U0190553

中 国 手 工 纸 文 库

Library of Chinese Handmade Paper

中国手工纸文库

Library of Chinese Handmade Paper

汤书昆
总主编

贵州卷

·下卷

Guizhou Ⅱ

汤书昆　陈　彪

主　编

中国科学技术大学出版社

University of Science and Technology of China Press

图书在版编目（CIP）数据

中国手工纸文库．贵州卷．下卷／汤书昆，陈彪主编.—合肥：中国科学技术大学出版社，2019.11

国家出版基金项目

“十三五”国家重点出版物出版规划项目

ISBN 978-7-312-04636-0

Ⅰ．中…　Ⅱ．①汤…　②陈…　Ⅲ．手工纸—介绍—贵州　Ⅳ．TS766

中国版本图书馆CIP数据核字（2018）第300904号

中国手工纸文库

贵州卷·下卷

出品人	伍传平
责任编辑	项赟飚
艺术指导	吕敬人
书籍设计	敬人书籍设计 吕　旻+黄晓飞
出版发行	中国科学技术大学出版社 地址 安徽省合肥市金寨路96号 邮编 230026
印　　刷	北京雅昌艺术印刷有限公司
经　　销	全国新华书店
开　　本	880 mm×1230 mm　1/16
印　　张	38.75
字　　数	1 124千
版　　次	2019年11月第1版
印　　次	2019年11月第1次印刷
定　　价	1980.00元

总 序

造纸技艺是人类文明的重要成就。正是在这一伟大发明的推动下，我们的社会才得以在一个相当长的历史阶段获得比人类使用口语的表达与交流更便于传承的介质。纸为这个世界创造了五彩缤纷的文化记录，使一代代的后来者能够通过纸介质上绘制的图画与符号、书写的文字与数字，了解历史，学习历代文明积累的知识，从而担负起由传承而创新的文化使命。

中国是手工造纸的发源地。不仅人类文明中最早的造纸技艺发源自中国，而且中华大地上遍布着手工造纸的作坊。中国是全世界手工纸制作技艺提炼精纯与丰富的文明体。可以说，在使用手工技艺完成植物纤维制浆成纸的历史中，中国一直是人类造纸技艺与文化的主要精神家园。下图是中国早期造纸技艺刚刚萌芽阶段实物样本的一件遗存——西汉放马滩古纸。

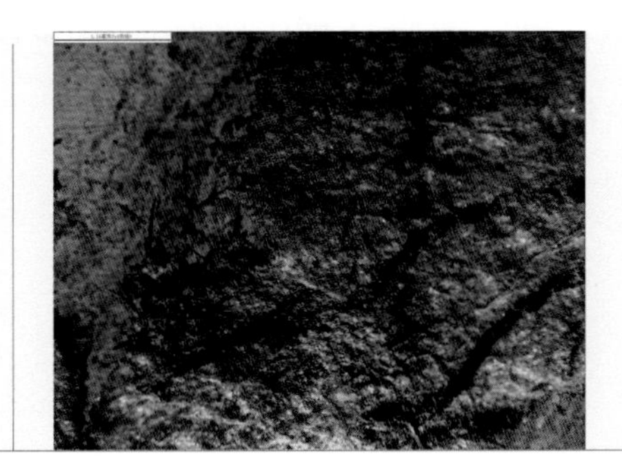

西汉放马滩古纸残片
纸上绘制的是地图
1986年出土于甘肃省天水市
现藏于甘肃省博物馆

Map drawn on paper from
Fangmatan Shoals
in the Western Han Dynasty
Unearthed in Tianshui City,
Gansu Province in 1986
Kept by Gansu Provincial Museum

Preface

Papermaking technique illuminates human culture by endowing the human race with a more traceable medium than oral tradition. Thanks to cultural heritage preserved in the form of images, symbols, words and figures on paper, human beings have accumulated knowledge of history and culture, and then undertaken the mission of culture transmission and innovation.

Handmade paper originated in China, one of the largest cultural communities enjoying advanced handmade papermaking techniques in abundance. China witnessed the earliest papermaking efforts in human history and embraced papermaking mills all over the country. In the history of handmade paper involving vegetable fiber pulping skills, China has always been the dominant centre. The picture illustrates ancient paper from Fangmatan Shoals in the Western Han Dynasty, which is one of the paper samples in the early period of papermaking techniques unearthed in China.

一

本项目的缘起

从2002年开始，我有较多的机缘前往东邻日本，在文化与学术交流考察的同时，多次在东京的书店街——神田神保町的旧书店里，发现日本学术界整理出版的传统手工制作和纸（日本纸的简称）的研究典籍，先后购得近20种，内容包括日本全国的手工造纸调查研究，县（相当于中国的省）一级的调查分析，更小地域和造纸家族的案例实证研究，以及日、中、韩等东亚国家手工造纸的比较研究等。如：每日新闻社主持编撰的《手漉和纸大鉴》五大本，日本东京每日新闻社昭和四十九年（1974年）五月出版，共印1 000套；久米康生著的《手漉和纸精髓》，日本东京讲谈社昭和五十年（1975年）九月出版，共印1 500本；菅野新一编的《白石纸》，日本东京美术出版社昭和四十年（1965年）十一月出版等。这些出版物多出自几十年前的日本昭和年间（1926~1988年），不仅图文并茂，而且几乎都附有系列的实物纸样，有些还有较为规范的手工纸性能、应用效果对比等技术分析数据。我阅后耳目一新，觉得这种出版物形态既有非常直观的阅读效果，又散发出很强的艺术气息。

1. Origin of the Study

Since 2002, I have been invited to Japan several times for cultural and academic communication. I have taken those opportunities to hunt for books on traditional Japanese handmade paper studies, mainly from old bookstores in Kanda Jinbo-cho, Tokyo. The books I bought cover about 20 different categories, typified by surveys on handmade paper at the national, provincial, or even lower levels, case studies of the papermaking families, as well as comparative studies of East Asian countries like Japan, Korea and China. The books include five volumes of *Tesukiwashi Taikan* ("*A Collection of Traditional Handmade Japanese Papers*") compiled and published by Mainichi Shimbun in Tokyo in May 1974, which released 1 000 sets, *The Essence of Japanese Paper* by Kume Yasuo, which published 1 500 copies in September 1975 by Kodansha in Tokyo, Japan, *Shiraishi Paper* by Kanno Shinichi, published by Fine Arts Publishing House in Tokyo in November 1965. The books which were mostly published between 1926 and 1988 among the Showa reigning years, are delicately illustrated with pictures and series of paper samples, some even with data analysis on performance comparison. I was extremely impressed by the intuitive and aesthetic nature of the books.

我几乎立刻想起在中国看到的手工造纸技艺及相关的研究成果，在我们这个世界手工造纸的发源国，似乎尚未看到这种表达丰富且叙述格局如此完整出色的研究成果。对中国辽阔地域上的手工造纸技艺与文化遗存现状，研究界尚较少给予关注。除了若干名纸业态，如安徽省的泾县宣纸、四川省的夹江竹纸、浙江省的富阳竹纸与温州皮纸、云南省的香格里拉东巴纸和河北省的迁安桑皮纸等之外，大多数中国手工造纸的当代研究与传播基本上处于寂寂无闻的状态。

此后，我不断与国内一些从事非物质文化遗产及传统工艺研究的同仁交流，他们一致认为在当代中国工业化、城镇化大规模推进的背景下，如果不能在我们这一代人手中进行手工造纸技艺与文化的整体性记录、整理与传播，传统手工造纸这一中国文明的结晶很可能会在未来的时空中失去系统记忆，那真是一种令人难安的结局。但是，这种愿景宏大的文化工程又该如何着手？我们一时觉得难觅头绪。

《手漉和纸精髓》附实物纸样的内文页
A page from *The Essence of Japanese Paper* with a sample

《白石纸》随书的宣传夹页
A folder page from *Shiraishi Paper*

The books reminded me of handmade papermaking techniques and related researches in China, and I felt a great sadness that as the country of origin for handmade paper, China has failed to present such distinguished studies excelling both in presentation and research design, owing to the indifference to both papermaking technique and our cultural heritage. Most handmade papermaking mills remain unknown to academia and the media, but there are some famous paper brands, including Xuan paper in Jingxian County of Anhui Province, bamboo paper in Jiajiang County of Sichuan Province, bamboo paper in Fuyang District and bast paper in Wenzhou City of Zhejiang Province, Dongba paper in Shangri-la County of Yunnan Province, and mulberry paper in Qian'an City of Hebei Province.

Constant discussion with fellow colleagues in the field of intangible cultural heritage and traditional craft studies lead to a consensus that if we fail to record, clarify, and transmit handmade papermaking techniques in this age featured by a prevailing trend of industrialization and urbanization in China, regret at the loss will be irreparable. However, a workable research plan on such a grand cultural project eluded us.

2004年，中国科学技术大学人文与社会科学学院获准建设国家“985工程”的“科技史与科技文明哲学社会科学创新基地”，经基地学术委员会讨论，“中国手工纸研究与性能分析”作为一项建设性工作由基地立项支持，并成立了手工纸分析测试实验室和手工纸研究所。这一特别的机缘促成了我们对中国手工纸研究的正式启动。

2007年，中华人民共和国新闻出版总署的“十一五”国家重点图书出版规划项目开始申报。中国科学技术大学出版社时任社长郝诗仙此前知晓我们正在从事中国手工纸研究工作，于是建议正式形成出版中国手工纸研究系列成果的计划。在这一年中，我们经过国际国内的预调研及内部研讨设计，完成了《中国手工纸文库》的撰写框架设计，以及对中国手工造纸现存业态进行全国范围调查记录的田野工作计划，并将其作为国家“十一五”规划重点图书上报，获立项批准。于是，仿佛在不经意间，一项日后令我们常有难履使命之忧的工程便正式展开了。

2008年1月，《中国手工纸文库》项目组经过精心的准备，派出第一个田野调查组（一行7人）前往云南省的滇西北地区进行田野调查，这是计划中全中国手工造纸田野考察的第一站。按照项目设计，将会有很多批次的调查组走向全中国手工造纸现场，采集能获

In 2004, the Philosophy and Social Sciences Innovation Platform of History of Science and S&T Civilization of USTC was approved and supported by the National 985 Project. The academic committee members of the Platform all agreed to support a new project, “Studies and Performance Analysis of Chinese Handmade Paper”. Thus, the Handmade Paper Analyzing and Testing Laboratory, and the Handmade Paper Institute were set up. Hence, the journey of Chinese handmade paper studies officially set off.

In 2007, the General Administration of Press and Publication of the People’s Republic of China initiated the program of key books that will be funded by the National 11th Five-Year Plan. The former President of USTC Press, Mr. Hao Shixian, advocated that our handmade paper studies could take the opportunity to work on research designs. We immediately constructed a framework for a series of books, *Library of Chinese Handmade Paper*, and drew up the fieldwork plans aiming to study the current status of handmade paper all over China, through arduous pre-research and discussion. Our project was successfully approved and listed in the 11th Five-Year Plan for National Key Books, and then our promising yet difficult journey began.

The seven members of the *Library of Chinese Handmade Paper* Project embarked on our initial, well-prepared fieldwork journey to the northwest area of Yunnan

取的中国手工造纸的完整技艺与文化信息及实物标本。

2009年，国家出版基金首次评审重点支持的出版项目时，将《中国手工纸文库》列入首批国家重要出版物的资助计划，于是我们的中国手工纸研究设计方案与工作规划发育成为国家层面传统技艺与文化研究所关注及期待的对象。

此后，田野调查、技术分析与撰稿工作坚持不懈地推进，中国科学技术大学出版社新一届领导班子全面调动和组织社内骨干编辑，使《中国手工纸文库》的出版工程得以顺利进行。2017年，《中国手工纸文库》被列为"十三五"国家重点出版物出版规划项目。

二
对项目架构设计的说明

作为纸质媒介出版物的《中国手工纸文库》，将汇集文字记

调查组成员在香格里拉县白地村调查
2008年1月

Visiting Baidi Village of Shangri-la County in January 2008

Province in January 2008. After that, based on our research design, many investigation groups would visit various handmade papermaking mills all over China, aiming to record and collect every possible papermaking technique, cultural information and sample.

In 2009, the National Publishing Fund announced the funded book list gaining its key support. Luckily, *Library of Chinese Handmade Paper* was included. Therefore, the Chinese handmade paper research plan we proposed was promoted to the national level, invariably attracting attention and expectation from the field of traditional crafts and culture studies.

Since then, field investigation, technical analysis and writing of the book have been unremittingly promoted, and the new leadership team of USTC Press has fully mobilized and organized the key editors of the press to guarantee the successful publishing of *Library of Chinese Handmade Paper*. In 2017, the book was listed in the 13th Five-Year Plan for the Publication of National Key Publications.

2. Description of Project Structure

Library of Chinese Handmade Paper compiles with many forms of ideography language: detailed descriptions and records, photographs, illustrations of paper fiber structure and transmittance images, data analysis, distribution of the papermaking sites, guide map

录与描述、摄影图片记录、样纸纤维形态及透光成像采集、实验分析数据表达、造纸地分布与到达图导引、实物纸样随文印证等多种表意语言形式，希望通过这种高度复合的叙述形态，多角度地描述中国手工造纸的技艺与文化活态。在中国手工造纸这一经典非物质文化遗产样式上，《中国手工纸文库》的这种表达方式尚属稀见。如果所有设想最终能够实现，其表达技艺与文化活态的语言方式或许会为中国非物质文化遗产研究界和保护界开辟一条新的途径。

项目无疑是围绕纸质媒介出版物《中国手工纸文库》这一中心目标展开的，但承担这一工作的项目团队已经意识到，由于采用复合度很强且极丰富的记录与刻画形态，当项目工程顺利完成后，必然会形成非常有价值的中国手工纸研究与保护的其他重要后续工作空间，以及相应的资源平台。我们预期，中国（计划覆盖34个省、市、自治区与特别行政区）当代整体的手工造纸业态按照上述记录与表述方式完成后，会留下与《中国手工纸文库》伴生的中国手工纸图像库、中国手工纸技术分析数据库、中国手工纸实物纸样库，以及中国手工纸的影像资源汇集等。基于这些伴生的集成资源的丰富性，并且这些资源集成均为首次，其后续的价值延展空间也不容小视。中国手工造纸传承与发展的创新拓展或许会给有志于继续关注中国手工造纸技艺与文化的同仁提供

to the papermaking sites, and paper samples, etc. Through such complicated and diverse presentation forms, we intend to display the technique and culture of handmade paper in China thoroughly and vividly. In the field of intangible cultural heritage, our way of presenting Chinese handmade paper was rather rare. If we could eventually achieve our goal, this new form of presentation may open up a brand-new perspective to research and preservation of Chinese intangible cultural heritage.

Undoubtedly, the *Library of Chinese Handmade Paper* Project developed with a focus on paper-based media. However, the team members realized that due to complicated and diverse ways of recording and displaying, there will be valuable follow-up work for further research and preservation of Chinese handmade paper and other related resource platforms after the completion of the project. We expect that when contemporary handmade papermaking industry in China, consisting of 34 provinces, cities, autonomous regions and special administrative regions as planned, is recorded and displayed in the above mentioned way, a Chinese handmade paper image library, a Chinese handmade paper technical data library, a Chinese handmade paper sample library, and a Chinese handmade paper video information collection will come into being, aside from the *Library of Chinese Handmade Paper*. Because of the richness of these byproducts, we should not overlook these possible follow-up

更多元的机遇。

毫无疑问，《中国手工纸文库》工作团队整体上都非常认同这一工作的历史价值与现实意义。这种认同给了我们持续的动力与激情，但在实际的推进中，确实有若干挑战使大家深感困惑。

三
我们的困惑和愿景

困惑一：

中国当代手工造纸的范围与边界在国家层面完全不清晰，因此无法在项目的田野工作完成前了解到中国到底有多少当代手工造纸地点，有多少种手工纸产品；同时也基本无法获知大多数省级区域手工造纸分布地点的情况与存活、存续状况。从调查组2008~2016年集中进行的中国南方地区（云南、贵州、广西、四川、广东、海南、浙江、安徽等）的田野与文献工作来看，能够提供上述信息支持的现状令人失望。这导致了项目组的田野工作规划处于“摸着石头过河”的境地，也带来了《中国手工纸文库》整体设计及分卷方案等工作的不确定性。

developments. Moving forward, the innovation and development of Chinese handmade paper may offer more opportunities to researchers who are interested in the techniques and culture of Chinese handmade papermaking.

Unquestionably, the whole team acknowledges the value and significance of the project, which has continuously supplied the team with motivation and passion. However, the presence of some problems have challenged us in implementing the project.

3. Our Confusions and Expectations

Problem One:

From the nationwide point of view, the scope of Chinese contemporary handmade papermaking sites is so obscure that it was impossible to know the extent of manufacturing sites and product types of present handmade paper before the fieldwork plan of the project was drawn up. At the same time, it is difficult to get information on the locations of handmade papermaking sites and their survival and subsisting situation at the provincial level. Based on the field work and literature of South China, including Yunnan, Guizhou, Guangxi, Sichuan, Guangdong, Hainan, Zhejiang and Anhui etc., carried out between 2008 and 2016, the ability to provide the information mentioned above is rather difficult. Accordingly, it placed the planning of the project's fieldwork into an obscure unplanned route,

困惑二：

中国正高速工业化与城镇化，手工造纸作为一种传统的手工技艺，面临着经济效益、环境保护、集成运营、技术进步、消费转移等重要产业与社会变迁的压力。调查组在已展开了九年的田野调查工作中发现，除了泾县、夹江、富阳等为数不多的手工造纸业态聚集地，多数乡土性手工造纸业态都处于生存的“孤岛”困境中。令人深感无奈的现状包括：大批造纸点在调查组到达时已经停止生产多年，有些在调查组到达时刚刚停止生产，有些在调查组补充回访时停止生产，仅一位老人或一对老纸工夫妇在造纸而无传承人……中国手工造纸的业态正陷于剧烈的演化阶段。这使得项目组的田野调查与实物采样工作处于非常紧迫且频繁的调整之中。

困惑三：

作为国家级重点出版物规划项目，《中国手工纸文库》在撰写开卷总序的时候，按照规范的说明要求，应该清楚地叙述分卷的标准与每一卷的覆盖范围，同时提供中国手工造纸业态及地点分布现

贵州省仁怀市五马镇
取缔手工造纸作坊的横幅
2009年4月

Banner of a handmade papermaking mill in Wuma Town of Renhuai City in Guizhou Province, saying "Handmade papermaking mills should be closed as encouraged by the local government". April 2009

which also led to uncertainty in the planning of *Library of Chinese Handmade Paper* and that of each volume.

Problem Two:
China is currently under the process of rapid industrialization and urbanization. As a traditional manual technique, the industry of handmade papermaking is being confronted with pressures such as economic benefits, environmental protection, integrated operation, technological progress, consumption transfer, and many other important changes in industry and society. During nine years of field work, the project team found out that most handmade papermaking mills are on the verge of extinction, except a few gathering places of handmade paper production like Jingxian, Jiajiang, Fuyang, etc. Some handmade papermaking mills stopped production long before the team arrived or had just recently ceased production; others stopped production when the team paid a second visit to the mills. In some mills, only one old papermaker or an elderly couple were working, without any inheritor to learn their techniques... The whole picture of this industry is in great transition, which left our field work and sample collection scrambling with hasty and frequent changes.

Problem Three:
As a national key publication project, the preface of *Library of Chinese Handmade Paper* should clarify the standard and the scope of each volume according to the research plan. At the same time, general information such as the map with locations of Chinese handmade

状图等整体性信息。但由于前述的不确定性，开宗明义的工作只能等待田野调查全部完成或进行到尾声时再来弥补。当然，这样的流程一定程度上会给阅读者带来系统认知的先期缺失，以及项目组工作推进中的迷茫。尽管如此，作为拓荒性的中国手工造纸整体研究与田野调查就在这样的现状下全力推进着！

当然，我们的团队对《中国手工纸文库》的未来仍然满怀信心与憧憬，期待着通过项目组与国际国内支持群体的协同合作，尽最大努力实现尽可能完善的田野调查与分析研究，从而在我们这一代人手中为中国经典的非物质文化遗产样本——中国手工造纸技艺留下当代的全面记录与文化叙述，在中国非物质文化遗产基因库里绘制一份较为完整的当代手工纸文化记忆图谱。

汤书昆
2017年12月

papermaking industry should be provided. However, due to the uncertainty mentioned above, those tasks cannot be fulfilled, until all the field surveys have been completed or almost completed. Certainly, such a process will give rise to the obvious loss of readers' systematic comprehension and the team members' confusion during the following phases. Nevertheless, the pioneer research and field work of Chinese handmade paper has set out on the first step.

There is no doubt that, with confidence and anticipation, our team will make great efforts to perfect the field research and analysis as much as possible, counting on cooperation within the team, as well as help from domestic and international communities. It is our goal to keep a comprehensive record, a cultural narration of Chinese handmade paper craft as one sample of most classic intangible cultural heritage, to draw a comparatively complete map of contemporary handmade paper in the Chinese intangible cultural heritage gene library.

Tang Shukun
December 2017

编撰说明

1

《中国手工纸文库·贵州卷》按六盘水市、黔西南布依族苗族自治州、安顺市、黔南布依族苗族自治州、毕节市、贵阳市、遵义市、铜仁市、黔东南苗族侗族自治州等九个市（州）区域划分一级手工造纸地域，形成"章"的类目单元，如第六章"毕节市"。章之下的二级类目以县为单元划分，形成"节"的类目，如第六章第一节"纳雍皮纸"。

2

本卷各节的标准撰写格式通常分为七个部分："××××纸的基础信息及分布""××××纸生产的人文地理环境""××××纸的历史与传承""××××纸的生产工艺与技术分析""××××纸的用途与销售情况""××××纸的相关民俗与文化事象""××××纸的保护现状与发展思考"。如遇某一部分田野调查和文献资料均未能采集到信息，将按照实事求是原则略去标准撰写格式的相应部分。

3

本卷设专节记述的手工纸种类标准是：其一，项目组进行田野调查时仍在生产的手工纸种类；其二，项目组田野调查时虽已不再生产，但保留着较完整的生产环境与设备，造纸技师仍能演示或讲述完整技艺和相关知识的手工纸种类。

Introduction to the Writing Norms

1. In *Library of Chinese Handmade Paper: Guizhou*, handmade papermaking sites are categorized in nine major regions, i.e. , Liupanshui City, Qianxinan Bouyei and Miao Autonomous Prefecture, Anshun City, Qiannan Bouyei and Miao Autonomous Prefecture, Bijie City, Guiyang City, Zunyi City, Tongren City, and Qiandongnan Miao and Dong Autonomous Prefecture. Each area covers a whole chapter, e.g. , "Chapter VI Bijie City". Each chapter consists of sections covering introductions to handmade paper in different counties. For instance, first section of the sixth chapter is "Bast Paper in Nayong County".

2. Each section of a chapter consists of seven sub-sections introducing various aspects of each kind of handmade paper, namely, Basic Information and Distribution, The Cultural and Geographic Environment, History and Inheritance, Papermaking Technique and Technical Analysis, Uses and Sales, Folk Customs and Culture, Preservation and Development. Omission is also acceptable if our fieldwork efforts and literature review fail to collect certain information.

3. The handmade paper included in each section of this volume conforms to the following standards: firstly, it was still under production when the research group did their fieldwork; secondly, the papermaking equipment and major sites were well preserved, and the handmade papermakers were still able to demonstrate the papermaking techniques and relevant knowledge, in case of ceased production.

4. Many handmade papermaking sites in Guizhou Province are inhabited by multiple minority groups. Accordingly, their official names include all the group names, e.g. , Wuchuan Gelo and Miao Autonomous County, which is comparatively complicated. In this volume, for the purpose of brevity, we employ a concise naming mode: the paper name, the ethnic group the papermaker

4

贵州省的很多手工造纸地为少数民族聚居区域，县名按中国地名使用规范应标注出全称，考虑到多民族地区的县名构成往往较复杂，如“务川仡佬族苗族自治县”，为兼顾地名的使用规范与简洁，本卷所有“节”的标题及“节”下一级类目的标题均直接标示为“××（县域）（+民族）+纸名”，而不出现多民族县名全称及“县”这一称谓，如“镇宁皮纸”（镇宁县的全称应为“镇宁布依族苗族自治县”）。

5

本卷造纸点地理分布不以测绘地图背景标示方式，而以示意图标示方式呈现。全卷每一节绘制三幅示意图：一幅以市州为单位，绘制造纸点所在市州的位置示意；一幅为从县城到造纸点的路线示意图；一幅为现存活态造纸点和历史造纸点在县境内的位置示意图。在标示地名时，均统一标示出县城与乡镇两级，乡镇下则直接标注造纸点所在村。本卷中涉及的行政区划名称，均依据调查组田野调查当时的名称，以尊重调查时的真实区划名称信息。

6

本卷对造纸点的地理分布按“造纸点”和“历史造纸点”两类区别标示。其中，历史造纸点选择的时间上限项目组划定为民国元年（1911年），而下限原则上为20世纪末已不再生产且基本业态已完全终止，1911年以前有记述的造纸点不进行示意图标示。因贵州省域历代造纸信息多样性、多变性突出，本卷示意图上的历史造纸点原则上以调查组通过田野工作和文献研究所掌握的信息为标示依据。

7

本卷珍稀收藏版原则上每一个所调查的造纸村落的代表性纸种均在书中相应章节附调查组实地采集的实物纸样。采样足量的造纸点代表性纸种于书中均附全页纸样，不足量的则附1/2、1/4或更小规格的纸样，个别因停产等原因导致采样严重不足的则不附实物纸样。

belongs to, and the county name, instead of using the complete name. For instance, “Bast Paper in Zhenning County” is used as a section name instead of using the complete name “Bast Paper in Zhenning Bouyei and Miao Autonomous County”.

5. In this volume, we draw illustrations instead of authentic maps to show the distribution of local papermaking sites. In each section of this volume, we draw three illustrations: the first one shows the location of papermaking sites in a specific city or prefecture; the second one draws roadmap from county centre to the papermaking sites; the third one shows the distribution of papermaking sites still in production or in history in a county. We provide county name, town name and village name of each site. All the administrative divisions' names in this volume are the ones in use when we made the field investigation, reflecting the real situation of the time.

6. In the distribution maps we cover both papermaking sites and the historical ones, which refer to papermaking sites that were active from the year of 1911 to the end of the 20th century, and were no longer involved in papermaking for present days. Papermaking sites before the year of 1911 are excluded in this study, together with the sites that had ceased production by the end of the 20th century. The papermaking sites marked in our maps are consistently based on our fieldworks and literature review, for the variety and variability of papermaking literature in Guizhou Province.

7. For each type of paper included in *Library of Chinese Handmade Paper: Guizhou* (Special Edition), we attach a piece of paper sample (a full page, 1/2 or 1/4 of a page, or even smaller if we do not have sufficient sample available) to the section. For some sections no sample is attached for the shortage of paper sample (e.g. , the papermakers had ceased production).

8. All the paper samples in this volume were tested based on

本卷对所采集纸样进行的测试参考了宣纸的技术测试分析标准（GB/T 18739—2008），并根据贵州地域手工纸的特色做了调适，实测或计算了所有满足测试分析足量需求已采样手工纸的厚度、定量、紧度、抗张力、抗张强度、白度、纤维长度和纤维宽度共8个指标。由于所测贵州省手工纸样的生产标准化程度不同，因而所测数据与机制纸或宣纸的标准存在一定差距。

（1） 厚度 ▸ 所测纸的厚度指标是指纸在两块测量板间受一定压力时直接测量得到的厚度。以单层测量的结果表示纸的厚度，以mm为单位。
所用仪器 ▸ 长春市月明小型试验机有限公司JX-HI型纸张厚度仪。

（2） 定量 ▸ 所测纸的定量指标是指单位面积纸的质量，通过测定试样的面积及质量计算定量，以g/m²为单位。
所用仪器 ▸ 上海方瑞仪器有限公司3003电子天平。

（3） 紧度 ▸ 所测纸的紧度指标是指单位体积纸的质量，由同一试样的定量和厚度计算而得，以g/cm³为单位。

（4） 抗张力 ▸ 所测纸的抗张力指标是指在标准实验方法规定的条件下，纸断裂前所能承受的最大张力，分纵向、横向测试，若试样无法判断纵横向，则视为一个方向测试，以N为单位。
所用仪器 ▸ 杭州高新自动化仪器仪表公司DN-KZ电脑抗张力试验机。

（5） 抗张强度 ▸ 所测纸的抗张强度指标一般用在抗张强度试验仪上所测出的抗张力除以样品宽度来表示，也称为纸的绝对抗张强度，以kN/m为单位。
本卷采用的是恒速加荷法，其原理是抗张强度试验仪在恒速加荷的条件下，把规定尺寸的纸样拉伸至撕裂，测其抗张力，计算出抗张强度。公式如下：

$$S=F/W$$

公式中，S为试样的抗张强度（kN/m），F为试样的绝对抗张力（N），W为试样的宽度（出于测试仪器的要求，为定值15 mm）。

the technical test and analysis standards of Xuan paper in China (GB/T 18739-2008), with modifications adopted according to the specific features of the handmade paper of Guizhou Province. Eight indicators of the samples were tested and analyzed, including thickness, mass per unit area, density, tension stress, tensile strength, whiteness, fiber length and width. Due to the various production standards involved in papermaking in Guizhou Province, the statistical data may vary from those of machine-made paper and Xuan paper.

(1) Thickness: the values obtained by using two measuring boards pressing the paper. In the measuring process, the result of a single layer represents the thickness of the paper, and its measurement unit is mm. The measuring instrument (Specification: JX-HI) employed is produced by Yueming Small Testing Instrument Co., Ltd., Changchun City.

(2) Mass per unit area: the values obtained by measuring the sample mass divided by area, with the measurement unit g/m². The measuring instrument employed is 3003 Electronic Balance produced by Shanghai Fangrui Instrument Co., Ltd.

(3) Density: mass per unit volume, obtained by measuring the mass per unit area and thickness, with the measurement unit g/m³.

(4) Tension stress: the maximum tension that the sample paper can withstand without tearing apart, when tested by the standard experimental methods. Both longitudinal and horizontal directions of the paper should be covered in test, or only one if the direction cannot be ascertained, with the measurement unit N. The testing instrument (Specification: DN-KZ) is produced by Hangzhou Gaoxin Automatic Instrument Co.

(5) Tensile strength: the values obtained by measuring the sample maximum tension stress against the constant loading, then divided the maximum stress by the sample width, with the measurement unit kN/m.

In this volume, constant loading method was employed to measure the maximum tension the material can withstand without tearing apart. The formula is:

(6) 白度▸所测纸的白度指标是指被测物体的表面在可见光区域内相对于完全白（标准白）物体漫反射辐射能的大小的比值，以%为单位，即白色的程度。所测纸的白度指标是指在D65光源、漫射/垂射照明观测条件下，纸对主波长475 nm蓝光的漫反射因数，表示白度测定结果。
所用仪器▸杭州纸邦仪器有限公司ZB-A色度仪。

(7) 纤维长度/宽度▸所测纸的纤维长度/宽度指标是指从所测纸里取样，测其纸浆中纤维的自身长度/宽度，分别以mm和μm为单位。测试时，取少量纸样，用水湿润，并用Herzberg试剂染色，制成显微镜试片，置于显微分析仪下，采用10倍及20倍物镜进行观测，并显示相应纤维形态图各一张。
所用仪器▸珠海华伦造纸科技有限公司XWY-VI型纤维测量仪。

9

本卷对每一种调查采集的纸样均采用透光摄影的方式制作成图像，显示透光环境下的纸样纤维纹理影像，作为实物纸样的另一种表达方式。其制作过程为：先使用计算机液晶显示器，显示纯白影像作为拍摄手工纸纹理透光影像的背景底；然后将纸样平铺在显示器上进行拍摄，拍摄相机为佳能5DIII。

10

本卷引述的历史与当代文献均一一注释，所引文献原则上要求为一手文献来源，并按统一标准注释，如“刘仁庆.我国少数民族地区的传统手工纸[J].纸和造纸,2007,26(5):89-91”“陈国生.明代云贵川农业地理研究[M].重庆:西南师范大学出版社,1997:7”。所引述的田野调查信息原则上要标示出信息源，如“2014年仍在从事手工纸生产的赵金妹老人描述……”“据造纸老人赵金学与赵金成回忆……”等。

$$S=F/W$$

S stands for tensile strength (kN/m), *F* is tension stress (N), and *W* represents width (as required by the testing instrument, width adopted here is 15 mm).

(6) Whiteness: degree of whiteness, represented by percentage (%), which is the ratio obtained by comparing the radiation diffusion value of the test object in visible region to that of the completely white (standard white) object. Whiteness test in our study employed D65 light source, with dominant wavelength 475 nm of blue light, under the circumstances of diffuse reflection or vertical reflection. The whiteness testing instrument (Specification: ZB-A) is produced by Hangzhou Zhibang Instrument Co., Ltd.

(7) Fiber length and width: Fiber length (mm) and width (μm) of paper sample were tested by dying the moist paper sample with Herzberg reagent, then the specimen was made and the fiber pictures were taken through ten times and twenty times lens of the microscope. The fiber testing instrument (Specification: XWY-VI) is produced by Zhuhai Hualun Papermaking Technology Co., Ltd.

9. Each paper sample included in *Library of Chinese Handmade Paper*: *Guizhou* was photographed against a luminous background, which vividly demonstrated the fiber veins of the samples. This is a different way to present the status of our paper sample. Each piece of paper sample was spread flat-out on the LCD monitor giving white light, and photographs were taken with Canon 5DIII camera.

10. All the quoted literature are original first-hand resources and the footnotes are used for documentation. For instance, “Liu Renqing. *Traditional Handmade Paper in Ethnic Regions of China* [J]. Paper and Papermaking, 2007, 26(5): 89-91” and “Chen Guosheng. *Research of Agricultural Geography in Yunnan, Guizhou and Sichuan During the Ming Dynasty* [M]. Chongqing: Southwest Normal University Press, 1997: 7”. Sources of information based on our fieldworks are also identified, e.g., “Zhao Jinmei, an old woman who was still making paper in 2014 introduced that...” “old

11

本卷所使用的摄影图片主体部分为调查组成员在实地调查时所拍摄，也有项目组成员在既往田野工作中积累的图片，另有少量属撰稿过程中所采用的非项目组成员的摄影作品。由于项目组成员在完成本卷过程中形成的图片的著作权属集体著作权，且在过程中多位成员轮流拍摄或并行拍摄为工作常态，因而对图片均不标示拍摄者。项目组成员既往积累的图片，以及非项目组成员拍摄的图片在后记中特别说明，并承认其个人图片著作权。

12

考虑到本卷中文简体版的国际交流需要，编著者对本卷重要或提要性内容同步给出英文表述，以便英文读者结合照片和实物纸样领略本卷的基本语义。对于文中一些晦涩的古代文献，英文翻译采用意译的方式进行解读。英文内容包括：总序、编撰说明、目录、概述、图目、表目、术语、后记，以及所有章节的标题、图题、表题与实物纸样名。"贵州省手工造纸概述"是本卷正文第一章，为保持与后续各章节体例一致，除保留章节标题英文名及图表标题英文名外，全章的英文译文作为附录出现。

13

《中国手工纸文库·贵州卷》的术语收集了本卷中与手工纸有关的地理名、纸品名、原料与相关植物名、工艺技术和工具设备、历史文化5类术语。术语选择遵循文化、民族、工艺、材料、历史特色表达优先，核心内容与关键概念表达优先的原则，力求简洁精练。各个类别的术语按术语的汉语拼音先后顺序排列。每条中文术语后都给以英文直译，可以作中英文对照表使用。由于本卷涉及术语很多且多次出现，以及因语境之异具有一定的使用多样性与复杂性，因此术语一律不标注出现的页码。

papermakers Zhao Jinxue and Zhao Jincheng recalled that..." etc.

11. The majority of photographs included in this volume were taken by the research team members when they were doing fieldworks of the research. Others were taken by our researchers in even earlier fieldwork errands, or by the photographers who were not involved in our research. We do not give the names of the photographers in this volume, because almost all our researchers are involved in the task. Yet, as we have claimed in the epilogue, we officially admit the copyright of all the photographers, including those who are not our research team members.

12. For the purpose of international academic exchange, English version of some important parts is provided, namely, Preface, Introduction to the Writing Norms, Contents, Introduction, Figures, Tables, Terminology, Epilogue, and all the headings, captions, and paper sample names. Among them, "Introduction to Handmade Paper in Guizhou Province" is the first chapter of this volume, and its translation is appended in the appendix part. For the obscure ancient texts included, we use free translation to present a more comprehensible version.

13. Terminology is appended in *Library of Chinese Handmade Paper: Guizhou*, which covers Places, Paper Names, Raw Materials and Plants, Techniques and Tools, History and Culture, relevant to the handmade paper research in this volume. We highlight cultural and national factors, as well as unique techniques, materials, and historic features, and make key contents and core concepts our priority in the winnowing process to avoid a lengthy list. All the terms are listed following the alphabetical order of the first Chinese character. As a glossary of terms, both Chinese and English versions are listed for reference. Different contexts may endow each term with various implications, so page number or numbers are not provided in this volume.

目 录
Contents

总 序
Preface
I

编撰说明
Introduction to the Writing Norms
XI

第七章 贵阳市
Chapter VII Guiyang City
001

002 第一节 乌当竹纸
Section 1 Bamboo Paper in Wudang District

第八章 遵义市
Chapter VIII Zunyi City
025

026 第一节 仁怀手工纸
Section 1 Handmade Paper in Renhuai City

070 第二节 正安手工纸
Section 2 Handmade Paper in Zheng'an County

102 第三节 务川仡佬族皮纸
Section 3 Bast Paper by the Gelo Ethnic Group in Wuchuan County

128 第四节 务川仡佬族竹纸
Section 4 Bamboo Paper by the Gelo Ethnic Group in Wuchuan County

144 第五节 余庆竹纸
Section 5 Bamboo Paper in Yuqing County

第九章 Chapter IX
铜仁市 Tongren City
161

162 第一节 石阡仡佬族皮纸
Section 1 Bast Paper by the Gelo Ethnic Group in Shiqian County

184 第二节 石阡仡佬族竹纸
Section 2 Bamboo Paper by the Gelo Ethnic Group in Shiqian County

204 第三节 江口土家族竹纸
Section 3 Bamboo Paper by the Tujia Ethnic Group in Jiangkou County

226 第四节 印江合水镇皮纸
Section 4 Bast Paper in Heshui Town of Yinjiang County

258 第五节 印江土家族手工纸
Section 5 Handmade Paper by the Tujia Ethnic Group in Yinjiang County

第十章 Chapter X
黔东南苗族侗族自治州 Qiandongnan Miao and Dong Autonomous Prefecture
279

280 第一节 岑巩侗族竹纸
Section 1 Bamboo Paper by the Dong Ethnic Group in Cengong County

306 第二节 三穗侗族竹纸
Section 2 Bamboo Paper by the Dong Ethnic Group in Sansui County

328 第三节 黄平苗族竹纸
Section 3 Bamboo Paper by the Miao Ethnic Group in Huangping County

346 第四节 凯里苗族竹纸
Section 4 Bamboo Paper by the Miao Ethnic Group in Kaili City

366 第五节 丹寨苗族皮纸
Section 5 Bast Paper by the Miao Ethnic Group in Danzhai County

398 第六节 榕江侗族皮纸
Section 6 Bast Paper by the Dong Ethnic Group in Rongjiang County

422 第七节 从江秀塘瑶族竹纸
Section 7 Bamboo Paper by the Yao Ethnic Group in Xiutang Town of Congjiang County

438 第八节 从江翠里瑶族手工纸
Section 8 Handmade Paper by the Yao Ethnic Group in Cuili Town of Congjiang County

470 第九节 从江小黄侗族皮纸
Section 9 Bast Paper by the Dong Ethnic Group in Xiaohuang Village of Congjiang County

492 第十节 从江占里侗族皮纸
Section 10 Bast Paper by the Dong Ethnic Group in Zhanli Village of Congjiang County

510 第十一节 黎平侗族皮纸
Section 11 Bast Paper by the Dong Ethnic Group in Liping County

附　录
Appendices

535

535 Introduction to Handmade Paper in Guizhou Province

559 图目
Figures

575 表目
Tables

576 术语
Terminology

后　记
Epilogue

583

第七章
贵阳市

Chapter Ⅶ
Guiyang City

第一节

乌当竹纸

贵州省
Guizhou Province

贵阳市
Guiyang City

乌当区
Wudang District

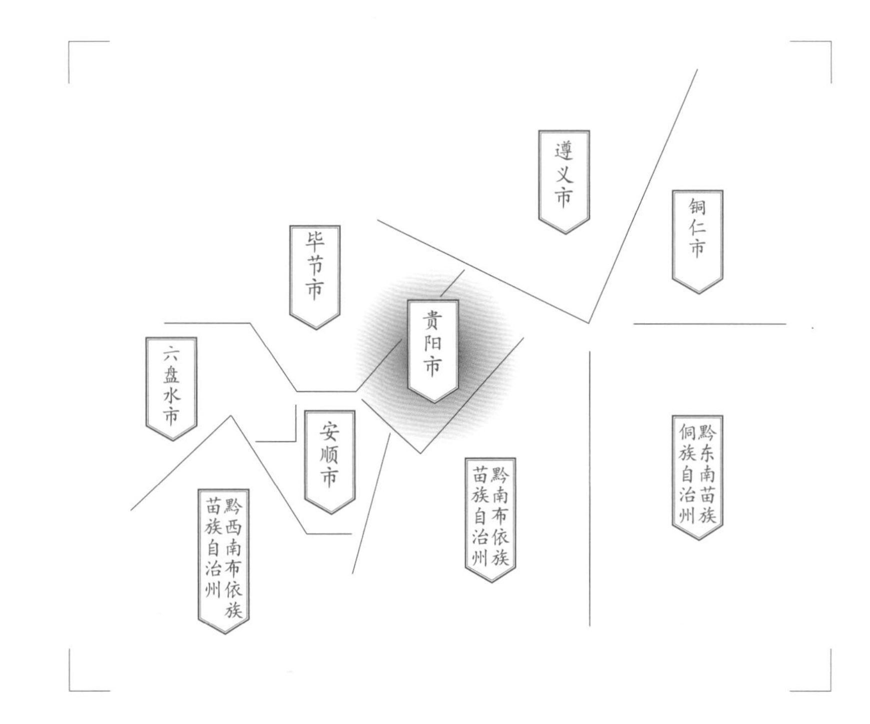

调查对象
新堡布依族乡
陇脚行政村
竹纸

Section 1
Bamboo Paper in Wudang District

Subject

Bamboo Paper in Longjiao Administrative Village of Xinbao Bouyei Town

一

乌当竹纸的基础信息及分布

1

Basic Information and Distribution of Bamboo Paper in Wudang District

乌当区新堡布依族乡陇脚行政村布依族、汉族的造纸村民用传统技艺制作的竹纸以苦竹与慈竹（当地村民也称之为青竹）为原料，成纸纸质绵韧，纸面平整，隐约有竹帘纹，色泽金黄，吸水性强。由于在造纸时掺入香叶，该纸作冥纸焚烧时呈灰白色，且带有淡淡的清香。乌当竹纸主要作为当地民间敬神祭祖时的焚烧用纸、卫生用纸和制作烟花、爆竹以及其他产品、物品的包装用纸。

陇脚行政村的上陇脚村民组、下陇脚村民组、香纸沟村民组和白水河村民组都保存着相对完整的手工古法造纸工艺。陇脚村古法造纸工艺于2006年被列入中华人民共和国国务院首批公布的国家级非物质文化遗产名录中，罗守全为该项技艺的代表性传承人。当地居民自誉该造纸工艺为“传承千年的布依族土法造纸工艺”。1997年，为保护这种造纸工艺流程，贵阳市政府下拨40万元资金，在香纸沟修建了一座面积约230 m^2的古法造纸博物馆，并新造了一座碾坊，为游客演示抄纸工艺。

⊙1

⊙1 香纸沟古法造纸博物馆 Traditional Papermaking Museum in Xiangzhigou Scenic Spot

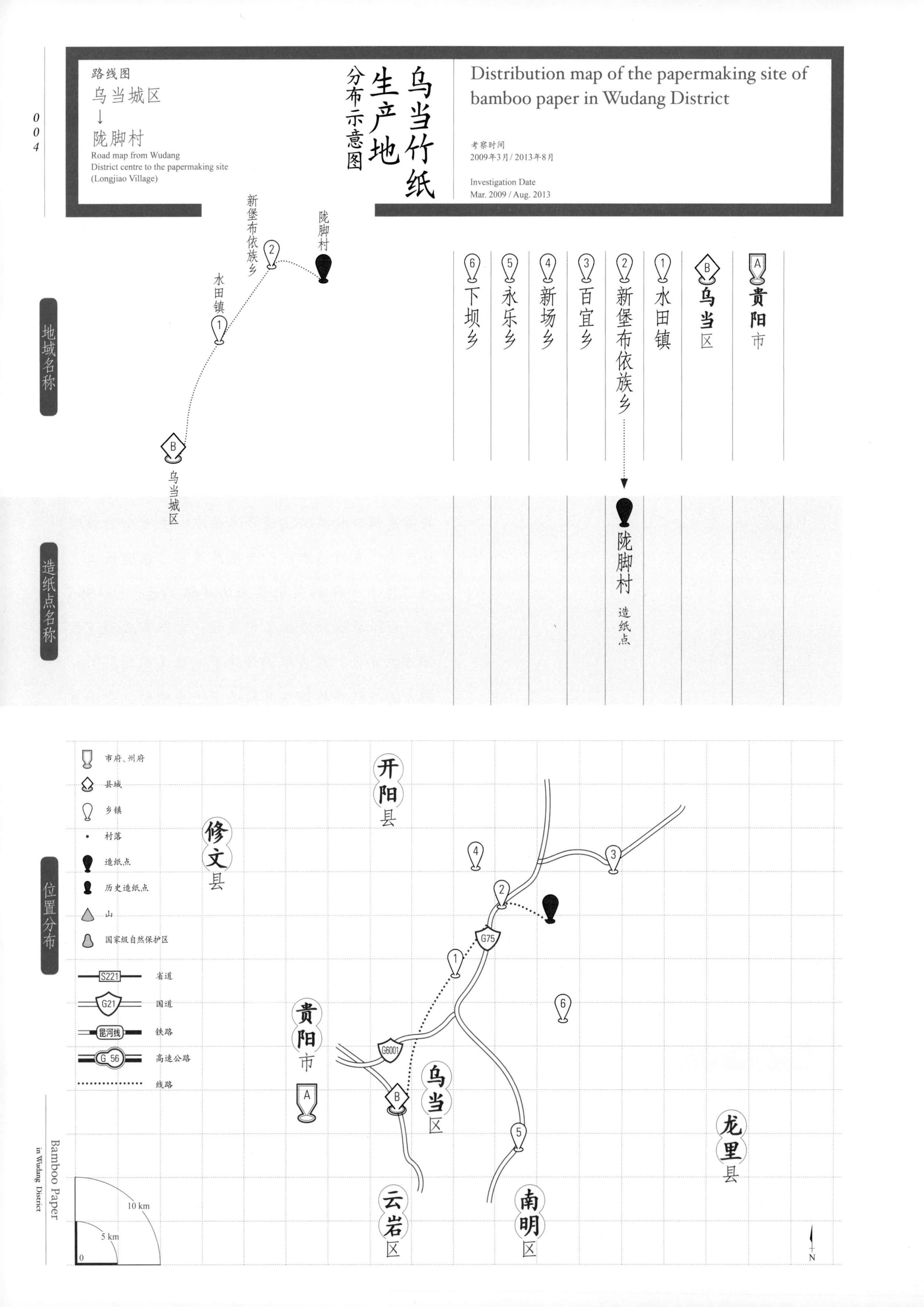
乌当竹纸
生产地
分布示意图
Distribution map of the papermaking site of bamboo paper in Wudang District
路线图
乌当城区
↓
陇脚村
Road map from Wudang District centre to the papermaking site (Longjiao Village)
考察时间
2009年3月 / 2013年8月
Investigation Date
Mar. 2009 / Aug. 2013
地域名称
造纸点名称
位置分布
新堡布依族乡
陇脚村
水田镇
乌当城区
贵阳市
乌当区
水田镇
新堡布依族乡
百宜乡
新场乡
永乐乡
下坝乡
陇脚村
造纸点
市府、州府
县城
乡镇
村落
造纸点
历史造纸点
山
国家级自然保护区
S221
省道
G21
国道
昆河线
铁路
G 56
高速公路
线路
开阳县
修文县
贵阳市
乌当区
龙里县
云岩区
南明区
G75
G6001
10 km
5 km
0
N

二 乌当竹纸生产的人文地理环境

2 The Cultural and Geographic Environment of Bamboo Paper in Wudang District

乌当区历史悠久，春秋时期属柯（音）国，战国时期为楚黔中地，秦属象郡，西汉时期属夜郎县，唐宋属矩州。清康熙二十六年（1687年），改贵州卫、贵州前卫为贵筑县，辖十七里。民国元年（1912年），贵筑县并入贵阳府。1958年2月，撤贵筑县，成立乌当区，属贵阳市辖区。

陇脚村造纸点地处香纸沟景区。香纸沟位于贵阳市北部，依山傍水，是贵州省著名的风景旅游胜地。悠悠古韵、民族风情与秀丽景色勾画出美丽的山乡旅游风景画卷。景区有一条三级公路与“104”省道相通，乡村公路网密布，交通便利。

陇脚村距贵阳市区42 km，距乌当区政府驻地18 km。该村地处东经106°53′24″～106°55′12″、北纬26°46′23″～26°47′30″。全村面积约20 km^2，由上陇脚、下陇脚、香纸沟、白水河、葫芦冲5个村民组组成，共有农户223户。

⊙1

⊙2

陇脚村的造纸作坊百年来一直都默默竖立在村中，直到2003年7月10日，该地遭遇了百年一遇的特大暴雨，引发了泥石流，大面积的滑坡、崩塌，大量的砂石淤积在上陇脚村民组、下陇脚村民组和香纸沟景区一带，河边的造纸作坊大部分被冲垮，只有小部分遗存下来，但调查时已废弃不用。据当地村民说，香纸沟现在已没有造纸户，只有白水河村民组的碾坊保存完好，并且还在生产古法草纸。

白水河村民组距下陇脚村民组5 km，是一个幽静的小村。调查组入村时，美丽的民居木楼沿山坡而筑，高低错落，比屋连甍。山上满是翠竹丛树，屋前平坦的田畴里开满了金黄的油菜花。河沿上散布着抄纸的碾坊，高岸上垒砌着成排煮竹的池子。村民三五成群，有的往池里撒石灰，有的扛着捶破好的竹子往池里码放。碾坊是由十几根木柱撑起的歇山式的瓦棚，正中间是碾盘，流动的河水冲击着水车，推动磙子转动，循环周

⊙1 洪水后废弃的造纸作坊 Abandoned papermaking mill after flood

⊙2 白水河村造纸现场 Papermaking site in Baishuihe Village

转，把煮好发酵的竹子碾成竹丝。抄纸池边，有人在添滑、搅浆、抄纸、压榨。各家屋子里的房梁上都晾满了香纸。调查组在白水河村民组完整地考察了乌当竹纸手工古法造纸的全部工艺。

⊙1

⊙2

⊙1
白水河村小景
View of Baishuihe Village

⊙2
香纸沟小景
View of Xiangzhigou Scenic Spot

三
乌当竹纸的历史与传承

3
History and Inheritance of Bamboo Paper in Wudang District

陇脚村竹纸生产大约始于明初。据调查组重点访谈对象——陇脚村70岁的老人汪长伦口述，他家祖先是随明朝军队来到香纸沟的，随后在此开始了古法造纸。

当地《彭氏家谱》同样记载："明洪武年间，朱元璋'调北征南'，彭氏祖先随越国汪公从军到此屯军，拓荒耕耘，伐竹造纸。"他们来自湖南，为祭祀军中阵亡将士的英灵，在此营造钱纸（冥纸）。因湖南简称"湘"，彭氏祖先为表示不忘自己是湘中弟子，故将此地称为"湘子沟"（亦名"乡思沟"），旧版《贵阳志·建置志》记载有"湘子沟"之名。后因这些钱纸又销给贵阳附近的寺庙烧香拜佛用，所以改称为香纸，"湘子沟"的地名年久日深，讹读成"香纸沟"。当地的彭氏宗祠里至今还供奉着"越国汪公彭氏宗祖之位"的神主。

《明史》中也能找到类似的线索："明洪武十五年（1382年），朱元璋在平定云贵高原的叛乱后，为了加强对叛乱区域的军事管辖，决定把军队留下，屯兵驻守，威慑四方。"这一军事行动被称作"调北征南"。之后还有迁徙百姓的"调北填南"的朝廷举措。这些移民主要来自江南和中原一带。据史书记载，明洪武年间进入贵州的移民在160万人以上，永乐年间在35万人以上，他们带来了内地的生产方式和手工百业。可见，手工造纸在此时传入陇脚村的传说，还是与这段历史相印证的。

相传在明洪武年间，陇脚村古法造纸只有彭氏一户，主要为军中提供祭品，到清初就发展到10余户。清乾隆年间，此地香纸被皇帝赐为御贡冥品。随着彭氏家族与当地胡、汪、罗三姓通婚，该技艺又传给胡、汪、罗三姓，随后该地造纸户发展到40余户。据汪长伦口述，汪氏从第一世祖来此地后就开始造纸，直到2003年发大水把碾子冲烂后才不再造纸，造纸技艺从汪氏祖上至

⊙3

⊙4

⊙3 调查组成员采访汪长伦
A researcher interviewing the local papermaker Wang Changlun

⊙4 『越国汪公彭氏宗祖之位』牌位
Memorial tablets of Wang Hua, lord of Yue State, and the ancestor of the Pengs in ancient China

今已传了14代，共300余年。因年代久远，汪长伦能确切记得的只有祖父汪法造过纸，更早的祖先名字还需查证，汪长伦的儿子汪勤闲、汪贵钱、汪勤林也都造过纸。

民国年间，香纸沟的竹纸除作冥纸之外，还可作为卫生纸用于日常生活，造纸户曾发展到60余户。20世纪六七十年代，国家提倡破除封建迷信，香纸沟就很少制作冥纸了，而是集体抽调技艺精干的村民专门制作土纸，所制作的土纸由供销社收购，并作为卫生纸批发到各地。1978年改革开放之后，各家各户又掀起了古法造纸的热潮，造纸作坊从原来的30个增加到60个，高峰时期有160户参与造纸。

由于现代机制竹纸的冲击，陇脚村古法造纸在人们的生产生活中，除用于冥品外，已无他用，造纸规模正在逐步萎缩。2003年发大水之后，香纸沟的古法造纸作坊几乎被毁灭殆尽，仅附近的白水河村还保存着古法造纸作坊。如今，香纸沟只有当地政府在1997年为了旅游而修建的一座面积约230 m^{2}的古法造纸博物馆。这是一个新建的碾坊，主要为游客演示抄纸工艺，介绍香纸沟土法造纸的工艺流程，其中包括介绍香纸沟造纸起源和72道造纸工序。调查时发现，这72道工序其实只是将造纸的工序逐一列出来，有凑数之嫌。

调查时发现，只有白水河村的碾坊在水灾后保存完好，还能继续按古法生产草纸。随着香纸沟旅游观光业的拓展，古法造纸又有恢复兴旺的迹象。

⊙1

⊙2

⊙1
香纸沟的抄纸展演
Papermaking show in Xiangzhigou Scenic Spot

⊙2
香纸沟土法造纸工艺流程展板
Flowchart of traditional papermaking techniques in Xiangzhigou Scenic Spot

四
乌当竹纸的生产工艺与技术分析

4
Papermaking Technique and Technical Analysis of Bamboo Paper in Wudang District

(一)
乌当竹纸的生产原料与辅料

生产乌当竹纸的主要原料为生长了两三年的苦竹、慈竹和钓鱼竹。正月砍苦竹、慈竹，二月砍钓鱼竹。如造纸户栽的竹子不够用，亦可到别家买。2009年调查时100 kg生竹子需16元，100 kg竹麻需32元。

生产乌当竹纸所需辅料为石灰和滑水。

当地人将纸药称为滑水。调查时造纸户所用纸药为滑根或糯叶。

(二)
乌当竹纸的生产工艺流程

明代宋应星的《天工开物》第十三卷《杀青》中关于生产竹纸的工艺有详尽的记载，书中还附有造纸操作图。

香纸沟土法造竹纸的过程基本沿用古法，但与《天工开物》记载的将煮好的竹子“取出入臼，舂成泥面状”不同，它是将处理过的竹子放入水碾槽里，碾成泥面状。

经访谈白水河、香纸沟等村民组多位造纸传人，总结乌当土法造竹纸的生产工艺流程如下：

壹	贰	叁	肆	伍	陆	柒	捌	玖	拾	拾壹	拾贰	拾叁
砍竹	修竹	破竹	晾竹	捆竹	打石灰水	煮竹	洗竹	装池	发酵	滤干	碾竹麻	上槽

拾肆	拾伍	拾陆	拾柒	拾捌	拾玖	贰拾	贰拾壹	贰拾贰	贰拾叁	贰拾肆
加水搅拌	放水	加清水	放水过滤	加入滑水	抄纸	压榨	揭纸	晾纸	理纸	裁纸

壹

砍竹

1 ⊙1

正月砍苦竹、慈竹，二月砍钓鱼竹，均为2～3年生的老竹子，长3～4 m，每根竹子的质量为3.5～4 kg，一天可砍300～350 kg。

贰

修竹

2 ⊙2

用柴刀修掉被砍倒的竹子的竹叶。

叁

破竹

3 ⊙3

用木槌将小竹子捶破即可，不需要破成若干片；将大竹子拿在手里用脚踩破。长的竹子可砍成3～4截，短的竹子可砍成2截。

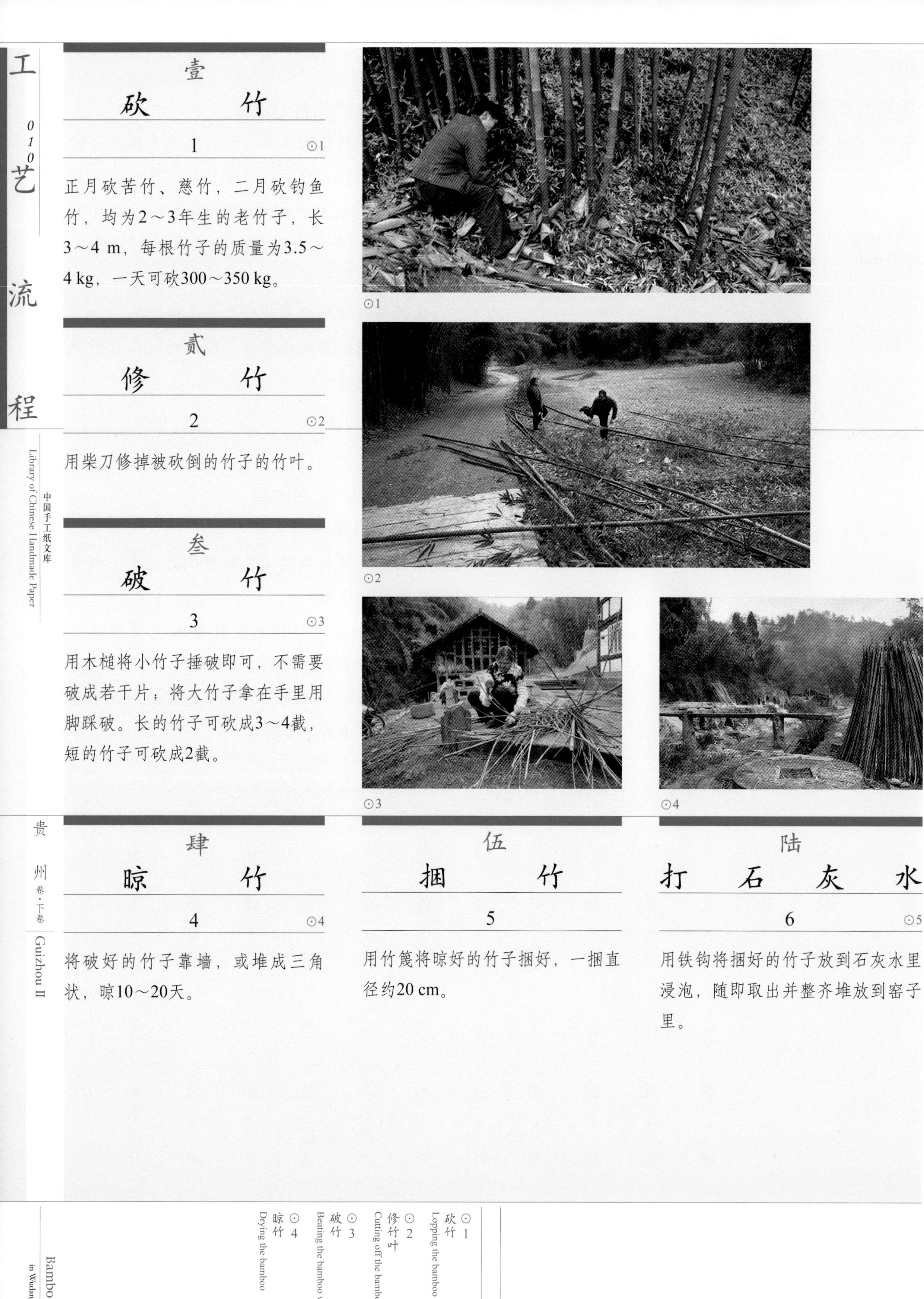

⊙1 ⊙2 ⊙3 ⊙4

肆

晾竹

4 ⊙4

将破好的竹子靠墙，或堆成三角状，晾10～20天。

伍

捆竹

5

用竹篾将晾好的竹子捆好，一捆直径约20 cm。

陆

打石灰水

6 ⊙5

用铁钩将捆好的竹子放到石灰水里浸泡，随即取出并整齐堆放到窑子里。

⊙1 砍竹 Lopping the bamboo

⊙2 修竹叶 Cutting off the bamboo leaves

⊙3 破竹 Beating the bamboo with a wooden mallet

⊙4 晾竹 Drying the bamboo

柒

煮竹

7

加水，顶部用泥巴盖住。一窑可放10 000 kg生竹子，需煮一个月，约用2 t煤。

⊙5

捌

洗竹

8

待竹子煮烂后，拿到河里将石灰洗掉，大约15个人一天可洗干净整池竹子。

玖

装池

9

将洗净的竹子再装到池子里，因池内还有余温，发酵会更快。

拾

发酵

10

在池子里自然发酵半个多月后，加清水浸泡。

拾壹

滤干

11

发酵后的竹子称为竹麻，用时取出，将水滤干。

拾贰

碾竹麻

12 ⊙6⊙7

将竹麻置于碾子里碾，一般一次可碾1 500～2 000 kg，具体根据碾子大小而定，一般一槽要碾七八个小时。若竹麻容易碾，一天可碾两槽。水车在碾子下，水流没有落差，只能靠水的冲力带动碾子，因此，碾竹麻前需有水先流进去。水板即是阻挡水流入的拦水板，一拉开水板，水便流进去，水车就转动起来。

⊙6

⊙7

⊙5 打石灰水 Soaking the bamboo in limewater

⊙6 碾竹麻 Grinding the bamboo materials with a grinder

⊙7 拉开水板 Removing the wooden water barrier

拾叁

上　槽

13

一槽料分3次放，150 kg料（包含水）可抄50～100 kg纸。

⊙8

拾肆

加　水　搅　拌

14　⊙8

上槽后，加水搅拌。

拾伍

放　水

15

放掉大部分水，使料逐渐沉淀下来，同时尽可能多地除去污水。

拾陆

加　清　水

16　⊙9

抄纸前，一边往槽里加清水，一边搅拌。

⊙9

拾柒

放　水　过　滤

17

适当放掉部分水，使大部分料沉淀。

拾捌

加　入　滑　水

18　⊙10⊙11

往槽里加滑水后，搅拌均匀，一般一天需要加3～4次滑水。

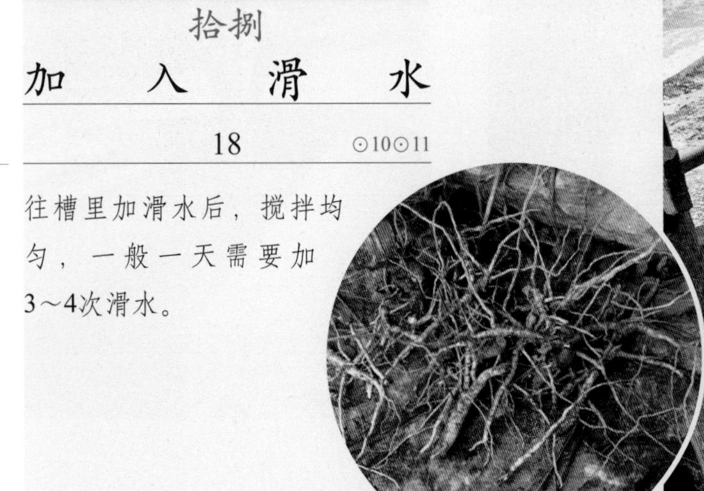
⊙10

⊙11

⊙8 搅料 Stirring the papermaking materials

⊙9 加清水 Adding in water

⊙10 滑根 Papermaking mucilage as adhesive

⊙11 搅纸浆 Stirring the paper pulp

拾玖
抄纸

19 ⊙12～⊙15

手持帘架，前后振荡两次，即可得一张湿纸膜。取下纸帘，转身倒扣盖于湿纸垛上。

⊙12 ⊙13 ⊙14 ⊙15 ⊙16

贰拾
压榨

20 ⊙16～⊙18

抄完纸后，将废旧纸帘盖在湿纸垛上，再盖上盖板、码子等，然后用榨杆缓慢压榨，压榨过程中需加两次码子。纸没抄完前，若需要做其他工作，可先用石头压榨。榨干后，一般在纸尾处打红，即用竹笔蘸上红色颜料在纸尾处点一下。

⊙17 ⊙18

⊙12/15 抄纸
Scooping and lifting the papermaking screen out of water and turning it upside down on the board

⊙16 石头压榨
Pressing the paper with stones

⊙17 石头上的红记
Red mark on stone

⊙18 打红用的竹笔
Bamboo stick for marking the paper

⊙19

⊙20

⊙21

贰拾壹 揭纸 21

用手一张张地从纸头揭纸，水从纸头流到纸尾，纸头比纸尾略厚。一次可揭17～18张。

贰拾贰 晾纸 22

⊙19

将揭好的纸放到室内竹竿上晾干。若天气好，一周可晾干；若天气不好，需要一个月，甚至更久才能晾干。

贰拾叁 理纸 23

⊙20

将晾干的纸取下，理齐。

贰拾肆 裁纸 24

⊙21

用裁纸刀裁纸时，需将一提纸头和另一提纸尾并放，以保证纸张厚度均匀。首先把毛边裁掉，然后用比子测量出纸的大小，最后再裁纸。

⊙19 晾纸 Drying the paper

⊙20 理纸 Sorting the paper

⊙21 裁纸 Trimming the paper edge

(三)

乌当竹纸生产使用的主要工具设备

壹 比 子 1

竹片，裁纸时用于测量纸张大小的工具，长约37 cm，柄宽约3 cm，上有两切口，用于确定香纸大小。

⊙22

贰 晾纸耙 2

将纸顶上和取下的工具。长柄长约92 cm，宽柄长约30 cm，一般由红籽树木制成。

⊙23

叁 鞋底板 3

布鞋底，上面放沙、柴油，可作磨刀工具。

肆 纸 帘 4

用竹子和丝线编织而成，长约 72.3 cm，宽约34.0 cm。

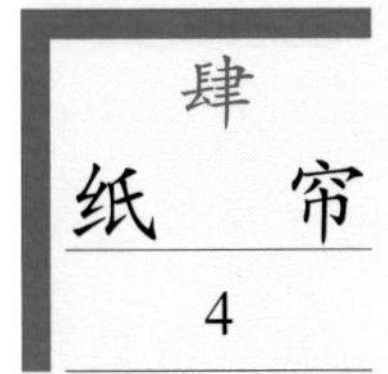

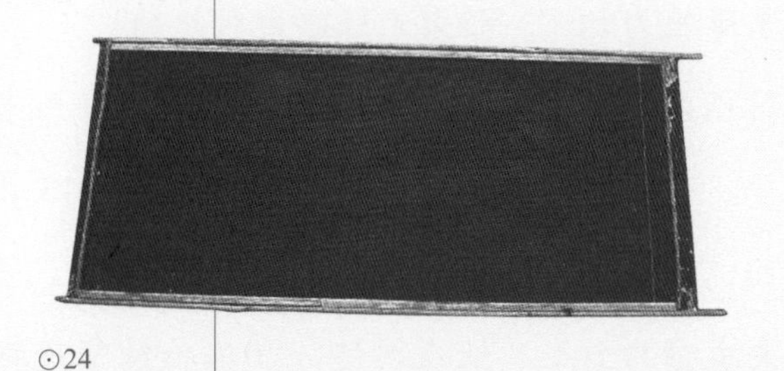

⊙24

伍 拍纸方 5

长约56.5 cm，宽约8.0 cm，高约6.5 cm。用于将捆好的纸拍整齐，前后左右四面都拍。

⊙22 造纸人用比子测量香纸的大小
A papermaker using a ruler to measure the size of Xiang paper

⊙23 晾纸耙
Rake for drying the paper

⊙24 纸帘
Papermaking screen

(四)

乌当竹纸的性能分析

所测乌当陇脚村竹纸的相关性能参数，见表7.1。

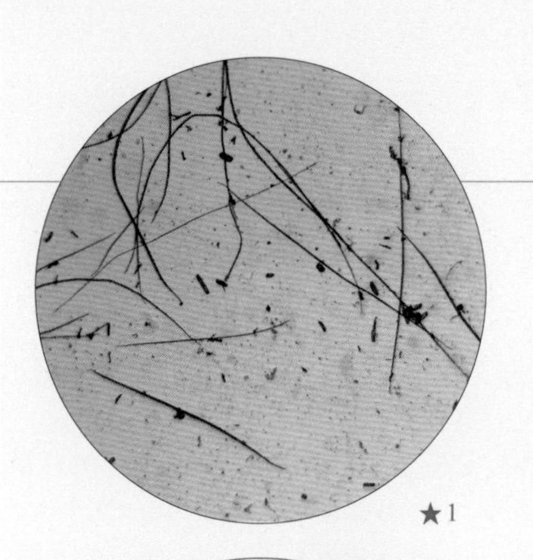

表7.1 陇脚村竹纸的相关性能参数
Table 7.1 Performance parameters of bamboo paper in Longjiao Village

指标		单位	最大值	最小值	平均值
厚度		mm	0.033	0.028	0.031
定量		g/m²	—	—	69.5
紧度		g/cm³	—	—	2.242
抗张力	纵向	N	6.6	4.7	5.5
	横向	N	5.0	3.4	4.6
抗张强度		kN/m	—	—	0.337
白度		%	12.7	12.5	12.6
纤维长度		mm	7.51	0.69	2.43
纤维宽度		μm	37.0	2.0	12.0

由表7.1可知，所测陇脚村竹纸最厚约是最薄的1.18倍，相对标准偏差为0.15%，厚薄较为均匀。竹纸平均定量为69.5 g/m²。所测竹纸的紧度为2.242 g/cm³。

经计算，其抗张强度为0.337 kN/m，抗张强度值较小，可能是因为乌当竹纸的主要成分为竹纤维，而竹浆纤维在所有禾本科造纸原料中纤维壁最厚，且很少有弯曲。

所测陇脚村竹纸白度平均值为12.6%，白度较低，白度最大值约是最小值的1.02倍，相对标准偏差为0.07%，差别相对较小，这可能是乌当竹纸竹浆未经漂白引起的。

所测陇脚村竹纸纤维长度：最长7.51 mm，最短0.69 mm，平均2.43 mm；纤维宽度：最宽37.0 μm，最窄2.0 μm，平均12.0 μm。所测竹纸在10倍、20倍物镜下观测的纤维形态分别见图★1、图★2。

★1
陇脚村竹纸纤维形态图(10×)
Fibers of bamboo paper in Longjiao Village (10× objective)

★2
陇脚村竹纸纤维形态图(20×)
Fibers of bamboo paper in Longjiao Village (20× objective)

五
乌当竹纸的用途与销售情况

5
Uses and Sales of Bamboo Paper in Wudang District

(一)
乌当竹纸的用途及演化

清乾隆年间，由于香纸沟声名远播，当地所造的纸被皇帝赐为御贡的“神圣”冥品。从此当地造纸业进入鼎盛时期，甚至发展到家家有造纸作坊、户户有香纸的盛况。由于成为御贡冥品，香纸的价格一度达到“万纸担米”的最高峰。

民国时期，香纸的用途有所增加，除了用于制作祭祀冥品外，部分大户人家将其用作卫生纸，此用途一直延续到20世纪60年代。

20世纪七八十年代，这种土纸除少部分继续用于制作冥品外，大部分用于制作烟花、爆竹及其他产品的包装材料。21世纪以来，随着机械与半机械竹纸品种增多，古法造的“香纸”在人们的生产生活中，除用于制作冥品外已别无他用。

(二)
乌当竹纸的销售渠道与价格

陇脚村草纸的传统销售市场是贵阳市和开阳、修文县以及相邻乡镇，一般是当地商人上门收购，不过也有造纸户自己运到贵阳和地方市镇去销售的。

在当地，每个造纸户造纸年产量为1 000~5 000 kg。抄出的纸在没打钱眼、没成香纸之前，每千克大约能卖8元；打了钱眼的成品，每千克大约能卖10元。

⊙1

⊙1 打钱眼 Perforating the paper

六 乌当竹纸的相关民俗与文化事象

6 Folk Customs and Culture of Bamboo Paper in Wudang District

⊙1

⊙1
『越国汪公』及蔡伦的牌位
Memorial tablets of Wang Hua, lord of Yue State, and Cai Lun the originator of papermaking

(一) 祭蔡伦与越国汪公

据调查中汪长伦老人口述，陇脚村每年6月6日都要用香、纸、肉、饭、酒祭拜蔡伦，只有磕头拜祭后才能吃东西。拜祭时要讲“祭念你，保佑我们”之类的吉祥话。

香纸沟的39户造纸人家中，除一户姓朱外，其他都姓彭。据彭氏最年长的老人彭庆尧（号俗卿，外号小诸葛，生于1915年）回忆，每年清明节后的第一个星期六，是彭氏祭祀蔡伦先祖、烧纸挂青的日子。届时，全族男性，包括那些已经在外成家立业的人，都要回来参加聚会。彭氏家族的神龛上供奉着“越国汪公彭氏宗祖之位”的牌位。他们称“越国汪公”为“宗祖”，而不以汪为姓，这是明代移民文化的一个特殊案例，调查时彭氏后人对此也语焉不详。

据安徽地方文献《新安文献志》所载，汪公原名汪世华，徽州歙县人，在徽文化中有着重要的地位。隋朝时，任徽州地方官，曾占据着江南六州之地，后来归顺唐，随唐太宗李世民征战有功，死后被追封为“越国汪公”，为避唐太宗讳，改名汪华。明代时，徽州较大的汪公庙就有26座。洪武四年（1371年），朝廷封越国汪公为神，命春秋祭祀，祭祀越国汪公进入国家正祀的行列，徽州汪姓都奉越国汪公为祖先。这也说明了彭氏家族与徽州移民有着密切关系，从而为香纸沟竹纸工艺的来源提供了想象空间。

此外，在汪氏的祠堂同时供着蔡伦的牌位。

(二) 柴刀砍竹习俗

由于香纸沟的纸用竹子作原料，所以香纸沟人对竹子有着敬畏之心。在砍竹前，先要举行庄重的仪式，在竹林前烧3炷香后再砍，并且只能用柴刀砍竹，忌用篾刀，这是因为用柴刀砍竹有“发财”

之意，而“篾”音同“灭”，有“灭竹”之意。

(三)

女性不得参与造纸

当地造纸的工匠必须是男性，女性不得参与，因为他们认为女性身有不净，若让其造纸，是对皇帝、神灵的不敬。这种风俗在其他造纸地区并不多见，是对女性的一种性别歧视。调查组推测，这或许与乌当竹纸承担庄严祭祀和供御的文化有关。

七 乌当竹纸的保护现状与发展思考

7 Preservation and Development of Bamboo Paper in Wudang District

(一)

乌当竹纸的生存与保护现状

长期以来，陇脚村古法造纸是当地农户的主要经济来源之一，为当地经济发展发挥了十分重要的作用，但是近年来却濒临生产危机。调查中，当地造纸人对此总结出如下原因：

(1) 由于土纸生产工序繁多，技术难度大，所以年轻人都不愿学，而选择外出打工。他们认为造纸不但劳动强度大、生产周期长、利润低，而且工作地点在乡下，生活单调乏味，不如到城里打工强，故土法造纸技艺已出现传承断代、后继无人的现象。

(2) 市场上，各种机制竹纸对土纸的冲击很大，部分从事手工纸生产的农户为了省时、省力，也将其手工生产转向半机械生产。

（3）由于受到了2003年洪灾的毁灭性冲击，香纸沟古法造纸作坊绝大部分已毁，维修成本高，且造纸的利润太低，造成许多作坊被荒废，当地的手工造纸业难以维系。如果不加强扶持，或许以后只能在香纸沟的造纸博物馆参观造纸表演了。

(二)

乌当竹纸的发展机遇探索

随着陇脚村的古法造纸工艺被列入首批公布的国家级非物质文化遗产名录，以及香纸沟被评为贵州省级风景名胜区，政府加大了投入力度，每年有数万人次的游客来到新堡乡陇脚村、马头村等处观光旅游。依托这些得天独厚的生态文化旅游资源进行开发，一方面促进了当地农村产业的结构调整，如陇脚村220多户农户中就有180多户从事与旅游业相关的职业；另一方面旅游业态的快速发展也为恢复古法造纸提供了重要的机遇。

旅游开发使当地传统的古法造纸作坊恢复了生机与活力，调查时，陇脚村已有40多户农户重新燃起了造纸的念头。但是，这因旅游业的发展而恢复的古法造纸究竟能有多大的发展前景，目前尚不明确。

政府目前对乌当竹纸的重视和投入有目共睹，但面对后继乏人的现实危机，建议重点扶持和补助传承有序且技艺出色的造纸户，留住那些外出打工的子弟，使古法造纸工艺得以延续。

在对外宣传策略上，当地一直自称是古法蔡伦造纸嫡传，并自称其所造纸是“传承千年的布依族土法造纸”“中国最古老的造纸活化石”。调查组建议当地能更平实地向外界宣传陇脚村的造纸工艺历史，因为，当地祭祀蔡伦并不意味着陇脚村的造纸工艺与明代的竹纸制造工艺相矛盾，同时，汉族造纸对乌当竹纸文化的贡献也不宜隐晦不提。

⊙1

⊙1 香纸沟的纸坊
Papermaking mill in Xiangzhigou Scenic Spot

乌当竹纸

Bamboo Paper in Wudang District

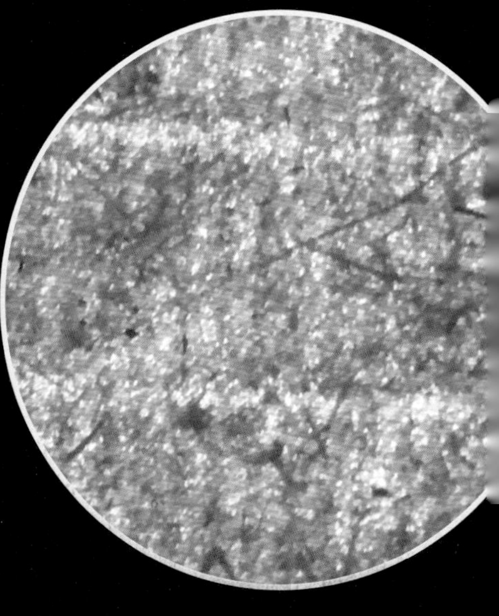

陇脚村竹纸透光摄影图

A photo of bamboo paper in Longjiao Village seen through the light

第八章
遵义市

Chapter Ⅷ
Zunyi City

第一节 仁怀手工纸

贵州省
Guizhou Province

遵义市
Zunyi City

仁怀市
Renhuai City

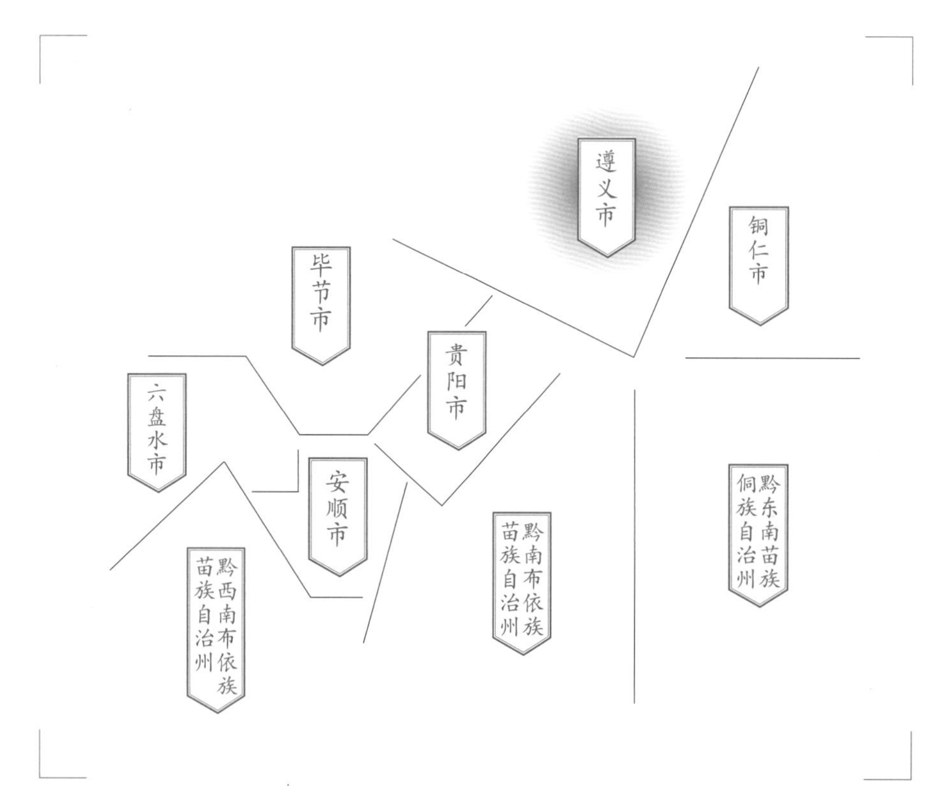

调查对象

五马镇
三元行政村
皮纸

调查对象

鲁班镇
竹纸

Section 1
Handmade Paper in Renhuai City

Subjects

Bast Paper in Sanyuan Administrative Village of Wuma Town

Bamboo Paper in Luban Town

一
仁怀手工纸的基础信息及分布

1
Basic Information and Distribution of Handmade Paper in Renhuai City

仁怀历史上出现过远近闻名的皮纸和竹纸的手工造纸业。手工竹纸的生产大约从清代初期开始出现，而皮纸则于清代中期开始出现，与贵州省一些明代就开始造纸的地方相比，仁怀手工造纸虽起步较晚，但发展规模却比较突出。

仁怀的慈竹、巴竹、寮竹、刺竹和各种山草资源丰富，当地以竹和山草混合制成的手工纸叫作草纸，且“随地可造”。鲁班纸厂坡、隆胜火石堂、新田长槽沟等处历史上都曾办过纸厂[1]，其中鲁班镇的生产规模较大，延续时间也较长。20世纪90年代以后，鲁班镇的草纸生产活动才逐渐消失。

仁怀皮纸主要产于五马镇境内五马河东岸三元村的岩头、沙坡、彭头山，五马河西岸茅坝镇的石关、一碗井、二合湾、马洞岩、新寨、石良坝等几个村落。皮纸生产的主要聚集地——三元村位于赤水河支流的五马河岸边，依山傍水，水源充足，交通便利，构树资源丰富，当地还产煤和石灰，因而成为生产手工皮纸的理想地点。2009年4月，因涉及赤水河水源保护问题，地方政府最终强行关停所有手工造纸作坊。2009年4月底调查组到达时，仁怀历史上一度十分红火的活态手工造纸业基本消失。

⊙1

⊙1 五马河边的造纸村落——三元村
Papermaking sites in Sanyuan Village alongside Wuma River

[1] 蒲宗尧.仁怀县志[M].贵阳:贵州人民出版社,1991:359.

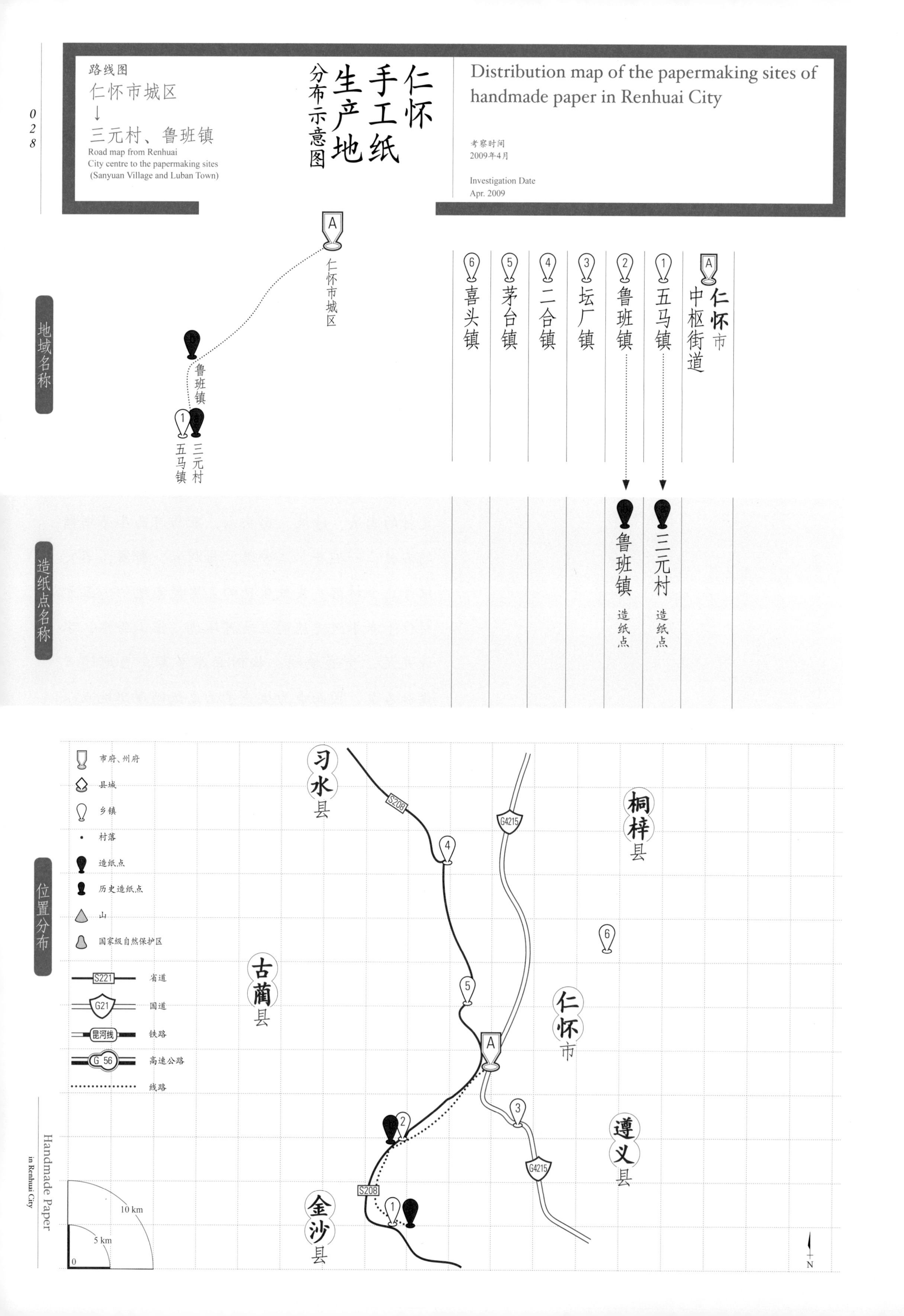
028
路线图
仁怀市城区
↓
三元村、鲁班镇
Road map from Renhuai City centre to the papermaking sites (Sanyuan Village and Luban Town)
仁怀手工纸生产地分布示意图
Distribution map of the papermaking sites of handmade paper in Renhuai City
考察时间
2009年4月
Investigation Date
Apr. 2009
地域名称
仁怀市城区
鲁班镇
五马镇
三元村
仁怀市
中枢街道
五马镇
鲁班镇
坛厂镇
二合镇
茅台镇
喜头镇
造纸点名称
三元村造纸点
鲁班镇造纸点
位置分布
市府、州府
县城
乡镇
村落
造纸点
历史造纸点
山
国家级自然保护区
S221 省道
G21 国道
昆河线 铁路
G56 高速公路
线路
习水县
桐梓县
古蔺县
仁怀市
遵义县
金沙县
S208
G4215
10 km
5 km
0
N
Handmade Paper in Renhuai City

二 仁怀手工纸生产的人文地理环境

2 The Cultural and Geographic Environment of Handmade Paper in Renhuai City

仁怀市隶属遵义市，地处贵州省西北部、大娄山脉西段、赤水河中游，地理坐标为北纬27°33′30″~28°10′19″，东经105°59′49″~106°35′50″，地域面积1 788 km²，是云贵高原向四川盆地过渡的典型山地，东南高、西北低，梯级分布，落差明显，地貌多样。

仁怀市北连习水县，东界桐梓县、遵义县，南临金沙县，西接四川古蔺县，自古以来就是黔北门户，是川盐入黔和盐马交易的要道。全市辖3个街道、12个镇、6个乡，居住着汉、苗、布依、仡佬、彝、壮、水、白、土家等11个民族，65万人，是地少人多的山区内陆地区。境域属中亚热带湿润季风气候，冬无严寒，夏无酷暑，年平均无霜期300天以上，年平均温度15.8 ℃，年平均降水量1 019.8 mm，适宜植物生长，森林覆盖率为26.47%，丰富的煤炭、石灰石和竹木资源为当地发展手工造纸业提供了良好的条件。

早在旧石器时代，仁怀辖地就已经有了古人类的活动[2]。1994年4月，考古工作者在仁怀云仙洞发掘出古人类居住用火的灰烬层以及商周时期的陶器、石器等。汉唐时期，仁怀属巴蜀荆楚的边缘地带，主要由少数民族定居开发。自宋代开始，大量汉族移民迁入，形成了多民族杂居的格局。北宋大观三年（1109年）置县，辖属关系几经更迭，明万历二十八年（1600年）改土归流，次年重置仁怀县。1995年11月30日，经国务院批准，仁怀撤县设市。

仁怀不仅是“四渡赤水”的重要战场，也是茅台酒的故乡，境内有多处自然风光和人文景观，观光旅游、红色旅游资源丰富。

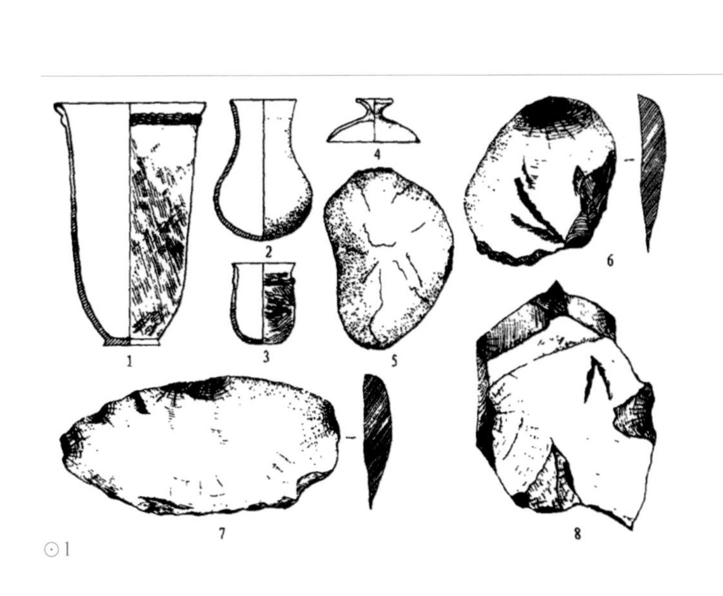

⊙1

⊙1 仁怀市云仙洞遗址出土的陶缸、陶圆底壶、陶杯、陶杯盖及石器
Earthenwares and stonewares unearthed from Yunxian Cave in Renhuai City

[2] 贵州省仁怀市地方志编纂委员会.仁怀市志(1978—2005)[M].北京:方志出版社,2013.

⊙1

五马镇是仁怀皮纸生产的集中区，位于市区西南方向，距城区直线距离19 km，沿“208”省道驾车行驶距离大约29 km。这里背靠摩天岭，前指五马河（古称吴江），文化底蕴深厚，因辖区内的唐代永安古寺以及古街、古桥、古墓群、摩天岭林场、田园风光等景观而声名远播。当地流传的《五马之歌》唱道：“巍巍摩天岭下有条河，流淌着古老文明的歌，歌声里依稀可见神仙足迹，歌声里早已收藏将士雕戈，古镇古街古桥古刹，骚人墨客盐夫胜多，五马河啊五马河，古韵悠悠千载颂，金光灿灿万顷波。巍巍摩天岭下有条河，流淌着现代文明的歌，流淌着和谐社会的歌。”

五马镇有一批诗词爱好者长年坚持创作诗词，他们把诗词文化建设与学校教育、“四在农家”（富在农家、学在农家、乐在农家、美在农家）的口号相结合，长期开展“村民院坝唱山歌”“逢年过节诗词朗诵”等活动。全镇诗词文化墙多达百面，除展示历代经典诗词外，还经常推出本地诗人反映五马变化的原创作品。2009年创办的半年刊《吴江颂》已经刊印多期，2012年3月五马镇被授予贵州省“诗词之乡”的称号。

⊙2

仁怀竹纸生产集中在鲁班镇，鲁班镇离市区直线距离13 km，沿“208”省道驾车行驶距离大

⊙1 五马古街
Ancient Street in Wuma Town

⊙2 五马乡村文化节
Rural Cultural Festival in Wuma Town

约15 km。南边有摩天岭山脉，北边有玉车屯山脉，下场口两山对峙，形成钳口状，迎向东面是开阔的丘陵地带；向西是一条长约10 km的狭长山沟，地形险要，易守难攻，是重要的军事要塞。1935年3月15日，红军为了三渡赤水河突破重围，打响了鲁班场战役。目前，鲁班镇内有保存完好的古碉遗址、鲁班烈士陵园以及鲁班场战役纪念碑。

鲁班辖区的陶家寨有一座省级文物保护单位“双魁田四合院”，这座典型的黔北穿斗式民宅院落建于清光绪二十二年（1896年）。因房屋主人蔦氏家族中有开甲、开仲两公参加县府两级考试并双双夺魁，乡民们遂将此地称为“双魁田”。建筑坐北朝南，占地约1 200 m²，建筑面积884 m²，有房屋34间，其中正房7间，左右厢房各2间，前面为下厅，形成四合天井。堂屋及客房悬挂多副对联，如“岳峙渊涵，学为世用；鸢飞鱼跃，心与天游”（严寅亮）等晚清名人手迹。

⊙3

20世纪30年代这里曾办过私塾，受其影响，房屋主人蔦显威自学自修成为乡村诗人，并于78岁高龄时出版了诗集《怒涛之声》（2000年）。其弟蔦显刚在71岁时出版了诗集《鳞爪集》（2004年）。

⊙4

⊙3 堂屋正面陈设 Furnishings of the living room

⊙4 双魁田四合院内院落 Shuangkuitian Courtyard

三 仁怀手工纸的历史与传承

3 History and Inheritance of Handmade Paper in Renhuai City

(一) 仁怀手工纸的历史

据地方史志文献记述，仁怀手工造纸业有300多年的发展历史。清代初年，在五马镇三元村地区开始出现了竹纸生产活动，与贵州一些明洪武、嘉靖年间就开始出现手工造纸的地区相比，仁怀的手工造纸起步相对较晚。在造纸村落的民间记忆中，仁怀造纸大约从清初开始出现。据五马镇三元村造纸人李文清介绍，李姓人家是迁居此地最久的家族，从祖先定居三元村开始李家就以造纸为业。村里的李姓子孙都熟知族谱："应茂先芝，父国庭正，永吉清安，天启太平，雨洪思万，上元华新。"从"应"字辈开始到现在的"清"字辈，已历11代，造纸技艺代代相传应该有接近300年的历史了。

⊙1

⊙1 五马河 Wuma River

清代中期，仁怀造纸业有了初步发展，成为新增的皮纸主要产地之一[3]。仁怀地方文化学者徐文仲发现三元村《程氏家谱》中记载了当地手工造纸历史的重要信息：清道光中叶，居住在仁怀县三元行政村岩头村民组的朱怀顺，看到村庄周边的林、水、煤俱全，宜于制造皮纸，便到当时造纸较发达的遵义县西安寨瓮花水（在岩头村东南方向 20 km处，现属遵义县泮水镇）学习造纸技术。学成归来的朱怀顺在五马河边的凉水井旁开设皮纸槽，在当地原来的造纸方法的基础上进行改进，开始制作皮纸，并收徒授艺。由于原料丰富、成本低、质量好、价格合理，皮纸销路愈来愈宽，四川古蔺县贩卖皮纸的商人和纸工不断迁来落户，修建了街房数十间，三元村逐渐成为新兴的皮纸贸易集镇，促进了皮纸生产的发展。到了清末，仅三元村一带造纸专业户就有百余户，皮纸的质量也明显提高[4]。

民国初年，仁怀手工造纸业开始呈现出繁荣景象。据民国四年（1915年）县政府建设科统计，全县有450户人家从事造纸业，其中，80户人家生产火纸（草纸），产量为70 200捆；370户人家生产皮纸，年产皮纸24 690担[5]。20世纪30年代开始，仁怀的竹纸和皮纸的质量和特色逐渐在遵义府辖区内崭露头角。据《续遵义府志》记载，造纸作坊生产的竹纸（毛边纸）“以金竹、水竹制成者，可抵川纸之红批毛边，仁怀制者与板桥（今遵义县板桥镇）同”[6]，“芦江水（仁怀境内五马河上游）专以构皮制成曰皮纸，再舀而成者曰夹皮纸，多行本属及四川川北一带”[6]。这说明当时仁怀所产的竹纸和皮纸都已经进入了黔北知名产品的行列。

抗日战争时期，大量机关、学校、工厂搬迁到贵阳，用纸需求激增，手工纸生产迅速发展。据1943年三元村蔡侯会（即造纸同业公会）统计，三元村顺河一带及茅坝、五马附近方圆约 7.5 km的地区内，已有造纸户500多家，拥有纸槽250余架[5]。抗日战争结束后，手工纸的需求回落，仁怀手工造纸业迅速衰退。1947年，仁怀生产皮纸2万刀，约7 t，占全省的比重分别为0.17%和0.18%；生产草纸10万刀，约24 t，占全省的比重分别为0.49 %和0.47 %[7]。

到了1949年，纸的产量又重新回升，仅皮纸一种年产量100 t左右。每天在这里从事经营活动的纸商、搬运工人不少于300人，专供他们食宿的旅店就有20多家，三元成为仁怀境内仅次于茅台的工商业集市[5]。

20世纪50年代以后，仁怀造纸业经历了“社会主义改造运动”和“改革开放”两次大的转折。1953年秋天，仁怀县政府投资5 000元，兴建地方国营仁怀三元皮纸厂。皮纸厂虽仍按老法生产，却改变了纸业私营的格局，首开仁怀历史上规模化生产手工纸的先河。三元皮纸厂有纸槽20架，焙子（烘纸用）60个，职工160人。1955年，生产53 cm × 53 cm的白皮纸8万多张。1956年合作化高潮期间，三元一带成立了5家皮纸生产合作社。此后，生产皮纸的原料由供销社收购供应，皮纸也由供销社包销。1958年，三元皮纸厂启动公私合营模式，对原有的合作社实行转厂过渡，并将私营皮纸作坊纳入国营仁怀皮纸厂，设立二分厂，职工迅速扩大到380多人。1958年，大规模

[3] 贵州省地方志编纂委员会.贵州省志·轻纺工业志[M].贵阳：贵州人民出版社，1993：53.

[4] 徐文仲.三元白皮纸[J].贵州文史丛刊，1986(2)：71.

[5] 蒲宗亮.仁怀县志[M].贵阳：贵州人民出版社，1991：359.

[6] 周恭寿，等.续遵义府志：卷十二[M].民国二十五年(1936年)刊本.

[7] 贵州省地方志编纂委员会.贵州省志·轻纺工业志[M].贵阳：贵州人民出版社，1993：54.

毁林开荒造成原本丰富的构树资源锐减，20多年后生态环境才逐渐恢复[8]，这对皮纸生产产生了严重的影响。1960年，皮纸产量达到第一个历史峰值（173 t），此后，皮纸生产开始了连续15年的低迷（表8.1）。自1964年起，三元皮纸厂对传统手工纸生产进行了一系列造纸工艺和技术的改造。

从1978年开始，仁怀手工皮纸生产又恢复为个体经营，生产规模不断扩大，沿五马河岸迅速形成了一批皮纸作坊集群。2005年，当地造纸户成立了以三元为核心的贵州省仁怀市五马镇手工造纸协会，共有会员200~300家。2008年，仁怀手工皮纸产业达到鼎盛，仅五马河沿岸的手工造纸户就达到279户，从业人员达724人。

与皮纸的情况相似，草纸制造业也进行了公有制改造。从20世纪50年代中期到20世纪70年代末，仁怀草纸属集体副业，生产经营归生产队或生产大队所有。1954年，长岗区供销社接管经营在民国年间由杨志义创办的桑树湾毛边纸厂（今在坛厂镇境内）。1957年，该造纸厂改用谷草、笋壳、巴竹、黄篾丝等为原料制造毛边纸，并获得了成功，当年产纸837担。1958年，增设纸槽4架，从业职工由原来的10人增加到45人。“大跃进”时期，县商业局提出“农产品加工一条龙”的口号，全县每个生产队“大食堂”都建了纸浆厂，用大麦秆、小麦秆打成堆，自然发酵，制浆舀黄板纸，结果造成大量积压，导致黄板纸生产最终结束。1959年，仁怀草纸产量在达到第一个历史峰值（245 t）后开始了近十年的低迷。桑树湾毛边纸厂也终因严重亏损于1962年停办[9]。竹纸生产早在20世纪70年代以后日渐式微，并在20世纪90年代最终退出人们的视野。

⊙1

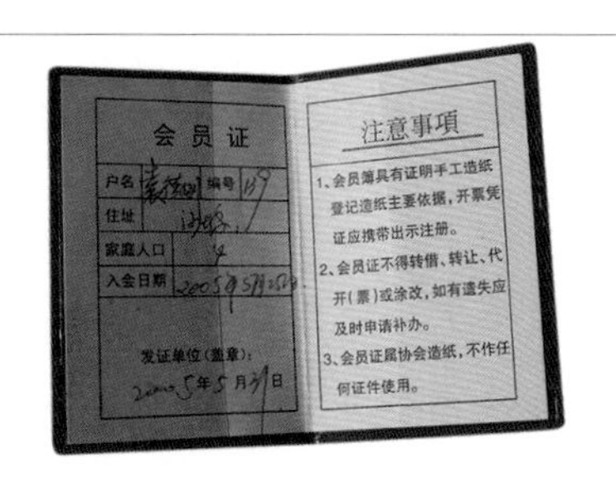
会员证
户名　编号
住址
家庭人口
入会日期
发证单位（盖章）：
年　月　日

注意事项
1、会员簿具有证明手工造纸登记造纸主要依据，开票凭证应携带出示注册。
2、会员证不得转借、转让、代开（票）或涂改，如有遗失应及时申请补办。
3、会员证属协会造纸，不作任何证件使用。

⊙2

⊙1 仁怀市五马镇手工造纸协会公章
Official seal of Handmade Papermaking Association in Wuma Town of Renhuai City

⊙2 手工造纸协会会员证
Membership card of Handmade Papermaking Association

杨存志老人20岁就开始造纸，经历了竹纸从繁荣到衰落的全过程，他认为，导致竹纸生产最终消失的原因主要有三个：一是水土流失造成水源不足，无法满足造纸用水的需要；二是造纸非常辛苦，仅仅依靠年龄大的人难以延续和发展；三是造纸经济效益差，年轻人不愿意从事造纸工作。

(二)
仁怀手工纸的传承

仁怀手工纸的传承主要有家族传承、师徒传承等形式。三元行政村岩头村民组的农民祖祖辈辈以造纸为主业，村民李文清祖上在清初时制作竹纸的技艺是通过父传子、兄传弟的方式一代一代传承的。掌握了手工纸制作技艺的家庭在子承父业的过程中也同时实现了技术的传递，这也是在仁怀手工造纸300多年的历史上，特别是在造纸技艺没有重大突破的时候，造纸技艺传承的主流

[8] 贵州省地方志编纂委员会.贵州省志·轻纺工业志[M].贵阳：贵州人民出版社，1993：52.

[9] 蒲宗亮.仁怀县志[M].贵阳：贵州人民出版社，1991：362，538-539.

表8.1 1949~1988年仁怀市构皮收购量与手工纸产量的变化
Table 8.1 Paper mulberry bark purchase and handmade paper output from 1949 to 1988 in Renhuai City

年份	构皮(t)	土纸(t)	皮纸(t)	年份	构皮(t)	土纸(t)	皮纸(t)	年份	构皮(t)	土纸(t)	皮纸(t)	年份	构皮(t)	土纸(t)	皮纸(t)
1949	-	109	36	1959	-	245	133	1969	-	215	21	1979	-	-	181
1950	-	82	30	1960	-	214	173	1970	-	87	21	1980	-	30	207
1951	-	91	30	1961	-	85	59	1971	-	79	25	1981	-	-	227
1952	5	90	35	1962	38	110	35	1972	-	532	53	1982	-	8.5	261
1953	-	128	30	1963	-	112	43	1973	-	182	55	1983	-	40	259
1954	-	77	134	1964	-	112	43	1974	-	323	62	1984	109	8	225
1955	-	100	98	1965	-	104	45	1975	88	617	110	1985	-	-	284
1956	-	68	88	1966	27	17	23	1976	-	121	173	1986	-	323	262
1957	20	73	43	1967	-	42	14	1977	-	41	200	1987	-	160	286.5
1958	-	107	129	1968	-	4	49	1978	40	47	240	1988	22	-	294.1

方式。

在造纸生产方式出现重大变革之前，仁怀的手工造纸从业者往往会选择外出学艺、请师上门或师徒传承的方式实现技术的革新与传播。“清道光中期，三元村的朱怀顺曾外出去瓮花水（地名）学习皮纸的生产技艺，后回到三元村，一边生产制作皮纸，一边收徒传艺。他收的第一个徒弟叫李廷尧，李廷尧又将造纸技术传给从四川太平渡迁来三元村落户的小商贩程绍先、程绍品兄弟，程氏兄弟又收徒传艺，这样辗转相传，互传技艺，几年间，开槽造纸的就有几十户人家了。”[10] 清咸丰、同治年间，朱怀顺去世，被安葬在三元村五马河边上。光绪六年（1880年），他的二徒弟李廷孝为师父立碑纪念[11]。几年前修建乡村公路时，原来墓地的位置由于在公路修建规划范围内，因此无法被看到，但是村里的许多造纸户至今还能指认出修路前坟茔的位置。

民国三十四年（1945年），长岗区桑树湾人杨志义开办了桑树湾毛边纸厂，雇请6个工人，并从遵义县请来2位造纸的技工师傅教授造纸技术。他们所产的毛边纸色泽淡黄、光滑抗水、呈半透明状，长127 cm，宽60 cm，当时主要有抄写公文、习字、裱褙、包装等用途。

⊙3

⊙3 三元村支书葛光远指认朱怀顺墓地遗址
Tomb site of Zhu Huaishun, a brilliant local papermaker during the Qing Dynasty (pointed out by the local secretary Ge Guangyuan of Sanyuan Village)

[10] 徐文仲.三元白皮纸[J].贵州文史丛刊,1986(2):70-71.

[11] 蒲宗亮.仁怀县志[M].贵阳:贵州人民出版社,1991:360.

四
仁怀皮纸的生产工艺与技术分析

4
Papermaking Technique and Technical Analysis of Bast Paper in Renhuai City

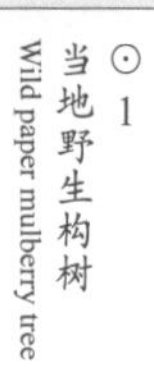

⊙1

⊙1 当地野生构树 Wild paper mulberry tree

(一) 仁怀皮纸的生产原料与辅料

1. 构树皮

仁怀皮纸所用原料为构树皮。尽管当地构树资源非常丰富，但仍难以满足大规模的手工造纸需求，对于长年专门从事手工造纸的人们来说，也没有足够的劳动力去采集和运输构树皮，特别是20世纪50年代末到70年代中叶，由于生态遭到破坏，构树皮资源更难以满足生产的需要。手工纸产业集群的规模生产进一步促进了社会分工的发展，三元村一带用于造纸的构树皮主要靠贞丰人从长顺、紫云等地采集后转运过来。2009年4月底，调查组入村调查时，干构树皮的售价是2 000 元/t。

2. 纸药

仁怀皮纸所用纸药包括玄麻、野生血藤、羊桃藤、三条筋、柁杉根等[12]。其中，玄麻也称作玄药、荨麻、荨药，是最主要的纸药。玄麻农历六七月开花，花瓣有黄、白、红三种颜色，叶子呈绿色，花蕊呈红色，根呈白色。三元村附近不生长玄麻，造纸户多从茅坝、云安、六板等地购买。以前茅坝镇安良村有人专门种玄麻卖给三元村造纸户，玄麻附生在小米地、高粱地等，一年种一次，收一次，1 m^2地可长1.5 kg，即使种了粮食1 m^2也可生长0.75 kg。2009年4月，玄麻的市场售价是0.5~0.7元/kg。先将玄麻浸泡透并用脚踩融，放入纸药槽后再加水调好浓度即可使用，一般准备好的纸药只能用一天。另一种主要的纸药是野生血藤，但它不如玄麻好用。从1999年开始，人们逐渐使用化学纸药——

[12] 徐文仲.三元白皮纸[J].贵州文史丛刊,1986(2):70.

聚丙烯酰胺，市场售价为20元/kg，可以捞4 000帘（12 000张）纸。

3. 石灰、煤和烧碱

仁怀手工皮纸生产的传统工艺中以石灰和煤炭浆煮原料，对环境的影响很小。从2000年前后开始，煤价飞涨，造纸户无力承担，不得已选择了烧碱。调查时，造纸户一致认为20世纪70年代生产的皮纸质量最好。当时人们用石灰作碱，用传统打碓方法打散构皮纤维，再用煤火蒸煮软化构皮纤维，这样造出来的纸纤维保存较为完好。后来改用烧碱，不但腐蚀性大，纤维被严重破坏，纸质变差，而且也严重污染了环境。

(二)
仁怀皮纸的生产工艺流程

调查组2009年4月底到三元村调查时，正值政府取缔造纸作坊，纸槽基本被拆除，已不再有抄纸活动，许多作坊抄出的纸垛尚未贴到墙上，还能看到少部分造纸工艺流程，大部分工艺流程仅能通过造纸户口述获得。

经现场记录整理，总结三元村皮纸的生产工艺流程如下：

壹	贰	叁	肆	伍	陆	柒	捌	玖	拾	拾壹	拾贰
砍构树	剥构皮	揭构皮	晒构皮	泡构皮	浆构皮	煮构皮	洗构皮	二次泡构皮	择构皮	打皮板	搅料子

拾叁	拾肆	拾伍	拾陆	拾柒	拾捌	拾玖	贰拾	贰拾壹	贰拾贰	贰拾叁	贰拾肆
滤料子	加荨药	攒槽	舀纸	吊垛子	退垛子	理垛子	揭纸	晒纸	撕纸	抖纸	捆纸

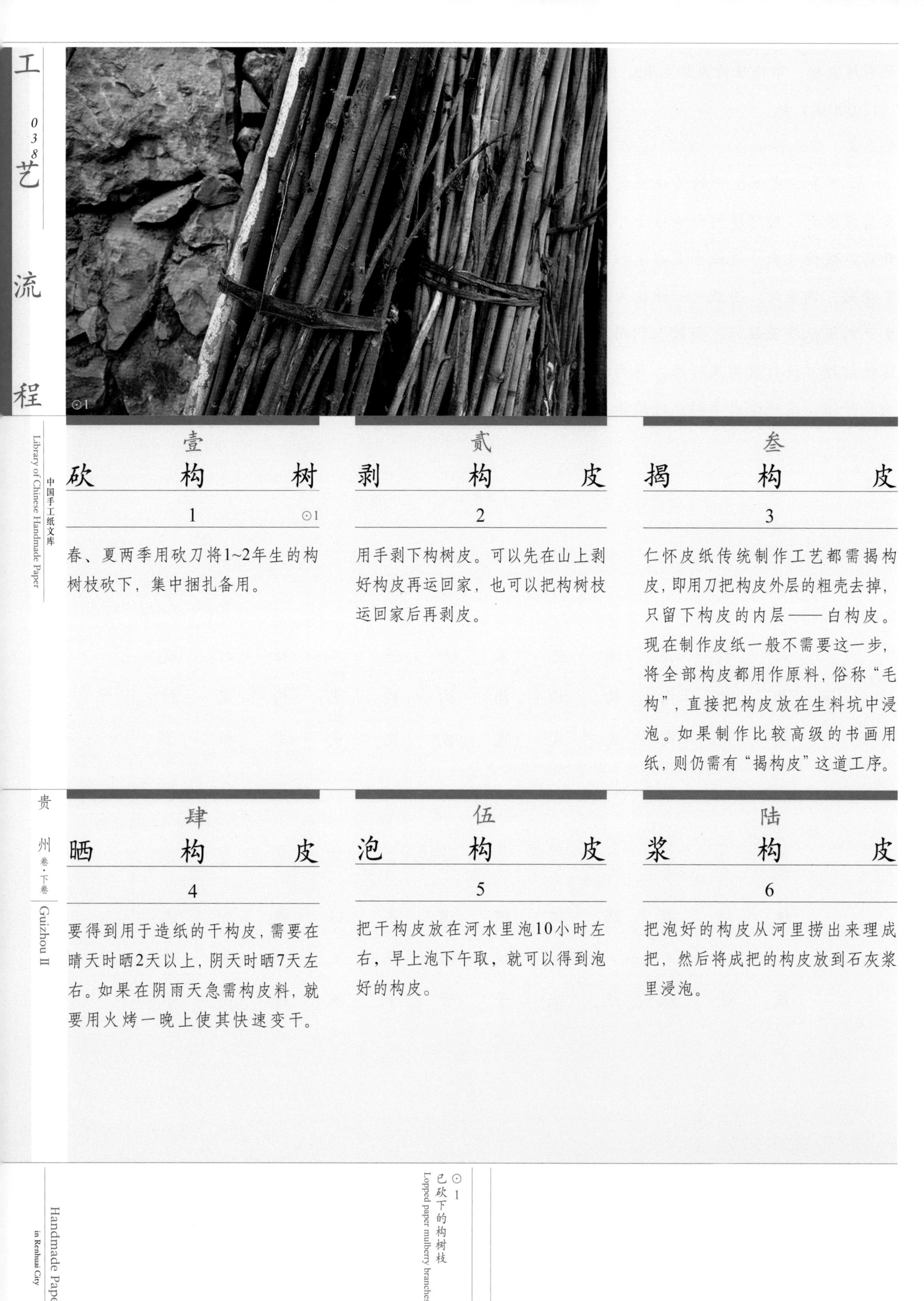

⊙1

壹 砍构树

1 ⊙1

春、夏两季用砍刀将1~2年生的构树枝砍下，集中捆扎备用。

贰 剥构皮

2

用手剥下构树皮。可以先在山上剥好构皮再运回家，也可以把构树枝运回家后再剥皮。

叁 揭构皮

3

仁怀皮纸传统制作工艺都需揭构皮，即用刀把构皮外层的粗壳去掉，只留下构皮的内层——白构皮。现在制作皮纸一般不需要这一步，将全部构皮都用作原料，俗称“毛构”，直接把构皮放在生料坑中浸泡。如果制作比较高级的书画用纸，则仍需有“揭构皮”这道工序。

肆 晒构皮

4

要得到用于造纸的干构皮，需要在晴天时晒2天以上，阴天时晒7天左右。如果在阴雨天急需构皮料，就要用火烤一晚上使其快速变干。

伍 泡构皮

5

把干构皮放在河水里泡10小时左右，早上泡下午取，就可以得到泡好的构皮。

陆 浆构皮

6

把泡好的构皮从河里捞出来理成把，然后将成把的构皮放到石灰浆里浸泡。

⊙1 已砍下的构树枝 Lopped paper mulberry branches

⊙2

⊙3

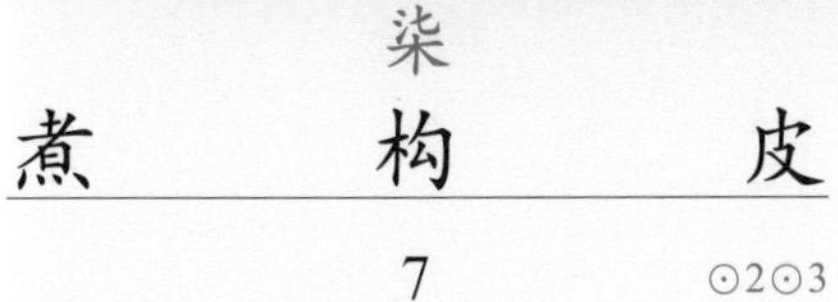

柒

煮构皮

7 ⊙2⊙3

传统工艺煮构皮所用的锅较小，每次可蒸煮干构皮约100 kg，需要加石灰50 kg。要将浸泡过石灰浆的构皮放在甑桥上，用铁盖盖好即可生火蒸煮，甑内的构皮需蒸煮一天一夜（24小时），中间不需再向甑内加水，蒸煮好的构皮叫作熟构皮。从操作方式看，这道工序应当叫作蒸构皮，但当地造纸户称之为煮构皮。当代蒸煮构皮所用的锅较大，一锅能蒸750 kg干构皮，需75 kg烧碱，先放一层构皮，再撒一层烧碱，也需蒸煮一天一夜，约需125 kg煤。造纸户们反映，近几年来，造纸户通常会加入纸花（废纸、废书），煮纸花所用的锅直径约100 cm，一次煮25 kg的废纸、废书需一个多小时，加约0.75 kg的烧碱，消耗木柴10~15 kg。

捌

洗构皮

8

传统工艺是把熟构皮从甑中取出，拿到河里用脚踩，让流动的河水将构皮上的石灰冲洗干净。一个人洗净一甑熟构皮，大约需要3小时。当代工艺是把蒸煮好的皮料放在熟料坑里泡，不用脚踩，第一次浸泡2小时，换水，第二次浸泡5小时，再换水，总共换4次水，大约需要24小时。

玖

二次泡构皮

9

将洗净的熟构皮放在河水里继续泡约一天时间。

拾

择构皮

10

把泡好的构皮从河水中捞起堆好，将构皮的细枝、黄筋、黑壳、废渣等择拣干净。一甑熟构皮一个人择拣，大约需要两天时间。

⊙2 待蒸煮的皮料 Papermaking materials ready for steaming

⊙3 桑皮 Paper mulberry bark

拾壹
打皮板
11 ⊙4

传统工艺是把择净的皮料放在石礅上用碓捶打，一次打一团约1.5 kg由干构皮做成的熟构皮，一个人打碓，一个人翻皮板，大约打7分钟，2天大约可以打好150 kg干构皮的纸料，打好皮板后就可以得到做皮纸的纸料。现在通常使用打浆机加工皮纸原料，一个人即可操作，同样150 kg干构皮的纸料，用7.5 kW电动机打40多分钟即可，省去一个人工及许多体力和时间。普通造纸户一年约需加工10 t干构皮，另需添加约0.5 t纸花。打浆的同时加入漂精，干构皮料和漂精的比例是30:1。前期用烧碱煮烂的纸花要再放到构皮中一起打散。有时造纸的用料还会应客户的要求做出调整。造纸户李文清家曾经应遵义一位画家客户的要求造过加棉花的皮纸，干构皮和棉花的比例是7:3。当时一天只能舀300张皮纸，棉花价格为20元/kg，每天要用2桶漂精（35元/桶），成本较高。最初舀的纸质量不是很好，每张纸卖0.8元，后来纸的质量好了，每张纸涨到1元。舀了半年纸以后，这位画家客户过世了，现在李文清家中还剩下部分当时造的纸没有销售出去。

⊙4

拾贰
搅料子
12 ⊙5

把打好的皮料背到纸坊，放入纸槽里，加入清水并用槽杆搅散。一次加20 kg左右干构皮，够一天抄纸所需。

⊙5

拾叁
滤料子
13

滤料子的目的就是清洗纸料和控制纸浆的浓度，要先把搅好的纸料舀起来装到料兜里滤干，再根据舀纸时纸浆浓度的需要逐渐把纸料加入纸槽，通常一天会补充6次左右。

⊙4 打浆机加工的皮纸料 Papermaking materials processed by electronic beater

⊙5 搅料子 Stirring the paper pulp

拾肆
加　荨　药
14

传统工艺使用荨麻作纸药，纸料装入纸槽后要加入准备好的荨药，正常生产的情况下，一天需消耗2.5~3 kg，天热时，一天需消耗3.5~4 kg。要配合纸料的使用分6次加入和补充荨药，每放一次纸料就加一次荨药。

拾伍
攒　槽
15

加入荨药后，要用槽杆从左右两边呈倾斜角度匀速推拉大约5分钟，名叫“攒（cuán）槽”，其目的是使纸料和荨药充分混合。

拾陆
舀　纸
16

⊙6~⊙10

舀纸者手持纸帘架，使远身一侧纸帘先入纸浆，由外往里舀纸，舀起纸浆到帘上后，前后左右倾斜纸帘，使纸浆均匀分布在纸帘上，再使多余的纸浆从近身一侧流出。近身一侧的帘棍往内翻转，纸头变为双层，揭纸时更容易分开。

⊙6

⊙7
⊙8

⊙9

⊙10

⊙6/10
舀纸
Scooping and lifting the papermaking screen out of water and turning it upside down on the board

拾柒
吊 垛 子

17 ⊙11

舀纸者参照托台上2根等高的扳桩位置，一张一张地把舀起的湿纸反扣在托台上。一天舀出的纸叫作一榨垛，包括一帘二纸、一帘三纸或一帘四纸的纸垛。纸垛上要放好旧纸帘，加上杠板（厚木板），并在杠板中间及两边分别压上70~80 kg的石头，质量共约220 kg，目的是利用石头的重力挤出湿纸垛中的水分。大约2小时以后，用穿过托台底下木棍的铁丝圈和上面的吊杆（木杠）形成一个支点，压在纸垛上，再于吊杆的另一头加上75 kg左右的石头，持续增加压力约一夜的时间，继续排出湿纸垛中的水分。

⊙11

拾捌
退 垛 子

18 ⊙12

经过一夜“吊垛子”工序，湿纸垛中的水分大多被排出，第二天早上要先搬去杠板左右两边的石头，再去掉吊杆上的石头，最后除去吊杆和杠板等。

⊙12

拾玖
理 垛 子

19 ⊙13⊙14

把纸垛搬回家里并放到纸凳上，用手捏松纸垛的左、右、下三边，做好揭纸的准备。为了让皮纸在墙上粘得更牢，一般还要在纸垛的上、左、右三边刷上米浆。

⊙13 ⊙14

⊙11
吊垛子
Pressing the paper with stones

⊙12
退垛子
Removing the stones

⊙13
理垛子
Sorting the paper

⊙14
刷米浆
Pasting rice pulp on the paper as adhesive

贰拾
揭　　纸

20　⊙15⊙16

先用左手从纸垛的右上角往左上角揭开湿纸，再由上往下撕，右手拿棕刷接住纸的左上角，双手将纸刷在墙上。墙上同一位置最多可以错位刷5~6张，这样的纸叫作“一重纸”。

⊙15

⊙16

贰拾壹
晒　　纸

21　⊙17

此工序也可以称为“焙纸”。早年造纸户多用火焙（火墙）烘干湿纸，每面墙可贴三排纸，如果使用火焙烘干5~6层湿纸，则需大火烘4小时，小火烘5~6小时。通常起早贪黑一天可焙3次，早、中、晚各一次。火焙虽然省时，却消耗燃料，一天要消耗20 kg煤炭。自从煤炭价格上涨以后，人们已经不用火焙而改为在墙上自然晒干或风干。晴天时晒一天即可，阴天时则要晒两天。

⊙17

贰拾贰
撕　　纸

22　⊙18

皮纸焙干或晾干以后就可以从纸焙或墙上撕下，从哪里入手撕纸并无明确规定，可以用左手将纸的左上角撕开，然后往右撕，右手接住纸的右上角，再由上往下撕；也可以先撕右上角，再由右往左撕。

⊙18

⊙15 揭纸角 Peeling the paper down from the upper corner

⊙16 揭纸 Peeling the paper down

⊙17 刷纸上墙 Pasting the paper on a wall for drying

⊙18 撕纸 Peeling the paper down

贰拾叁 抖纸

23 ⊙19~⊙21

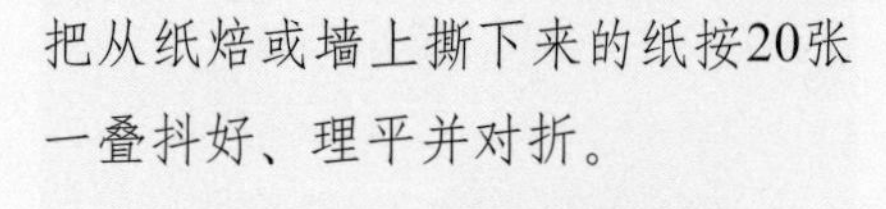

把从纸焙或墙上撕下来的纸按20张一叠抖好、理平并对折。

⊙19

⊙20

⊙21

贰拾肆 捆纸

24 ⊙22⊙23

把对折以后的皮纸按50叠1驮扎成一小捆，再把4~6驮扎成一大捆。早先人们通常会用皮纸搓成纸绳捆扎，大约从20世纪90年代开始改用塑料绳捆扎，这样会更加节约时间，也更加结实和方便运输。

⊙22

⊙23

仁怀皮纸制作的传统工艺和20世纪80年代以后逐渐采用的新工艺的对比如表8.2所示。

表8.2 仁怀皮纸制作的传统工艺和20世纪80年代以后逐渐采用的新工艺的对比

Table 8.2 Contrast of traditional and modern papermaking techniques after 1980s of bast paper in Renhuai City

工序	传统工艺	新工艺
揭构皮	需揭构皮	只有生产高级书画纸时需揭构皮，其余直接用“毛构”，一般纸甚至加了纸花
煮构皮	锅小，一甑可蒸煮干构皮100 kg，加石灰50 kg	锅大，一锅可蒸煮干构皮750 kg，加烧碱75 kg
洗构皮	拿到河里用脚踩、水冲3小时	放在熟料坑里泡，不用脚踩，总共需要换4次水，大约24小时
打皮板	用碓打料，一个人打碓，一个人翻皮板，一次放1.5 kg干构皮做成的料，1把构皮大约打7分钟	用打浆机一次可打150 kg干构皮做的纸料，用7.5 kW电动机打40多分钟即可；打一池（约750 kg）干构皮料需放25 kg漂精（次氯酸钠）
加萆药	用萆麻作纸药（萆药），加一次料，补一次“萆药”	抄1 000帘纸需250 g化学萆药（聚丙烯酰胺），加一次料，补一次纸药
焙纸或晒纸	用火焙，一次5小时左右	自然晒干，晴天时一天即可晒干，阴天时则要晒两天

⊙19 抖纸 Shaking the paper

⊙20 理平 Flattening the paper

⊙21 对折 Folding the paper

⊙22 捆纸 Binding the paper

⊙23 运输 Transporting the paper

(三)

仁怀皮纸生产使用的主要工具设备

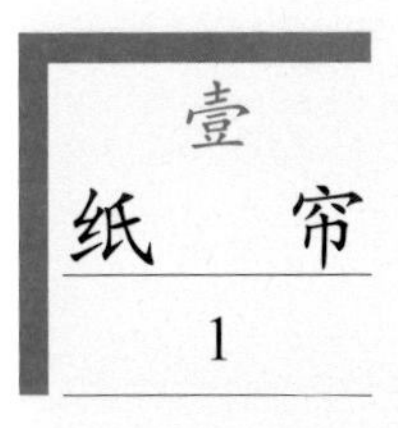

壹 纸帘 1

⊙24

用苦竹或慈竹的竹丝制成的舀纸工具。纸帘要配合木质的纸帘架一起使用。纸帘的尺寸不统一，最大的约170 cm × 50 cm，最小的约130 cm × 50 cm。

贰 泡料池 2

尺寸不统一，调查组现场实测的2个泡料池的尺寸分别约为340 cm × 330 cm × 70 cm和530 cm × 330 cm × 50 cm。

⊙25

叁 蒸灶 3

用于蒸煮构皮。蒸灶内有铁锅，锅内有甑桥，用来摆放将要蒸煮的构皮，锅上有铁盖。调查组现场实测的一座蒸灶内径约170 cm，外径约230 cm，高约300 cm，底部有砖块，上部被毁，现可测其高度约160 cm。

⊙26

肆 火焙 4

传统火焙是中空生火、焙干湿纸的设备。火焙两侧各用4根直径约20 cm的棕树粗木棒作为火墙的垂直支撑；上面再放两根木棒，作为火墙内部的框架；然后用苦竹竹篾编成席子作为内层；最外面用石灰敷平，形成内空外平的结构。两侧长方形大面积墙面长约4 m，高约1.7 m，用于贴纸焙干，大小为53 cm × 53 cm（1.6尺 × 1.6尺）的皮纸能焙3行7列。较窄的梯形面共有2面：一面是火灶口，顶上宽约50 cm，底下宽约100 cm，用于加煤生火；另一面是出烟口。火焙生火烧热以后即可贴纸焙干，一个传统火焙可以使用20多年。1997年三元村开始生产捆钞纸时，曾有人采用砖砌火焙干燥皮纸。用火焙烘干的手工纸很光滑，在20世纪60年代“广交会”上，此类产品获得广泛好评。20世纪90年代以后村里的房屋质量逐渐提高，平整的可用墙面变多，由于煤炭价格上涨，造纸户大多选择把纸直接晒在墙上。

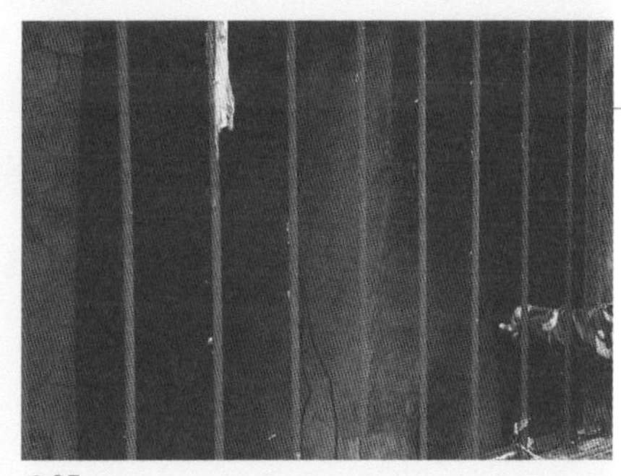

⊙27

⊙24 纸帘及纸帘架 Papermaking screen and its supporting frame

⊙25 泡料池 Pool for soaking the papermaking materials

⊙26 废弃的蒸灶 Abandoned kiln for steaming the papermaking materials

⊙27 火焙 Drying wall

(四)

仁怀皮纸的性能分析

所测仁怀三元村皮纸的相关性能参数见表8.3。

表8.3 三元村皮纸的相关性能参数
Table 8.3 Performance parameters of bast paper in Sanyuan Village

指标		单位	最大值	最小值	平均值
厚度		mm	0.110	0.070	0.088
定量		g/m²	—	—	20.0
紧度		g/cm³	—	—	0.227
抗张力	纵向	N	15.8	11.7	14.4
	横向	N	5.8	4.0	4.8
抗张强度		kN/m	—	—	0.640
白度		%	62.4	61.8	62.3
纤维长度		mm	9.49	1.07	4.37
纤维宽度		μm	34.0	4.0	15.0

由表8.3可知，所测三元村皮纸最厚约是最薄的1.57倍，相对标准偏差为0.13%，厚薄差异相对较小。皮纸的平均定量为20.0 g/m^2。所测皮纸紧度为0.227 g/cm^3。

经计算，其抗张强度为 0.640 kN/m，抗张强度值较小。

所测三元村皮纸白度平均值为62.3%，白度最大值约是最小值的1.01倍，相对标准偏差为0.24%，白度差异相对较小。

所测三元村皮纸纤维长度：最长9.49 mm，最短1.07 mm，平均4.37 mm；纤维宽度：最宽34.0 μm，最窄4.0 μm，平均15.0 μm。所测皮纸在10倍、20倍物镜下观测的纤维形态分别见图★1、图★2。

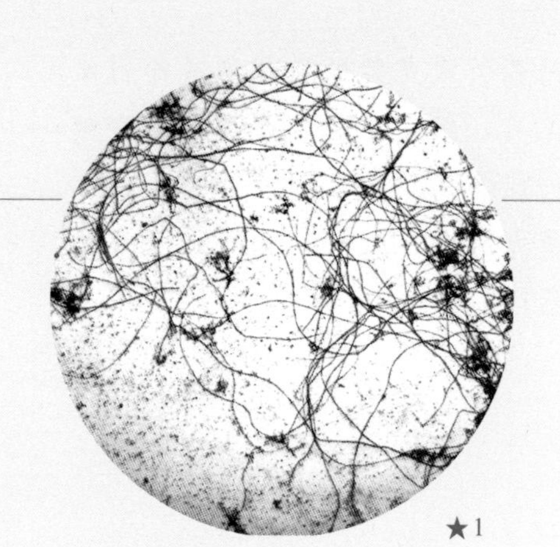

★1
三元村皮纸纤维形态图(10×)
Fibers of bast paper in Sanyuan Village (10× objective)

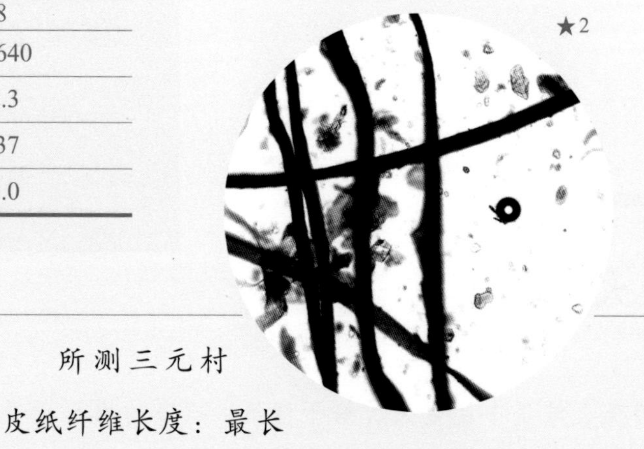

★2
三元村皮纸纤维形态图(20×)
Fibers of bast paper in Sanyuan Village (20× objective)

五
仁怀竹纸的生产工艺与技术分析

5
Papermaking Technique and Technical Analysis of Bamboo Paper in Renhuai City

⊙1

(一)
仁怀竹纸的生产原料与辅料

1. 竹

仁怀市鲁班镇竹纸所用原料主要是慈竹，调查中造纸户认为其他竹子也可以，关键是要用当年生的嫩竹子，不能用老竹子。

2. 纸药

鲁班镇制造竹纸需用苎麻根作纸药，制备方法是把苎麻根锤融后装入粗布袋，再放到水里，将挤出来的黏液搅匀，使用的时候舀起加入纸槽。

3. 石灰

仁怀竹纸的造纸工艺中，石灰兼有分解、软化和漂白竹纤维的作用，竹子和石灰的比例为2:1。仁怀多处出产和销售石灰，如大坝石龙场沿公路一带和交通乡磨槽湾公路一带成为生产石灰的主要基地[13]。

(二)
仁怀竹纸的生产工艺流程

据重点调查对象鲁班纸工杨存志老人介绍，鲁班镇竹纸的生产工艺流程如下：

壹	贰	叁	肆	伍	陆	柒	捌	玖	拾	拾壹	拾贰	拾叁	拾肆
砍竹	破竹	捆竹	泡竹	搽料子	沤竹	洗料子	发酵	碾料	落槽	放水	打槽	加苎药	舀纸

拾伍	拾陆	拾柒	拾捌	拾玖	贰拾
压榨	松榨	撕纸	晾纸	切纸	捆纸

⊙1 苎麻
Nettle as adhesive

⊙2 造纸工杨存志老人讲述竹纸制作工艺
Yang Cunzhi, an old papermaker introducing the papermaking procedures

⊙2

[13] 蒲宗亮.仁怀县志[M].贵阳：贵州人民出版社，1991:358.

壹 砍竹 1

农历腊月的时候用柴刀砍下当年生的竹子，一人一天可砍150余千克竹料。

贰 破竹 2

用柴刀将竹子砍成段，每段长约2 m，粗的竹子用刀破成4瓣，细的竹子用木槌捶破。

叁 捆竹 3

用竹篾将破好的竹子扎成捆后背回家，一捆50~75 kg。

肆 泡竹 4

将成捆的竹子整齐地摆放在纸塘里，待摆放到1 m高时，加一层石灰；再向上码放1 m，加一层石灰；最上面一层石灰用量要最大。竹子码放好后即可加入清水。

伍 搽料子 5

加满竹料、石灰和水后，用锄头将竹料上面的石灰溶于水中，使石灰、水和竹料充分混合。

陆 沤竹 6

让竹麻在石灰水里沤泡2个月以上，竹麻吸收石灰水以后变重，不会浮起来。

柒 洗料子 7

先用锄头把沤好的竹麻从纸塘里捞出来，并将纸塘里的石灰清洗干净；再将料子放回去，加入清水；最后将竹麻洗干净，把水放掉。

捌 发酵 8

在竹麻上面用谷草盖严，发酵40天，并加水浸泡20天，再将竹麻清洗干净。

玖 碾料 9

将发酵好的竹麻取出来，并用石碾把竹麻轧融，一次可放75 kg左右的竹麻。碾料时要不停地翻转竹麻，竹麻碾细以后要于其上面加些水，这样可提高工效。碾料时间的长短根据竹麻情况而定，石灰重（即石灰加得较多）或发酵充分的竹麻碾料时间较短，一般4小时左右可以碾完750 kg竹麻。

拾

落　槽

10

将一次碾的料子全部放到纸槽里，加水，先用捅耙打散，再用捞筋杆将粗的竹纤维捞出，最后用铲槽杆铲匀，整个工序需半小时左右。

拾壹

放　水

11

将水放掉，静置一晚，等到第二天抄纸时使用。

拾贰

打　槽

12

取出纸槽中部分纸料，留下适量纸料后加入清水至纸槽高度的1/2，先用捅耙将沉在槽底的纸料翻起来，再用铲槽杆把纸浆搅和均匀。

拾叁

加　芽　药

13

加水至纸槽高度的4/5左右，将芽药放入槽里，并用捞筋杆搅开。

拾肆

舀　纸

14

舀纸时需舀三盘水，依次是由内往外、由外往内、由右往左，然后将湿纸倒扣在底板上。一帘一纸，一天可舀2 000多张纸。

拾伍

压　榨

15

舀完纸后，在纸垛上依次加纸壳、杠板、马凳（也叫木方）和吊杆，用绳子将吊杆和滚筒系牢，并用手扳动滚筒卷起绳子，吊杆压低支点缓慢增加压力。吊杆压到底一次后就放开绳索，再加一次马凳继续压榨，整个压榨过程约需半小时。

拾陆

松　榨

16

压榨工序完成以后把木榨吊杆、杠板、马凳、纸壳等移开，先掰去纸垛左边的纸头，后将榨好的纸垛搬回家。

拾柒

撕　纸

17

用刀将纸的两个长边刮松，后将纸垛翻转过来，用手把纸头搓松，由左向右揭纸到大约1/3处，再一张张地揭下，每2张纸之间错开1 cm左右距离，20张为一托，双手揭开放在木凳上。

拾捌

晾　纸

18

揭开若干托纸后，将纸一托托地晾在屋内竹竿上。夏天阳光强烈、气温高，3天可晾干；冬天时间长些，需一周左右。如果晾纸的地方过于阴凉，则需在纸下生火烤干。

拾玖

切　纸

19

将5托纸合成一叠放在纸凳上，用弯刀修边，后用杠板压住，切成3份。

贰拾

捆　纸

20

50托为1捆，用竹篾扎紧纸的两端。

1991年版的《仁怀县志》中也记载了仁怀手工竹纸的生产流程：『春天收料，泡料，夏秋冬季洗料、碾料、舀纸。一人每天可舀一个纸墩子，可揭二三万张纸，等晾（焙）干后，方分成二十、三十、五十、六十张不等的刀数，十刀为一捆出售。』[14]

[14] 蒲宗尧.仁怀县志[M].贵阳：贵州人民出版社，1991：359.

(三)仁怀竹纸生产使用的主要工具设备

壹 泡料塘 1

浸泡竹料的水塘。塘子大小不同，可放竹子的数量也不同。大的塘子可放5 000 kg竹料，小的塘子只能放750 kg竹料。

⊙1

贰 平面碾 2

碾轧纸料的石质设备，由碾盘和碌子组成。调查组现场拍到的碾盘直径约3 m，外周有由高出碾盘平面近20 cm的石条围成的圆圈，以阻挡纸料流出。碌子为一头粗一头细的圆柱体，纵向刻有24条深沟，细的一头靠近圆盘中心，粗的一头靠近圆盘边缘。使用石碾时，先把纸料铺在碾盘上，再一圈一圈地转动碌子，将纸料轧融，直到可以用于舀纸。

⊙2

⊙3

叁 纸槽 3

舀纸时盛放纸浆的容器。最初的纸槽是椭圆形的木桶，后来逐渐改为用砖块、水泥制成的水池。与木桶相比，水泥池更结实、耐用。

⊙1 废弃的泡料塘 Abandoned soaking pool

⊙2 掩映在草丛中的平面碾 Grinder covered by grasses

⊙3 废弃的纸槽 Abandoned papermaking trough

肆
捞纸棚和纸榨
4

捞纸棚是指竹纸作坊的纸槽和纸榨上方遮阳挡雨的简易小棚，纸榨是指捞出湿纸后压出水分的木质设备。

⊙4 废弃的捞纸棚 Abandoned papermaking shed

⊙5 纸榨 Pressing device

伍
裁　刀
5

造纸师傅用来裁切竹纸的工具。

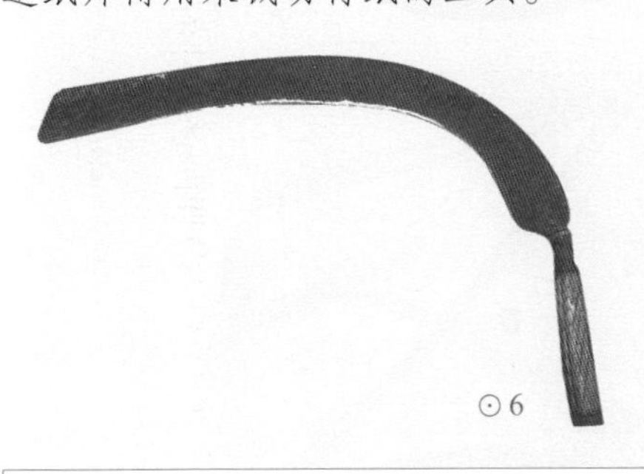

⊙6 裁刀 Sickle for trimming the paper

2009年4月，调查组到仁怀市鲁班镇进行实地调研时，采访了已76岁高龄的杨存志等制作竹纸的师傅。据介绍，过去这里竹纸生产十分繁荣，20世纪90年代以后逐步停业，至今已有十几年没有再造过纸。由于停业时间太久，纸帘、捅耙、捞筋杆、铲槽杆、杠板、马凳、吊杆等工具现在已经很难看到。

⊙7 调查组成员与杨存志合影 A researcher and Yang Cunzhi

(四)

仁怀竹纸的性能分析

所测仁怀鲁班镇竹纸的相关性能参数见表8.4。

表8.4 鲁班镇竹纸的相关性能参数
Table 8.4 Performance parameters of bamboo paper in Luban Town

指标		单位	最大值	最小值	平均值
厚度		mm	0.150	0.100	0.127
定量		g/m^2	—	—	28.3
紧度		g/cm^3	—	—	0.223
抗张力	纵向	N	11.0	6.1	7.4
	横向	N	5.8	1.8	3.7
抗张强度		kN/m	—	—	0.370
白度		%	22.6	21.7	22.2
纤维长度		mm	3.43	0.72	1.70
纤维宽度		μm	30.0	4.0	12.0

由表8.4可知，所测鲁班镇竹纸最厚是最薄的1.5倍，相对标准偏差为0.15%，厚薄差异相对较小。竹纸的平均定量为28.3 g/m^2。所测竹纸紧度为0.223 g/cm^2。

经计算，其抗张强度为0.370 kN/m，抗张强度值较小。

所测鲁班镇竹纸白度平均值为22.2%，白度最大值约是最小值的1.04倍，相对标准偏差为0.33%，白度差异相对较小。

所测鲁班镇竹纸纤维长度：最长3.43 mm，最短0.72 mm，平均1.70 mm；纤维宽度：最宽30.0 μm，最窄4.0 μm，平均12.0 μm。所测竹纸在10倍、20倍物镜下观测的纤维形态分别见图★1、图★2。

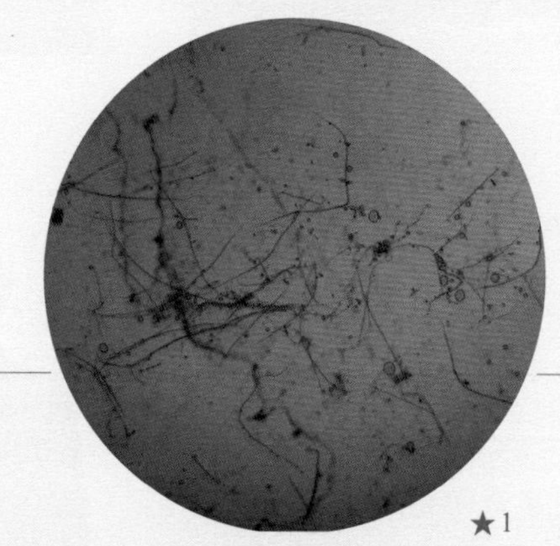

★1 鲁班镇竹纸纤维形态图(10×)
Fibers of bamboo paper in Luban Town (10× objective)

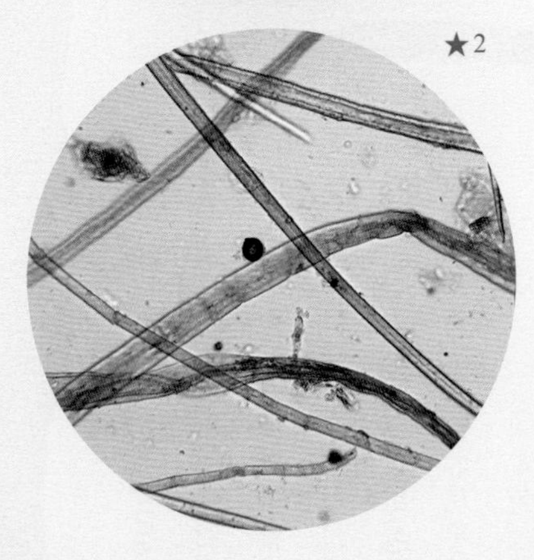

★2 鲁班镇竹纸纤维形态图(20×)
Fibers of bamboo paper in Luban Town (20× objective)

六 仁怀手工纸的用途与销售情况

6 Uses and Sales of Handmade Paper in Renhuai City

(一) 仁怀手工纸的基本用途

1. 书画艺术用纸

仁怀皮纸，尤其是以传统工艺生产的仁怀皮纸，其纸质细腻、拉力强，是很好的书画用纸。据三元村的造纸户介绍，过去集体性质的仁怀皮纸社（后来改成仁怀皮纸厂）用纯手工方法抄过由构皮制成的“宣纸”，所用的帘子、帘架与现在通用的尺寸都不一样。

2. 抄经和文档用纸

由于仁怀皮纸的纸质结实、保存时间长，当地抄写经文、医书、家谱、账簿、公文等时都会用到它。用仁怀皮纸抄写的经文和文档存放几十年甚至上百年都不会坏。

3. 养殖用纸

仁怀地方曾将皮纸用于孵化春蚕和家禽。“民国年间，皮纸就曾作为孵蚕的种纸使用。”[15] 在传统的家禽孵化过程中，人们先把谷子炒热，后将鸭蛋放在其上面，再用仁怀皮纸包好，这样可以长时间保温。孵化家禽卵时使用仁怀皮纸包裹的成活率要高于使用毛布包裹，贵州在四川周边的几个县市用仁怀皮纸孵化春蚕和家禽的情况很普遍。

4. 制作生活品用纸

仁怀及其周边的厂家、作坊在制作纸伞、斗笠、纸扇、烟花、爆竹等生活民俗用品时常使用三元皮纸作为原料。赤水县雨伞厂曾用仁怀皮纸制作油纸伞，方法是先在皮纸上印花再刷上桐油，这样可以防水、遮阳。苗族青年结婚时，

⊙1

⊙2

⊙3

⊙1 书写在仁怀皮纸上的《见病知方汤头歌》 Medical book written on bast paper in Renhuai City

⊙2 在仁怀皮纸上进行绘画创作 Painting on bast paper in Renhuai City

⊙3 用仁怀皮纸抄写的账簿 Account book on bast paper in Renhuai City

[15] 周恭寿修，等.续遵义府志：卷十二[M].民国二十五年(1936年)刊本.

男女双方都要各买一把皮纸花伞，故纸伞的销量一直很好。有的乡村市场的作坊在竹编斗笠上糊皮纸、刷桐油，用来挡风遮雨，既经久耐用，又轻巧方便。过去烟花、爆竹作坊多用皮纸制作引线，手工纸引线较粗，易燃不回潮，现在的鞭炮引线多采用细的机制纸来制作。

5. 擦机器、枪炮用纸

皮纸去污力强，无划痕和纤维残留，与棉布、绸缎相比，皮纸擦机器、枪炮效果更好。遵义电厂从20世纪70年代开始，每年都要购买30多万张仁怀皮纸用于擦拭机器。

6. 包装用纸

仁怀皮纸曾被长期、广泛用于酒瓶、糖果、饼干等的包装。茅台酒用皮纸包装的时间很长，至今还有商人用三元村的黑皮纸（没漂白的皮纸）作为酒瓶的内包装。现在，糖果和盐等的包装用纸已被其他新材料代替。从2003年开始，三元村造纸人李吉祥陆续到云南联系业务，将仁怀皮纸用作云南普洱茶的包装，后来村里大部分人都开始制作45 cm × 45 cm尺寸的茶叶包装纸。2008年茶叶包装纸开始滞销，手工纸协会希望抓住2008年第2届中国国际茶叶及茶文化（深圳）博览会这一机会，拓展皮纸的茶叶包装市场，决定将纸样送到疾病防控部门检测，并取得作为茶叶包装用纸的合格证明。

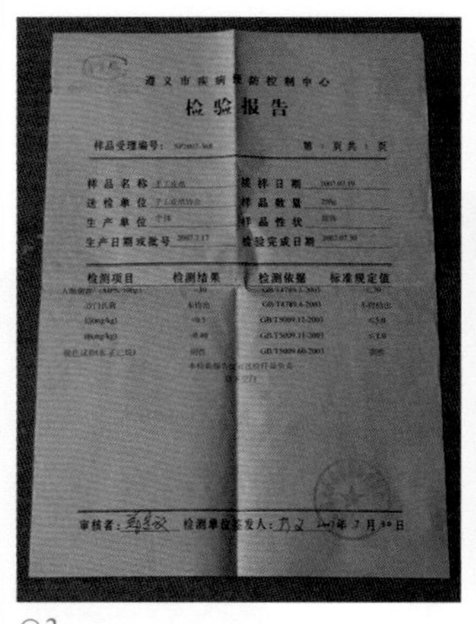
遵义市疾病预防控制中心
检验报告

⊙3

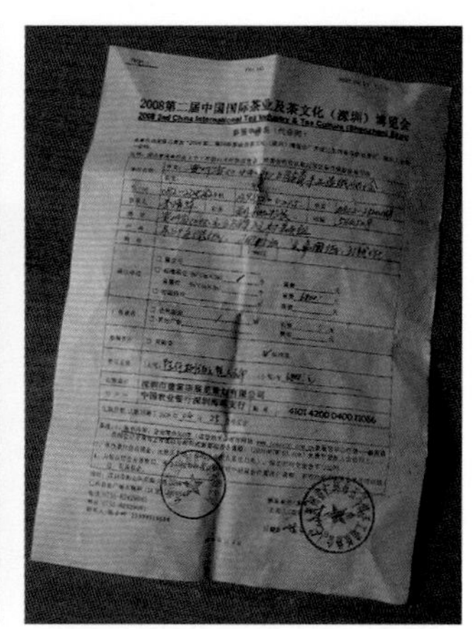

⊙4

7. 捆钞用纸

捆钞用纸是银行专用纸，与一般皮纸相比，明显厚得多，通常是双层、三层夹纸，制作方法是晒纸时2~3张一起晒，这种纸厚且结实，三元村曾有多家造纸户生产捆钞纸销售。

8. 民间工艺用纸

普通百姓制作风筝、糊窗户、做油桶等都会选用仁怀皮纸。每年农历正月末至二月间当地有放风筝的习俗，用皮纸制作风筝的人很多，这种现象至今还很普遍。传统民居中通常会在木质窗户格上糊一层透光的皮纸，糊一次可用一年多。过去百姓在制作盛放油的竹篓桶时，会在竹篾兜上糊一层皮纸，刷一层桐油，一层层地糊上去，晾干之后做成油桶，可以使用多年。

⊙1

⊙2

⊙1 用仁怀皮纸包装的地方白酒
Local liquor bottle wrapped by bast paper in Renhuai City

⊙2 1972年用仁怀皮纸包装的茅台酒
Maotai bottle wrapped by bast paper in Renhuai City in 1972

⊙3 仁怀皮纸检验报告
Test report of bast paper in Renhuai City

⊙4 第2届中国国际茶叶及茶文化（深圳）博览会参展申请表
Application form of the Second International Tea Culture Exposition (Shenzhen)

9. 祭祀与丧葬用纸

祭祀、丧葬用纸的数量大，对纸质要求不高，价格也较低。地方上，传统的丧葬方式需要使用大量皮纸，如设道场、灵堂时需要用皮纸来裱褙，入殓时要把皮纸叠成三角形以固定尸体在棺材内的位置，这些习俗至今在遵义还保存着。

10. 打经簿

老人过世的时候，家人通常会请人写一本经簿，名为打经簿，经簿上记载过世者的生卒年月、生平德业、过世原因（如病故、伤亡等）、家族源流、始祖以来的所有祖先、后代及五服亲戚名号，丧葬实况及道场上道士所念经书，以及后辈哀思。一本经簿通常要写数百甚至上千人的名字，需要100多张皮纸。用皮纸抄写的经簿存放几十年甚至上百年都不会坏，可以一代代保留下来，如有损毁需要重新誊写。1995年，仁怀鲁班镇双魁田村蔼氏后人看到记载祖先蔼永庆“白手创业之维艰”的经簿已过于破旧，“不克阅览”，于是请当时年届八秩的方思广老人重新抄录，并增补其后裔若干人，在县府两级考试中夺魁的蔼开甲，出版诗集的蔼显威、蔼显刚都在所列后裔之中。

仁怀竹纸的用途也十分广泛，质量好的毛边纸（也称水纸）质地细、吸水性好，且呈微黄色，书写容易干，字迹经久不变，是当年写公文、习字、裱褙、包装、印刷书籍报刊用纸。质地疏松的火纸主要用于祭祀，制作鞭炮、纸捻、卫生纸等[16]。逢年过节、各种祭祀活动，尤其是每到农历七月十五当地百姓都要祭祀祖先，为过世者做斋，冥纸用量很大。

(二)
仁怀手工纸的销售情况

仁怀手工纸历史上一度是仅次于茅台酒的重要轻工业品。其中，仁怀皮纸“具有洁白细腻、厚薄均匀、柔软绵韧、拉力强、吸水好、无黄疤黑点等特点，是同类产品中的佼佼者”[17]，市场反应很好。仁怀皮纸依据不同的用途分为不同的规格：按厚薄和复舀的次数分为单纸和二夹、四夹等品种；按大小分为“方帘纸”（53 cm×53 cm）、“联二纸”（50 cm×40 cm）、“云皮纸”（尺寸稍大于“联二纸”）、“小纸”（25 cm×20 cm）。其中，方帘纸有单皮纸和夹皮纸两种，主要用于书法和绘画创作。方帘皮纸以千张为“一把”，其余皮纸均以千张为“一捆”，皮纸捆有齐捆、毛捆之分，齐捆销往远处，毛捆销往本地。

20世纪40年代以前，仁怀外销皮纸除供应本省的贵阳、毕节等地外，主要销往重庆、成都、泸州等地。曾经有一段时间，本省及四川、云南等外地客商纷纷慕名到仁怀三元村购买、定制不同用途的皮纸。每逢赶场，三元村销售的皮纸都在50挑（50万张）以上。

20世纪50年代中期到70年代末期，仁怀皮纸由供销社统购统销。1957年，三元皮纸厂产品质量大幅提高，开始生产国家批准的出口纸。当时，出口皮纸的白度达到70%，纸质细、光滑、平整、油嫩、纤维组织均匀，略有光泽、拉力强、无水眼、无破烂，最高年产量达65 t，全部由外贸部门包销到马来西亚、新加坡等地。1978年以后因原料价格上涨，生产成本过高而停止生产[18]。

20世纪80年代以后，开始有商贩上门收购，或者由造纸户自己运纸出去销售。皮纸市场形势最好的时候，销售区域一度扩展到全国14个省的46个县市。未经漂白的仁怀大黑皮纸曾销往西藏，用来装裱菩萨画像。20世纪80年代末，曾有

[16] 蒲宗亮.仁怀县志[M].贵阳：贵州人民出版社，1991：262，359.

[17] 徐文仲.三元白皮纸[J].贵州文史丛刊，1986(2)：70.

[18] 蒲宗亮.仁怀县志[M].贵阳：贵州人民出版社，1991：359.

日本客商采购仁怀皮纸用于书画的创作和鞭炮的生产。从1990年手工纸市场疲软开始到2009年春，在强行取缔三元造纸作坊的近20年时间里，在贵州许多地方的手工纸停业、大量造纸户另谋生路的情况下，三元皮纸一直维持生产和销售，并有一定的发展。

三元行政村岩头村民组人多地少，世代以手工造纸为业，近200户人家，仅13 000 m^2土地，粮食、青菜、肉蛋等不能自足，都要在市场上购买。据造纸人李吉祥介绍，三元村普通造纸户一年大约消耗10 000 kg干构皮，抄67槽皮纸，每槽舀10 000~11 000张纸，按2009年的皮纸价格0.1元/张计算，造纸户年收入约7万元，若不计劳动力的投入，去除成本开支约3万元（表8.5），每年可赢利约4万元，尚可维持生计。

仁怀竹纸的销售市场主要在仁怀本地，20世纪40年代，桑树湾纸厂所产的127 cm × 60 cm毛边纸曾一度销往遵义、金沙等周边县城。由于冥纸、卫生纸和火纸需求量大，仁怀境内长年都有竹纸的生产和销售。竹纸通常由造纸户赶场时拉到镇上或者运到附近农村进行销售。市场上的竹纸通常以“托”为单位计价，20世纪90年代一托（20张）竹纸卖0.5元（调查时可卖1.2元）。50 kg竹料可舀100托纸，可卖50元，除去竹料的成本，基本没有利润空间，仁怀竹纸手工业因此陷入难以为继的局面。

表8.5 三元村普通造纸户一年造纸开支表
Table 8.5 Papermakers' annual papermaking cost in Sanyuan Village

类 别	开支情况	备 注
干构皮	10 000 kg × 1.8元/kg=18 000元	毛构1.6~2元/kg，取1.8元/kg
烧碱	1 000 kg × 4元/kg=4 000元	
漂精	35元/槽 × 67槽=2 345元	一次打150 kg干构皮（需打67槽），一槽一桶，一桶35元
煤	125 kg/锅 × 13锅 × 0.7元/kg=1 138元	10 000 kg干构皮需蒸13锅
煮构皮	150元/锅 × 13锅=1 950元	请人煮构皮一锅需支付工钱150元
打浆	12元/槽 × 67槽=804元	
化学莩药	20元/kg × 0.75 kg/槽 × 67槽=1 005元	抄一槽纸需0.75 kg，20元/kg的化学莩药聚丙烯酰胺
帘子折旧	320元	一个可用一年半，单帘250元，双帘500元，折中算
帘架折旧	125元	150元，一般用一年，好的一年半，折中算
纸槽折旧	80元	2 000元，可用25年
其他	200元	
合计	29 967元	

七
仁怀手工纸的相关民俗与文化事象

7
Folk Customs and Culture of Handmade Paper in Renhuai City

(一)
祭蔡伦敬工艺

调查组在仁怀的鲁班和五马一带看到，完全以手工业为生的人们对于工艺和先师的尊敬尤其看重，每个造纸户的中堂之上都有蔡伦的牌位。在“天地国（君）亲师”牌位的两边，“蔡伦先师（祖师）”与“门中宗祖”“福禄财神”“灶王府君”一道接受家里香火的供奉。通常在逢年过节或新的一年开始舀纸的时候，都要焚香化纸祭祀蔡伦。

鲁班镇的地名也与这里人们敬重工艺的传统有关。相传当地的大户孟氏一族在场口修建石牌坊，掌墨师傅费尽心思都无法将牌坊的横梁安放好，恍惚之间，他走到附近的小山洞躺下休息，很快就睡着了。梦里，一位风度翩翩的白胡子老者向他走过来，传授技艺，掌墨师傅醒来后如法行事，果然轻松地安放好横梁。之后，掌墨师傅将此事说出，众人认定是鲁班先师显圣托梦真传。掌墨师傅与众人一道备办香烛钱纸和红羊供果前去拜献。从此，这个山洞就被称为鲁班洞，鲁班场和鲁班镇的名称也由此而来，在鲁班洞前祭拜鲁班先师也成为当地的一项传统民俗活动。

(二)
正月闹花灯

正月闹花灯是仁怀的传统民俗，活动从正月初八开始到正月十六结束，共9天。先用竹篾扎成植物、动物、车船等骨架，外面糊以皮纸作皮肤，再安好蜡烛就做成了各式花灯。清光绪《增修仁怀厅志》记载：“正月初九日，为上九日，凡乡场村庄先于初八日竖灯杆谓之‘试灯’，或满树（23盏灯）、或九皇（9盏灯）、或三官（3盏灯），每夜燃点到十五夜为满，十六日倒灯杆，治酒沿村聚会，谓之‘灯会’。”[19]

（三）

清明节漂清

在仁怀清明节民俗中，每一个家庭都要从一世祖开始，依次漂清祭祀。漂清就是烧纸、放鞭炮，每座坟上都放皮纸，数量不拘多少，质地也无需很好。仁怀历史上相关习俗古已有之，清光绪《增修仁怀厅志》记载："清明节前后十日，人各上坟挂纸于祖坟上，谓之‘挂青’。”[19]

（四）

中元节烧袱包

按照遵义民间习俗，农历七月十五中元节要给祖先烧袱包，意在寄钱给祖先亡灵，让他们在阴间有钱用，以更好地庇佑子孙。烧袱包在五马一带也叫“打望山钱”。袱包也称作袱子，是烧给祖先的大信袋，里面装着纸钱。信袋的正面要写上对先人的称呼、自己的名字与烧袱包的时间，比如“今逢某某之期，虔具信袱共某包共某圓，奉上第某包，奉故显考（妣），讳某公（母）老大（安）人正魂（性）名下收用，孝男某某叩。天运某年某月某日，公元某年某月某日火化”。写袱要查阅家里记载祖先名号的金簿本，从一世祖开始到最近过世的亲人，逐一用毛笔小楷工整填写。接下来是在草纸纸钱上打21个眼，每排7个，共3排。一个袱子里装2叠纸钱，每叠40张，装好后在信袋反面写上一个“封”字。然后在房屋附近将所有袱子堆成一堆，天黑后焚化。据造纸户介绍，在五马镇民间有专门打望山钱的店铺，中元节时普通农家仅打望山钱就要用去1 000多张皮纸和竹纸（草纸）。

（五）

十月一送寒衣

清光绪《增修仁怀厅志》记载：“十月一日，俗以纸作衣送往墓间焚之，曰‘送寒衣’。”[19]如今，在仁怀地方送寒衣的民俗已看不到。

（六）

化钱买山

早在宋代，仁怀地方就有烧化纸钱向阴界买坟穴的习俗。仁怀市三合镇两岔村宋墓中发掘出土的石刻《阴地卷》记载：“大宋夔路播州石粉栅居黎二氏将钱万余贯贴，买山下坟穴一所建立寿堂。”[20]这与现在为过世的人在下葬前化纸钱“买山”或“买坟穴”的丧葬民俗基本相似。

⊙1

⊙1 蔡伦先师牌位 Memorial tablet in memory of Cai Lun, the originator of papermaking

[19] [清]崇俊，等.增修仁怀厅志[M].光绪二十八年(1902年)刻本.

[20] 蒲宗尧.仁怀县志[M].贵阳：贵州人民出版社，1991：358.

八
仁怀手工纸的保护现状与发展思考

8
Preservation and Development of Handmade Paper in Renhuai City

（一）仁怀手工纸发展问题与生态的冲突

仁怀发展手工纸的条件很好，面临的生态冲突和发展问题也非常典型。2003年，仁怀市人民政府曾采取措施治理五马河流域手工造纸作坊的污染问题，但是，由于涉及面广，造纸户抵触态度强烈，工作难以深入。2004年，贵州省和遵义市领导现场办公，制定了“集中制浆、分散造纸、达标排放、总量控制”的治理手工造纸业污染问题的基本原则。政府有关部门对五马河沿岸手工作坊的小造纸制浆污染采取了多种治理措施，希望保留手工造纸的传统手艺和特色产业。当年年底，当地政府投入资金58万余元，在三元行政村岩头村民组修建集中制浆设施，修建了6个集中制浆蒸灶和1座废水处理设施，并于2005年6月投入运行。由于设备设计、运营管理的缺陷和造纸户认知的不足，五马河手工造纸集中制浆设施建成使用以后，反而加剧了对五马河的污染。环境部门监测结果表明，小造纸集中制浆工艺产生的废水未经处理直接进入五马河，所排废水应控制的指标和下游水体化学需氧量超标，直接污染了五马河，也对国酒基地的生态环境和茅台酒的生产环境造成了威胁。2008年6月，政府决定关停五马河流域279家从事个体手工皮纸生产的作坊，并制定了取缔小造纸作坊的工作方案和助学、退耕还林等帮扶政策。

2009年4月4日，五马镇政府公布的《取缔五马河流域个体手工造纸作坊补助标准》中规定，补助范围包括造纸设备补助和过渡期的生活补助。对上缴工具设备的造纸户给予一定的补助：每架纸槽（含四副以上可用帘子、帘架及附属设施）给予4 000元补助，每台打浆机（含打浆房、浆槽等设施）给予1 500~6 000元补助。4月5～22日，五马河流域的五马镇、茅坝镇279户造纸户（表8.6）全部完成自拆工作，共计拆除纸槽242

⊙2
⊙3

⊙2
感谢支持传统造纸工艺标语
Slogan thanking the government's support for traditional papermaking technique

⊙3
呼吁自动拆除造纸作坊的标语
Slogan advocating removal of papermaking mills

表8.6 2009年仁怀市取缔小造纸作坊前五马河沿岸手工造纸户基本情况
Table 8.6 Basic information of papermakers alongside the Wuma River before papermaking mills are banned in 2009

	造纸户数(户)	人口总数(人)	<18岁人口数(人)	18~40岁人口数(人)	40~55岁人口数(人)	>55岁人口数(人)	从业人员数(人)
五马镇	245	1 036	279	481	167	109	640
茅坝镇	34	190	63	61	37	29	84
合 计	279	1 226	342	542	204	138	724

架、打浆机42台，兑现造纸设施设备拆除后补助资金1 365 800元。另外，自行拆除造纸设备的造纸户给予两年过渡期内每人每月165元的生活补助（不含有正式工作的人员）。

调查组于2009年4月24日到达三元村时，正值政府全力取缔手工皮纸生产，造纸作坊群中的造纸设备基本被拆光，只有大面积的碎砖和瓦砾还能够让我们想象这里手工造纸业昔日的繁荣景象。曾是当地百姓重要生计的仁怀手工造纸行业就此消失。

（二）关于保护仁怀传统手工造纸文化遗产的思考

仁怀手工纸正处在保护文化遗产与保护自然生态激烈的历史性冲突之中。一方面，仁怀手工纸制造技艺对于保护黔北文化生态、发展地方传统特色产业，以及开发新兴文化旅游产业都是不可多得的宝贵资源；另一方面，五马河、赤水河的生态安全又事关国家环保法治、国酒茅台的水源生态，以及沿岸生产、生活用水的安全。有效和妥善地解决这个冲突不仅对于仁怀的包容性发展具有重要意义，同时也将对其他面临类似问题的地方具有重要的借鉴和启示价值。破解保护手工纸生产的文化生态与保护自然生态之间的矛盾，需要解决下列几个关键问题。

⊙1

第一，可考虑恢复传统的石灰制浆，杜绝化学制浆。造纸户普遍反映，石灰制浆和化学制浆各有利弊，以前用石灰作为分解植物纤维的辅料，其优点是对环境的影响很小，对植物纤维损伤不大，纸的内在品质和保存时间方面都有明显优势；缺点是生产成本高、劳动强度大、生产周

⊙2

⊙1 五马镇取缔岩头小造纸指挥部 Head office for abolishing papermaking mills in Wuma Town

⊙2 拆除后的造纸作坊 A demolished papermaking site

期长。现在使用工业烧碱和化学漂白剂造纸，优点是可以减少劳动量、缩短生产周期，让皮纸看上去更白，同样一槽纸可比传统工艺多产10%的皮纸；缺点是纸的质量和保存时间明显下降，制浆生产的污水直接排入河水中，破坏环境和威胁生态安全。如果能在制浆工艺中恢复使用传统的石灰辅料，保护传统工艺遗产与守护绿水青山之间看似不可调和的矛盾有望在很大程度上得到缓解，甚至解决。

第二，在建设污水处理设施的同时引进集中制浆设备。近年来，手工纸产业的环境压力越来越大，一些地方引进了集中制浆的做法。目前普遍存在着重硬件建设、轻制度建设的问题。这也是仁怀第一轮手工纸治污失败的原因。手工纸行业协会等民间力量弱小、缺乏预见性和调控手段，以及行动滞后，导致造纸户在这场文化与生态的博弈中败下阵来。如果能够在投入硬件的同时，加强与有关各方利益格局的整合和调整，培育和发挥第三方在缓冲生态矛盾中的作用，加强设备运营效果的制度设计，实现造纸户与社会公众利益的共赢，在手工纸产业的去留问题上政府就不会陷于不相容的行为选择。

第三，推进手工纸的产品结构调整，促进产业高值化。在仁怀皮纸的众多市场中，祭祀所用的冥纸销量远远超过书画、包装等用纸量，但价格却最低。以2009年为例，53 cm × 53 cm（1.6尺见方）的冥纸价格为0.1元/张，同样品质的45 cm × 45 cm的普洱茶包装纸价格为0.3元/张，尺寸为60 cm × 60 cm用于染制民族服装的皮纸依薄厚不同分别为0.4元/张、0.8元/张。另外，仁怀皮纸以初级产品为主，深加工产品品种少、比例低。

第四，明确手工纸制造业在非物质文化遗产保护框架下的定位。许多造纸户希望地方政府学习丹寨的做法，深入挖掘仁怀手工纸制作技艺所涉及的相关民俗文化，丰富仁怀手工纸的文化内涵，将其纳入国家非物质文化遗产保护范围，让手工纸制作技艺进入乡土文史教育，并进行特色体验旅游开发，和茅台酒一样得到保护、规范、传承和发展，重现茅台酒和手工纸相互依存、共同发展的和谐图景。把延续300年的手工纸制作技艺纳入博物馆保护体系，调查、记录和保护仁怀手工纸在发展中积累的生产原料、配方、工艺和满足不同需求的皮纸品种，手工纸生产中形成的种植、生产、销售分工协作体系，仁怀皮纸“被消亡”以后逐渐消失的相关风俗习惯，为子孙后代保存一份仁怀手工纸历史、传播、技术演变，以及相关文化、经济、民俗等方面的宝贵记忆。

皮纸 加厚

Bast Paper (Extra Thick) in Wuma Town of Renhuai City

三元村皮纸（加厚）透光摄影图
A photo of bast paper (extra thick) in Sanyuan Village seen through the light

皮纸 带丝

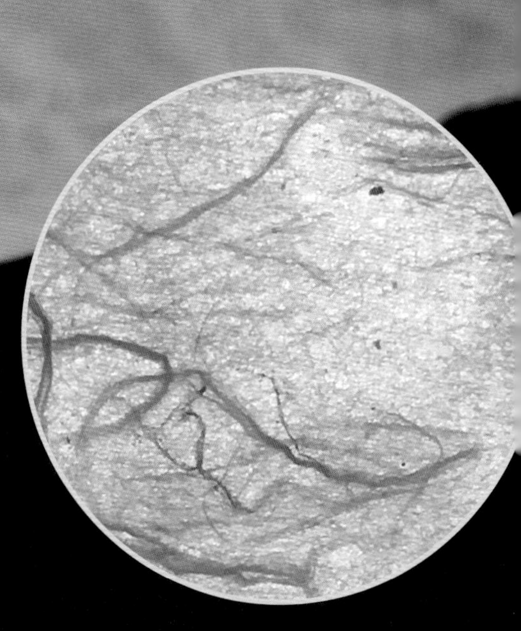

三元村皮纸（带丝）透光摄影图
A photo of bast paper (with threads) in Sanyuan Village seen through the light

皮纸

三元村皮纸透光摄影图
A photo of bast paper in Sanyuan Village seen through the light

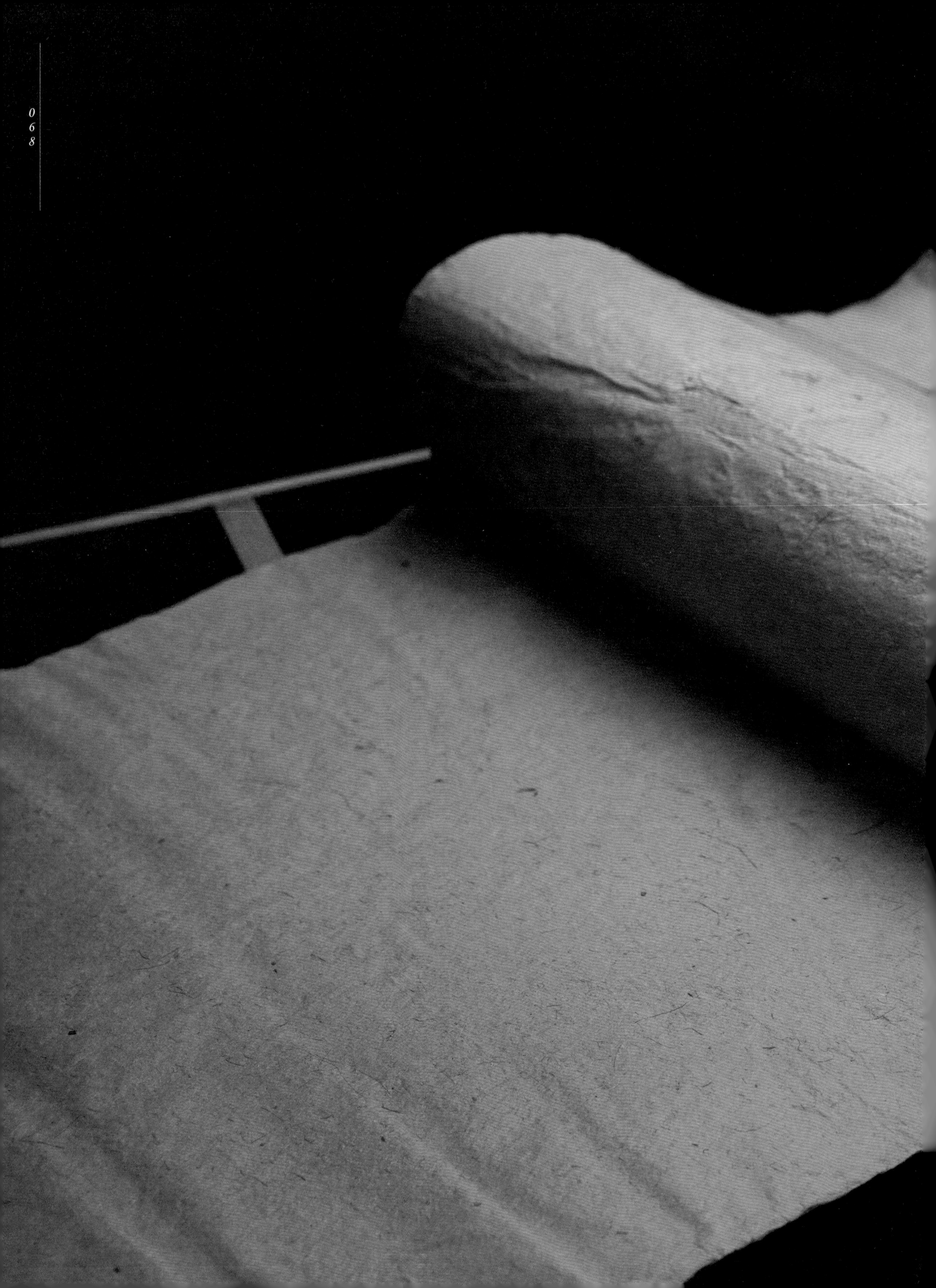

竹纸

鲁班镇竹纸透光摄影图
A photo of bamboo paper in Luban Town seen through the light

第二节

正安
手工纸

贵州省
Guizhou Province

遵义市
Zunyi City

正安县
Zheng'an County

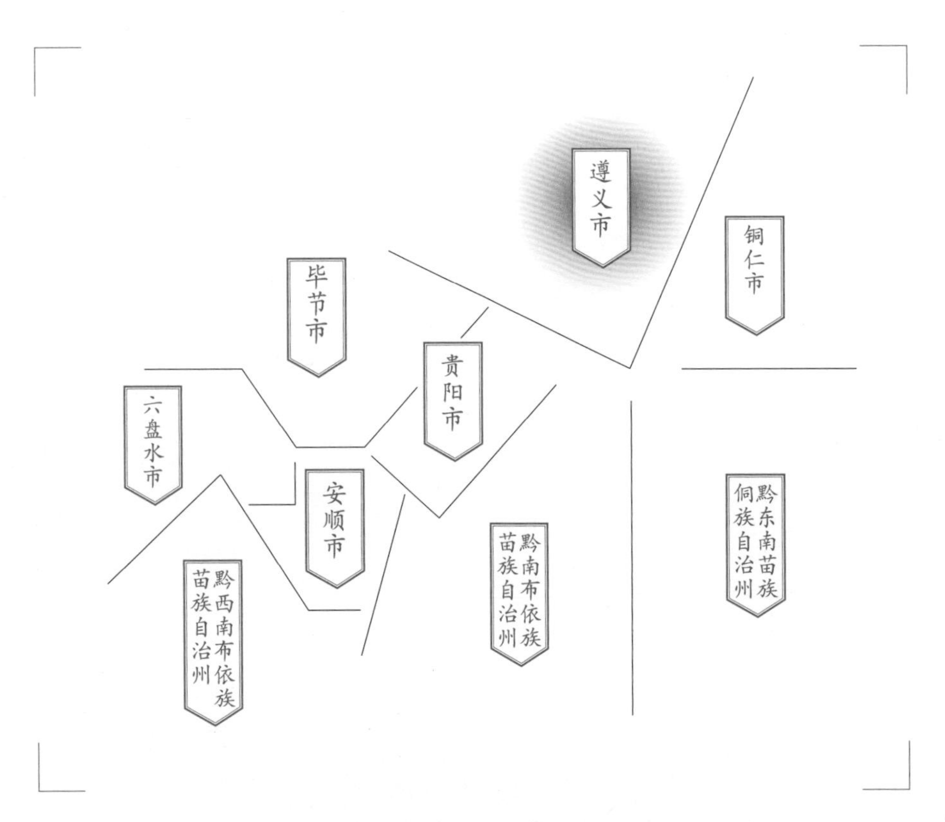

调查对象

和溪镇杉木坪行政村
凤仪镇梨坝行政村
手工纸

Section 2
Handmade Paper in Zheng'an County

Subject

Handmade Paper in Liba Administrative Village of Fengyi Town, Shanmuping Administrative Village of Hexi Town

一
正安手工纸的基础信息及分布

1
Basic Information and Distribution of Handmade Paper in Zheng'an County

从目前文献调研和实地考察掌握的资料可判断出，正安县从清代到20世纪90年代这段时间曾有过相当普遍的手工纸生产现象，产品涉及皮纸、竹纸、草纸等多种类型。

正安的皮纸以构树皮为原料，主要产于清溪河流域的凤仪镇南部的梨坝、桑坝一带。竹纸以金竹、水竹为原料，民国年间曾在县城北部的夫烟坪（今在桴焉乡政府西北方18 km有"扶烟坪"一地，"夫烟"与"桴焉"同音）生产过质量优良的竹纸。20世纪50年代曾经在桴焉乡以水竹、杂竹为原料生产过毛边纸。另外，乡间以慈竹为原料生产的火纸和火炮纸也属正安竹纸的一类。正安的手工竹纸有"新州滩""杨兴滩"之分，新州滩的纸张薄于杨兴滩的纸张，但长于杨兴滩的纸张。

草纸由稻草沤制抄造而成。《道光遵义府志》记载："以竹、杂草为者曰草纸，供冥镪粗用。"正安方言中所称的"纸壳"则是一种质地粗糙、厚实的包装草纸。

正安县历史上手工纸生产活动非常普遍，有记录可查的曾经生产过手工纸的村落就有10处以上，然而调查时仅有2处手工纸生产点仍存活态（表8.7）。2009年3月，调查组第一次去正安调查时，在凤仪镇梨坝村看到了粗制竹纸的生产活动。2011年2月，调查组外请成员冯其伟进行田野调查时发现，和溪镇杉木坪村道角村民组也存在手工竹纸的生产活动。

⊙1

⊙1 清代正安竹纸制成的土坪镇郑氏家谱
Genealogy of the Zhengs in Tuping Town written on bamboo paper in Zheng'an County during the Qing Dynasty

表8.7 正安历史上手工纸生产的地点及现状
Table 8.7 Handmade papermaking sites in history and their current status in Zheng'an County

	手工纸生产的地点	纸品种类	现状	年份
1	桴焉乡桴焉坪村	竹纸	停产	民国《续遵义府志》[21]
2	凤仪镇梨坝村	竹纸	生产	调查组2009年3月田野调查
		皮纸	停产	1999年《正安县志》
3	和溪镇杉木坪村道角村民组	皮纸	停产	1999年《正安县志》
		竹纸	生产	调查组2011年2月田野调查*
4	瑞溪镇三把车村堡上村民组	竹纸	停产	调查组2011年2月田野调查
5	流渡镇百花村白石村民组	竹纸	停产	调查组2011年2月田野调查
6	班竹乡旦坪村其麻坝村民组	竹纸	停产	调查组2011年2月田野调查
7	小雅镇桐梓坪村	草纸	停产	调查组2012年3月田野调查
8	市坪乡市坪村	竹纸	停产	调查组2012年3月田野调查
9	桴焉乡红岩村河马沟村民组	草纸	停产	调查组2012年3月田野调查

[21] 周恭寿,等.续遵义府志:卷十三[M].民国二十五年(1936年)刊本.

* 2011年2月和2012年3月两次补充田野调查由冯其伟主持。

路线图
正安县城
↓
梨坝村、杉木坪村
Road map from Zheng'an County centre to the papermaking sites (Liba Village and Shanmuping Village)

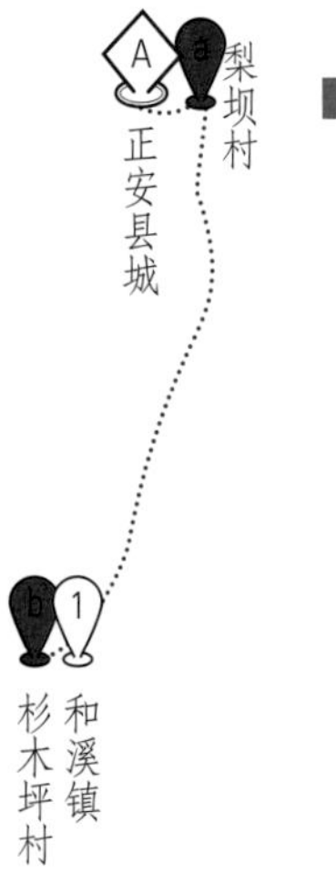

正安手工纸生产地分布示意图

Distribution map of the papermaking sites of handmade paper in Zheng'an County

考察时间
2009年3月/2011年2月

Investigation Date
Mar. 2009/Feb. 2011

A 正安县城
a 梨坝村
1 和溪镇
b 杉木坪村

地域名称

A 正安县 凤仪镇
1 和溪镇
2 流渡镇
3 庙塘镇
4 瑞溪镇
5 碧峰乡

造纸点名称

a 梨坝村 造纸点
b 杉木坪村 造纸点

位置分布

市府、州府
县城
乡镇
村落
造纸点
历史造纸点
山
国家级自然保护区
S221 省道
G21 国道
昆河线 铁路
G56 高速公路
线路

重庆市
桐梓县
道真仡佬族苗族自治县
正安县
务川仡佬族苗族自治县
绥阳县
湄潭县
凤冈县
S303
S207
10 km
5 km
0
N

二 正安手工纸生产的人文地理环境

2 The Cultural and Geographic Environment of Handmade Paper in Zheng'an County

正安县地处贵州高原北部、大娄山脉东麓、芙蓉江上游。这里气候温和、四季分明、雨量充沛，年平均温度16.14 ℃，无霜期平均290天，平均年降水量1 076 mm，属于中亚热带常绿阔叶林植被带，动植物资源非常丰富。

正安地处贵州东北部，位于东经107° 4′ ~107° 42′、北纬28° 08′ ~28° 51′。春秋战国时期，其西南为鳖国，东北与巴、楚接壤（现在县域的北面与重庆市南川区交界），是黔、川、湘等地经济交流的重要通道，也是巴蜀文化、楚文化与古黔北文化相互渗透、交融发育的重要区域。1996年12月，正安县杨兴乡农民胡明冲、郑继光上山挖药时，从石缝中挖出蟠璃纹青铜甬钟。据贵州省博物馆鉴定，该钟系春秋晚期至战国的宫廷乐器，定为二级文物。从其形制和纹饰进行推断，该钟与楚文化的关系较为密切[22]。2002年6月，正安县小雅镇出土了青铜虎钮錞于。錞于是古代打击乐器，既可用于战争中指挥军队进退，也可用于贵族举行的祭祀、朝会、宴飨等礼仪活动，其所处的时代应当在战国到西汉间。有专家认为虎钮錞于"为巴族遗物均无可疑"[23]。两件文物的出土说明，早在东周到秦汉的这段时间，正安的土地上就已经出现了巴蜀、楚和黔文化之间的交流。

正安古为"濮人"和"九黎"等少数民族世居繁衍的地方。仡佬族人（古称濮人）遍布全县乡村。苗族人则在秦汉以后由湖南等地迁入。明万历年间，实行"改土归流"*，中央政府派军队屯兵留守。大批汉民迁入，也把汉族的文化带入正安。根据1990年第五次人口普查统计数据，正安县总人口为593 960，其中仡佬族、苗族、土家

⊙1

⊙ 1 尹珍像 Statue of Yin Zhen

[22] 张伟翼.正安出土春秋战国青铜甬钟[J].贵州文史天地，1999(1).

[23] 徐中舒.四川涪陵小田溪出土的虎钮錞于[J].文物,1974(5).

* "改土归流"是指改革原先民族首领土司制度为由中央政府委派官员的流官制。改土归流有利于消除土司制度的落后性，同时加强中央对西南地区的统治。

族、布依族、回族、壮族、彝族等23个少数民族的人口为42 399，占总人口的7.1%。

正安建置最迟在唐贞观三年（629年），至今已有近1 400年历史。明万历二十九年（1601 年）实行“改土归流”，至今也已超过400年。正安地域文化积淀深厚，是东汉儒学大师和教育家尹珍的故里。据《后汉书》记载，东汉桓帝时，牂牁郡毋敛人尹珍（字道真）跋涉千山万水到洛阳求师于经儒大师许慎，“学成还乡里教授，于是南域始有学焉”[24]。今天，正安县新州镇毋敛坝仍保留着尹珍当年讲学的草堂“务本堂”的遗迹。2009年，正安县获得文化部“中国民间文化（小说）之乡”的荣誉。

调查时，正安手工纸的生产还能够在凤仪镇梨坝行政村大田坝村民组及和溪镇杉木坪行政村道角村民组看到。2009年3月，调查组成员出县城向南，沿“207”省道正安往绥阳方向大约行驶9 km出公路，向西步行约2 km就到了大田坝村民组。这里地处天子山脉的余脉，海拔约660 m。梨坝和大田坝都属于云贵高原上一种称为“坝子”的局部平原地形。坝子主要分布于山间盆地、河谷沿岸和山麓地带。

三
正安手工纸的历史与传承

3
History and Inheritance of Handmade Paper in Zheng'an County

在正安手工纸生产的文献调查中所得到的资料十分有限。历史上正安曾有过六部方志：宋代的《珍州图志》、明代的《正安志》、清乾隆年间的《正安旧志》、清嘉庆年间的《正安州志》、清咸丰年间的《正安新志》和清光绪年间的《续修正安州志》。其中，前三部均已毁于战火，无法看到其中的内容；后三部正安方志中关于手工纸的信息比较少。

《道光遵义府志》记载：“纸，州县同出。”可见，清代整个遵义地区都有手工纸的生

[24] [宋]范晔,等.后汉书:卷八十六·南蛮西南夷列传第七十六[M].北京:中华书局,1956:2845.

产，正安的造纸业应该此时已经出现。然而，当时方志中记载的几个造纸名声较高的地方中并没有特别提到正安县。比如，“皮纸出遵义者，以上溪场为上；出绥阳者，以黄泥江为上，白腻坚绵，更胜上溪”“竹纸唯绥阳专制，其上者曰厚水纸，料专用水竹，粉白细腻，极佳者胜上皮纸”“草纸，随地可制，然亦分佳、恶，以绥阳茅丫制者为上，曰茅丫纸”[25]。民国时期，正安的造纸业开始崭露头角。民国年间《续遵义府志·物产》中对正安县的竹纸大加赞赏：“正安夫烟坪以金竹、水竹制者，可抵川纸之红批、毛边。”[26]清代皮纸质量原本很好的绥阳县受到批评：“今唯绥阳造者劣，将必谋制法乃可畅行。”[26]

清代及民国时期，正安县的手工造纸业与炼铁、酿酒、榨油、制砖、缫丝、纺织、印染、采煤等行业一样，多是以家庭为单位的独户经营。手工土纸生产是农民的一种家庭副业，并没有专门的从业者。调查组在田野调查中了解到，正安县南部的市坪苗族仡佬族乡市坪村历史上曾经生产过火纸，至今境内还保留着“纸厂沟”（又名“严家沟”“鸡公蹬”）的地名。凤仪镇梨坝村夏氏祖先400多年前从四川涪陵移民过来，从曾祖时期就开始造纸，至今已经有100多年的历史了。

20世纪40年代末，正安县白皮纸年产量相对较小，为300余捆（40张为1合，10合为1捆）。相比之下，火炮纸的年产量非常惊人，有520万捆之多。

20世纪50年代初期，农民土纸的生产经营仍然以“户营户管”的家庭副业方式存在。1955年开始“合作化运动”以后，土纸的生产经营改由集体管理。全县普遍成立了合作性质的纸厂，家庭副业转为生产大队、生产队或公社的集体副业。1958年，在桴焉乡建纸厂一间，用土法造毛边纸，生产一年多后停产。到1977年底，全县512个社、队办企业中已有多家土纸厂。社办纸厂的经营决策、用工方式、经济核算、财务开支、规章制度、劳动分配等，均由管委会讨论决定。大队、生产队办的纸厂，除务工人员实行记工分、参加生产核算分析外，其他管理形式与社办纸厂相同。1978年改革开放初期，这些纸厂仍然由社办社管或队办队管，纸厂的一切资产归社、队集体所有。

1981年，正安农村实行联产承包责任制，伴随土地承包到户，社、队企业资产也逐渐分解到户，社、队纸厂锐减。到1983年，社、队土纸厂大多关闭，土纸的产品产量已无数据上报。1984年，县政府号召大力兴办乡镇企业，要求“乡乡都要有企业”，原来单一的社、队办企业的格局拓展为乡办、村办、联户办、个体办多种形式并存的局面。政府对新办乡镇企业实行信贷优惠、减税让利、税前还贷和财政贴息等政策，一度停滞的土纸厂再次出现发展的生机。1987年，乡镇企业土纸的产量最高时达到1 300 t（表8.8）。到1990年底，仍有造纸企业分布在县域各地，产量

表8.8 1982~1990年正安县乡镇企业土纸及石灰产量变化情况*
Table 8.8 Local handmade paper and lime output in Zheng'an County from 1982 to 1990

	土纸（t）	石灰（万吨）
1982	32	-
1983	-	-
1984	-	-
1985	-	10.10
1986	101	5.73
1987	1 300	3.70
1988	98	5.30
1989	643	2.50
1990	282	2.37

[25] [清]平翰，等.道光遵义府志：卷十七[M].上海：上海古籍出版社，1995:28.

[26] 周恭寿，等.(民国)续遵义府志：卷十二[M].民国二十五年(1936年)刊本.

* 本表根据《正安县志》(贵州人民出版社，1999年9月)有关表格整理。

也有282 t。根据冯其伟等人的田野调查，瑞溪镇三把车村堡上村民组的竹纸生产活动一直延续到2002年才逐步停止。班竹乡旦坪村其麻坝村民组（小地名西村沟）在鼎盛时期全组有50多户人家生产用于祭祀和制作爆竹的竹纸，大约在2003年停止生产。

调查时，竹纸生产仍然是凤仪镇梨坝村部分村民补贴家用的收入来源。梨坝村委会副主任夏忠鹏说，1978~1990年是梨坝村竹纸生产的兴盛时期，造纸户的收入有六成来自造纸。自20世纪90年代以后，遵义、绥阳和南川的机制纸大量进入城乡居民的消费领域，皮纸无法与机制纸抗衡，首先从正安淡出。竹纸的情况强于皮纸，仍然可以保住自己的部分市场。

虽经文献调查和田野作业多方求证，正安手工纸最早生产技艺源自何处现在尚无从考察。土纸的生产者亦农亦工，属季节性生产，造纸的手艺代代相传。

四 正安竹纸的生产工艺与技术分析

4 Papermaking Technique and Technical Analysis of Bamboo Paper in Zheng'an County

(一) 正安竹纸的生产原料与辅料

1. 慈竹或刺竹等

凤仪镇梨坝村村民造纸大多用慈竹。慈竹喜丛生，内实节疏，性柔，可以代藤用，五六月出笋，次年长成，竹高可达6 m。往往一丛至少生竹几十根，即使长出数百根竹，其笋也只向丛内生出。老竹、新竹高低相倚，形同一家人老少相守，在我国传统文化中常用其比喻父母的慈爱，所以取名“慈竹”。宋代文学家乐史（930~1007年）曾写过《慈竹》诗，咏叹其是有灵性的植物。人们常用“慈竹长春”祝愿做寿者父慈子孝，人丁兴旺。根据朱惠方、腰希申所做的“33种竹材制作纸浆适宜性研究”（1964年）[27]，竹材按制造纸浆质量的优劣顺序可分为四级，其中慈竹列在第一级。用慈竹作造纸原料时需要等第一年生出的竹材长到第二年农历十至十二月才可以砍下。调查组还了解到，正安县城西北方的瑞溪镇三把车村的村民曾砍下洋山慈竹、拐脚慈竹、米筛慈竹、甜慈竹等种类的竹材造纸。

⊙1

⊙2

⊙1 瑞溪镇三把车村李红全老人指认竹原料 Choosing the papermaking materials by the local papermaker Li Hongquan in Sanbache Village of Ruxi Town

⊙2 造纸作坊的水源 Water source for papermaking

梨坝村的村民中也有将刺竹用于造纸的。头年生刺竹当年冬天就可以用作造纸原料，但冬天砍的刺竹太嫩，损耗大，不适合造纸。次年正月到二月砍的竹材比较适合造纸。

2. 纸药

20世纪80年代以来，正安竹纸生产所用的纸药至少有三种：第一种是仙人掌。第二种是被村民称为“滑药根”的植物。春天滑药根的毒性较大，长时间浸泡在加入纸药的纸浆里容易腐蚀造纸人的手部。所以，村民通常是冬天抄纸，接触纸浆后一般用白矾水洗手消毒。瑞溪镇三把车行政村堡上村民组李红全和陈昌文两位老人介绍了一种被当地村民称为“华芍”的多年生草本宿根植物。搓出这种植物的汁液涂在手上或加入纸浆可以保护捞纸者长时间接触冷水的手部皮肤不致皲裂。第三种是山杨柳树的叶子，和溪镇杉木坪行政村道角村民组的造纸户多把这种叶子当作滑药使用，有和匀纸浆和黏结竹纤维的功能。梨坝村的村民早年造皮纸的时候，用山杨柳树皮或构树皮作原料，也用山杨柳树叶作滑药。滑药加水放入容器内煎煮，当容器里液体能够被挑起，并从上到下流成一根线时就可以使用。

历史上，当地用作滑药的植物还有多种，不同滑药的作用也有细微的差别。据《道光遵义府志》记载，遵义府境内的“造纸者用一种料，名滑药。分三种，一即羊桃，俗呼羊桃梗。一其树大一围叶尖被有白毛，俗名三条筋。一其茎如指大，结黑子，大如小豆，叶似桑，角似罂粟，名大滑药。乡人自种，春种秋生。凡造纸用三条筋，纸易用。羊桃梗则久逾佳。大滑叶更在二种

[27] 朱惠方，腰希申．国产33种竹材制浆应用上纤维形态结构的研究[J]．林业科学，1964(4).

之上”[28]。其中，“羊桃”亦称“阳桃”，即猕猴桃，是手工造纸常用的一种滑药；“三条筋”亦称“华草”，即中药柴桂，可能是李红全和陈昌文两位老人所说的“华芍”。

3. 生石灰

生石灰是造竹纸的辅助材料，其作用是通过腐蚀和分解竹材，形成适宜于造纸的竹纤维。石灰和竹料的配比是100 kg的竹料要用30 kg刚烧出来的石灰。

4. 水

在手工造纸过程中，水主要用于浸泡、冲洗竹材，以及搅拌纸浆。造纸时需要把山泉水引到纸槽。经测定，梨坝村造纸所用的水，其pH为6.3。

(二) 正安竹纸的生产工艺流程

通过与梨坝行政村大田坝村民组、瑞溪镇三把车行政村堡上村民组以及和溪镇杉木坪村多位造纸人的访谈，总结正安竹纸的生产工艺流程如下：

壹	贰	叁	肆	伍	陆	柒	捌	玖	拾	拾壹
砍竹	断竹	摔料	捆把	晾竹	下池	洗料	洗池	发汗	淹水	碾料

拾贰	拾叁	拾肆	拾伍	拾陆	拾柒	拾捌	拾玖	贰拾	贰拾壹
淘料	打槽	做滑	舀纸	扳垛边	榨纸	分垛	揭纸	晾纸	捆纸

[28] [清]平翰，等.道光遵义府志：卷十七[M].上海：上海古籍出版社，1995:28.

壹
砍竹
1

用柴刀把一年生的竹材砍下来备用。若竹子较小，一人一天可砍250 kg左右竹材；若竹子较大，一个人一天可砍500 kg左右竹材。

贰
断竹
2

把竹材砍断成每截2 m左右统一的长度，便于破竹和捆扎。

叁
摔竹
3

手握断好的竹材，往石头或水泥地上猛烈摔打，一头打开之后再摔打另外一头。细竹的顶端比较结实，要用木槌将其破开。

肆
捆把
4

也叫“捆浆把”。用竹篾把竹材捆成把，当地叫“个”，每把10~15 kg。

伍
晾竹
5

将捆好的竹材放在山上晾干，这样便于将其运到山下的纸棚中。500 kg湿竹子晾干后只有约200 kg，提高了工作效率。但是，若将湿的竹子直接下塘，则造出来的纸颜色微黄带绿，比较好看。用晾干后的竹子造出的纸没有用湿的竹子造出的纸好看。

陆
下池
6

把竹料整齐地放到池子里，铺一层料子，撒一层石灰，再铺一层料子，如此反复，直至装满整个池子。石灰的用量比较大，每100 kg竹料需要用30 kg刚烧出来的石灰。如果第二天下池，当天就要把石灰化好备用。竹料装好以后，最上面需要放2~3组木棒，每组2根。木棒上面加放若干块石头，数目不限，其作用是不让竹料浮起来。接下来加水直至漫过竹料。泡料一般需要100天左右。

柒
洗料
7

先在池子里把竹料上的石灰刷洗干净，再将竹料拿到水沟里清洗。通常情况下，2~3人一天可以洗完一整池竹料。

捌
洗池
8

洗料结束后要把池子洗干净，并将石灰水全部洗掉。

玖

发汗

9

重新把料子放到池子里，上面用干谷草铺好，两头用木棒盖住，最上面用石头压好。发汗时间的长短要根据气候而定。如果下雨且气温高，则发汗最快，12~13天后竹料即可使用。一般情况下，则需要25天左右。如果一直是晴天，则要用水把竹料的表面泼湿。如果下雨且气温低，则发汗最慢，需要一个月甚至更长的时间。竹料发汗的环节需要十分小心，每天早上都要到池子边上观察。判断发汗结束的标准：一是用手搓料子，感觉料子绵软；二是看到池子里的料子最上面一层普遍长满了菌。若发汗不足，料子不容易碾烂，“碾料”工序的操作就比较费劲；若发汗过度，竹纤维被破坏得多，竹料损失也相应增多，“舀纸”工序需要反复进行，比较费工费时。有一年，梨坝村陈长登家的料子发汗太慢，陈长登把蒸豆腐的泔水洒在竹材上后，很快就完成了发汗，于是这就成了大家借鉴的“陈家经验”。

拾

淹水

10

“淹”当地方言念“ā n”，意思就是向池子里加满水，把完成发汗的竹料泡上。碾多少竹料就从池子里取多少。

拾壹

碾料

11

从池子里把竹料取出来，铺在碾盘上。牛拉磙子在碾盘上碾压竹料，每碾一次200 kg。每当牛拉磙子过后，人要立即上去用铲子把纸料翻起，再铺匀，这样可以提高碾压的效率。若竹料容易碾，则碾2~3小时即可；若竹料不容易碾，则需要碾半天，甚至一天。“碾料”是非常费力的一种工序，即使是一头耐力很强的牛一天也只能碾两盘料，否则牛的力气也会不济。这道工序过去叫“碓料”，就是把竹料放到石质的碓臼里，用木质的碓梃碓舂。需要两个人合作，一人踩碓梃，一人翻料。一次可碓一个浆把，7~8 kg，一天可碓5个浆把，劳动效率较低。大约从1995年开始改为牛碾后，工效提高很多。

拾贰

淘料

12

把料子放到料坑里，一次淘一槽料（50 kg左右）。将料坑装满水，用捅耙和料，将淘洗后的浊水放掉，再加水。如此循环4~5次，直到淘洗干净方可。

拾叁

打槽

13

把一槽料全部放到纸槽里，用槽杆捶打料子，直到把料子打成“汤汤”（即糊状），即将料子打到均匀为止。一般每槽料要打1 000杆。每打200杆后再用捅耙把下面打不到的料翻上来。打槽一次通常需半小时。打好槽后把多余的水放掉，重新淹水。如果第二天抄纸，则头一天就要把槽打好。

拾肆

做滑

14

抄纸的前一天要把滑根放入石碓中碓破，再放到滑药坑里加水浸泡。用槽杆打70~80杆，让滑根中的汁液充分溶入水里。舀纸加滑药的时候还需要再打20杆左右。若全天舀纸，则要加5~6次滑药。滑药的兑制没有严格的标准，全凭经验判断。夏、秋季节温度偏高，滑药极易变质。

⊙1

舀纸之前要把纸帘放到纸帘架上，从右往左挑水，水从右边进入，还由右边流出，再由外往里舀纸。如此最多走两道水。通常，舀出来的纸左边会厚一些。一次加料35~40 kg，可以舀1 500帘纸。刚舀出来的纸尺幅为60 cm×47 cm。现在一帘一般可以出3张纸，过去也有一帘出4张纸的，这根据纸帘上的粗线走向而定。通常情况下，舀纸从农历十月开始，十二月结束，最晚可以到次年正月。春耕农忙时，造纸户又会以农耕为主。

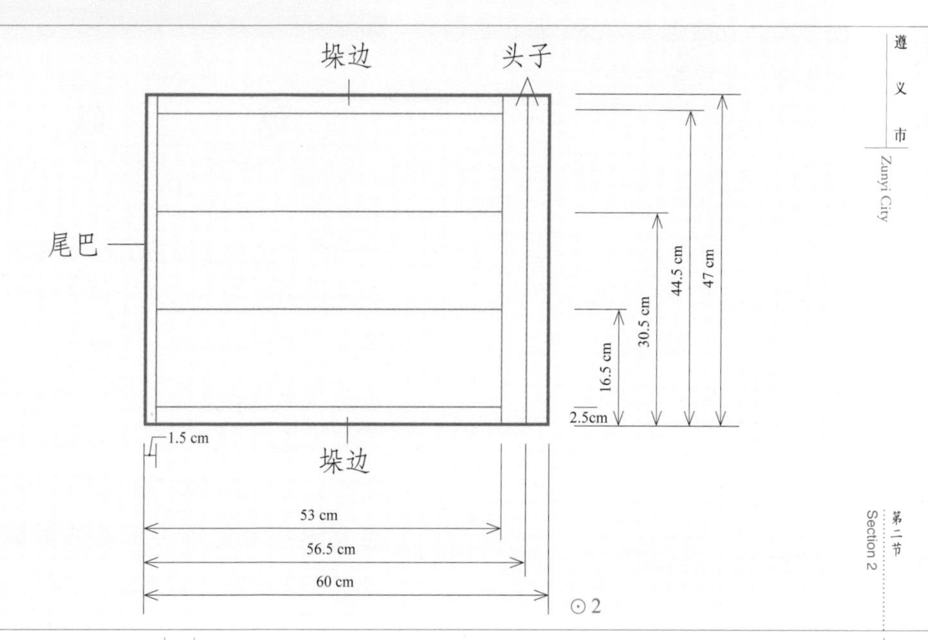

⊙2

⊙1
和溪镇杉木坪村蒋礼昌老人在舀纸
Jiang Lichang, the local papermaker in Shanmuping Village of Hexi Town, lifting the papermaking screen out of water

⊙2
目前使用的纸帘示意图
Schematic diagram of the papermaking screen currently in use

拾陆
扳垛边
16

将纸舀起来以后，把不需要的“边边”去掉。

拾柒
榨纸
17

将舀起的纸放在木榨上，达到一定高度后，借助垛板、底码、码子、吊杆、绳索等工具，运用杠杆原理把纸里原有的水分挤出去。榨纸一般需要半小时左右，若人力扳不动了则说明纸里面的水分已经很少了，这时就可以停止加压。也曾有人用过千斤顶施压，但千斤顶的使用寿命短，一般只有2~3年。

拾捌
分垛
18

松开纸榨以后就可以沿纸帘上粗线经过的地方把纸分成数垛，并将其搬回家。

拾玖
揭纸
19

用“压纸锤”沿着从“垛脑壳”到“垛尾巴”的方向划过，使纸与纸之间形成空隙，这样纸才能更好地分开。揭纸的时候要从纸相对较厚的左边开始，慢慢揭起。第一张纸揭到一半时，开始揭第二张，每两张纸相距一定距离，这样可以提高工效。分开的纸垛中，多的一叠有上百张，少的也有二三十张，它们都需要一张张地被揭开。

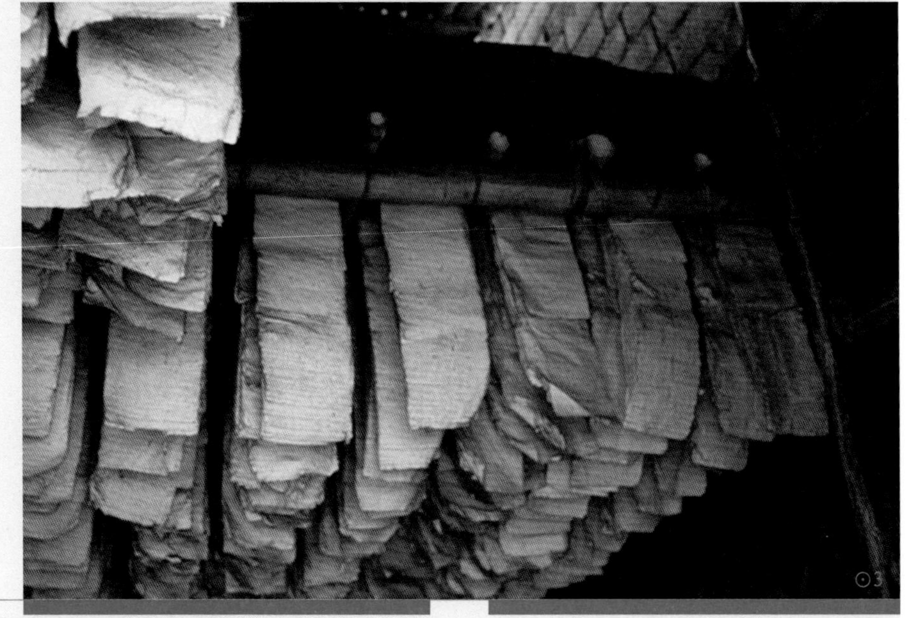

⊙3

贰拾
晾纸
20 ⊙3

将揭开的纸放在屋内竹竿上晾干。天气好的情况下十几天即可晾干，若天气不好则需要一个多月。遇到冬天气温低且没有太阳，则一个多月都不能晾干，只能一直晾在屋内阴干。虽然阴干纸的时间比较长，但也不能用火烤，否则纸会卷起来，因不平整而不好卖。

贰拾壹
捆纸
21

把晾干的纸整理好并捆扎起来。过去是5~8张，或10张整理成1叠。目前的做法普遍是4张为1叠，10叠为1捆，10捆为1担，10担为1挑。每担用竹篾捆扎为一个包装单元。每担的底部和上部各放一竹片，再用竹篾从中间捆两道。1 500 kg料子可造出100多担纸。

⊙3
凤仪镇梨坝行政村大田坝村民组屋内晾纸
Drying the paper in a room in Datianba Villagers' Group of Liba Administrative Village in Fengyi Town

(三)

正安竹纸生产使用的
主要工具设备

壹
舀纸棚
1

舀纸棚是造纸的简易作坊，棚内会设置纸槽、引水槽、滑药坑、料坑、纸榨、石臼、踩席等设施。

⊙4

⊙5

贰
滑药坑
纸　槽
料　坑
2

舀纸棚里从左到右排列着滑药坑、纸槽和料坑，尺寸大小不完全相同。陈长登家的滑药坑的长、宽、高分别约为115 cm、90 cm、70 cm，纸槽的长、宽、高分别约为150 cm、125 cm、80 cm，料坑的长、宽、高分别约为125 cm、80 cm、70 cm。纸槽有倾斜角度，便于操作。靠人的一侧有塑料布遮挡，防止弄湿或弄脏衣服。

⊙4/5
和溪镇杉木坪村的舀纸棚
Papermaking sheds in Shanmuping Village of Hexi Town

叁
纸　榨
3

纸榨是榨出纸中水分的设备。包括左高右矮、大小各一的两对垛桩，大的叫“高立人”，高约170 cm，小的叫“矮立人”，高约100 cm；垛桩上的横梁叫“千斤担”，高约125 cm；连接两对垛桩的横木叫“榨桥”，长约130 cm，高约50 cm。垛板放在两对垛桩之间，用于摆放舀出来的竹纸，垛板的尺寸约是100 cm × 65 cm × 5 cm。

⊙6

肆
踩　席
4

放在纸槽和料坑下面用来和料的设施。

伍
纸帘和纸帘架
5

梨坝村制作竹纸的纸帘尺寸通常为65 cm × 43 cm或50.5 cm × 43 cm。比较特别的是，这里的纸帘被较粗的线分成四个板块。捞纸的时候粗线经过的部分竹纤维沉淀得比较少，一次捞上来的纸可以比较方便地沿粗线形成线条的地方把纸掰开，分成55.5 cm × 26 cm、49 cm × 17 cm、17 cm × 6.5 cm、43 cm × 6.5 cm等几个部分。根据纸的尺幅大小不同，分别叫作菜板纸、七折纸和五折纸等不同的名称。纸帘在抄纸的时候会配合纸帘的架子使用。纸帘架的外框边缘尺寸是68 cm × 63 cm，内框边缘尺寸是60 cm × 54 cm。左手处有一方便手握的半圆形固定把手，右手处有一木棍作为活动把手。

⊙7

⊙8

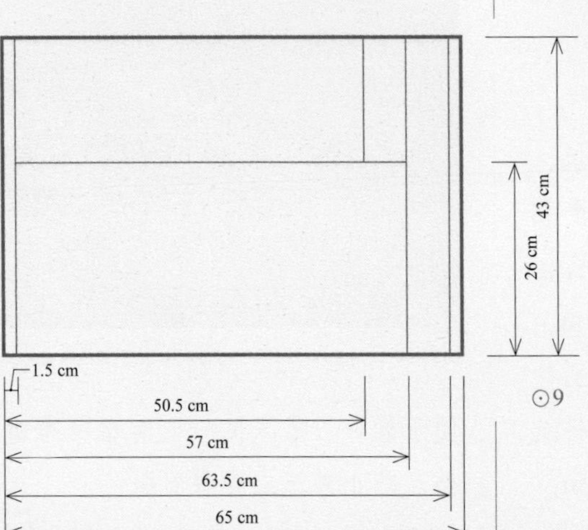

⊙9

⊙6
三把车行政村堡上村民组的舀纸池
Papermaking trough in Baoshang Villagers' Group of Sanbache Administrative Village

⊙7
梨坝村现在使用的纸帘及纸帘架
Papermaking screen and its supporting frame currently in use in Liba Village

⊙8/9
梨坝村过去使用的纸帘及其示意图
Papermaking screen and its schematic diagram formerly used in Liba Village

陆

捅耙和纸筋刷

6

捅耙是用于搅拌纸料的木质工具，其一端为40 cm × 18 cm的木板，木板上安装一根长95 cm的木棍，作用是把投入纸槽的料搅拌成均匀的纸浆。纸筋刷是捞纸筋的竹质工具，由数根长40 cm的细竹捆扎而成，根部捆一道方便手握，中间再捆一道以固定数根细竹张开的位置。其作用是在舀纸前把搅匀后的纸浆里没有打碎的粗纸筋捞出来，以免影响纸的质量。

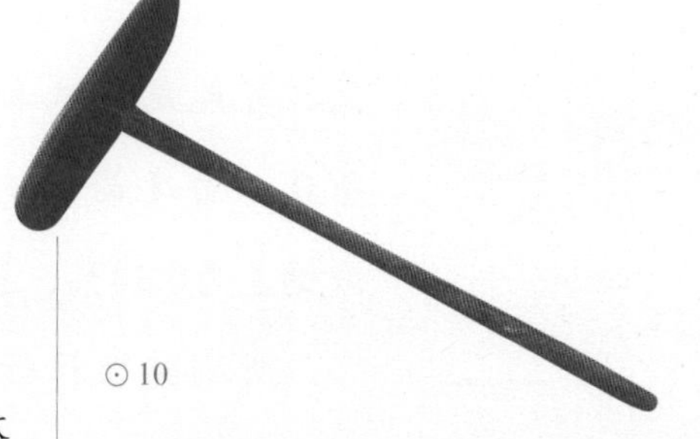

⊙10

⊙11

柒

泡料池

7

用来浸泡竹料的设施。一般设在舀纸棚的周边，造纸量大的造纸户往往将数个池子，甚至数十个池子密集分布，连成一片。池子的深度为2.3~3.5 m，有些池子的一侧底部会留一个排水孔。池子的大小通常依地形情况而定，容量的差距可能有几倍之多。

⊙12

捌

碾盘和磙子

8

碾盘是用来轧碎竹材的设施，通常是一处石块或水泥围成的圆盘。碾槽的内径约300 cm，外径约320 cm。磙子的直径约57 cm，高约38.5 cm。使用时先把浸泡好的竹材平铺在碾盘上，然后用牲口拉着磙子绕碾盘中间的一根石柱做圆周运动，直至把竹材碾碎为止。

⊙13

⊙10
梨坝村使用的捅耙
Rake for stirring paper pulp used in Liba Village

⊙11
梨坝村的纸筋刷
Sticks for picking out the impurities in Liba Village

⊙12
三把车行政村堡上村民组李红全老人介绍池子的使用方法
Li Hongquan,the local papermaker in Baoshang Villagers' Group of Sanbache Administrative Village, introducing the use of soaking pool

⊙13
碾竹材用的碾盘
Grinder for grinding bamboo materials

(四)

正安竹纸的性能分析

所测正安梨坝村竹纸的相关性能参数见表8.9。

表8.9 梨坝村竹纸的相关性能参数
Table 8.9 Performance parameters of bamboo paper in Liba Village

指标		单位	最大值	最小值	平均值
厚度		mm	0.370	0.190	0.343
定量		g/m²	—	—	51.0
紧度		g/cm³	—	—	0.149
抗张力	纵向	N	9.4	6.0	7.6
	横向	N	5.9	3.7	4.6
抗张强度		kN/m	—	—	0.407
白度		%	17.0	16.0	16.5
纤维长度		mm	4.15	0.81	1.88
纤维宽度		μm	45.0	4.0	14.0

由表8.9可知，所测梨坝村竹纸最厚约是最薄的1.95倍，相对标准偏差为0.66%。竹纸的平均定量为51.0 g/m²。所测竹纸紧度为0.149 g/cm³。

经计算，其抗张强度为0.407 kN/m，抗张强度值较小，可能是因为正安竹纸的主要成分为竹纤维，其纤维壁厚且很少有弯曲。

所测梨坝村竹纸白度平均值为16.5%，白度较低，可能是竹纸未经过漂白，加上表面比较粗糙所致，白度最大值约是最小值的1.06倍，相对标准偏差为0.34%，差异相对较小。

所测梨坝村竹纸纤维长度：最长4.15 mm，最短0.81 mm，平均1.88 mm；纤维宽度：最宽45.0 μm，最窄4.0 μm，平均14.0 μm。所测竹纸在10倍、20倍物镜下观测的纤维形态分别见图★1、图★2。

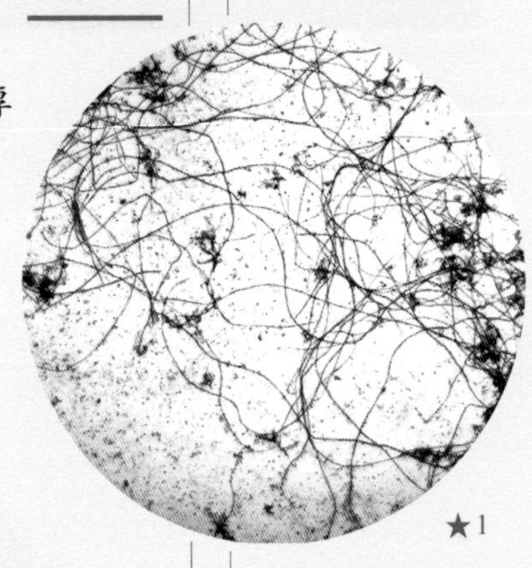

★1

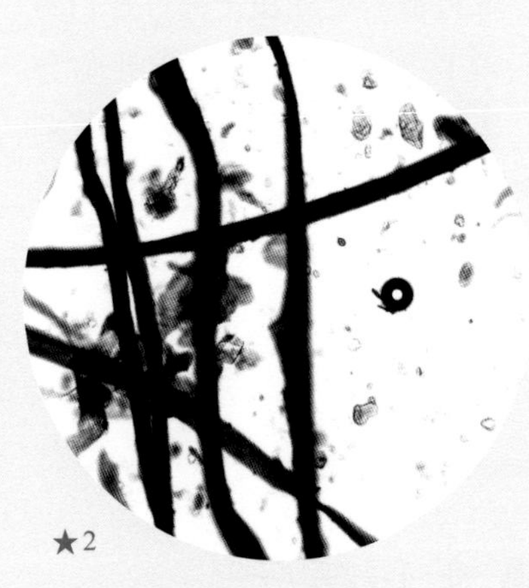

★2

★1
梨坝村竹纸纤维形态图(10×)
Fibers of bamboo paper in Liba Village (10× objective)

★2
梨坝村竹纸纤维形态图(20×)
Fibers of bamboo paper in Liba Village (20× objective)

五
正安皮纸的生产工艺与技术分析

5
Papermaking Technique and Technical Analysis of Bast Paper in Zheng'an County

2009年4月下旬，调查组进入正安县凤仪镇梨坝村调查时，该地皮纸生产业态已完全终止，但通过访谈多位老造纸人，初步记录了颇具特色的梨坝村皮纸生产工艺。

(一)

正安皮纸的生产原料与辅料

据造纸人陈大明（男，时年60岁）、李进文（男，时年57岁）介绍，当地制造皮纸所用的原料是生长于当地的山杨柳和杨树的树皮。通常在农历三至五月间上山砍山杨柳树的树枝，因为此时的树皮容易剥开。若到了农历六月，树皮就不易从树枝上刮开剥离了。砍杨树树枝的时间要比砍山杨柳早些，结束的时间也相应早些，一般都会提前半个月左右。

梨坝村皮纸生产使用的纸药主要有两种。一种纸药是山杨柳叶。先将采集来的山杨柳叶打融，再放入水中浸泡，待泡出黏汁液后即可马上使用。但是这种纸药的药性保存时间有限，仅限于当天使用，时间长了就失去了纸药的作用。另一种纸药是松根。松根要先放在石碓上打融，再浸泡在水中待用。这种纸药的药性保存时间较长，只要出来的水不干，就可以连续使用几天。

(二)
正安皮纸的生产工艺流程

据陈大明和李进文等村民口述，正安皮纸的生产工艺流程如下：

壹	贰	叁	肆	伍	陆	柒	捌	玖	拾
砍树	火漂	剥皮	打『把把』	浆『把把』	蒸料	洗料	二次蒸料	漂料	榨料

拾壹	拾贰	拾叁	拾肆	拾伍	拾陆	拾柒	拾捌	拾玖	贰拾
打皮板	切皮板	淘料	打槽	加滑	舀纸	榨纸	揭纸	焙纸	包装

壹
砍树
1

造纸户通常会使用斧头、弯刀砍伐山杨柳和杨树。小的山杨柳也可以用手拔出来。山杨柳不管是老枝还是嫩枝都可以用作造纸的原料。两年生的杨柳树枝出纸量高，同一棵山杨柳树上的树枝一般是两年砍一次。

贰
火漂
2

将采集来的树枝捆成直径大约为40 cm的“把把”，质量在20 kg左右，点上明火烤热树枝，一个“把把”大约要烤十几分钟。

叁
剥皮
3

经过火漂之后，要尽快趁热把树皮从树枝上剥离下来。

肆
打“把把”
4

将剥下的树皮以把为单位整理好，每把1~1.5 kg，用杨树皮捆扎起来备用。

伍
浆“把把”
5

按照每3 kg“把把”用1 kg石灰的比例，将纸料浸入石灰浆里，为纸料上浆。

陆
蒸料
6

把浆好的纸料放入纸甑里，最上面用一层干草覆盖，再用泥巴封严，大火蒸24小时左右。

柒
洗料
7

取出蒸好的纸料，拿到河里冲洗，用脚踩压，尽可能挤出纸料中的石灰水。

捌
二次蒸料
8

把洗净的纸料放入纸甑，进行二次蒸煮，这次蒸煮无须加入石灰浆，大火烧2小时，当看到纸甑上面有蒸气冒出时就可以停火。

玖
漂料
9

从纸甑里取出完成二次蒸煮的纸料，放入流动的小溪里漂洗三天，其间，经过一天漂洗以后，要把纸料从上到下翻理一遍，以便透彻地漂洗纸料。将整理好的纸料再次放入小溪的水流中漂洗两天。

拾 榨料
10

从小溪中捞出漂洗好的纸料，将其放到木榨上压榨以使纸料中的水分被压干。

拾壹 打皮板
11

从木榨上取下压干水分的纸料，将其放入脚碓里，一人踩碓，一人翻料，把杨树皮打成皮板。通常两个人花费约3小时可打好35~40 kg纸料。

拾贰 切皮板
12

用皮刀将皮板切成段，每段长1~2 cm。

拾叁 淘料
13

将成段的纸料装入淘筐里淘洗，淘洗时要上下搅动纸料，同时，把其中的粗纤维和杂质拣出来。10 kg干料要淘1小时左右。

拾肆 打槽
14

纸料淘洗好以后需放到纸槽里，加水，用竹竿搅匀。在打槽的过程完成之后，无须再放水淘洗纸料。

拾伍 加滑
15

在纸槽中加入事先制备好的“滑药”（也称“纸药”）。通常舀一槽纸要用掉大约15 kg纸药。

拾陆 舀纸
16

用纸帘抄起纸浆，通常要过两道水，先往左舀一次，再往右舀一次，然后将有纸浆的一面向下，反扣在木榨上被称作“水盘”的厚木板上，揭起纸帘。一张纸帘舀出一张皮纸，一个纸工每天可舀1 100张左右皮纸。

拾柒 榨纸
17

水盘上的纸达到一定的数量后，就要使用木榨压干其中的水分。

拾捌 揭纸
18

将挤干水分的纸垛翻转过来，使纸垛原来贴在水盘的一面向上，逐张揭起皮纸。揭纸时，先从左上角开始，轻轻地往右下角拉，一张张地撕开。

拾玖 焙纸 19

左手徒手、右手持棕刷，双手把揭起的纸贴到纸焙上焙干。贴纸时须按从上到下、从左到右的顺序依次贴齐。一面纸焙可贴25张：从上到下有3张，从左到右有8张，第25张皮纸则是贴在靠火灶的一头。将焙干的皮纸从火焙上撕下来时，要先将纸的左上角揭起，从左上角向右下角撕。一个工人一天可晒（焙）一捆（1 000张）纸。一个人舀纸，一个人晒纸，两个人配合，操作的进度基本相当。

贰拾 包装 20

将揭下的皮纸理齐，以100张为单位把纸折三道为一合。皮纸的质量小于火纸的质量，一合火纸的质量约为350 g，一合皮纸的质量约为250 g。10合为一捆，用竹篾或棕条捆住两头，即可以成捆对外出售。

（三）正安皮纸生产使用的主要工具设备

壹 纸甑 1

蒸煮皮纸原料的设备。多以石块、水泥砌成，内有铁锅，锅上架着木棒，纸料码放在木棒上。一个纸甑一次可以蒸煮100 kg左右的纸料。

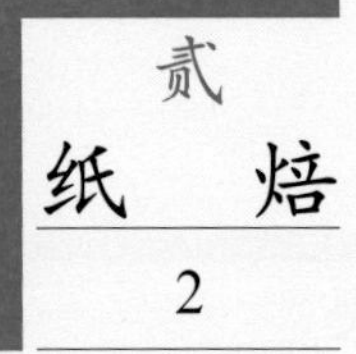

贰 纸焙 2

焙干湿纸的设备。长约510 cm，高约150 cm，火堂上面有根长杉木，两侧有两排人字形木架支撑，在木架上扎竹篾席，外刷石灰和纸筋，形成两面均可焙干湿纸的墙面。使用时在火堂的一头生火，另一头排烟，整个纸焙的温度就会上升。如果焙纸时湿纸无法粘上墙，则需要在墙上刷些米汤。一般刷夹层纸都会先在纸焙墙面上刷米汤，单层纸则不需要这道工序。

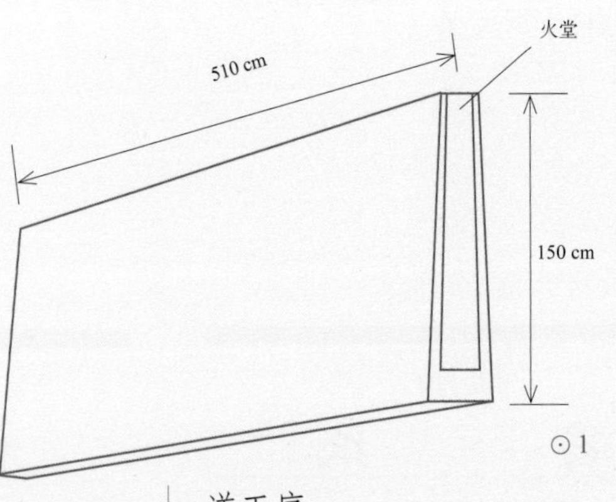

⊙1 纸焙示意图 Schematic diagram of the drying wall

六
正安手工纸的基本用途与销售情况

6
Uses and Sales of Handmade Paper in Zheng'an County

⊙2

⊙3

⊙2
梨坝村道士用皮纸书写的《永言孝思》
Book in Memory of the Deceased written by Taoist priest on bast paper in Liba Village

⊙3
用正安皮纸绘制的神像
Buddha paintings on bast paper in Zheng'an County

(一)
正安手工纸的基本用途

正安手工纸的应用历史十分悠久，使用范围非常广泛，至今还有非常精彩的纸工传承。主要有以下几种用途：

1. 书写用纸

《道光遵义府志》引述当时贵州名儒郑珍（1806~1864年）所撰《田居蚕室录》中关于手工纸用途的记载："遵义之纸以构皮制者曰皮纸，以竹制者曰竹纸，皆宜书。"[29]这道出了正安纸用于书写这一用途。传统习俗是，正安百姓遇到重要的事情时还要郑重地买来皮纸订成本子认真记载。比如老人过世时，要请来道士写一本百余页的《永言孝思》，以记载逝者的世系、生平及祭祀的盛况，留给后人作为纪念。

2. 绘画用纸

正安的绘画多用皮纸，在民间，人们对于皮纸和宣纸的区别不明显，他们更愿意分清竹纸和皮纸。旧时道士做法事所用的神像须请青丹师绘制，一般是先用皮纸裱褙，再用浆糊将多层画纸粘贴在一起。晾干后用铅笔勾影，绘出神像轮廓，然后着色。阴干后开光安位，便告成功。

3. 印刷用纸

正安县境内小场镇用木刻板印刷迷信文书，春节期间印刷财神、春联供应市场。民国初年，县城南街杨家置石印板一台，以供政府印布告之用。

4. 祭祀用纸

祭祀用纸有朝皇纸、黄裱纸、草纸、烧纸、草纸、火纸等多种名称。其中，朝皇纸的名称源

[29] [清]平翰，等. 道光遵义府志：卷十七[M].上海：上海古籍出版社,1995:28.

于法事活动中道士烧给天皇大帝、玉皇大帝的纸。道士在道场上从事法事活动的时候烧的“袱纸”是七折纸。袱纸一般需要盖上冥币的戳印，封成小包，上书“孝男××，奉送冥币于某府故先考先妣神前，袱乞正魂收用，外鬼勿争”。根据正安县的风俗，清明节前后十天都可以扫墓挂青。挂青是另一种给过世者“送钱”以寄托思念之情的方式，其用纸量也非常大。

5. 工艺用纸

工艺用纸主要用来扎制龙灯、狮子、风筝、剪纸、花圈、灵屋、孔明灯等工艺作品，可以用竹纸，也可以用皮纸。彩扎工艺在正安有比较明确的分工。以安场镇为例，周明鑫、黎定成、冯道俊、彭三元（聋哑人）4位精通彩扎的民间艺术高人，擅长的手艺各不相同。周明鑫的国徽纸扎艺术制品曾经在安场镇里的大会会场悬挂多年。黎定成的灵屋彩扎工艺在当地及其周边一带很有名气。冯道俊制作的龙灯彩扎是安场舞龙民俗文化活动中最抢眼的作品。彭三元的麻老鹰、花蝴蝶风筝是许多正安人童年时代奢侈的玩具。

虽然4位老一辈的民间纸扎艺术家均已过世，但是如今仍然有许多人在做纸艺。正安仡佬族民间艺术家周开智用各种纸张制作折纸仿真昆虫、立体山水画、盆景等，其以精美的纸艺作品于2006年和2008年两次在“多彩贵州”能工巧匠选拔大赛中脱颖而出，获得“贵州名匠”的称号。刘朝毅师从黎定成，学习彩扎，利用业余时间探索纸艺，2009年创作完成了全县春节游行中的安场镇龙灯彩扎。

金光神呪
天地玄宗 萬炁本根
廣修萬劫 証吾神通
三界內外 惟吾獨尊
體有金光 覆映吾身
視之不見 聽之不聞
包羅天地 養育群生
受持萬遍 身有光明

⊙1

⊙2

⊙3

⊙1 用正安手工纸印制的道家经书《金光神咒》
Taoist classics named *Golden Light Mantra* on handmade paper in Zheng'an County

⊙2 祭祀牌位前的火纸
Huo paper in front of memorial tablets

⊙3 灵屋和花圈
Paper house and wreath used for sacrificial ritual

6. 制鞋用纸

正安的手工纸还有一个独特的用处是做“布壳”，即当作鞋底或鞋垫使用。用糨糊将皮纸或草纸糊裱在布上，晒干了就是布壳。布壳是做布鞋和鞋垫的原料。在童谣“老婆婆（儿），打鞋底（儿），巴（儿）的一针锥过去，巴（儿）的一针锥过来”里，老婆婆们给布壳加上针线就做成了鞋垫，或是纳成了鞋底。

7. 包装用纸

传统包装食盐、糖、点心、果饼和药材用的就是质地比较粗厚的纸壳。

8. 建筑用纸

正安手工纸另一个独特的用途是用作建筑材料。当地有大量干栏式建筑，木房隔墙需要用一种粗制纸壳作为原料。将草和竹等造纸材料用水浸泡后捶打，提取竹纤维纸壳，再混合石灰粉糊墙壁，这种墙壁被当地百姓称作“石灰篾杆壁”。

9. 照明用纸

正安有一种用皮纸糊制而成的照明用具，被称为“糍粑灯笼”。这种灯笼打开时呈圆柱形，内可点燃蜡烛照明，上面有根线，方便挑着行走；不用时可折叠收起形如糍粑，所以取名为“糍粑灯笼”。

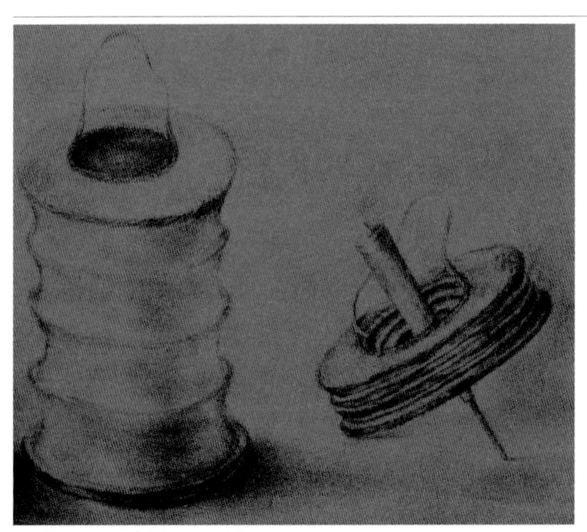

⊙4

⊙4 糍粑灯笼（周信绘画，1999年版《正安县志》）
Lantern made of bast paper in Zheng'an County (drawn by Zhou Xin, included in *The Annals of Zheng'an County*, 1999)

10. 制作鞭炮用纸

在机制鞭炮尚未大量生产之前，正安曾有很多手工鞭炮作坊。这些作坊都会使用手工纸包裹火药和药捻原料。

此外，正安的手工纸还有作为土枪引线用纸、妇女经期卫生用纸、擦拭机器用纸等多种用途。在20世纪80年代以前，大量的皮纸用来擦拭机器，这给梨坝村的造纸户带来了可观的财富。

（二）正安手工纸的销售情况

早在清道光年间至20世纪50年代，正安的皮纸不像质量上乘的遵义、绥阳两县所产的纸销路广阔，绥阳皮纸“极佳者贩入蜀中”[30]，“（遵义县）芦江水专以构皮制成曰皮纸，再舀而成者曰夹皮纸。多行本属及四川川北一带”[31]。即使正安夫烟坪的竹纸“可抵川纸之红批、毛边”，也只是受到本地市场的认可和欢迎，并没有改变“卖之本郡”的基本情势。当时，生产土纸仅仅是农业生产劳动及收入的一种补充。土法生产的白皮纸、火炮纸、草纸及纸壳等产品中有相当一部分还处于自给自足、自产自销的自然经济状态，除自用以外，大部分就近出售，满足本地市场需求。

民国年间，正安县城和安场镇有两家文具店，所售商品限于笔墨纸张之类。民间所用纸张

[30] [清]平翰，等.道光遵义府志：卷十七[M].上海：上海古籍出版社，1995:28.

[31] 周恭寿，等.(民国)续遵义府志：卷十二[M].民国二十五年(1936年)刊本.

多为县内所产皮纸和竹纸。从20世纪中叶以后，钢笔替代了毛笔，传统的文房四宝仅保留在少数爱好书法、绘画的人家中。竹纸和皮纸逐步被毛边纸、打字纸、书写纸及各种色纸替代，当地生产的纸类逐渐从市场中淡出。

1952年，正安县供销合作社开始经营日用杂货，从本地收购土纸等十余种商品在各供销门市部销售。也正是从这时开始，正安县有了纸品销售数量的统计（表8.10）。在有统计数据的9年中，1978年的纸张销售量最高，达到76 t；1956年的纸张销售量最低，仅为15 t。在土纸的产量有统计数据的6年中，1982年的土纸产量最低，为32 t；1978年的土纸产量最高，竟有1 300 t之多。对比全县乡镇企业土纸产量和本地纸品销售的数据可以得出结论：第一，自20世纪80年代以后，正安本地土纸的产量逐渐与本地纸的销量拉开距离；第二，唯一同时有产销数据的1990年，生产土纸282 t，销售纸品32 t，这一年正安县有250 t土纸没有在本县销售。

正安手工纸生产的经济环境近年来发生了巨大的变化。以梨坝行政村大田村民组为例，在土纸生产最兴盛的1978~1990年，全组80户人家中有16户长年造纸，占到当时户数的20%。一个村民组全年产量最高时有数十吨。造纸户超过60%的收入为造纸所得。

但在调查组进村调查时，全组160户人家中只有4户造纸，仅占村民户数的2.5%。若按平均每户生产纸品300 kg计算，一个村民组一年的产量也仅有1 200 kg。造纸户造纸的收入仅占其全部收入的20%左右。现在全组村民造纸产量加起来还不及过去一户人家的产量。如今，大田村民组村民主要以耕种和务工为生，少数经商，造纸早已不是村民最重要的生活来源。由于市场萎缩、劳动力外出打工，以及老手艺人年事渐高，许多乡村的土纸生产活动逐步停止，仅存的只有纸槽的遗迹和闲置已久的工具了。

表8.10　1952~1990年正安县部分年度全县手工纸销售数量*
Table 8.10　Annual sales of handmade paper in Zheng'an County between 1952 and 1990 (parts of the data)

年份	纸张（t）
1952	-
1956	15
1958	20
1961	21
1963	41
1970	40
1978	76
1980	50
1985	34
1990	32

手工纸生产的衰落对于石灰生产等相关行业的发展造成了一定的影响。正安县的石灰石资源比较广泛，各个乡镇都有分布。氧化钙含量普遍在48%以上。安场镇有高纯度的石灰岩，氧化钙含量达到53.93%。20世纪80年代中后期的数据对比表明，手工纸产量与石灰产量大体上呈现出同步变化趋势。

* 本表根据《正安县志》(贵州人民出版社，1999年9月)有关表格整理。

七 正安手工纸的相关民俗与文化事象

7 Folk Customs and Culture of Handmade Paper in Zheng'an County

(一) 谚语

正安有一句与纸有关的谚语："烧钱化纸，以缅阳人。"这句话明示人们，一切丧葬礼俗都不过是做给活人看的仪式。

(二) 习俗

1. 祭祖师

每年第一次舀纸之前，造纸户都要备好肉、鱼、鸡、酒、香、烛、纸等供品，或摆在家中蔡伦先师的神龛牌位前，或摆在纸槽内，或摆在纸槽附近合适的地方。家中负责造纸的男子上前祈祷蔡伦先师或老辈的师傅保佑这一年舀纸平安顺利，揭纸不破。祷告的话语并没有固定的程式，比如，一位年届70岁的老造纸师傅说，他在祭槽的时候心中会祷告："老辈儿的老师傅们，保佑我这一年舀纸顺顺序序（顺利）啊！"现在造纸户中仍然普遍存在这样的祭祖师活动，当地村民称之为"祭槽"。

2. 开天门、传弟子

正安县的老人去世后一般仅享受道士安排的3天法事活动。但是，造纸工匠和纸扎工艺师傅等手艺人在去世后却有资格享受5天以上法事活动的待遇，以便有时间举行一场"开天门、传弟子"的仪式。在这个仪式上要把承载和象征老一辈造纸师傅和纸扎师傅手艺的加工和制造工具交给继承者。没有人能够说清楚这个传统仪式起源于何时。但是，可以肯定的是，"开天门、传弟子"仪式和对这一仪式的时间预留反映了历史上正安先民对于拥有技艺并传承技艺的人及其所创价值的尊重。

3. 放孔明灯

正安县本地举行的丧礼安排在法事活动结束前一天夜里12点以后，一般要放用手工纸制作的孔明灯，数量为2~8盏。放孔明灯的目的是给逝者照亮升天的路。

⊙1

⊙ 1
正安班竹乡的旧字库塔
Ancient Paper-Burning Tower in Banzhu Town of Zheng'an County

4. 敬惜字纸

正安县主要有两处字库塔，一处位于俭坪乡合作行政村繁荣村民组，另一处位于班竹乡旦坪完小附近。班竹乡字库塔有石碑嵌于塔基，碑文如下：

创修字库小引。尝闻仓颉圣人制字，雨粟来□*。上天点画，形象既成，悲号闻于夜鬼。可知一字千斤之重，万古昭然。而褙屏糊窗之轻，见而□□。然遍觅当世，一粟片赀之末，往往失则知求。六经字纸之文在在，视犹敝屣。岂知埋字纸而五世登科，扬全善之善根非浅，葬字纸而一身显贵。□材之种福，有由是字，固宜惜然，必有□□之所者。今有正婺壤接地名祥云寺者，文人肄业之区，字纸凑积之所矣。是以予也，目击心伤，素体惜字之典难为，弗若修化字之炉而为功。以故，立念多年，无奈独立难支。赞劝必赖多人，蒙绅士人仁共出囊内金钱，不惜有余珍宝，庶乎文教振兴。不负帝君惜字一千，延寿一纪之明训矣。领首增生周光仁，男应田，孙德慎、德浩，门生周尚纲、郑作梁各助钱五千。生员雷登谱乙千二百文。张子俊、张子周、叶兴□各乙千二。助领首僧月有。道光玖年，已丑岁，仲冬月上完吉旦。

有专家考证，我国敬惜字纸的传统始于宋，盛于明清。明清时期民间多有托名文昌帝君的《惜字宝诰》《惜字律》《惜字新编》等书籍刊印。要求人们“不轻笔乱写，涂抹好书”“不以书字放湿处霉烂，并扯碎践踏”。《敬字纸功例》中明示，“敬惜字纸有富贵福寿之报”。《慢字纸功例》则告知，“不敬惜字纸有穷苦夭寿天诛之报”，不能“以字纸经书放船舱底并马上，令人骑坐”，违之要记“二十罪”“生毒疮，受人欺凌”；不能“以经书枕头”，违之要记“十五罪”，遭“穷苦，受杖刑”。民国年间李幼芝曾著有《课孙竹枝词》，告诫晚辈要珍惜纸张：“蔡伦造纸费神功，遂使教化普天穹。寸纸如金应珍爱，说与儿孙勿看轻。”

从班竹乡字库塔的碑文可见，当地也有类似《敬字纸功例》的俗信。字库塔附近村民记得老辈人传说，废弃的字纸不得任意抛撒、踩踏，而须统一焚化，凡人眼瞎者，搜集字纸入内烧之，不久复明。“敬字惜纸”的传统和民俗饱含着古人对传播知识的字纸的敬重和爱惜之情，在文化教育资源相对稀缺的古代有着积极的意义。

⊙2

⊙2 创修字库碑
Memorial monument of the ancient Paper-Burning Tower

* “□”符号表示碑文中字迹模糊无法认读的文字。

八 正安手工纸的保护现状与发展思考

8 Preservation and Development of Handmade Paper in Zheng'an County

今天，正安手工纸生产技艺正濒临失传。究其原因，首先是造纸手艺的新老交替出现了断裂。老工匠年事渐高，许多技艺高的造纸师傅还没来得及把手艺传给后人就过世了。新一代村民外出打工现象比较普遍，愿意从事手工纸生产的人不多。其次是手工纸生产从业者的生产热情减退。一方面是手工纸原材料和用工成本的增加，另一方面是市场价格提升的空间有限，双向挤压造成手工纸的利润渐薄，使得许多手工纸生产者改行从事种植业和养殖业。最后是面向现代生活的手工纸市场尚未培育起来，一些原先用量较大的手工纸市场逐步被机制纸替代。民俗用途的手工纸市场空间十分有限，其可替代性较强。

传承正安手工纸生产技艺的关键是对历史文化资源的保护。利用统筹规划，官方和民间多方协调，文化、旅游等部门同步推进，共同拓展新的市场空间。

(1) 将手工纸的保护纳入文房用品的恢复、开发和集成保护的整体计划中。正安历史上生产过宜书宜画的手工纸。目前，内地厂家普遍使用化学原料生产书画纸，高端古法手工纸生产已较稀缺。开发和生产高端手工纸将会吸引更多的村民重操手工造纸的旧业，也应当能够为村民带来可观的经济收入。

乡土资料显示，贵州省历史文化名镇正安县安场镇在明清时期就盛产制作墨条（烟墨）的优质油桐、卷子（乌桕）。自清代中期开始，安徽“胡开文墨庄”就在安场镇设立“下烟房”，点烟制作油烟墨和松烟墨的原料。目前，这里还有当年为胡开文墨庄点烟的房屋，“下烟房”的地名也沿用至今。正安县的民间艺术家周开智先生近年也制作了石雕砚台、水盂、笔筒等文房高雅玩具。如果能够面向中高端书画和文玩收藏市场，把纸、墨、砚等文房用品整合开发，形成集成优势和品牌效应，就有可能吸引更多的人关注和支持手工纸的保护。

（2）将手工纸的保护与旅游产品开发和旅游经济发展相结合。目前，正安手工纸的主要用途与民俗，特别是与人生礼俗有关的各种活动结合较密切。如果能够在民俗纸工艺品的开发中融入更多地域文化的内涵，比如传播正安尊老敬老的孝道文化、祈福纳祥的吉祥文化、敬惜字纸的传统文化等，面向各个年龄段的旅游者、户外运动爱好者设计并生产方便携带的纸艺产品，则将会开辟出新的产品市场。

（3）将手工纸的保护与家庭纸艺装饰市场的开发相结合。正安的先民开创了手工纸在民居建筑中应用的先例。如果能够把这个传统向前推进，将纸艺产品与当代居家的皮纸艺术灯罩、室内装饰纸艺摆件、节庆吉祥对联等市场对接，就有望开辟出新时代的纸产品市场。

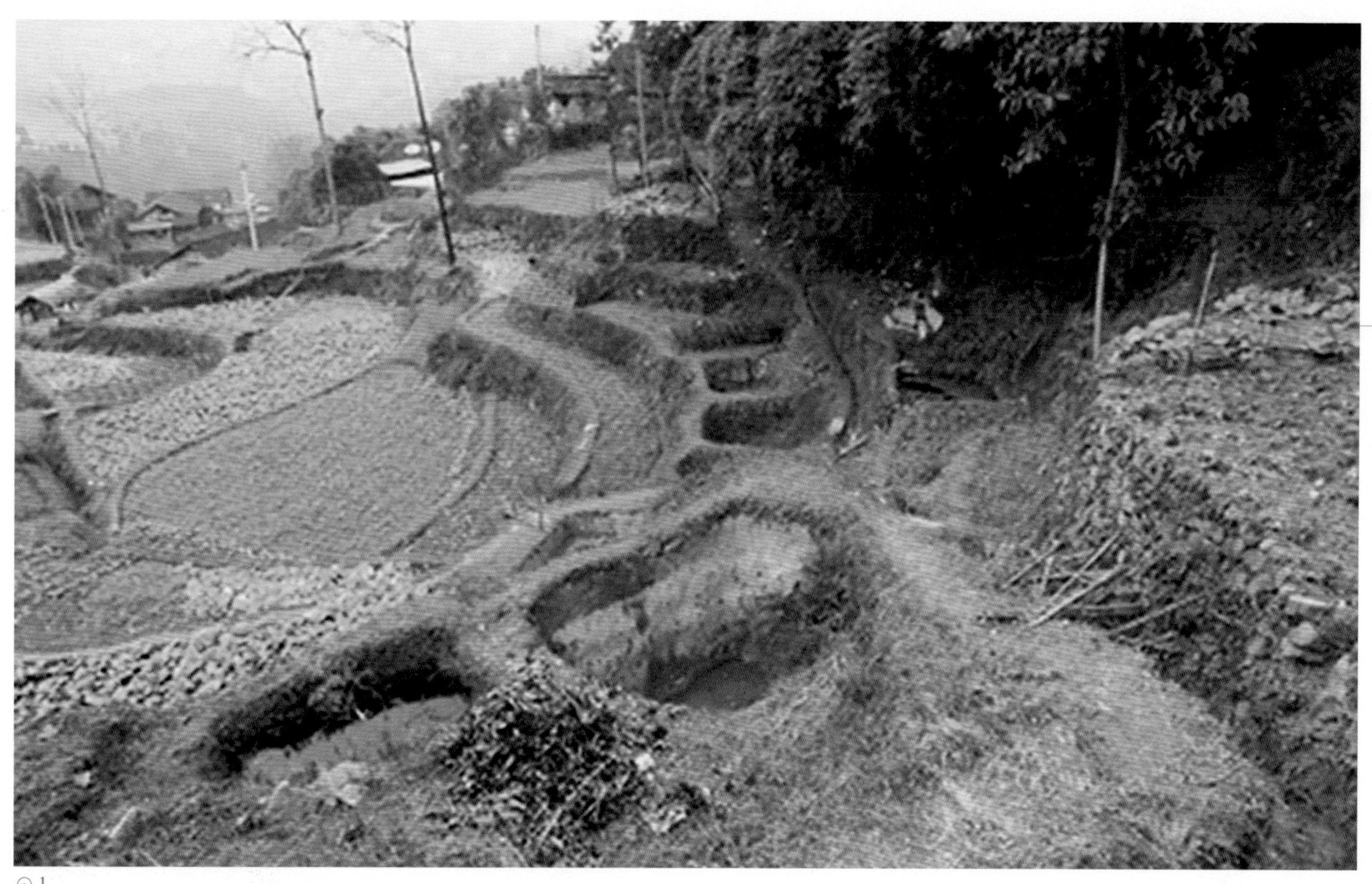

⊙1

⊙1
和溪镇杉木坪村的泡料池群
Soaking pools in Shanmuping Village of Hexi Town

正安竹纸

101

Bamboo Paper in Zheng'an County

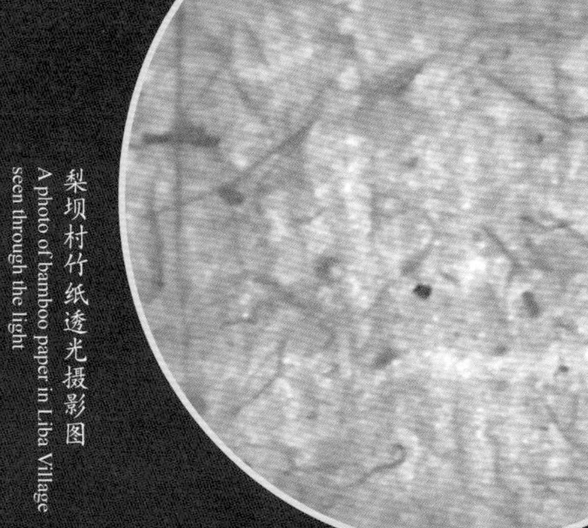

梨坝村竹纸透光摄影图

A photo of bamboo paper in Liba Village seen through the light

第三节

务川仡佬族皮纸

贵州省
Guizhou Province

遵义市
Zunyi City

务川仡佬族苗族自治县
Wuchuan Gelo and Miao Autonomous County

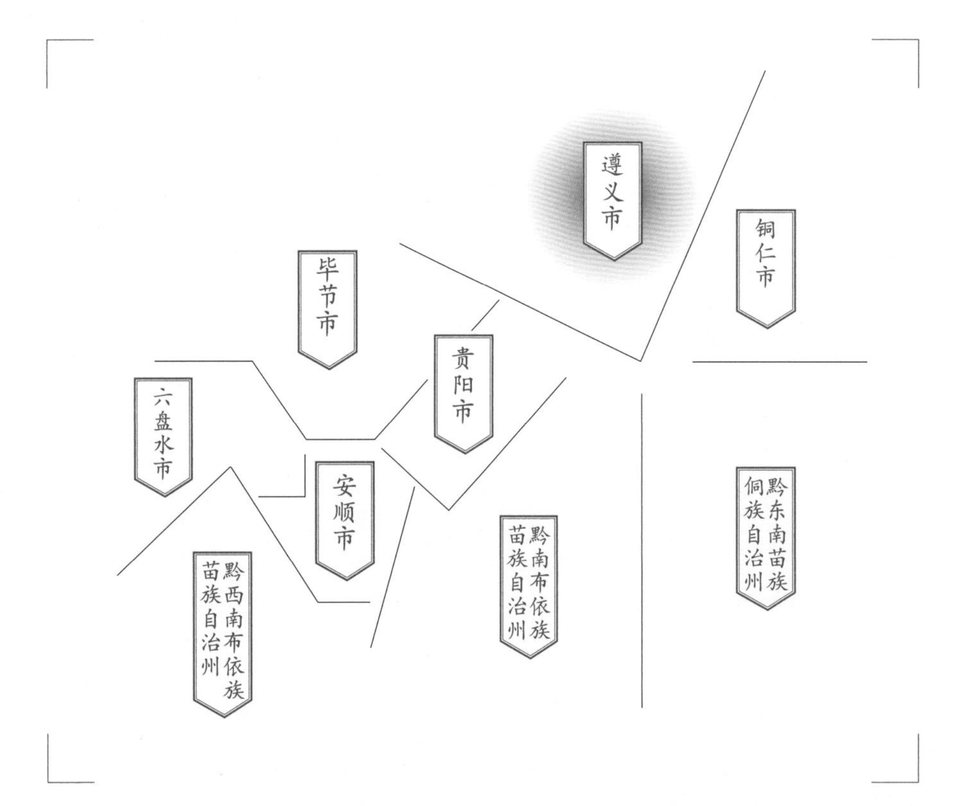

调查对象

丰乐镇新场行政村造纸塘村民组仡佬族皮纸

Section 3
Bast Paper by the Gelo Ethnic Group in Wuchuan County

Subject

Bast Paper by the Gelo Ethnic Group in Zaozhitang Villagers' Group of Xinchang Administrative Village in Fengle Town

一 务川仡佬族皮纸的基础信息及分布

1 Basic Information and Distribution of Bast Paper by the Gelo Ethnic Group in Wuchuan County

务川仡佬族皮纸当代的代表性产地是务川县丰乐镇新场行政村的造纸塘村民组。历史上，务川皮纸曾经是地方流行的文化与用纸，除了作为书写用纸这一主要用途外，还大量用于油纸伞、纸扇等乡土民生用品的制作。

务川仡佬族皮纸生产的主要原料是当地生长的野生构树皮，造纸塘村是经典的务川皮纸代表性产地，清道光年间皮纸生产技艺由印江县传入，直至调查组入村调查时的2009年4月，该村一直是务川仡佬族苗族自治县构皮纸的中心生产地。造纸塘村是一个偏僻的深山小村，道路崎岖，距务川县城35 km。

造纸塘村所造构皮纸具有韧性高、吸水吸湿性强、透气性好，以及无毒、不易变色、保存年代久等特点，其工艺完整地传习了古法手工造纸的原生态。2009年，“造纸塘皮纸制作技艺”凭借其良好的活态保护和传承状态，被贵州省人民政府列入第三批非物质文化遗产名录。

务川仡佬族皮纸生产地分布示意图

Distribution map of the papermaking site of bast paper by the Gelo Ethnic Group in Wuchuan County

考察时间
2009年4月

Investigation Date
Apr. 2009

路线图
务川县城
↓
造纸塘村
Road map from Wuchuan County centre to the papermaking site (Zaozhitang Village)

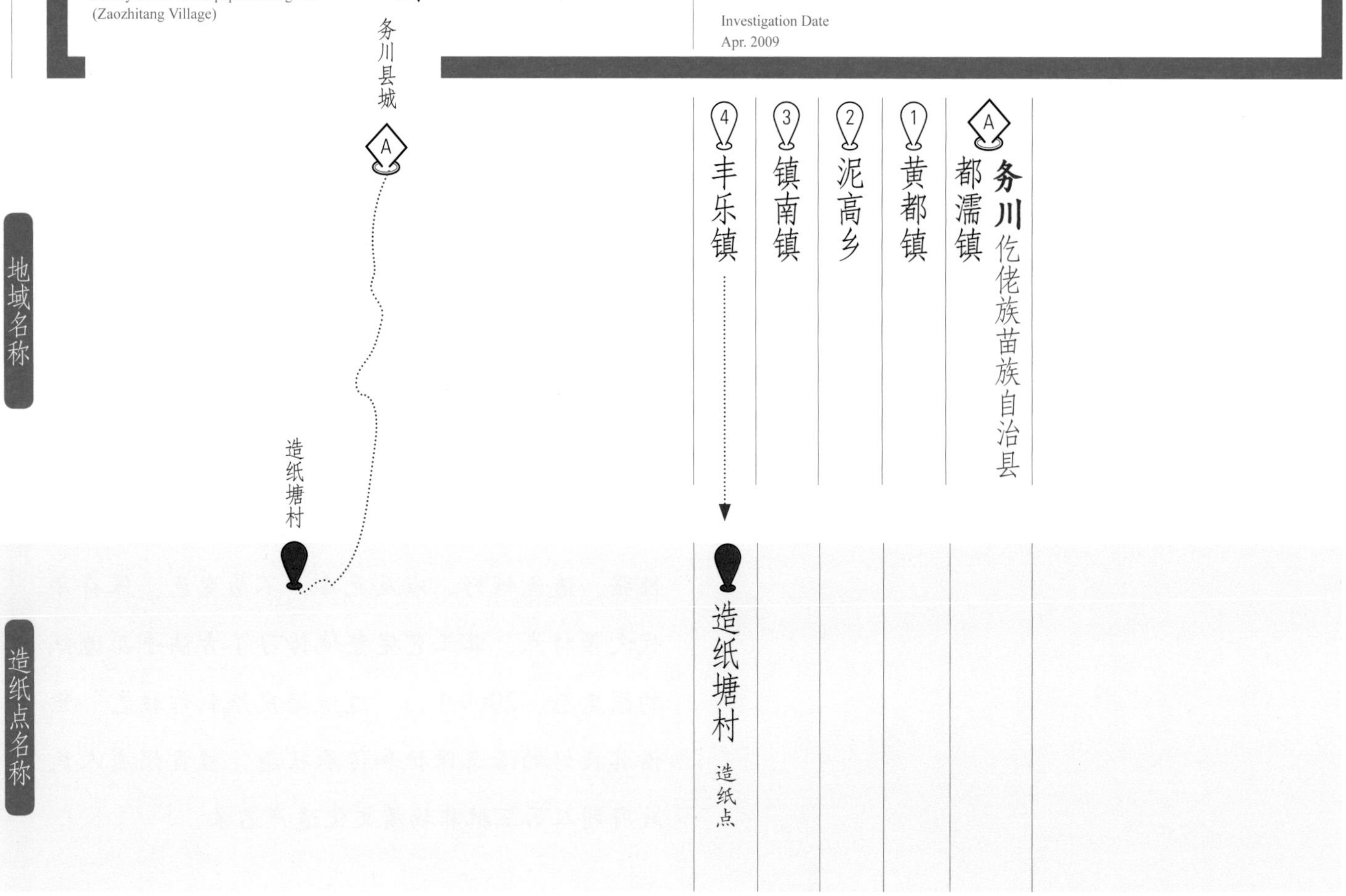

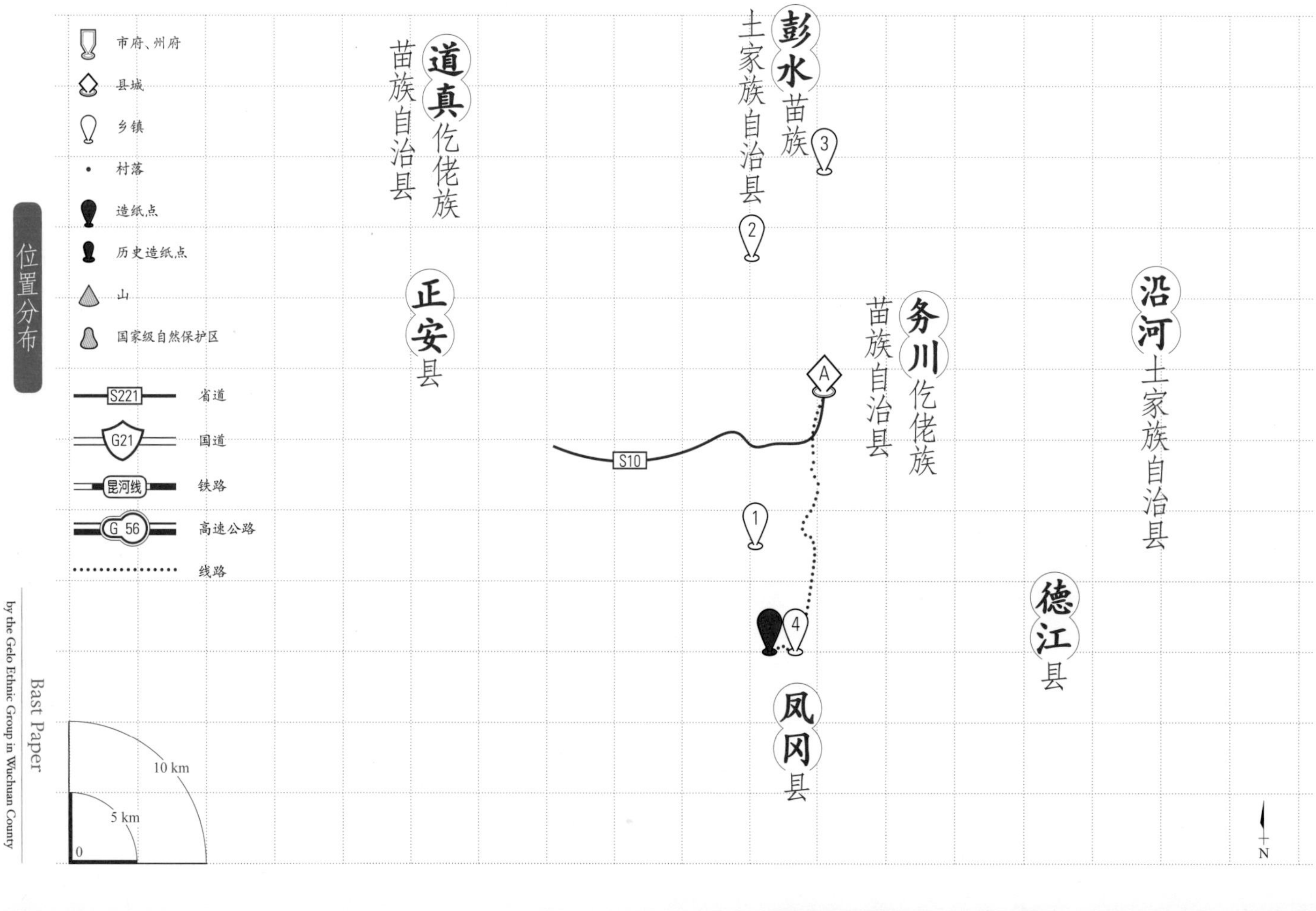

务川县城

造纸塘村

地域名称

务川仡佬族苗族自治县

都濡镇

黄都镇

泥高乡

镇南镇

丰乐镇

造纸点名称

造纸塘村 造纸点

位置分布

市府、州府
县城
乡镇
村落
造纸点
历史造纸点
山
国家级自然保护区
S221 省道
G21 国道
昆河线 铁路
G56 高速公路
线路

道真仡佬族苗族自治县

彭水苗族土家族自治县

正安县

务川仡佬族苗族自治县

沿河土家族自治县

德江县

凤冈县

S10

10 km
5 km
0
N

二
务川仡佬族皮纸生产的人文地理环境

2
The Cultural and Geographic Environment of Bast Paper by the Gelo Ethnic Group in Wuchuan County

务川仡佬族苗族自治县（以下简称务川县）位于贵州省遵义市北部，东与沿河土家族自治县、德江县接壤，南与凤冈县相连，西接道真仡佬族苗族自治县、正安县，北与彭水苗族土家族自治县相邻。地处东经107° 30′~108° 13′、北纬28° 11′~29° 05′。全县面积2 777 km^2，辖15个乡镇，114个行政村，人口43.2万，其中仡佬族人口18.5万，占总人口的42.8%，苗族人口15.1万，占总人口的35.0%。

据当代《务川仡佬族苗族自治县志》记载，春秋战国时期，务川地域属巴地，时归属秦黔中郡管辖，汉代归属武陵郡。隋开皇十九年（599年），招慰蛰僚奉诏置务川县。唐武德四年（621年），置务川郡，旋改为务州。贞观四年（630年），改务州为思州。北宋政和八年（1118年），移思州治于务川县都濡故地。元至元年间，以"婺星飞流化石"改"务"为"婺"，婺川县属思州安抚司。明永乐十二年（1414年）改土归流，婺川县属思南府。1959年，改"婺川县"为"务川县"。1986年，中华人民共和国国务院批准"撤销务川县，设立务川仡佬族苗族自治县"[32]。

务川是一个历史悠久的多民族自治县，早在2 000多年前，仡佬族的先民——百越中的濮人就在这块土地上采掘冶炼朱砂、烧炼水银。自隋开皇十九年（599年）置县，已有1 400余年历史，素有"丹砂古县，仡佬之源"之称。流经境内的洪渡河是"仡佬族的母亲河"，孕育了仡佬族千百年源远流长的灿烂文化，务川因此被誉为"仡佬之源"和"世界仡佬文化中心"。县境内

[32]务川县地方志编纂委员会.务川仡佬族苗族自治县志[M].贵阳：贵州人民出版社,2000:2.

有中国唯一遗存完整的仡佬民族文化村落，有史前仡佬族遗址——陀子箐，有仡佬族祭天拜神圣地——天祖坳，这是标准的仡佬族发源地。仡佬族分布于黔、滇、桂三省区，其中97%集中在贵州，而贵州仡佬族的聚集地则在黔北的务川、道真两个民族自治县。仡佬族有自己的民族语言，属汉藏语系，但语族、语支尚无学术归类定论，仡佬族无自己的文字，共用汉字。

务川为典型的亚热带高原干湿季风气候，具有“冬无严寒、夏无酷暑，四季如春、干湿分明”的特点。雨季降水量占全年降水量的80%~88%。务川地形为典型的喀斯特地貌，地势高低差异大，切割深，地面破碎，垂直地带性明显，具有强烈的自然景观垂直差异，“对山喊得应，走路要半天”是对这种地理状况的真实写照。由于山大、坡陡、谷深，因此人们的生产生活受到了较大影响。县境内有务（川）凤（冈）公路连接“326”国道，有务（川）彭（水）、道（真）德（江）出入境等级公路，交通相对深山区尚属便利。务川县是连接贵阳至重庆、四川的重要交通要塞，为黔渝、黔川的中转地之一。

造纸塘村民组隶属务川县南部丰乐镇新场行政村，距务川县城约35 km，其交通便利，入村公路通至村寨前。造纸塘属低山丘陵地形，光照充足，植被良好，溶洞、山泉众多，造纸河在村寨前蜿蜒而过。村寨两岸青石古道犹存，民居主要为传统木建筑，大部分修建于清末民国时期，建筑工艺较为讲究，属于历史感较鲜明的古老村寨。造纸塘全村71户，318人，耕地面积3.3×10^5 m^2，其中水田8.5×10^4 m^2、旱地1.8×10^5 m^2。村内以卢、王两姓为主要姓氏，其他尚有冉、秦、胡、阮、石、严等姓，村民多为仡佬族，苗族与土家族次之。

造纸塘原名枣子塘，据调查中查阅的《卢氏经单薄》记载，清道光年间，卢姓从铜仁地区的印江县迁入，以造纸为业。光绪至民国初年，因卢家在枣子塘的造纸生意兴隆，冉、胡、石、阮等其他姓氏家族也从印江迁来枣子塘。20世纪二三十年代，枣子塘造纸业兴盛，家家造纸。20世纪50年代，因枣子塘成立集体性质的新场造纸厂，寨名始改为“造纸塘”。

⊙1
⊙2

⊙1 村里的造纸作坊 Local papermaking mills

⊙2 造纸水碾 Hydraulic grinder for papermaking

三

务川仡佬族皮纸的历史与传承

3

History and Inheritance of Bast Paper by the Gelo Ethnic Group in Wuchuan County

据造纸塘村中卢氏家族所藏《卢氏经单薄》（民国十六年（1927年）抄本）记载，造纸塘原名枣子塘，清道光年间，卢定一随家人从铜仁印江县一甲洞竹林角迁来枣子塘，以造纸为业，开务川造纸之先河。随后冉、胡、石、阮等其他姓氏家族也从印江迁来枣子塘。清光绪年间，以枣子塘皮纸为主的务川皮纸生产兴盛。据清光绪二十五年（1899年）《婺川县备志》记载，“婺川所产皮纸、百合粉二物颇佳，畅销外属”。民国年间，枣子塘造纸业兴盛，家家造纸。

1955年，由地方政府推动，在造纸塘建立务川县第一个手工业生产合作社——新场造纸生产合作社。1958年，新场造纸生产合作社扩大生产规模，修建厂房3间，同时更名为新场造纸厂，1987年，厂房毁于山洪，纸厂自动解散。20世纪80年代以前，造纸塘皮纸生产一直兴盛，几乎家家都生产皮纸，鼎盛时，全寨有手工作坊40多家，一年可产皮纸400多万张。20世纪90年代以来，由于工业化造纸的普及以及农村产业结构的调整，造纸塘皮纸生产日趋衰落。从整体上看，造纸塘的皮纸生产业呈收缩态势，2009年4月，调查组入村调查时，全村只剩约10户村民生产手工皮纸，而且有多户属断续生产，即农闲时造纸，农忙或有其他事务时歇业的模式，另有1户改进为半机械化生产，即用竹原料制造火纸。

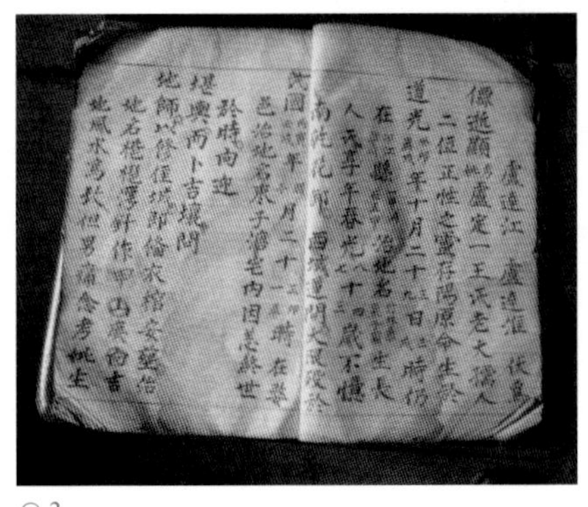

⊙3

⊙3 造纸塘村旧藏《卢氏经单薄》 Book of the Lus collected in Zaozhitang Village

四
务川仡佬族皮纸的生产工艺与技术分析

4
Papermaking Technique and Technical Analysis of Bast Paper by the Gelo Ethnic Group in Wuchuan County

(一)
务川仡佬族皮纸的生产原料与辅料

务川仡佬族皮纸制作的主原料是构树皮，传统使用的辅料包括石灰、草木灰和滑料（沙松树根，当地习称为沙根子），现代使用的辅料包括烧碱、漂精（即次氯酸）和滑石粉。

构树属桑科落叶乔木，又名楮树，高可达16 m，树皮纤维韧而细长，是优质的造纸原料。

造纸塘传统方式制作皮纸必须用石灰、草木灰，20世纪80年代以后逐渐改用烧碱和漂精。烧碱主要用于“蒸扑灰料”工序。漂精可增加构皮的白度，主要用在“洗毛料”和“打槽”这两道工序中。

造纸塘造纸传统工艺使用沙松根作滑，用打皮板的石碓舂融，在水中泡3~5天即可，7.5 kg沙松根干料可以浸泡制成纸药180 kg，大约需要20天。滑药原料一般从正安县的流渡镇或道真县的旧城镇购买，2006年左右的原料价格约为6元/kg。

(二)
务川仡佬族皮纸的生产工艺流程

据调查组2009年4月的入村考察及后续若干次补充调查掌握的信息，造纸塘皮纸的当代生产工艺流程如下：

壹	贰	叁	肆	伍	陆	柒	捌	玖	拾	拾壹	拾贰	拾叁
砍构树	剥构皮	晒构皮	泡构皮	浆构皮	蒸毛料	洗毛料	踩毛料	蒸扑灰料	出锅	洗扑灰料	踩扑灰料	拆料子

拾肆	拾伍	拾陆	拾柒	拾捌	拾玖	贰拾	贰拾壹	贰拾贰	贰拾叁	贰拾肆	贰拾伍	贰拾陆
打皮板	切皮板	淘料	挤料	打槽	刮料	掺滑水	舀纸	榨纸	晒纸	揭纸	理纸	打捆

壹 砍构树

1 ⊙1

造纸塘村民一般在农历二至四月份砍1~3年生的野生构树（熟土里生长的构树质量比生土里的好）。通过调查了解到，村里造纸户认为，纤维越嫩越好，一年生且树干直径为20 cm左右的构树皮质量最好。若树龄太长、纤维太老，则造出的纸质量不好。农历五至六月份时也可剥皮，但因那时的构皮含水量少、皮层薄，故质量较差。

⊙1

贰 剥构皮

2 ⊙2

用刀刮下或用手撕下鲜树皮，从根部刮到尾部，树叶不要，当地春分以前构树尚未长出树叶，即砍时没有树叶，当天砍当天刮，通常一个人一天可砍树并刮鲜皮100 kg左右。

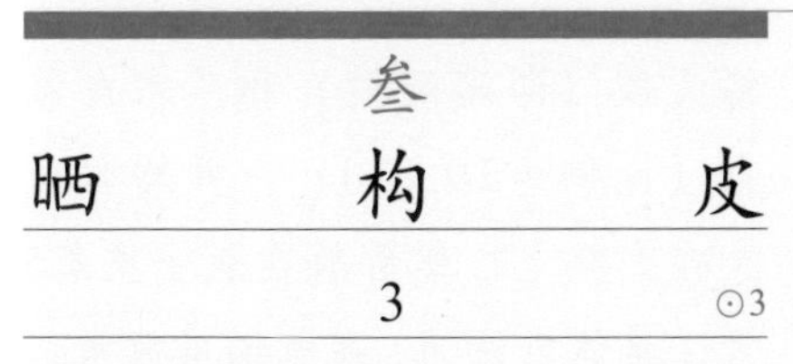

叁 晒构皮

3 ⊙3

即把构树皮晾干或晒干。如在室外经太阳晒，大约需要5天时间。如遇阴雨天气，收到室内，搭在梁上，要20天左右方能阴干。晾干后一捆大约3.5 kg。若天气不好，阴干后的皮料质量差，则蒸料后，皮不白，造出的纸质量不好。发霉的料造出的纸韧性差，纤维一拉就断，因此，造纸时一般不用霉变皮料。

⊙3

⊙2

⊙1 砍构树 Lopping paper mulberry tree

⊙2 剥构皮 Stripping paper mulberry bark

⊙3 阴干构皮 Drying paper mulberry bark

肆

泡 构 皮

4 ⊙4⊙5

将构皮放在河里泡3~5天，得到毛构皮，当地造纸户也称其为毛料。

⊙4

⊙5

伍

浆 构 皮

5

用捞钩将毛料放入石灰溶液中浆浸，100 kg干构皮需要40 kg石灰，将已浆浸的毛料堆放一天。毛料浆得越好，造出来的纸就越白。

陆

蒸 毛 料

6 ⊙6⊙7

据造纸户介绍，以前用祖辈传下来的土窑加水密封蒸，现在则改用锅炉，其上面要用塑料布盖住。实际上叫煮毛料更合适，但当地都叫蒸毛料。据造纸户们回忆，传统方式是窑顶用铁盖子盖住，大约从1987年开始改用塑料布，因为塑料布更加便利，但调查中被调查户们也说不清弃用铁盖子的原因。通常一炉一次可蒸750 kg干构皮，蒸七八天，不用加水。用煤炭作燃料，一天耗煤50 kg。

⊙7

⊙6

柒

洗 毛 料

7

用捞钩将蒸熟的毛料捞出锅炉并拉到河里洗，一个人钩毛料，另一个人在河里洗毛料。两个人一天可将全部毛料钩完并洗完。

捌

踩 毛 料

8

将洗好的毛料捞起来，就地在河边“钢板”（即木板）上用脚踩干，通常750 kg干构皮需要5个人踩一天。

玖

蒸 扑 灰 料

9

将大碱（即碳酸钠）和草木灰混合（比例为20:100），并均匀地涂在毛料上，然后放在甑子里蒸5天。具体方法为：在甑锅里放水，水上放甑桥（15根左右木质棍，10 cm粗、150 cm长，2根间隔3 cm左右），甑桥上堆放涂抹好碱料的扑灰料，一层一层码放，放好后在最上面用塑料布盖上封闭，然后从进水口加水，每天加2次，每次加2桶，每桶25 kg左右。蒸3天后，要翻锅一次，再蒸2天。

⊙4 泡构皮 Soaking paper mulberry bark

⊙5 泡好的构皮 Soaked paper mulberry bark

⊙6/7 蒸毛料 Steaming the papermaking materials

拾 出锅

10

将蒸好的扑灰料从甑子里用捞钩捞出来，通常750 kg料两个人一天可捞完。

拾壹 洗扑灰料

11

在河水里先用手洗，再用脚踩，两个人一天可洗净750 kg干构皮制成的扑灰料。

拾贰 踩扑灰料

12

将半干的扑灰料拿到岸上，三个人在“钢板”上踩一天，得到的粉团（或饼）叫熟料，熟料绞成一团，称为“一绞”。一绞熟料大小不等，质量1.5~2.5 kg。

拾叁 拆料子

13

将一绞熟料拆成3个小料团。

⊙8

⊙9

拾肆 打皮板

14 ⊙8⊙9

将小料团拿到石碓里并用水车舂，只需要一个人转料，打薄后将四周卷到中间再打成饼状，然后对折，如此反复3次，即对折3次，打4次。最后延长边折成5折，打平后对折就得到一个皮板，通常造纸塘熟练的纸工半小时可打一个皮板。

拾伍 切皮板

15 ⊙10

⊙10

将皮板放在马凳上，先用绳子套好，固定，再用皮刀将皮板切成刀口纸，一般12个皮板一起切，大约要切一小时。如果皮刀锋利且纸工操作熟练，半小时就能切完。

拾陆 淘料

16 ⊙11

⊙11

将12个皮板切成的刀口纸放于浸入水中的圆形竹箩兜里，并用手淘洗，其手法是用右手顺时针搅，中间形成漩涡，把所有污水淘掉，淘一次大约需15分钟。淘好后，当地纸工才称其为料子。

⊙8 打皮板 Beating the papermaking materials

⊙9 皮板 Folded papermaking materials

⊙10 切皮板 Trimming the papermaking materials

⊙11 淘料 Cleaning the papermaking materials

拾柒

挤　料

17

将淘洗干净的料子在竹箩兜里用手挤干。

拾捌

打　槽

18　⊙12

将淘洗好并挤干的料子放在纸槽里，先加入清水，然后用槽棍搅打均匀。

⊙12

拾玖

刮　料

19

把纸槽的两块插板取开，然后用插板把多余的料刮到小仓（小格，又叫二仓）中，需要刮两次，一般2分钟即可完成。

贰拾

掺　滑　水

20　⊙13⊙14

传统的纸药是在纸槽里加入滑水（沙松根浸泡液，加入量与纸浆的比例为1:4），用木耙搅打和匀，使纸浆液与滑水黏液混合成糊状。

2008年以后开始用滑石粉代替沙松根，50 g滑石粉兑水75 kg。

⊙13　⊙14

贰拾壹

舀　纸

21　⊙15

双手持抄纸竹帘两端，远离人的一侧叫纸头，靠近人的一侧叫进水，也叫纸尾。在纸槽内前后轻轻晃荡，使纸浆均匀附着在竹帘上。胚纸的质量与舀纸时的手法、力度密切相关，力度不匀、舀荡幅度过大，都会造成胚纸厚薄不一。若槽内浆水里料子纤维多，则舀2~3次，否则舀5~6次。调查时，多使用一帘三纸的纸帘。

⊙15

⊙12 打槽 Stirring the papermaking materials

⊙13 掺滑水 Adding in papermaking mucilage

⊙14 滑石粉 Talcum powder

⊙15 胚纸 Preliminary paper

贰拾贰 榨 纸

22 ⊙16

⊙16

纸全部舀完后，直接在榨板的湿纸垛上加炕板等，旋即压榨。由下到上，依次为木质榨板（当地又称为钢板）、把草、把帘、纸垛、炕板、四块木方（码子）、大码。压榨总共需一小时左右，应缓慢压，太快易因压力过大而压破湿纸，通常要压8次，每次压2分钟左右，再静置几分钟，大约压4次（半小时）后，需要松榨一次，加一块木方。压榨完后，将半干纸垛运回家中。

贰拾叁 晒 纸

23 ⊙17⊙18

先用手或夹子将半湿纸垛的四边打松，这样纸更容易揭开。然后右手拿刷子（棕刷），左手从纸右上角往下撕，右手再拿住右上角纸头。通常在揭湿纸时将纸架上纸的上面叫头纸，下面叫尾纸。由于抄纸

⊙17

时头纸靠近人的一侧，因此它比尾纸厚。撕、刷纸上墙及烘焙干一面墙的纸约需15分钟，白天一面墙一次可刷50张，晚上一面墙一次可刷100张或150张。调查时发现都是以50为单位计数的，理由是这样好计算。但纸焙上一次刷纸过多容易造成纸的质量不好，干得不透，硬度不够。按照造纸塘的习惯，造纸户要把当天抄的纸全部晒完，一帘三纸的，有2 000张左右。

⊙18

贰拾肆 揭 纸

24 ⊙19⊙20

在纸焙上烘焙干后，需要从纸焙上揭取下来，揭取方式是右手拿针揭纸的右上角，当地叫“用针揭角角”，然后将纸逐一从纸焙上撕揭下来。

⊙19

⊙20

贰拾伍 理 纸

25

揭了半刀纸（50张）后，抖齐，三对折叠好并理齐。

⊙16 榨纸 Pressing the paper

⊙17 撕纸 Peeling the paper down

⊙18 刷纸 Pasting the paper on the wall for drying

⊙19 揭角角 Peeling the paper down from the upper corner

⊙20 揭纸 Peeling the paper down

贰拾陆

打　　捆

26　　⊙21⊙22

1 000张为一捆，调查时得知，传统方式是用纸绳在纸头和纸尾各捆一道。21世纪初则一般用竹篾条、塑料绳捆，也有用纸绳捆的。

⊙22

⊙21

综合整个工艺的演化过程，与20世纪中后期的相对传统工艺相比，造纸塘皮纸生产工艺发生了一定的变化，主要如表8.11所示。

表8.11　传统和现代造纸塘皮纸生产工艺对照表

Table 8.11　Contrast of traditional and modern papermaking techniques of bast paper in Zaozhitang Village

工序	传统工艺	现代工艺
蒸料	用木甑蒸料，一次可蒸150~250 kg，需三四天	用锅炉蒸料，一次可蒸750 kg，需七八天
纸药	用沙松根，每7.5 kg可配制纸药175 kg	用滑石粉，每0.05 kg可配制纸药75 kg

(三)

务川仡佬族皮纸生产使用的主要工具设备

壹

甑　子

1

用于蒸熟料的设备，直径约190 cm，深约170 cm，外用石头砌成，内用泥巴砌糊。旁边有煤槽，由石头砌成，用于点燃煤。

⊙23

贰

锅　炉

2

用于蒸毛料，钢质圆柱形，直径约100 cm，深约250 cm，外用石头围砌（造纸塘村只有一个，20世纪80年代初从武昌买来，当时购买价为2 500元）。

⊙24

⊙21/22 理纸和捆纸 Sorting and binding the paper

⊙23 甑子 Utensil for steaming the papermaking materials

⊙24 锅炉 Wok for steaming the papermaking materials

叁

碓

3

用于舂料，实测碓石长约54 cm，宽约47 cm，碓杆长约150 cm，碓脑壳直径约24 cm。

⊙25

肆

纸　　槽

4

大小不完全一致，调查时所测纸槽内部尺寸约203 cm × 100 cm，高约56 cm，厚约5.0 cm，由水泥制成。四个角有些弧度，但不是直角。分为两格：大格，称为“头仓”，尺寸约203 cm × 73.5 cm；小格，称为“二仓”，尺寸约203 cm × 24.5 cm。两格之间有隔板，隔板上开两个缺口，用插板分隔大小格。

⊙26

伍

纸　　帘

5

一帘三纸的长方形纸帘。

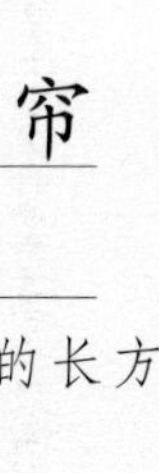

⊙27

陆

纸　　焙

6

大小不一，实测纸焙长约350 cm，高约280 cm。每面墙可焙2排，每排10檀，每檀3~4张，一面墙可同时烘焙60张纸。

⊙28

⊙25
打碓示意
Showing how to beat the papermaking materials with a pestle

⊙26/27
一帘三纸纸帘
Papermaking screen that can make three pieces of paper simultaneously

⊙28
纸焙
Drying wall

(四)

务川仡佬族皮纸的性能分析

所测务川造纸塘村皮纸的相关性能参数见表8.12。

表8.12 造纸塘村皮纸的相关性能参数
Table 8.12 Performance parameters of bast paper in Zaozhitang Village

指标		单位	最大值	最小值	平均值
厚度		mm	0.150	0.070	0.090
定量		g/m^2	—	—	19.6
紧度		g/cm^3	—	—	0.218
抗张力	纵向	N	10.1	6.6	8.2
	横向	N	4.5	2.3	3.3
抗张强度		kN/m	—	—	0.383
白度		%	56.7	55.7	56.1
纤维长度		mm	11.93	0.66	3.57
纤维宽度		μm	47.0	4.0	15.0

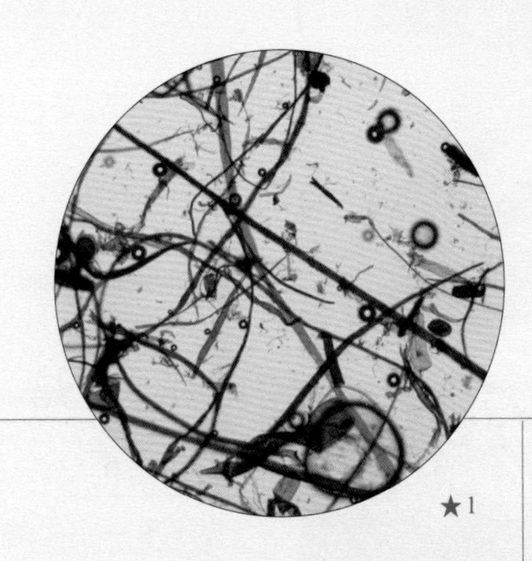

★1

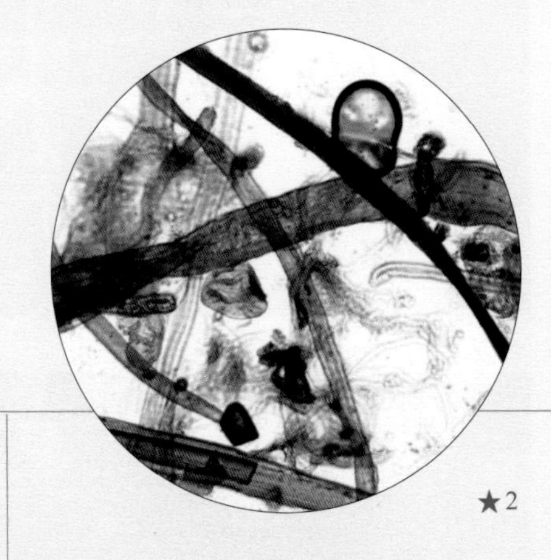

★2

由表8.12可知，所测造纸塘村皮纸最厚约是最薄的2.14倍，相对标准偏差为0.24%，纸张厚薄较为一致。皮纸的平均定量为19.6 g/m^2。所测皮纸紧度为0.218 g/cm^3。

经计算，其抗张强度为0.383 kN/m，抗张强度值较小。

所测造纸塘村皮纸白度平均值为56.1%，白度较高，白度最大值约是最小值的1.02倍，相对标准偏差为0.29%，白度差异相对较小，可能是因为皮纸在加工时进行了一定程度的漂白。

所测造纸塘村皮纸纤维长度：最长11.93 mm，最短0.66 mm，平均3.57 mm；纤维宽度：最宽47.0 μm，最窄4.0 μm，平均15.0 μm。所测皮纸在10倍、20倍物镜下观测的纤维形态分别见图★1、图★2。

★1 造纸塘村皮纸纤维形态图（10×）
Fibers of bast paper in Zaozhitang Village (10× objective)

★2 造纸塘村皮纸纤维形态图（20×）
Fibers of bast paper in Zaozhitang Village (20× objective)

五
务川仡佬族皮纸的用途与销售情况

5
Uses and Sales of Bast Paper by the Gelo Ethnic Group in Wuchuan County

造纸塘皮纸是当地民众日常生活用纸，传统上用于抄写、绘画、裱糊面具和制作各种民间宗教习俗用品等，也用于小商品包装、银行捆钞、制作档案文件封条等。

(一)
务川仡佬族皮纸的基本用途

1. 书画创作

在造纸塘皮纸上进行绘画的门神案图及过桥图，尽管因年代已久及保存不善，画面受损较严重，但观其背纸，老纸的质地至今仍相当良好。

2. 抄家谱、写契约

调查组见过不少用造纸塘皮纸抄写的清代至民国时期的家谱、契约。由此可见，传统的造纸塘皮纸的质量较优异，书写流畅，且防腐防蛀，因而当地人抄家谱、写契约习惯用当地出产的质地优良的皮纸。

⊙1

⊙2

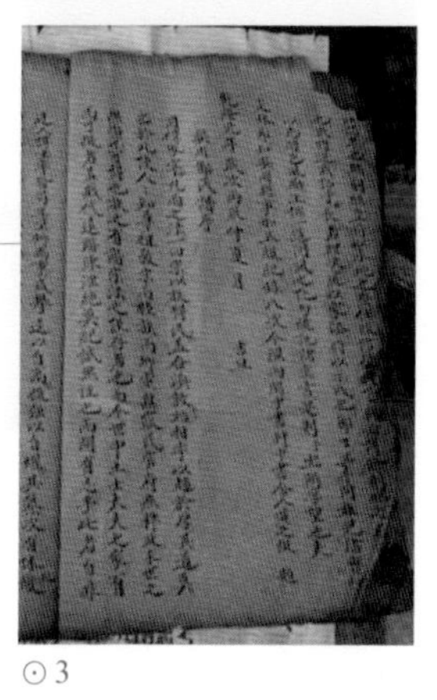
⊙3

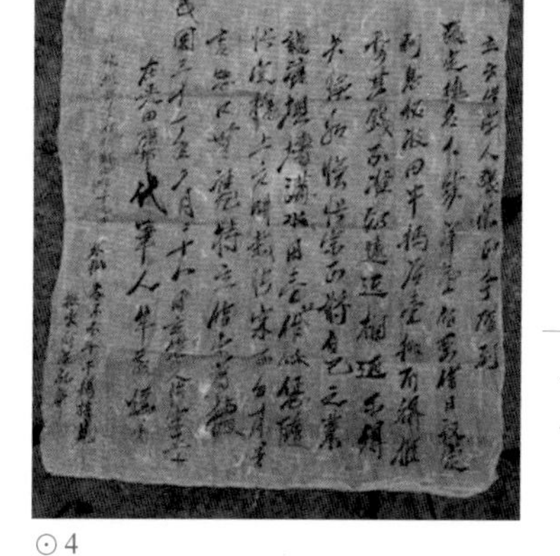
⊙4

3. 抄经

务川当地盛行道教，道士所用经书及符咒主要用造纸塘皮纸抄写。

4. 书写用纸

调查组调查时发现了一份民国时期当地的甲

⊙1 门神案图 Painting of Door Gods

⊙2 过桥图 Local traditional painting

⊙3 清代邹氏族谱 Genealogy of the Zous in the Qing Dynasty

⊙4 民国契约 Contract during the Republican Era of China

⊙1

⊙2

⊙3

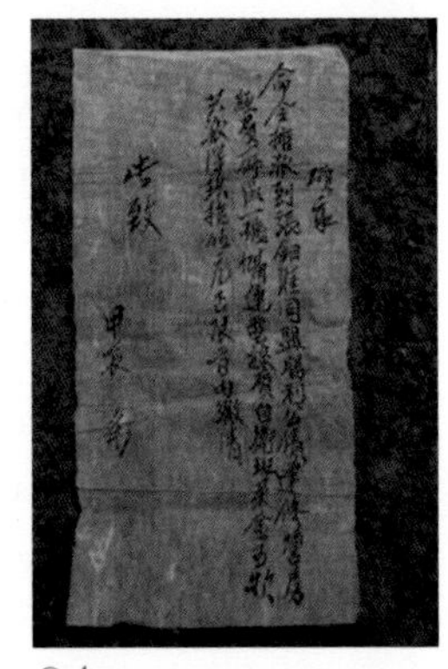

⊙4

⊙5

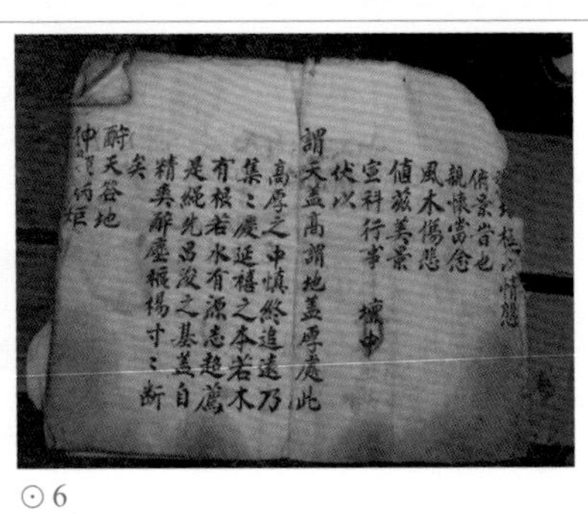

⊙6

⊙7

⊙8

长用皮纸书写的派款通知。随着机制纸的大量普及，这一用途已不断萎缩，民间仅少数人有时还作日常记事用。

5. 民间祭祀

主要包括制作清明时期上坟用的“青纸”，即将皮纸剪成纸串用于“挂青”，以及道场中抄写的经单簿。

6. 裱糊用纸

裱糊用纸包括面具裱糊纸、花轿裱糊纸等。

(二)

务川仡佬族皮纸的销售状况

造纸塘皮纸作为一种手工生产的民间用纸，柔韧性好是其主要特点，而且用途广泛，深受各地

⊙1 清光绪三十二年(1906年)抄经
Scriptures during Guangxu Reign of the Qing Dynasty (1906)

⊙2 清咸丰十年(1861年)抄经
Scriptures during Xianfeng Reign of the Qing Dynasty (1861)

⊙3 周相清老人抄写的花灯书
Book of Lantern Festival transcribed by Zhou Xiangqing

⊙4 民国时期的派款通知
Levy during the Republican Era of China

⊙5 『青』纸
Grave marker

⊙6 民国三十五年(1946年)申氏经单簿
Genealogy of the Shens during the Republican Era of China (1946)

⊙7 笑和尚、孙猴子面具
Masks of smiling monk and Sun Wukong

⊙8 花轿
Palanquin

客商的喜爱。20世纪50年代以前，造纸塘皮纸不仅覆盖了务川及其周边乡县的皮纸市场，还远销重庆市南川、涪陵等地。调查时，造纸塘皮纸的销售范围已有明显收缩，主要在本县销售，相邻的正安、凤冈县也有少量销售。除县城的纸是造纸户自己送货外，其余的都是就地销售。纸价一直维持在300~320元/捆，每捆1 000张，就地销售的价格为15元/刀，每刀50张。

造纸塘村普通造纸户一年造纸平均约需1 000 kg干构皮，大约可以生产50捆纸，按调查时就地销售的价格300元/捆计算，年收入约15 000元。这在黔北少数民族山区村落里算是一项较显著的家庭收益。

六 务川仡佬族皮纸的相关民俗与文化事象

6 Folk Customs and Culture of Bast Paper by the Gelo Ethnic Group in Wuchuan County

(一) 造纸塘的造纸起源传说

据造纸塘当地卢姓传说，卢家先祖来务川时很穷，只带了一双鞋垫，从印江县逃难到此，到时天已黑，于是就在河边安锅准备做饭。这时一条鱼竟然跳到了锅里，卢氏祖先觉得这真是一个好地方，于是就在这里落了脚，安了家，并利用其在印江县时已掌握的技艺开始造纸谋生。因为当地有一颗大枣子树，所以就把落脚的地方称为“枣子塘”，后因为造纸业态繁盛，就叫成了“造纸塘”。

(二) 叫花子的手艺

当地村民戏谑地说，舀纸是叫花子的手艺，

⊙1

讨口饭吃而已。说装料的木桶是"叫花子的米桶"，舀纸的帘子是"叫花子的筛子"，和料棒是"叫花子的杵二棒（即打狗棒）"，帘子上两头的竹片（称为顶砖）是"叫花子的莲花板"。又说舀纸的人家"不怕春官（儿）怕叫花子"。"春官（儿）"就是报春的人。舀纸的人家在做帘子时，如果遇到春官说春，就在帘子上放1.2元钱（称为"利师钱"），春官懂得造纸的，就要说造纸的"根生"（即造纸的来历），说对了才可以得到这个"利师钱"，不会说或说不对的，就不能拿"利师钱"了，实际上这是一种手艺人对手艺人的考校。而对于叫花子的乞讨，不论多少都要给一点钱或粮食，不会去为难他们。

(三)
清明会

造纸塘村及其周边村庄在清明节的前一天或当天，会操办宴席请客吃饭，邀请的客人除了村寨内的本家人以外，还有前来扫墓插"青"的亲戚，这种习俗被称作"清明会"。当地人认为，用当地皮纸做成的"青"是一面令旗，有了这面令旗，祖先的魂灵就可以在阴间畅通无阻，参加阴间的清明节。

⊙2

(四)
祭蔡伦

调查中，据村民们回忆，以前当地人要祭蔡伦，现在造纸的村民都在30~50岁范围，已说不清楚旧俗，只是逢年过节时在堂屋烧纸钱，要烧四五堆，但也不知道是烧给谁的。

⊙1
祭"将军柱"示意
Offering sacrifices to the General Pillars

⊙2
清末民国时期的旧栈房
Ancient house during the late Qing Dynasty and the Republican Era of China

(五)

祭“将军柱”

当地造纸的人家会在年三十夜里到造纸厂房里祭“将军柱”。“将军柱”是指木榨上两根起到主要受力作用的木柱子。祭“将军柱”的目的是祈愿来年造纸一切顺利。

祭“将军柱”由家里的男人主祭，祭品有长钱一束、烛一支、香三根、酒三杯、刀头一份（肉、豆腐、米粑）等，祭祀时要燃放鞭炮。

(六)

女人不单独造纸

当地一种特别的习俗是，只有男人在家时，女人才能参与造纸，女人一般情况下不单独造纸。寡妇一般也不允许造纸。当调查组向造纸户们询问缘由时，他们也说不出个所以然来，只是表明女人单独在家是做不出来好纸的。

(七)

合作使用造纸工具的传统

造纸塘村里的水碾一般由两三户造纸户共同修建；甑子里的铁锅是由使用者们共同出钱购买的；锅炉是20世纪80年代初集体造纸时买的，谁都可以使用。这些造纸工具的维修、更换也由造纸户共同出钱或出工来做。

造纸户少时，谁先“下河”（即谁最先开始造纸）谁就可以先使用这些工具。造纸户多时，则要排序确定。

造纸厂房一般是一户一个，也有两三户搭伙的。没有厂房的造纸户可以使用闲置的厂房，只要跟厂房的主人打个招呼就可以了，也不需要租金。

(八)

相关造纸俗语

1.“一帘三个人”

这句俗语生动地说明了造纸时的辛苦程度及投入状况。也就是说，用一张纸帘造纸，至少需要三个人完成。当代农村家庭一户多时也就五六口人，大多数家庭的孩子不是在上学读书，就是外出打工去了。因此，要正常生产皮纸，造纸的村民们就不得不付出更多的劳动，后继乏人，非常辛苦。

2.“新场卖不下的到造纸塘”

这句俗语是说在新场（附近的一个集市）卖不掉的肉、酒等商品拿到造纸塘来卖就可以卖完。对造纸塘过去的兴盛，当地老人至今仍津津乐道。清末至民国时期，造纸塘纸业兴隆时，每天卖“杯杯酒”就要卖100 kg。村子里有专门卖米粑等小吃的摊子，有栈房，重庆一带来买纸的客商都住在造纸塘村中，可见当地旧日的兴旺与发达。

七 务川仡佬族皮纸的保护现状与发展思考

7 Preservation and Development of Bast Paper by the Gelo Ethnic Group in Wuchuan County

(一) 造纸塘手工造纸传承与保护的机制

1. 政府扶持和地方立法保护

2009年，造纸塘的“皮纸制作技艺”被列入贵州省非物质文化遗产名录后，得到了国家和地方相关政府部门的扶持。务川县人民政府已在2008年提出了用五年时间初步建成“中国人文旅游新区、渝黔仡佬文化中心”的发展目标。《中共务川自治县委关于进一步加快以仡佬文化为重点的文化旅游发展的决定》（务党发〔2008〕11号）明确指出应加强仡佬族婚俗、高台舞狮、造纸塘皮纸等非物质文化遗产的保护和传承，制定了《务川自治县非物质文化遗产项目代表性传承人认定与管理暂行办法》，公布了首批务川县非物质文化遗产项目代表性传承人，其中卢朝辉、卢朝松两人被评为造纸塘皮纸制作技艺首批传承人。

地方政府在法规与政策方面的强力推动，无疑为务川仡佬族皮纸的生产性保护与传承提供了较强的动力，也显著地拓展了造纸塘皮纸当代发展的现实与想象空间。

2. 旅游带动

造纸塘古法造纸技艺被列入贵州省非物质文化遗产名录后，其品牌价值与生产要素价值开始

⊙1

⊙2

⊙3

⊙1 『非遗』传承人卢朝辉(左四)、卢朝松(右三)全家合影
A family photo of Lu Zhaohui (fourth from the left) and Lu Zhaosong (third from the right), inheritors of intangible cultural heritage

⊙2 新建的造纸厂房
Newly built papermaking factory

⊙3 新建的水碾
Newly built hydraulic grinder

凝聚传播；同时造纸塘村落一带具有优美的自然环境，共同构成了独具内涵的村落文化景观，促进了文化遗产保护与旅游产业发展需求的融合。调查时，当地旅游主管部门已经开始组织游客参观古法造纸作坊，并将造纸塘等村落文化景观与旅游产业有机融合的问题列入了自治县“十一五”文化旅游发展规划。

3. 已实施的保护措施

2008年以来，县文体广电旅游局已征集造纸工具13件（套）并陈列展览。2009年，县有关部门投资新建了造纸厂房250 m^2，将17个水槽、1个纸焙、4个水碾无偿提供给造纸塘的造纸户使用。

（二）造纸塘手工造纸面临的挑战及应对该挑战的对策

1. 传统的皮纸生产

造纸塘村寨所在地属低山丘陵地形，青山绿水，溶洞、山泉众多。村落内传统民居保存较为完整，大部分修建于清末民国初期，建筑工艺较为讲究。但是，当代造纸塘村落文化景观与自然生态景观也存在着山体、河流遭受破坏，传统民居部分损毁，村寨卫生条件差的情况。因此，造纸塘皮纸技艺的保护应纳入县级旅游规划，整合旅游、新农村建设、文化传媒等资金，以造纸塘传统的皮纸生产工艺活态传习展示为核心，实现村落文化、生态环境的整体改造。

2. 传统的皮纸保护与传承

建立保护与传承专项资金，着力解决扶持皮纸制作传承人以及复原传统工艺技术和产品的资金短缺瓶颈。

（1）扶持传承人是非物质文化遗产保护的重要措施，鼓励和支持非物质文化遗产项目代表性传承人开展传习活动，是保护和传承非物质文化遗产的一个重要方面。目前，能够掌握一整套造纸塘皮纸生产技术的村民大多数在40岁以上，年轻一代村民很少愿意学习造纸这门收益不高又费工费时的传统手艺。因此，设立专项的保护资金，鼓励和支持传承人开展传习活动，是务川仡佬族皮纸技艺传承发展的当务之急。

（2）纯正传统产品复原是造纸塘古法造纸工艺传承的大事。从20世纪晚期开始，传统的手工造纸技艺由于引进了一定的工业化要素，如化学漂洗药品等，致使纸质及工艺流程已不如从前。复原优质、传统的造纸塘皮纸技艺是独具民族文化内涵的务川仡佬族皮纸保护中非常直接的需求。然而，复原这些纸及纯正工艺不仅需要对现在的诸多造纸工具进行改良，还要选择适当的生态环境和场地，这对村民来说并非一件有动力的事，在专项资金的支持下才有可能推动。

⊙4

⊙5

⊙4 访谈造纸传人 Researchers interviewing papermakers

⊙5 洪渡河 Hongdu River

务川仡佬族皮纸

Bast Paper
by the Gelo Ethnic Group
in Wuchuan County

造纸塘村皮纸透光摄影图
A photo of bast paper in Zaozhitang Village
seen through the light

务川仡佬族皮纸

Bast Paper by the Gelo Ethnic Group in Wuchuan County

造纸塘村皮纸透光摄影图
A photo of bast paper in Zaozhitang Village
seen through the light

第四节

务川仡佬族竹纸

贵州省
Guizhou Province

遵义市
Zunyi City

务川仡佬族苗族自治县
Wuchuan Gelo and Miao Autonomous County

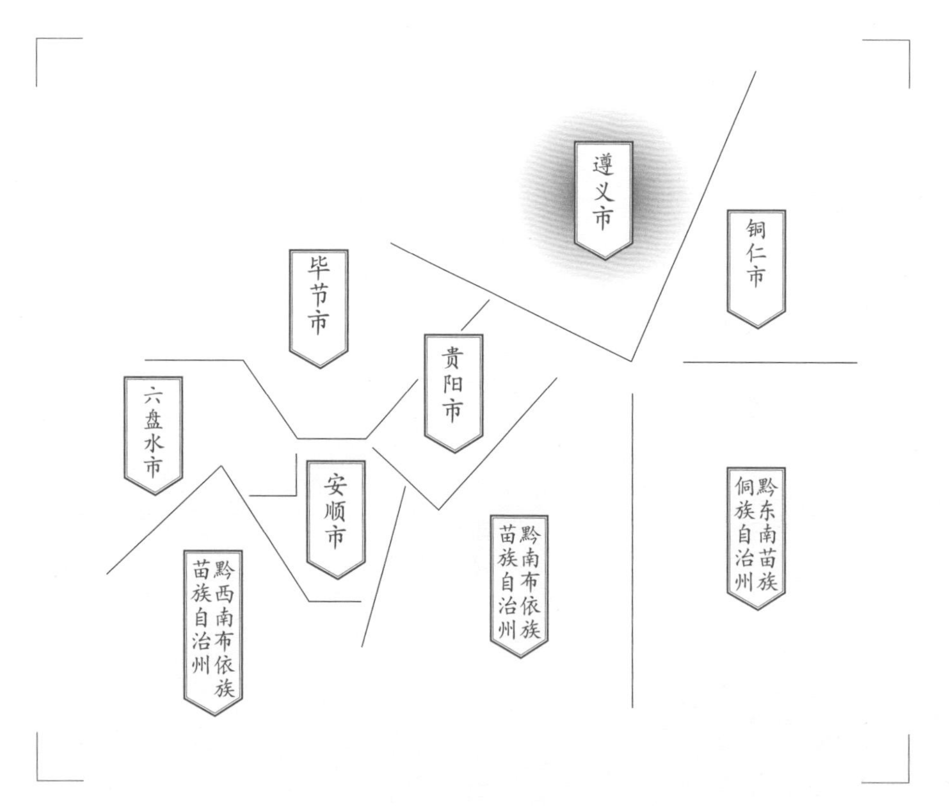

调查对象

黄都镇三合行政村学堂坡村民组仡佬族竹纸

Section 4
Bamboo Paper by the Gelo Ethnic Group in Wuchuan County

Subject

Bamboo Paper by the Gelo Ethnic Group in Xuetangpo Villagers' Group of Sanhe Administrative Village in Huangdu Town

一
务川仡佬族竹纸的基础信息及分布

1
Basic Information and Distribution of Bamboo Paper by the Gelo Ethnic Group in Wuchuan County

务川仡佬族竹纸的当代生产地以黄都镇三合行政村学堂坡村民组为经典代表。务川仡佬族苗族自治县历史上竹纸制作分布广泛，但当调查组2009年4月第一次进入县域调查时，发现业态已经萎缩得非常明显。在调查所能获知的信息中，学堂坡村已是正常维系竹纸生产仅存的村落了。以学堂坡村为代表的务川仡佬族竹纸所用原料主要是本地生长的慈竹。历史上曾经采用过上甑蒸煮的熟料法造纸，调查时则已改用不蒸煮的生料法造纸。学堂坡所造竹纸较为粗糙，以制作冥纸为主要用途，当地百姓也习称为“草纸”。

二
务川仡佬族竹纸生产的人文地理环境

2
The Cultural and Geographic Environment of Bamboo Paper by the Gelo Ethnic Group in Wuchuan County

务川县域历史文化与自然环境在第三节“务川仡佬族皮纸”中已有介绍，此节从略。

黄都镇位于务川的西南境，地处洪渡河流域，辖9个行政村，镇政府驻地为黄都坝村。早在隋唐年间，该地即为古代高富县的地域，已有相关文明的发育。1942年设黄都乡，1992年撤乡建镇。黄都镇为贵州省十大坝区之一，田多山少，恰恰与贵州省整体特色——山多田少形成反差，其万亩大坝别具特色，农业生态发育良好，有“黔北粮仓”之誉，主产水稻、油菜、烤烟、茶叶等，其中茶叶有较强的集成度，有“黔北茶叶大镇”的称号，已建茶园1.47 km^2。黄都镇野生大树茶群落也非常著名，极具历史价值。

黄都镇属典型的中国西南喀斯特地貌，地质形态奇特，旅游资源丰富，有绿海苍茫的燕龙山林

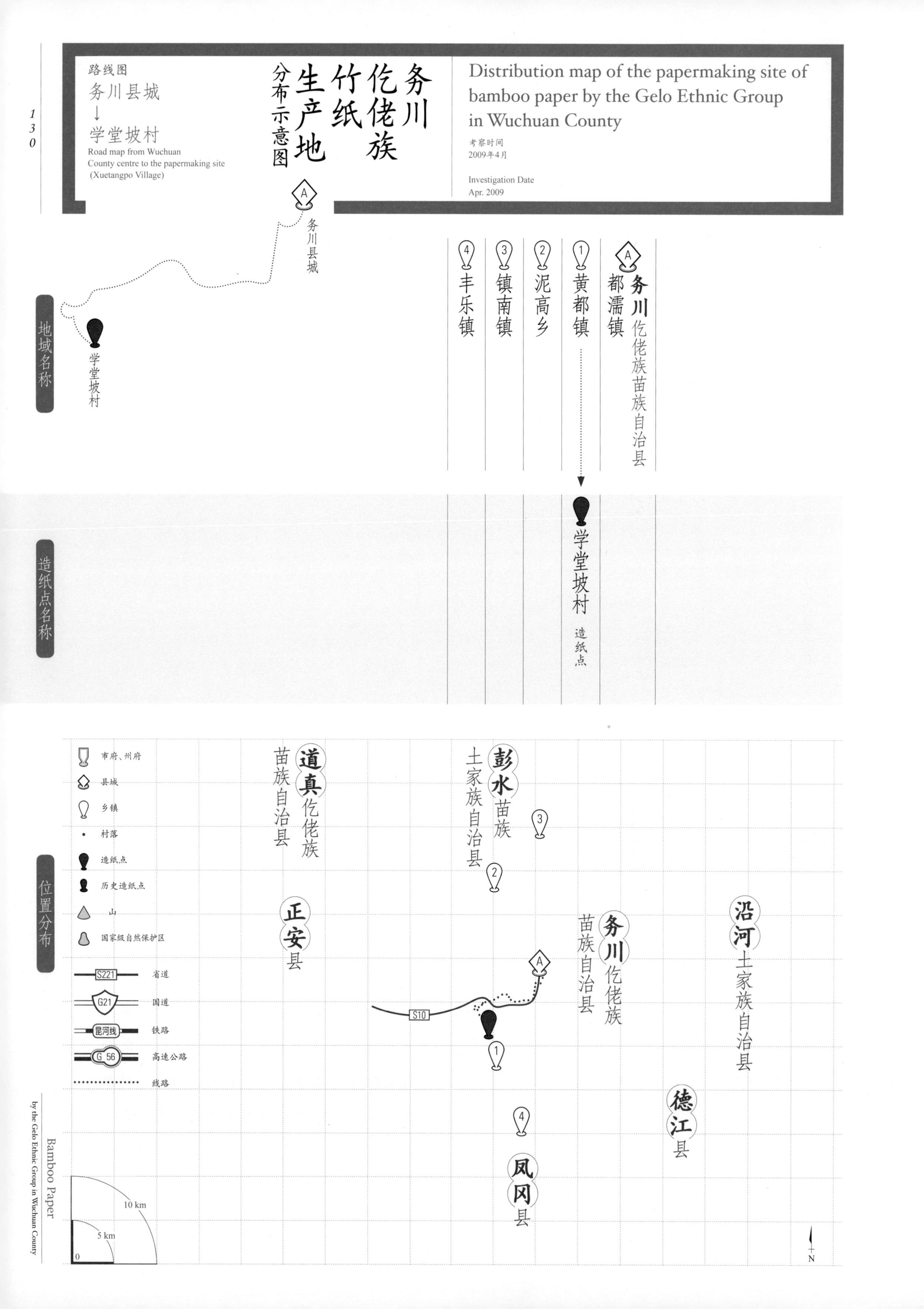

路线图
务川县城
↓
学堂坡村
Road map from Wuchuan County centre to the papermaking site (Xuetangpo Village)
务川仡佬族竹纸生产地分布示意图
Distribution map of the papermaking site of bamboo paper by the Gelo Ethnic Group in Wuchuan County
考察时间
2009年4月
Investigation Date
Apr. 2009
务川县城
学堂坡村
地域名称
务川仡佬族苗族自治县
都濡镇
黄都镇
泥高乡
镇南镇
丰乐镇
造纸点名称
学堂坡村
造纸点
位置分布
市府、州府
县城
乡镇
村落
造纸点
历史造纸点
山
国家级自然保护区
S221
省道
G21
国道
昆河线
铁路
G56
高速公路
线路
道真仡佬族苗族自治县
彭水苗族土家族自治县
正安县
务川仡佬族苗族自治县
沿河土家族自治县
德江县
凤冈县
S10
10 km
5 km
0
N

⊙1
⊙2

海、钟乳奇秀的大型溶洞——银狮洞景区、鬼斧神工的七柱山景区。另有沈家坝村的陈氏古宅，精彩木雕技艺遍布旧屋，门楣上悬挂翼王石达开的幕僚甘棠手书的“耕读人家、林泉山水”旧匾。

学堂坡村民组隶属务川县黄都镇三合行政村，距务川县城约56 km，务川县城至黄都镇的公路临村而过，交通便利；属山沟峡谷地形，一条小河顺沟谷流过，两山植被葱郁，学堂坡村建在南坡半山。2009年4月调查组入村时，学堂坡全村36户，152人，全部为张姓仡佬族居民。

三 务川仡佬族竹纸的历史与传承

3 History and Inheritance of Bamboo Paper by the Gelo Ethnic Group in Wuchuan County

据2009年调查组前往学堂坡村的调查及溯源访谈，获知该村的造纸均在张姓仡佬族村民中传习，但其技艺来源并非张家世传，而是由于外来人的迁入才开始造纸的。据调查中张明学、张明贵等造纸人自述，学堂坡村张氏先祖并不会造纸。张家掌握造纸技艺并形成谋生业态，源于张正香的继父王居臣从正安县鹿池沟村到张家当上门女婿，把祖传的造纸手艺带了过来，传给张正香、张正禄等人后，学堂坡村才开始造竹纸，距调查时已有约120年，是源自正安的输入性技术。

⊙1 学堂坡村远景 View of Xuetangpo Village

⊙2 学堂坡村民居小景 View of residences in Xuetangpo Village

20世纪六七十年代的学堂坡村曾一分为二，分别称为学堂坡生产队和文学生产队，两个生产队同时成立了造纸社，由生产队集体造纸，统一购买原料和销售竹纸。1982年后土地等资产下放给农户，学堂坡生产队集体所有的造纸厂房及设备被张明芳、张明学等人买下；文学生产队集体所有的造纸厂房及设备则被张光玉、张光能、张光彩三兄弟买下。据了解，调查时整个三合行政村造纸的农户也只有学堂坡生产队的张明学、张明芳、张明贵和文学生产队的张光玉、张光能、张光彩、张光贵两条传承脉系。由于两脉的工艺、材料等几乎一致，因此调查组本次主要以学堂坡生产队为考察对象。

四 务川仡佬族竹纸的生产工艺与技术分析

4 Papermaking Technique and Technical Analysis of Bamboo Paper by the Gelo Ethnic Group in Wuchuan County

(一) 务川仡佬族竹纸的生产原料与辅料

学堂坡村竹纸的生产原料以慈竹为主，其他竹子，如荆竹，因当地生长的数量少，只是偶尔会砍下来配用。传统造竹纸用的辅料包括石灰和滑料。

学堂坡村附近的慈竹资源丰富，通常会长到10余米高，四季常绿，以学堂坡村目前的小规模手工作坊生产用量来看资源供给充足。

滑药，即通常所说的纸药。据张明学介绍，

学堂坡造纸以前用从附近山上挖来的一种土名叫“三条茎”的植物作滑。21世纪初，开始种植卖纸帘子的师傅从四川带过来的植物种子。每年冬季下种，六月收获，作滑药时先用叶子，若叶子用完了，则用秆或根。做法是用布袋子将滑药装好，放在纸药槽子里浸泡，用时揉一下以挤出已浸出的汁液。一大把叶子通常就够一天用的了，一两棵这种植物就能采下一大把叶子。夏天用一天即换叶子，冬天用两天换叶子，夏天、秋天可用叶子，冬天无叶子时则可用秆或根。根据现场观察，这种由四川引入的植物长有类似棉花的骨朵，茎秆带刺，即俗称的“野棉花”，学名“黄蜀葵”，是中国西南地区流行的纸药植物。至于“三条茎”，调查中造纸户无法说清到底是什么植物，但应该与黄蜀葵不是一类。

⊙1

⊙2

（二）务川仡佬族竹纸的生产工艺流程

据调查组2009年4月入村的实地调查以及对张明学和张明芳等人的重点访谈，总结学堂坡村竹纸的制作工序如下：

壹	贰	叁	肆	伍	陆	柒	捌	玖	拾	拾壹	拾贰	拾叁	拾肆	拾伍	拾陆
砍竹	破竹	捆浆把	臜料	泡料	洗料	发汗	二次泡料	碾料	下槽	舀纸	榨纸	揭纸	晒纸	理纸	打捆

⊙1 四川引种的滑药植株黄蜀葵 Papermaking mucilage (*Abelmoschus manihot*) introduced from Sichuan Province

⊙2 已用过的滑药植物残草 Used plant for making papermaking mucilage

壹 砍竹

1 ⊙1

当地村民一般在农历四月砍一年生的新生慈竹，当年生的竹子嫩，没有枝叶，高5~6米。若竹龄太长，则竹纤维会变老，造出的纸产量和质量都会降低。由于劳动力等原因，调查时造纸户们除了砍自己家山上的竹子外，还会到附近当阳、打鼓坡等村寨收购。

⊙1

贰 破竹

2

先砍掉竹子的细竹枝，然后用柴刀把竹子主茎砍成段，每段长2 m，并用木槌将其捶破。一人一天可以捶破500 kg左右的新鲜竹料。

叁 捆浆把

3

把破好并捶好的竹段捆成普通碗径粗的浆把，50 kg料可捆成6把。

肆 腊料

4

这是村里造纸户的方言土语，即泡料的上一工序。其方法是将竹浆把放在塘子里，一层竹子撒一层石灰，一般一塘可放竹料2 500 ~3 000 kg，最大的塘子可放5 000 kg（这种塘子多为几家造纸户联合使用），竹料与石灰配放比例为10:3。放满水后，最上面用石头压紧、压实。

伍 泡料

5 ⊙2

竹料在泡料池里至少要泡4个月。当然，泡的时间再长也不会使竹料作废，据张明芳介绍，曾有一塘竹料泡了一年多。只是泡的时间长了以后，竹子的纤维会减少，产出的纸就要少些。泡好后的竹子被村里造纸户们称为毛料。

陆 洗料

6

用手将泡料池里已浸泡好的毛料翻起来，拿到河边洗去石灰残液和其他杂质。2 500~3 000 kg毛料，一般两三个人一天可以洗完。

⊙2

⊙1 村寨边的慈竹丛
Neosinocalamus affinis forest alongside the village

⊙2 泡料
Soaking the papermaking materials

柒

发 汗

7 ⊙3

⊙3

毛料洗好后，将泡料塘子用水清洗干净，再将毛料放进去，不用加水，堆放发汗40天左右，最上面用青草或干草（两者只用其一）盖上。据张明学、张明贵介绍，以前的传统工艺是用甑子蒸料，20世纪晚期改用更简便的堆放发汗法之后，即可由熟料法转为生料法了。

捌

二 次 泡 料

8

发汗完成后，将竹料放入池中，加入清水，继续泡，这时竹料已成熟，称为熟料，生产竹纸时可随用随取。

玖

碾 料

9 ⊙4⊙5

把熟料放在碾盘里，用牛拉碌子碾，一次可碾150 kg竹料，每次碾8小时左右，边碾边用手翻料。

⊙4

⊙5

⊙6

拾

下 槽

10 ⊙6⊙7

将已碾好的竹料放到抄纸槽里，一次放50 kg，加入清水，用槽棍打槽后，再加一些清水。抄纸前用手揉一下已泡好的装有滑药液的袋子，再将纸药水液挤入容器内（可用盆、桶等作为容器），然后将其倒入纸槽里搅匀。剩余的竹料则暂时放在边上的踩槽里备用。

⊙7

⊙3 熟料 Fermented papermaking materials

⊙4 碾料 Grinding the papermaking materials

⊙5 碾好的料 Processed papermaking materials

⊙6 槽棍 Sticks for stirring the papermaking materials

⊙7 挤滑 Squeezing the papermaking mucilage

拾壹 舀纸

11 ⊙8⊙9

一个熟练技工一天可舀1 200帘纸。“头帘水”由外往里舀，使纸纤维均匀入帘；“二帘水”再由纸头挑水，往纸尾冲，目的之一是使纸面平滑，二是方便湿纸撕揭开。舀一张湿纸后即转身将帘上的湿纸倒扣在榨床的底板上，然后一张张倒扣覆盖，形成湿纸垛。一槽纸浆料约可舀1 200帘，但需加七八次滑药液，加一次滑药液则用撸筋耙和一次料。

⊙8 ⊙9 ⊙10 ⊙11

⊙13 ⊙12

拾贰 榨纸

12 ⊙10~⊙13

一槽纸浆全部舀完后，即可直接加压板等迅速压榨。由下到上的顺序依次为底踩板、把草、纸垛、盖纸板、四块木方（码子）、大码。把长杆一端放在码子和千斤顶中间，另一端用绞索套在滚筒上，然后将扳子插在滚筒的孔中并向下扳，压榨湿纸垛。压榨约需要一小时，慢慢压，压到约一半厚度的时候（耗时约0.5小时）松榨，然后再加一块木方。压一小时后松榨，将半干的纸饼运回家中。

⊙8 舀『头帘水』
Scooping the papermaking screen for the first time

⊙9 舀『二帘水』
Scooping the papermaking screen for the second time

⊙10 待榨的纸
Paper to be pressed

⊙11/12 榨纸过程
Procedures of pressing the paper

⊙13 榨好的纸
Pressed paper

拾叁
揭　　纸
13

在家中将湿纸一张张揭开，每揭10张为一次工序，每2张之间形成一些间隔，揭完10张后，全部拖起来往竹竿上晾挂。

⊙14

拾肆
晒　　纸
14　⊙14

在屋内竹竿上晒纸，天气好时，约20小时即可晒干，冬天则要四五天。晒纸时，每张纸从尺寸短的一边翻折，折后留出一些距离以使每张之间不完全重叠。

⊙15

拾伍
理　　纸
15　⊙15

从竹竿上取下晾干的纸，每4张为一叠，每张纸的短边依次向下留出一截不重叠，理齐后再对折起来。

⊙16

拾陆
打　　捆
16　⊙16

分为两种规格：一种是20叠（80张）为一把；另一种是15把为一捆，即1 200张为一捆。调查时一般用棕叶、竹篾、塑料绳等捆扎。

与20世纪七八十年代相比，调查时，学堂坡村竹纸生产工艺发生了若干变化，具体情况见表8.13。

表8.13　传统和现代学堂坡村竹纸生产工艺对照表
Table 8.13　Contrast of traditional and modern papermaking techniques of bamboo paper in Xuetangpo Village

工序	传统工艺	现代工艺
做熟料	用甑子蒸	窖封发汗
碾料	用石碓舂（人力）	用石碾碾（畜力）
纸药	野生的“三条茎”	四川引种的纸药黄蜀葵

⊙14 晒纸 Drying the paper

⊙15 理纸 Sorting the paper

⊙16 捆好的纸 Bundles of paper

（三）

务川仡佬族竹纸生产使用的主要工具设备

壹 纸槽 1

大小并不完全一致，所测张明学家纸槽尺寸约152 cm × 105 cm × 80 cm，厚约10 cm，由水泥材料制作而成。紧邻纸槽旁有踩槽（尺寸约70 cm × 80 cm × 53 cm）一个，据造纸户介绍，过去它是用来踩料的，调查时则已被用来装一次舀纸打槽用不完的余料。

⊙1

贰 纸帘 2

一帘二纸的改进型抄纸竹帘，尺寸约75.5 cm × 31.5 cm。

⊙2

叁 木榨 3

用于压榨纸垛。由将军柱、千斤顶、懒板凳（即榨桥）、矮子、滚筒等构成，全部用硬实木料做成。平面呈梯形，长约150 cm，短边约20 cm，长边约80 cm；将军柱一端高约167 cm，矮子一端高约105 cm。

⊙3

⊙4

肆 石碾 4

用于碾熟料。由板子杆、碾盘、磙子、将军柱构成。实测张明学家使用的石碾各部件尺寸为：板子杆长约258 cm，离地面约25 cm；碾盘直径约340 cm，四周用石块围砌，碾盘上也用石块铺设，中间一块呈圆形；磙子直径约40 cm，长约65 cm；将军柱一根，竖立在碾盘正中间，用以固定板子杆、磙架，高约55 cm。

⊙1 纸槽 Papermaking trough

⊙2 一帘二纸纸帘 Papermaking screen that can make two pieces of paper simultaneously

⊙3 木榨 Wooden presser

⊙4 石碾 Stone grinder for grinding the papermaking materials

(四)
务川仡佬族竹纸的性能分析

所测务川学堂坡村竹纸的相关性能参数见表8.14。

表8.14 学堂坡村竹纸的相关性能参数
Table 8.14 Performance parameters of bamboo paper in Xuetangpo Village

指标		单位	最大值	最小值	平均值
厚度		mm	0.220	0.115	0.161
定量		g/m²	—	—	34.8
紧度		g/cm³	—	—	0.216
抗张力	纵向	N	6.7	4.7	5.8
	横向	N	4.0	1.7	3.1
抗张强度		kN/m	—	—	0.297
白度		%	21.1	19.3	20.4
纤维长度		mm	11.93	0.66	3.57
纤维宽度		μm	28.0	1.0	13.0

由表8.14可知，所测学堂坡村竹纸最厚约是最薄的1.91倍，相对标准偏差为0.33%，纸张厚薄差异较大。竹纸的平均定量为34.8 g/m^2。所测竹纸紧度为0.216 g/cm^3。

经计算，其抗张强度为0.297 kN/m，抗张强度值较小。

所测学堂坡村竹纸白度平均值为20.4%，白度较低，白度最大值约是最小值的1.09倍，相对标准偏差为0.23%，白度差异较小。

所测学堂坡村竹纸纤维长度：最长11.93 mm，最短0.66 mm，平均3.57 mm；纤维宽度：最宽28.0 μm，最窄1.0 μm，平均13.0 μm。所测竹纸在10倍、20倍物镜下观测的纤维形态分别见图★1、图★2。

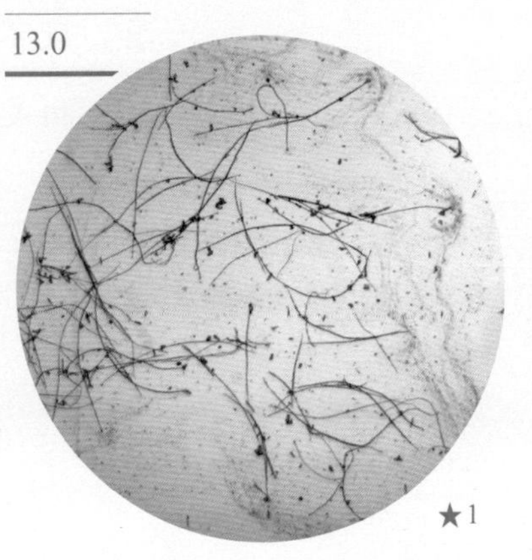
★1

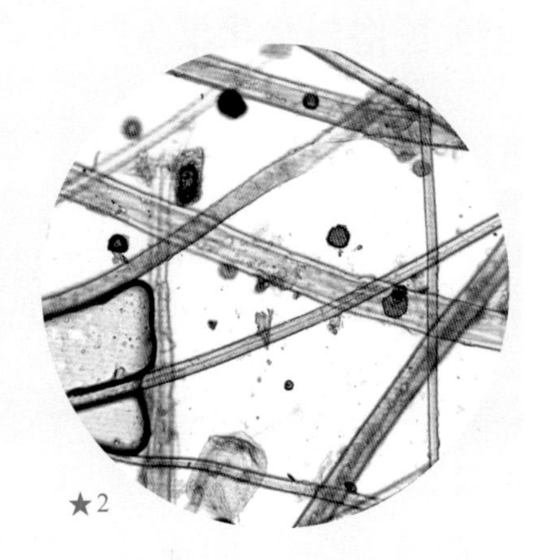
★2

★1 学堂坡村竹纸纤维形态图(10×)
Fibers of bamboo paper in Xuetangpo Village (10× objective)

★2 学堂坡村竹纸纤维形态图(20×)
Fibers of bamboo paper in Xuetangpo Village (20× objective)

五
务川仡佬族竹纸的用途与销售情况

5
Uses and Sales of Bamboo Paper by the Gelo Ethnic Group in Wuchuan County

据调查时获知的信息及实地考察所见，务川竹纸在当代属于生料法制作，纸质偏粗偏黄。调查时发现，学堂坡村所造的竹纸用途单一，仅作为祭祖祭亲时焚烧的冥纸，通常打制成纸钱的形式使用，未发现有其他用途。

学堂坡村当代所造竹纸主要在本县当阳、涪洋等乡村集散地进行销售，其方式以造纸户自己背到集市上贩卖为主，也有附近村民到造纸户家中来购买的。调查时，竹纸的售价为4元/把。

学堂坡村造纸户一年造纸平均约需600 kg竹子，可以生产400~500把纸，按4元/把计算，毛收入约2 000元。张明芳家是村中生产竹纸最多的造纸户，一年约生产3 000把，毛收入在1万元以上。据调查，张明芳家一年造纸平均开支信息如表8.15所示。

表8.15　张明芳家一年造纸开支表
Table 8.15　Zhang Mingfang's annual papermaking cost

类别	开支情况	备注
竹子	5 000 kg × 0.4元/kg =2 000元	
石灰	1 000 kg × 0.4元/kg =400元	
帘子	170元	250元/架，可使用1.5年
厂房折旧	100元	500元/间，可使用5年
木榨折旧	15元	250元/个，可使用20年
合计	2 685元	

如果不计入长年的劳力投入，张明芳家一年造纸纯利润约9 000元，这在黔北少数民族村落中还算是一笔较大的经济收入。

六　务川仡佬族竹纸的相关民俗与文化事象

6　Folk Customs and Culture of Bamboo Paper by the Gelo Ethnic Group in Wuchuan County

在对学堂坡村手工造纸的调查中发现：造纸无禁忌。学堂坡造纸户在生产过程中没有什么禁忌，既不祭师父，也不祭蔡伦，有着较大的随意性。相关习俗的采集也未有收获。

七　务川仡佬族竹纸的保护现状与发展思考

7　Preservation and Development of Bamboo Paper by the Gelo Ethnic Group in Wuchuan County

(一)　务川仡佬族竹纸传承与保护机制的构建

在对学堂坡村手工造纸技艺业态的实地考察中，我们切身地感受到对传统造纸工艺“活化”的保护和发展才是激发传统手工造纸工艺活力的根本。然而，对于像学堂坡村竹纸这种用途单一的低端纸品而言，这种“活化”的保护仅仅靠传承人勉强维持生产是不够的，迫切需要地方政府和相关公共机构的“活化”促进政策支持。

从务川已推动的保护与促进政策来看，确有系列的地方条规已经颁布，比较具有代表性的有：① 2008年，务川县人民政府正式提出用五年的时间，初步建成“中国人文旅游新区、渝黔仡佬文化中心”的发展目标。同时出台了《中共务川自治县委关于进一步加快以仡佬文化为重点的文化旅游发展的决定》（务党发〔2008〕11号），明确提出了要加强非物质文化遗产的保护和传承，并配套制定了《务川自治县非物质文化遗产项目代表性传承人认定与管理暂行办法》。② 学堂坡村造纸作坊已经被务川县人民政府公布为第三批县级文物保护单位，地方政府已在规制层面加强了对学堂坡村造纸业态的保护和管理。

⊙1

⊙1 山坡上的造纸作坊 Papermaking mills on a hill

（二）务川仡佬族竹纸面临的困境及思考

保护传承人是非物质文化遗产保护的重要措施，鼓励和支持非物质文化遗产项目代表性传承人开展生产性传习活动，是保护和传承非物质文化遗产的关键内容。目前，掌握整套务川竹纸技艺的村民多在50岁以上，年轻一代的村民很少愿意学习、传承手工造纸技艺。因此，设立专项的传习促进资金，提高该技艺业态的收益水平，鼓励和支持传承人开展传习活动，是务川仡佬族竹纸发展的当务之急。

产品质量的提升是学堂坡古法造纸工艺传承的大事。调查时发现，学堂坡村竹纸由于用途单一且很低端，加上造纸户对纸的品质要求不高，纸质较为粗糙，很难扩展其他的生活或文化用途，从而大大制约了该竹纸的发展空间。历史上，学堂坡村竹纸曾经用甑蒸料，制成纸质更优的熟料法竹纸产品。如若设立专项资金，试验性恢复熟料法生产，对现行设备工艺进行若干改良，或许能探索出更丰富的当代发展与传承活态。

务川仡佬族竹纸

Bamboo Paper by the Gelo Ethnic Group in Wuchuan County

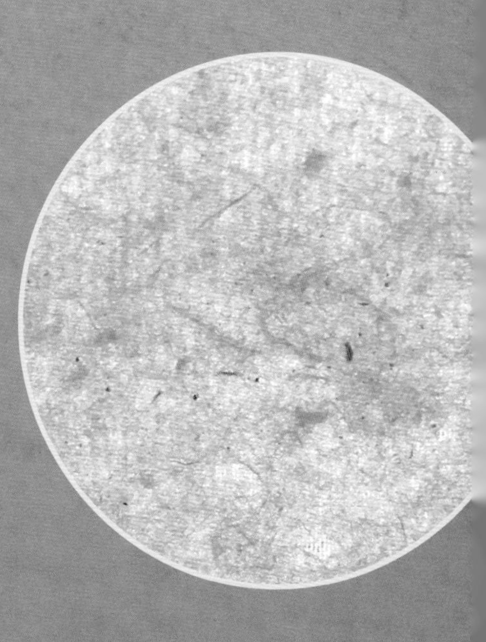

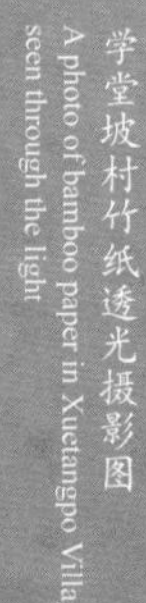

学堂坡村竹纸透光摄影图
A photo of bamboo paper in Xuetangpo Village seen through the light

第五节

余庆竹纸

贵州省
Guizhou Province

遵义市
Zunyi City

余庆县
Yuqing County

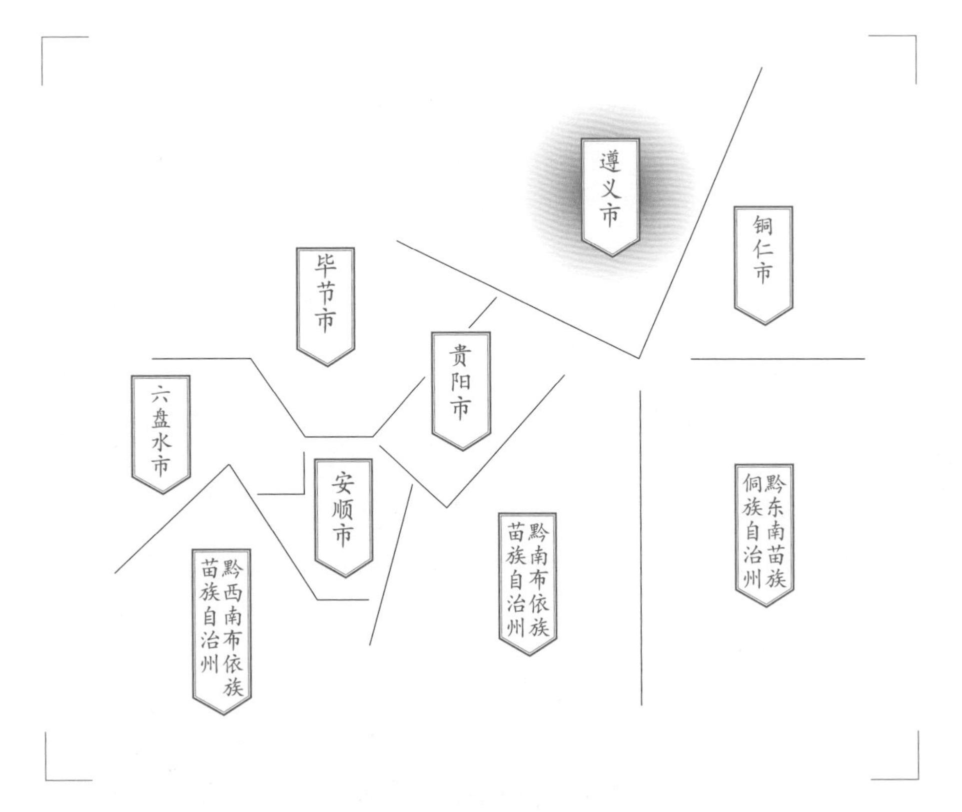

调查对象

大乌江镇乌江行政村竹纸

Section 5
Bamboo Paper in Yuqing County

Subject

Bamboo Paper in Wujiang Administrative Village of Dawujiang Town

一
余庆竹纸的基础信息及分布

1
Basic Information and Distribution of Bamboo Paper in Yuqing County

余庆竹纸是传统余庆手工纸的一种，据1992年版的《余庆县志》记载，明清时期余庆就有造纸业。种类有用慈竹作原料生产的土纸（又名火纸），主要用于祭祀；用构皮作原料生产的白纸（又称皮纸），主要用于习字、记录等。余庆竹纸主要是指用慈竹作原料生产的手工纸，主要用于祭祀。

民国时期，余庆造纸业较为兴旺，主要分布在现熬溪镇的箐口、后坝（今银坝）村，松烟镇的台上村等地。2009年4月，调查组进入余庆县境调查时，所获知的信息是目前仅剩大乌江镇乌江村仍有活态的竹纸生产，皮纸生产业态已经完全消失。

二
余庆竹纸生产的人文地理环境

2
The Cultural and Geographic Environment of Bamboo Paper in Yuqing County

余庆地处黔北南陲，系遵义、铜仁、黔东南、黔南四地州（市）的结合部。北与凤冈，东与石阡，南与黄平、施秉，西与湄潭、瓮安诸县接壤。余庆县属亚热带温润季风气候，四季分明，冬无严寒，夏无酷暑，气候温和，年平均气温16.4 ℃，年平均最高气温21.3 ℃，年平均最低气温12.9 ℃，无霜期长达300天。雨量充沛，年平均降水量1 056 mm。

余庆山清水秀，土壤类型多样，森林覆盖率达49%。水能资源蕴藏量大，开发前景广阔，境内河流密布，贵州第一大江——乌江横穿县境中部，已开发的地方电力达到1×10^4 kW装机容量，2003年已全面开工建设的国家级大型水电站——构皮滩水电站，位于余庆境内，装机容量达3×10^6 kW，2008年第一台机组发电，该电站

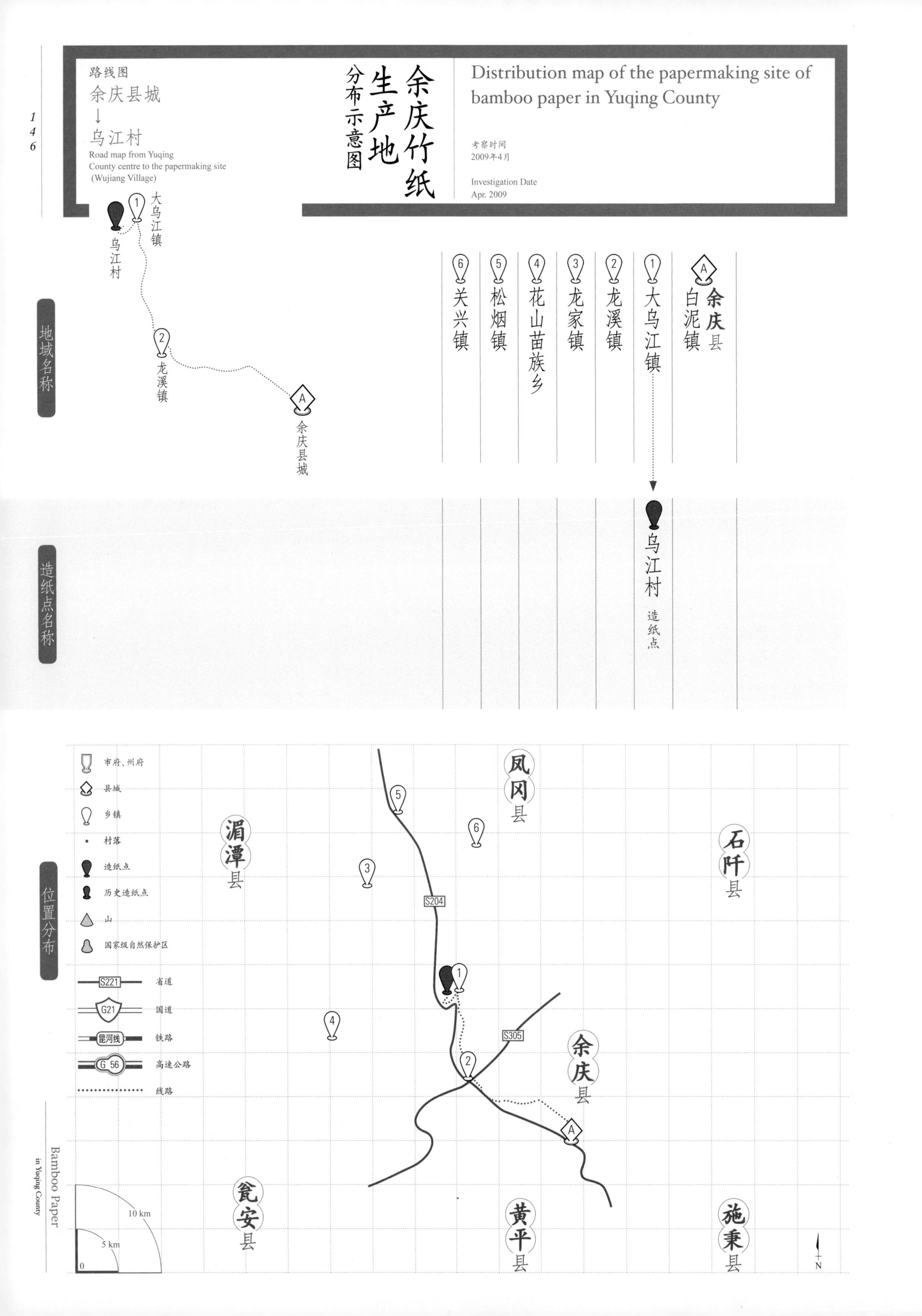

路线图
余庆县城
↓
乌江村
Road map from Yuqing County centre to the papermaking site (Wujiang Village)
余庆竹纸生产地分布示意图
Distribution map of the papermaking site of bamboo paper in Yuqing County
考察时间
2009年4月
Investigation Date
Apr. 2009
乌江村
大乌江镇
龙溪镇
余庆县城
地域名称
余庆县 白泥镇
大乌江镇
龙溪镇
龙家镇
花山苗族乡
松烟镇
关兴镇
造纸点名称
乌江村 造纸点
位置分布
市府、州府
县城
乡镇
村落
造纸点
历史造纸点
山
国家级自然保护区
S221 省道
G21 国道
昆河线 铁路
G 56 高速公路
线路
凤冈县
湄潭县
石阡县
S204
S305
余庆县
瓮安县
黄平县
施秉县
10 km
5 km
0
N

⊙1

的建设使余庆成为贵州水电能源供应的中心区域之一、贵州“西电东送”的重要电源点。县内矿产资源有煤、陶土、磷、高钙石灰岩、重晶石等。已探明的陶土矿仅黄土坡矿储量就已达400万吨，可开发生产系列别具风格的绿陶制品。

全县幅员面积1 623.7 km^2，有汉、苗等21个民族，2003年末的统计数据显示，全县总人口为29.36万，少数民族以苗族、土家族人口较多。

在调查中，余庆当地口耳相传中国第一幅对联——“新年纳余庆，佳节号长春”中说的余庆就是贵州的余庆县，但正史及流行文献中均未见记述。西晋初期，分汉代且兰县地置万寿县，为牂牁郡治，有今余庆县地。隋代以万寿县扩置牂牁县，为牂牁州牂牁郡治，今余庆县地在此境内。唐武德三年（620年），以牂牁地置牂州，余庆属牂州境。南宋绍定元年（1228年），今余庆隶绍兴府所领羁縻小州辖境。元至元十四年（1277年），属播州安抚司辖地。明洪武五年（1372年），思州、播州归入明朝，余庆、白泥二州随归，隶四川行省，万历二十九年（1601年），合余庆、白泥两土司地置余庆县，属平越军民府，隶贵州布政使司。1949年，余庆隶镇远专区，1956年改隶遵义地区，1958年撤余庆县，并入湄潭县，1961年恢复余庆县，建制至今未变。

⊙2

⊙1 构皮滩水电站 Goupitan Hydropower Station
⊙2 乌江村边的泡料塘 Soaking pool alongside Wujiang Village

三
余庆竹纸的历史与传承

3
History and Inheritance of Bamboo Paper in Yuqing County

关于余庆手工纸的情况，新版《余庆县志》记载：“明清时期，余庆有造纸业。民国时期，造纸业较为兴旺，主要分布在敖溪的箐口、后坝（今银坝）及松烟的台上等地。用慈竹作原料生产土纸（又名火纸），主要用于迷信活动。用构皮作原料生产的白纸（又称皮纸），主要用于习字、记录等。”《贵州传统工艺研究》里同样也提到了余庆有造纸业。

据余庆县文物管理所周登琳所长分析，乡间口述记忆认为余庆在土司时代就已经开始造纸，那么余庆至少有400年的造纸历史了。

2009年4月，调查组重点访谈了造纸人游成贵。据游成贵回忆，他们家从祖父起就开始造纸，到现在至少有四代了，然而他能记起来名字的，只有他父亲游福成。父亲将造纸技术传给游成贵，他又教会了时年仅16岁的儿子游德福，调查时游德福还在读书，但14岁就已经学会造纸，并帮父亲一起造过竹纸。

中华人民共和国成立初，全县有100余户生产土纸和白纸。1956年后，大多停业。1979年开放市场后，农村个体户造纸复业，仅箐口张家沟一带就有30余户。1980年，岩门纸厂生产包装纸约22 t、机制纸约8 t。1984年岩门纸厂迁往白泥镇草坪村，1987年生产机制纸358 t。

⊙1

⊙1 调查组成员在料塘边访谈
Researchers interviewing papermakers by a soaking pool

四
余庆竹纸的生产工艺与技术分析

4
Papermaking Technique and Technical Analysis of Bamboo Paper in Yuqing County

2009年4月，调查组在余庆县文物管理所周登琳、王勇军、高军等的协助下，到熬溪镇、大乌江镇等地进行实地调查。据现场考察，熬溪镇没有发现手工皮纸生产业态，大乌江镇乌江村则仍有竹料火纸生产。

(一)
余庆竹纸的生产原料与辅料

余庆竹纸生产所用的原料是一年生的慈竹，又称糍竹。

辅料为石灰。乌江村造纸所用纸药当地称为沙根滑，主要使用其根，也可以使用其叶子。早上先用木槌将沙根滑捶融，再用袋子将其装好后即可用，一天大约用1.5 kg。

(二)
余庆竹纸的生产工艺流程

经重点访谈造纸人游成贵和游天云，总结乌江村竹纸生产工艺流程如下:
壹 砍竹
贰 下节
叁 捆竹
肆 下塘
伍 洗竹麻
陆 晒竹麻
柒 发汗
捌 碾料
玖 打槽
拾 加滑
拾壹 抄纸
拾贰 去垛边
拾叁 压榨
拾肆 分垛子
拾伍 刮纸
拾陆 拆纸
拾柒 晾纸
拾捌 收纸
拾玖 捆纸

壹

砍竹

1

每年农历十二月农闲时，造纸户都会上山用柴刀砍一年生的慈竹，一天一般可砍350~400 kg，多时超过500 kg。

贰

下节

2

将竹子破成段，每段长约2 m，一般一根竹子可破成4段。

叁

捆竹

3

用木棒将竹子捶破，然后用竹篾捆成捆，每捆直径约25 cm，质量25~30 kg。

肆

下塘

4

用4根木棒平行垫在麻塘底部，并将竹子垂直放在木棒上，一般麻塘高约2 m，可放5层竹子。竹子上压石头，再撒石灰，竹子和石灰质量比约为100：42，后加水，泡4个月。麻塘大小不等，可放竹子数量也不等，最小的麻塘可放2 500 kg竹子，最大的可放6 000 kg。

伍

洗竹麻

5

用手在麻塘里将竹麻上的石灰及其他杂质洗干净，四个人洗一天即可洗完一塘竹麻。

陆

晒竹麻

6

将3根木棒放在第二个麻塘底部，并把洗好的竹麻放在木棒上晒干，一般需晒一个月。

柒

发汗

7

竹麻晒干后，往麻塘里加入清水，再泡一个月。

⊙1 泡竹麻 Soaking bamboo materials

捌 碾料

8 ⊙2

用钉耙将竹麻钩起来，挑到碾槽中。以前用牛碾，一次碾3挑料，湿料质量共约180 kg，需碾10小时；现改用打浆机，碾1挑料仅需半小时。

玖 打槽

9 ⊙3

把一挑料一次性放在槽子里，加水，先用手搅，后用拱耙将料拱散，再用捞筋杆以“8”字形搅拌，将料搅匀，同时把筋捞出来。

拾 加滑

10 ⊙4

乌江村造纸所用纸药当地称为沙根滑，主要用根，也可以用叶子。早上先用木槌将沙根滑捶融，再用袋子将其装好后即可用，一天大约用1.5 kg。抄纸前用手挤装滑的口袋，使滑流到槽内，再用捞筋杆搅匀后即可抄纸，大约每半小时加一次滑。

⊙4

⊙2

⊙3

拾壹 抄纸

11 ⊙5⊙6

抄纸时，将帘子置于帘架上，左手握帘把手压在头子垛边，右手持帘杆，先由外往里舀水，再由右往左挑水。这样得到的纸，左边的厚些叫头子，右边叫尾子，一天抄1 500帘左右，约3 000张。

⊙5

头子垛边
怀垛
背垛
尾子垛边
10 cm
背垛垛边

⊙6

⊙2 打浆机碾料 Grinding the papermaking materials with a machine

⊙3 打槽 Stirring the papermaking materials

⊙4 沙根滑 Shagenhua, papermaking mucilage as adhesive

⊙5 抄纸 Scooping and lifting the papermaking screen out of water

⊙6 纸帘示意图 Schematic diagram of papermaking screen

拾贰

去 垛 边

12

抄完纸后，用手将头子垛边、尾子垛边、背垛垛边去掉。

拾叁

压 榨

13 ⊙7

在纸垛上依次放上旧帘子、炕板、码子，在码子的凹口处先放纸壳，再放榨杆，纸壳的作用是使其不打滑。用手杆缓慢压，压榨10余分钟后，纸垛高度下降，再放一颗码子，继续压榨。一般需加5次码子，最后一次人脚踩在手杆上，直到纸垛不再流出水来，就认为纸垛已经干了，后用毛巾将四边擦干，再松榨。整个过程需一个半小时左右。

⊙7

拾肆

分 垛 子

14

松榨后，将纸垛搬到凳子上，用手将怀垛、背垛往两边掰，使之一分为二。

拾伍

刮 纸

15 ⊙8

用刨子由下往上刮纸垛两长边及头子，后翻转垛子，再刮尾子。

⊙8

拾陆

拆 纸

16 ⊙9

先用左手由下往上将纸头揉松，由头子往尾子撕起约15 cm，然后每2张相隔1 cm，每10张为一合（叠），拆了一合纸后，撕开置于纸垛旁。

⊙9

拾柒

晾 纸

17

拆了10余合后，一起拿到室内竹竿上晾，天热晾3天即干，阴天要晾一周，下雨天要晾10天或更长时间。

拾捌

收 纸

18

将晾干的纸收下来，10合为1把。

拾玖

捆 纸

19

将相邻的两把纸头尾相错，10把为一捆，用包装带在中间捆一道，以前用棕树叶捆。

这就完成了整个造纸过程，可以拿到余庆县的大乌江镇、龙溪镇、敖溪镇等地的市场去卖。如果做成纸钱出售，则价格会高些，也更好卖。

⊙7 压榨示意 Showing how to press the paper

⊙8 刮纸 Scraping the paper

⊙9 拆纸 Spliting the paper layers

(三)

余庆竹纸生产使用的主要工具设备

壹 纸帘 1

活动式抄纸帘，实测造纸户的尺寸：长约73 cm，宽约47 cm。

⊙10

贰 槽子 2

略有差异，实测造纸户的尺寸：长约224 cm，宽约120 cm，高约100 cm。

⊙11

叁 木榨 3

实测造纸户的尺寸：将军柱高约169 cm，榨杆长约220 cm，手杆长约187 cm。

乌江村用专门的纸钱架来打纸钱，将造好的火纸放在纸架和纸钱模之间，并将纸钱模固定好。手持钱錾沿着纸钱模逐一按，即可在火纸上打出钱印，按一定要求打完后，得到的就是纸钱。

(四)

余庆竹纸的性能分析

所测余庆乌江村竹纸的相关性能参数见表8.16。

表8.16 乌江村竹纸的相关性能参数
Table 8.16 Performance parameters of bamboo paper in Wujiang Village

指标		单位	最大值	最小值	平均值
厚度		mm	0.250	0.200	0.224
定量		g/m²	—	—	56.3
紧度		g/cm³	—	—	0.251
抗张力	纵向	N	9.0	7.7	8.1
	横向	N	7.6	5.5	6.7
抗张强度		kN/m	—	—	0.493
白度		%	29.6	27.9	28.7
纤维长度		mm	3.79	0.51	1.37
纤维宽度		μm	25.0	1.0	10.0

⊙10 打纸钱 Making joss paper

⊙11 纸钱架 Frame for making joss paper

由表8.16可知，所测乌江村竹纸最厚是最薄的1.25倍，相对标准偏差为1.70%，纸张厚薄较为一致。竹纸的平均定量为56.3 g/m^2。所测竹纸紧度为0.251 g/cm^3。

经计算，其抗张强度为0.493 kN/m，抗张强度值较小。

所测乌江村竹纸白度平均值为28.7%，白度较低，白度最大值约是最小值的1.06倍，相对标准偏差为0.57%，白度差异相对较小，可能是因为余庆县竹纸加工时没有经过较强的漂白。

所测乌江村竹纸纤维长度：最长3.79 mm，最短0.51 mm，平均1.37 mm；纤维宽度：最宽25.0 μm，最窄1.0 μm，平均10.0 μm。所测竹纸在10倍、20倍物镜下观测的纤维形态分别见图★1、图★2。

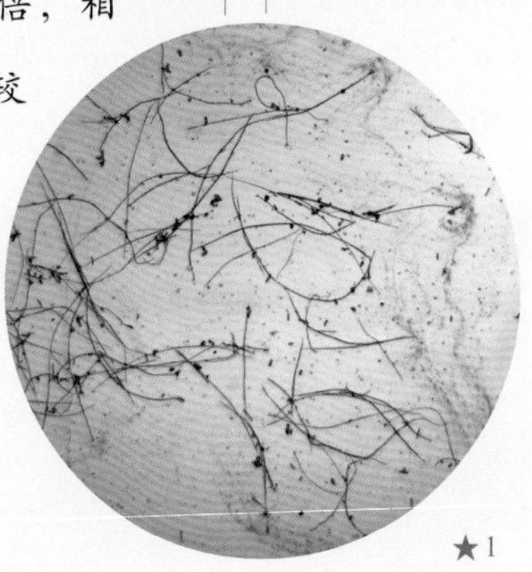

★1

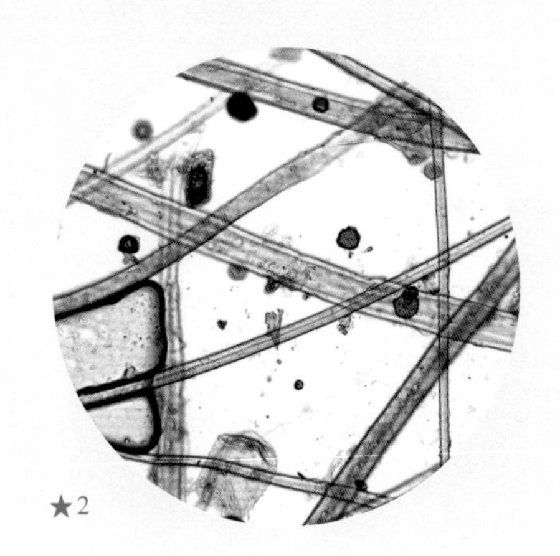

★2

★1
乌江村竹纸纤维形态图(10×)
Fibers of bamboo paper in Wujiang Village (10× objective)

★2
乌江村竹纸纤维形态图(20×)
Fibers of bamboo paper in Wujiang Village (20× objective)

五 余庆竹纸的用途与销售情况

5 Uses and Sales of Bamboo Paper in Yuqing County

乌江村火纸调查时的单一用途即为民间祭祀，在每年的“三节”（清明节、中元节、春节）祭祀时都要烧火纸。调查组调查时，火纸价格为3.5元/把，如果做成纸钱后则按千克算，价格为8元/kg，据造纸村民游天云所说，做成纸钱后会更好卖。

乌江村改用打浆机碾料后，除降低了劳动强度，提高了工作效率外，还提高了产量，增加了效益。据游天云统计，用打浆机碾料，5 000 kg竹子可抄3 000 kg纸，而用牛碾则只能抄2 250 kg左右。游天云夫妇一年造纸需生竹子5 500 kg，可造纸3 300 kg，按打成纸钱后的价格8元/kg来算，则销售额约为26 400元。其支出主要是石灰，用的是刚烧的广子灰，价格为1.6元/kg，约需2 310 kg，共计约3 700元。如不计劳力等成本，一年造纸收入约22 700元。

六
余庆竹纸的相关民俗与文化事象

6
Folk Customs and Culture of Bamboo Paper in Yuqing County

1. 纸钱有讲究

余庆竹纸的钱引为三排七列，多了少了都不行。

2. 婚姻礼仪用纸习俗

余庆汉族在1949年前，重在明媒正娶，讲究门当户对，结婚有放话、插香、开庚、过礼和迎娶等环节。各环节基本都要用到手工纸，或书写或烧化。

3. 丧葬祭祀用纸习俗

老人弥留之际，请阴阳先生看吉地，做好寿衣、寿帽、寿鞋，备好棺木。以前衣帽鞋都用手工纸制作。

治丧中多处要用纸，传统都用手工纸。如气绝之际，焚“倒头钱”；设灵堂时，堂屋门前悬长纸幡，用黄纸书挽联贴于大门两侧；出殡时，焚香烧纸，沿途丢买路纸钱；棺木抬到墓穴旁，阴阳先生焚纸钱买地。

春节、清明、端午、中元、中秋、重阳等传统节日，都要以酒馔、果品、香、烛、纸、爆竹等，在祖龛或祖坟前祭祀，形式大同小异。

游天云夫妇一年造纸收入约22 700元。相比之下，造纸收入并不算太多。这也是造纸人数逐渐减少的重要原因。据游天云介绍，张家沟村2000年左右还有20多家造纸户，2009年时只有5家。游天云夫妇坚持造纸的原因主要是可以照顾家。

乌江村造纸，除碾料时不再用牛而改用打浆机，捆纸时不再用棕树叶而改用包装带外，其余均保留了传统手工造纸工序，为研究西南地区传统手工造纸方法提供了鲜活的样本。

乌江村造纸各个工序、所用各种设备及工具都有单独的名称，典型的像纸帘，各部分都有名称；同时造纸户总结出相当丰富的造纸经验，如由于抄纸时力度不平衡，怀垛纸比背垛纸更平整、光滑且不易烂。这些都充分体现了乌江村造纸工艺承载着大量的传统手工造纸知识。

但乌江村手工造纸的业态传承并不理想，在当前社会急剧变迁的背景下，乌江村的手工纸生产已至濒危的边缘，亟待保护。这从近年来造纸户数量的急剧减少可见一斑。

乌江村造纸现已改用打浆机碾料，不但降低了劳动强度，提高了工作效率，而且增加了效益。改用打浆机，虽然对传统的工艺有了一定程度的改变，但这是造纸户针对火纸主要用于祭祀、对纸的质量要求不太高的现状而进行的有意识的自我保护。正是通过这种自我有意识地降低成本的保护，竹纸制造技艺才得以艰难地保存下来。

⊙1

⊙2

⊙ 1/2
乌江村景
View of Wujiang Village

余庆 竹纸

Bamboo Paper in Yuqing County

乌江村竹纸透光摄影图

A photo of bamboo paper in Wujiang Village seen through the light

第九章
铜仁市

Chapter IX
Tongren City

第一节 石阡仡佬族皮纸

贵州省
Guizhou Province

铜仁市
Tongren City

石阡县
Shiqian County

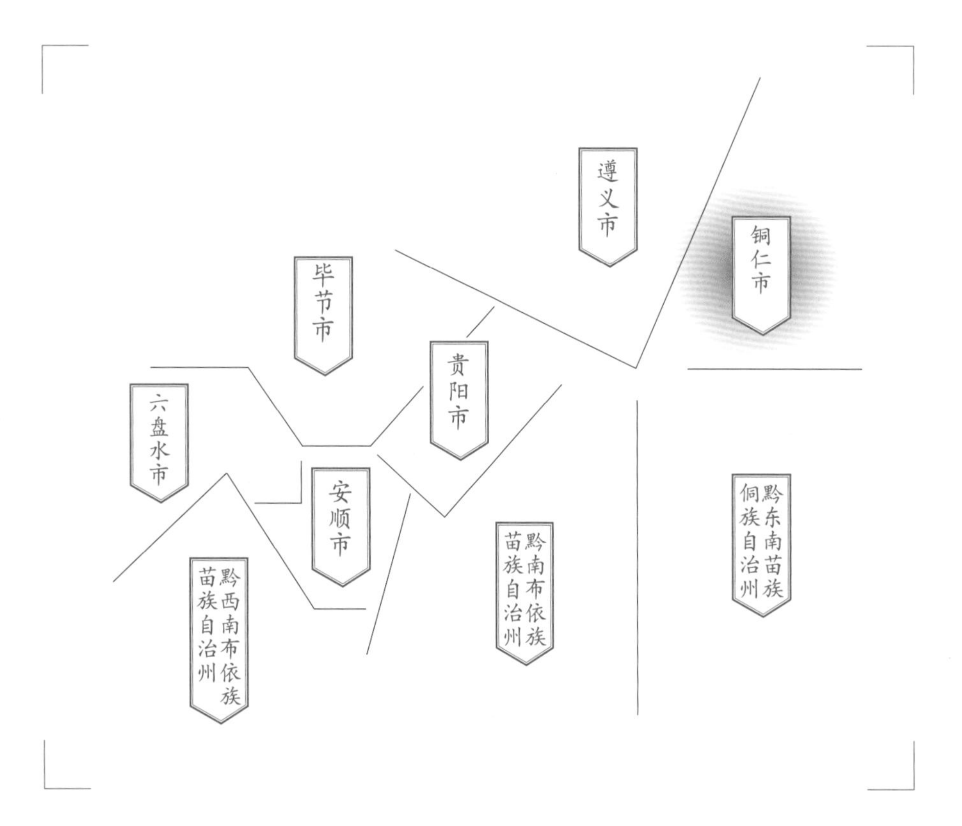

调查对象

汤山镇香树园行政村仡佬族皮纸

Section 1
Bast Paper by the Gelo Ethnic Group in Shiqian County

Subject

Bast Paper by the Gelo Ethnic Group in Xiangshuyuan Administrative Village of Tangshan Town

一
石阡仡佬族皮纸的基础信息及分布

1
Basic Information and Distribution of Bast Paper by the Gelo Ethnic Group in Shiqian County

在当代，石阡皮纸的生产地主要集中在汤山镇香树园村，这里是仡佬族聚居区，造纸户多以仡佬族为主。制作石阡皮纸的主要原料为当地产的构树皮。这种纸的主要用途有：(1) 祭祀，这是香树园村皮纸最主要的用途；(2) 书写，包括地方文书文献的抄录、日常生活记事和学生书写用纸；(3) 捆钞，石阡县有些银行专门到香树园村订制皮纸用来捆钞票。

石阡皮纸的造纸工序多达30道，工艺复杂，而且保留了传统的造纸技艺，其间有多次蒸煮、踩料、舂料、洗料、选料等工序，这充分体现了传统工艺精益求精的标准，为研究传统的造纸方法提供了鲜活的样本。

二
石阡仡佬族皮纸生产的人文地理环境

2
The Cultural and Geographic Environment of Bast Paper by the Gelo Ethnic Group in Shiqian County

石阡县位于贵州省东北部、铜仁市的西南部，地处东经107°44′55″~108°33′47″、北纬27°17′5″~27°42′50″，以及湘西丘陵向云贵高原过渡的梯级大斜坡地带。属中亚热带湿润季风气候区，日照充足，气候温和，雨量丰沛，无霜期长。

石阡古称山国，历史悠久，建置较早。秦始皇帝二十八年（公元前219年），置夜郎县于今县境西部，属象郡。元至元年间，置石阡军民长官司，石阡之名始于此。明永乐十一年（1413年），置石阡府，分辖龙泉县及石阡、苗民、葛彰葛商3个长官司。清康熙二年（1663年）废葛彰葛商长官司，乾隆七年（1742年）三月，石阡府分设七里，即江外迎仙里、江内迎仙里、水东里、苗民里、在城里、苗半里、龙底里。直至清末，石阡府直隶于省，仍领龙泉县。民国十六

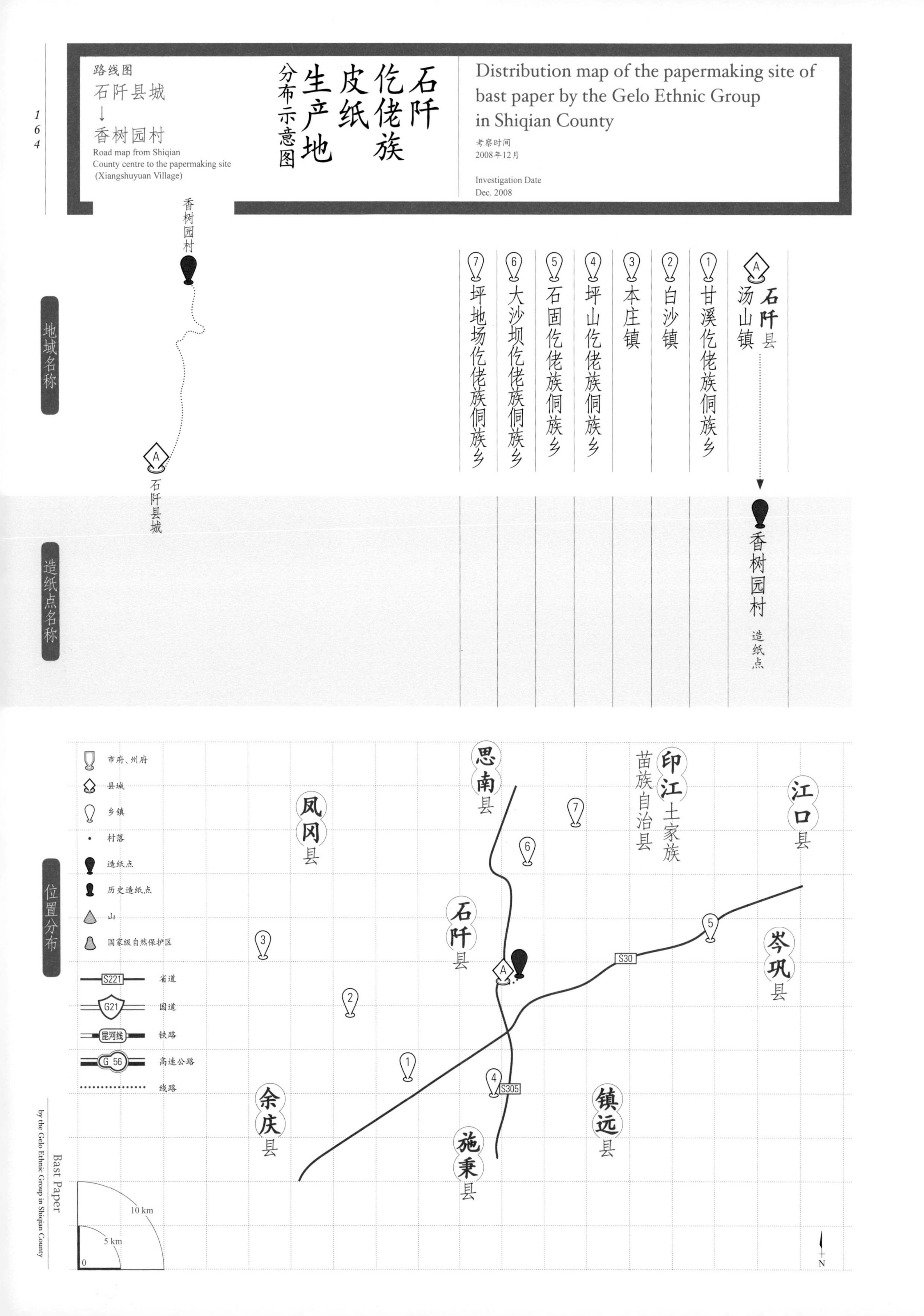

164
石阡仡佬族皮纸生产地分布示意图
Distribution map of the papermaking site of bast paper by the Gelo Ethnic Group in Shiqian County
考察时间
2008年12月
Investigation Date
Dec. 2008
路线图
石阡县城
↓
香树园村
Road map from Shiqian County centre to the papermaking site (Xiangshuyuan Village)
香树园村
A
石阡县城
地域名称
石阡县
A 汤山镇
1 甘溪仡佬族侗族乡
2 白沙镇
3 本庄镇
4 坪山仡佬族侗族乡
5 石固仡佬族侗族乡
6 大沙坝仡佬族侗族乡
7 坪地场仡佬族侗族乡
造纸点名称
香树园村
造纸点
位置分布
市府、州府
县城
乡镇
村落
造纸点
历史造纸点
山
国家级自然保护区
S221 省道
G21 国道
昆河线 铁路
G 56 高速公路
线路
思南县
印江土家族苗族自治县
江口县
凤冈县
石阡县
岑巩县
余庆县
镇远县
施秉县
S30
S305
10 km
5 km
0
N
Bast Paper
by the Gelo Ethnic Group in Shiqian County

年（1927年），石阡县直隶贵州省管理，民国三十二年（1943年），实施乡镇保甲式的新县制。1950年2月，成立石阡县人民政府，1970年起，归属原铜仁地区至今。

2012年的统计数据显示，石阡县总人口约46万，包括汉族、仡佬族、侗族、白族、傣族、壮族、苗族等29个民族，其中少数民族人口占总人口的68%，总面积2 173 km^2，辖7个镇、2个乡、9个民族乡，以仡佬族乡与侗族乡为主体。

尧上仡佬族文化村是石阡县境内的著名文化景点，位于坪山仡佬族侗族乡佛顶山村尧上村民组，该文化村坐落在佛顶山下、包溪河畔。该地距县城38 km，从县城出发，大约1小时车程。该村居住着67户仡佬族居民，所有的房屋都是两层的。仡佬族是一个能歌善舞的民族，每逢节日，都要在一个土坑里燃起篝火，男女老幼齐聚于篝火旁欢歌起舞。尧上仡佬族最重大的节日就是每年农历二月初一的敬雀节，每当这天，家家户户都要做糍粑，宰杀猪、牛、羊、马4种牲口祭祀神鹰和12只彩凤，祈祷家业兴旺和五谷丰登。除此之外，还要表演本民族特有的艺术节目——舞毛龙、傩戏、薅草锣鼓、木偶戏等，民族文化特色十分浓郁。

石阡县，1992年被列为贵州历史文化名城；2009年获“中国矿泉水之乡”誉称；2012年获“中国温泉之乡”“中国最佳休闲旅游目的地”“中国苔茶之乡”等誉称。石阡县最具代表性的风景名胜及特色文化与产品包括佛顶山自然保护区、凯旋河原始生态漂流区、泉都地热矿泉水、石阡苦丁茶、石阡茶灯表演等。

⊙1

⊙2

⊙1 尧上文化村祭祀活动 Sacrificial ceremony in Yaoshang Cultural Village

⊙2 石阡茶灯表演 Shiqian Tea Show during the Lantern Festival

三 石阡仡佬族皮纸的历史与传承

3 History and Inheritance of Bast Paper by the Gelo Ethnic Group in Shiqian County

石阡县盛产构皮、毛竹、梭草等造纸原料，手工纸的生产历史悠久。历史上该地民间产出的烧纸（即草纸）、舀茶罐的皮纸、青山棚的火纸久负盛名。据《（民国）铜仁府志》记载，石阡县的手工纸有三种：皮纸、竹纸和草纸。三种纸的质量和功用不尽相同：皮纸、竹纸质量较好，主要用于书写；草纸质地粗糙，主要用于制作百姓祭祀等活动中焚烧的冥镪（纸钱）。

最早对石阡手工纸进行记载的是清代田雯所撰的《黔书》。该书载："石阡纸，极光厚，可临帖。"《黔书》成书于康熙二十九年（1690年），可见当时石阡手工纸很早就有了。

田雯在1687~1691年任贵州巡抚期间，治黔有方，深受百姓崇敬。《黔书》二卷记载了他治理贵州的宝贵经验和对黔地山川地理、风土人情的介绍，为后世了解清初贵州石阡县的造纸状况留下了珍贵的资料。

1992年版《石阡县志》曾记载：同治八年（1869年），有造纸技工徐志文、杨志全由印江县到石阡的白沙田沟开设白皮纸作坊，所造纸销往黔东南一带。在其带动下，村里40多户人家有一半设槽买帘，办造纸作坊，小村造纸规模可观。至1949年，全县共有造纸作坊6个，从业者30多人，年产纸15 t左右。直至1985年，仍有合作、联营、个体造纸厂共15个，从业者71人，品种有烧纸、粗纸、白皮纸，年产量约15.7 t，年产值3万元。

随着现代纸业经济的发展，石阡手工造纸业受到了一定的冲击。以当地造纸户田家为例，调查时，据61岁的田儒忠介绍，他们家族100多年前从印江搬过来时就开始造纸，造纸技艺代代相传。但如今由于造纸的经济效益不理想，他的下一代传人已经放弃了这一祖业。

四
石阡仡佬族皮纸的生产工艺与技术分析

4
Papermaking Technique and Technical Analysis of Bast Paper by the Gelo Ethnic Group in Shiqian County

（一）
石阡仡佬族皮纸的生产原料与辅料

石阡县香树园村生产皮纸的主原料为当地生长5年以下的构树。生长时间过长的构树通常白皮纤维较少，出料率低，一般不用。

（二）
石阡仡佬族皮纸的生产工艺流程

根据调查组对田儒忠、田儒德等造纸人的调查，归纳香树园村皮纸的生产工艺流程如下：

壹	贰	叁	肆	伍	陆	柒	捌	玖	拾
买构皮	泡构皮	理构皮	浆构皮	堆构皮	蒸构皮	踩洗料	静置	泡料	洗料

拾壹	拾贰	拾叁	拾肆	拾伍	拾陆	拾柒	拾捌	拾玖	贰拾
二次静置	踩皮料	打散料	踩料	煮料	二次洗料	晾料	选料	舂料	切料

贰拾壹	贰拾贰	贰拾叁	贰拾肆	贰拾伍	贰拾陆	贰拾柒	贰拾捌	贰拾玖	叁拾
舂融	三次洗料	打槽	加滑根水	抄纸	压榨	晒纸	撕纸	理纸	捆纸

壹
买 构 皮
1

每年的农历三四月份，造纸户都会从附近村民处购买生长5年以下并已经晾干的构皮。因为构皮生长时间太长的话，其白色纤维含量少，出料率低，会影响纸的质量。

贰
泡 构 皮
2

农历七八月农闲时，将干构皮放在水塘里浸泡3天左右。

叁
理 构 皮
3

将泡好的构皮理成小捆，一捆构皮的质量约7.5 kg。

肆
浆 构 皮
4

先将石灰放入石灰池并加水搅拌使其相融，然后用皮钩钩起捆好的构皮，一捆捆地浆石灰水。

伍
堆 构 皮
5

将浆好的构皮堆放在空地上，并盖上塑料布，堆放四五天。以前造纸户用谷草、竹席覆盖，近10年来改用塑料布，主要目的是避免石灰被雨水淋湿，使构皮更好地发酵。

陆
蒸 构 皮
6

将构皮整齐地码入石甑里，用稻草将构皮盖严实后烧火蒸。待石甑上气后，熄火，用木槌将松烂下去的构皮捶实，然后烧火继续蒸。蒸1个月时间，每天需将构皮捶实一次，一次捶一个多小时。此外，每天还需要往石甑内加水一次，约100 L。一般一甑可蒸4 t构皮，需要4 t煤。

柒
踩 洗 料
7

将蒸好的构皮挑到河里，先用脚踩，然后用手搓洗。两个人两天可踩洗完一甑料。

捌
静 置
8

将踩洗好的皮料放在岩坎上，沥去皮料里的水分。

玖
泡 料
9

将沥干的皮料放到清水塘里继续浸泡10天左右。

拾
洗料
10

直接在塘里再次用手清洗已泡好的皮料，然后换掉塘里的水，再加入清水浸泡。泡料、洗料各反复3次，才可洗净皮料里的石灰，共需约一个月的时间。

拾壹
二次静置
11

将洗好的皮料放在岩垛上，再次沥水三四天。

拾贰
踩皮料
12

首先将一块木板（当地称作钢板）放在地上，然后于木板上放洗净的皮料，最后用脚将皮料踩干。四个人两天可踩好一甑皮料。

拾叁
打散料
13 ⊙1

将踩好的皮料拿回家，用木棍将其打散，并搭放在房子的梁架上备用。

⊙1

拾肆
踩料
14

从梁架上取下适量的皮料放在水池里略浸泡后，再放在木板上，用脚踩一两个小时，使皮料松软。

拾伍
煮料
15

将约50 kg皮料放入铁锅里，加入1.5 kg盐料（即纯碱），放满水煮10小时。

拾陆
二次洗料
16

将煮好的皮料拿到水池里清洗。一个人一天能洗完一锅料。

拾柒
晾料
17

将洗好的皮料放在平地上晾干。

拾捌
选料
18 ⊙2

待皮料晾干后，用手将皮料里的硬壳、杂质等拣掉。一个人两天可拣选完一锅料。

⊙2

⊙1 搭在梁架上的皮料 Papermaking materials hanging on the beams

⊙2 选料 Picking out the impurities

拾玖 舂料

19 ⊙3

选料完成后，取一团皮料并用脚碓舂。舂料时，一人翻料，一人踩碓，两人要专心、默契，否则翻料人的手会很容易被木碓舂伤，每翻动一次料，用脚碓舂一次，将皮料舂实舂平成一薄饼状，此时扁平的皮料被称为皮盘，随后再牵动皮盘，反复舂捶数次。将皮盘对折，再次舂捶，使皮盘更加平实，如此反复数次，直至将皮料舂融舂细。舂好后，将皮盘折叠成长条状，再舂几下，使其结实，随即放入筐中备用。一般一锅料要舂5天。

⊙3

贰拾 切料

20 ⊙4

⊙4

将皮料放在皮凳（当地把用于切皮料的板凳称为皮凳）上，并用钢丝绳套住，钢丝绳的两端下垂，交结后，取一根木棒搭在上面。切料时，人两脚分跨在凳的两旁，左脚踩着木棒，使钢丝绳勒紧皮料。两手持刀，从左上向右下斜切，速度大小保持不变，不能太快也不能太慢，将皮料切细。切得好的师傅，既能将皮料切细，又不伤刀口，一个人一天可切100 kg皮料。

贰拾壹 舂融

21 ⊙5

将切细的皮料放入碓窝，加少量水，两人配合，一人踩木碓，另一人站在碓窝边，不断地将料拨到碓窝中间，直至将皮料舂融。两个人一天可舂75~100 kg皮料。木碓不用时，为了保护好碓头和碓窝，在碓窝上横搭一块桥板，并用布盖住，然后再将碓头置于布上，这样做不仅可以避免碓头上的碓碗及碓窝损伤，还可以防止灰尘进入碓窝。

贰拾贰 三次洗料

22 ⊙6

将舂好的皮料放入簸箕里，加水，用手搓揉，将料中的杂质洗去。反复多次，将料洗净后，把水挤掉，捏成料团，放进盆或桶中备用。

⊙6

⊙5

⊙3 舂料 Beating the papermaking materials

⊙4 切料 Trimming the papermaking materials

⊙5 舂融 Beating the papermaking materials until they're melted

⊙6 洗料 Cleaning the papermaking materials

贰拾叁
打　槽

23　⊙7⊙8

将洗净的皮料按每槽10 kg的量放入捞纸槽里，同时加入1 kg已准备好的纸筋，然后加入水，用槽棍打槽，搅打一个多小时，将浆料纤维打散，均匀分布于水中。

贰拾肆
加滑根水

24　⊙9

将杉松树根放到碓窝里舂碎，然后再放入装有水的染缸（即纸药槽）里浸泡2天，滑汁即可浸出。

一般一次将5~6 kg杉松树根舂融，浸泡在加满水的染缸中，出滑，滑根水的黏性一般可保持10天左右，若滑根水黏性下降，再往槽里加杉松树根。平均一槽浆料需要0.5 kg杉松树根。加滑根水时，首先取一布袋，将布袋口用铁圈固定，再将布袋置于抄纸槽中，铁圈口搭在细竹竿与纸槽沿上，然后用盆从染缸盛出滑根水，倒入布袋口，静置片刻，去掉铁圈，拎起布袋，用手将剩余的滑根水挤入抄纸槽中，再取拱耙搅槽，使滑根水与皮料纤维混合均匀。

⊙7

⊙8

⊙9

贰拾伍
抄　纸

25　⊙10⊙11

抄纸前，先将帘架置于撑杆上，再放上纸帘，并使纸帘头与帘架挡板对齐，后用两个头子夹紧纸帘。抄纸时，手持帘架，把撑杆推开，由外往里挖水，再由里往外送水，如此反复两次，使之均匀。抄好纸后，把撑杆拉回来，并把帘架放在上面，将头子挪到两边。随后一手拎纸帘头，一手拿纸帘尾，转身，将纸帘头的挂钉对准钉柱，随后将湿纸盖于纸垛上，用手压下纸头，然后缓缓揭起纸帘。抄纸时，帘头处的纸面抄得稍厚些，以便于揭纸时可一张张地将纸揭开。一个人一天可抄500~600张纸。

⊙10

⊙11

⊙7 纸筋 Processed papermaking material ball

⊙8 打槽 Stirring the papermaking materials

⊙9 加滑根水 Adding in papermaking mucilage

⊙10/11 抄纸 Scooping and lifting the papermaking screen out of water and turning it upside down on the board

贰拾陆
压　榨
26　⊙12⊙13

⊙12

⊙13

抄好纸后，将纸垛静置一晚上。第二天早上，先在纸垛上均匀地铺一层稻草，再放上木板、方木、码子、榨杆，用榨绳将榨杆和滚筒连起来，将扳棍插进滚筒洞里，通过扳动滚筒，带紧榨绳，连动榨杆，缓慢将纸垛中的水分榨出，一般一次可榨500张纸，需要1.5~2小时。榨好后，松榨。

贰拾柒
晒　纸
27　⊙14

将榨好的纸垛搬到纸架上，纸头朝上。左手执纸夹将纸垛四周划松，右手执棕刷，撕开纸垛右上角，左手跟上，沿着上边，两手同时从右上往左下斜撕，待纸头全部撕开后，将纸从上往下撕离纸垛。然后用棕刷将撕下的纸从中间往四周刷在墙或木板上。若有皱痕，用棕刷慢慢将其刷平。晒纸时，每5张为一贴，贴于墙上，每2张之间错开约0.5 cm。晴天时半小时即可晒干，阴天时则需要一两天，甚至更长时间。

⊙14

⊙12/13 压榨 Pressing the paper

⊙14 揭纸 Peeling the paper down

贰拾捌

撕　　纸

28　⊙15⊙16

将晒干的纸从墙上揭下来，撕贴时，一人牵纸头，另一人揭纸，两人配合将一贴纸一张张撕开。此时不能太用劲，否则容易将纸撕烂。一个人也可将纸从墙上一张张撕下来。

⊙15

⊙16

贰拾玖

理　　纸

29　⊙17

每100张纸为一提，双手抓住一提纸的纸头，上下抖纸。发现纸张不齐时，一只手执纸，另一只手将纸理齐。然后将一提纸贴靠在墙上，一条腿站立，另一条腿就势将纸抵于墙上，将纸折成3折。

叁拾

捆　　纸

30

每10提为一捆，先用塑料带（以前为用纸搓成的纸条）捆好，再将其压实。

⊙17

这就完成了整个造纸过程。

调查组在调查中了解到，香树园村造的白皮纸，有很多掺入了新闻纸的纸筋，但无论是否掺入纸筋，白皮纸的价格都是一样的。两者的区别只是掺入纸筋的纸会更白更厚。香树园村从20世纪60年代开始在造纸时掺入纸筋，所加的纸筋是石阡县印刷厂的废旧白纸，价格为1.6元/kg。制作纸筋时，一般先将5~6 kg废旧白纸单独放在铁锅里煮三四个小时，再将其放在碓窝里舂一天，直至舂融，否则造出的白皮纸里就会起白点。

⊙15/16 撕纸 Peeling the paper down

⊙17 理纸 Sorting the paper

(三)
石阡仡佬族皮纸生产使用的主要工具设备

壹 帘架 1

木质，有不同的尺寸，实测帘架长约62 cm，宽约55 cm。

⊙1

贰 纸帘 2

用细苦竹丝编制而成，有不同的尺寸，实测纸帘长约58 cm，宽约48 cm。

⊙2

⊙1
帘架与头子
Frame and sticks for supporting the papermaking screen

⊙2
纸帘
Papermaking screen

叁
头　子
3

木质，两根，抄纸时置于纸帘上，手持帘架并握紧头子，从而夹紧纸帘。纸张的长度取决于两头子之间的距离。头子与相应帘架组合使用。

肆
碓
4

碓杆、碓头用橡子树等硬树制成，实测尺寸为：碓杆长约174 cm；碓头直径约19 cm、高约49 cm。碓头下部有一铁质有齿碓碗。碓盘用石头制成，形状不规则。

伍
木　榨
5

靠墙的两根大柱子称为立人，高约180 cm；在两立人之间有一横担；与地面平行的两根木桩称为顺木，长约280 cm；榨杆长约275 cm。

陆
纸　架
6

木质，中间有一搁板，用于搁放纸垛；纸架中间有三根与搁板垂直的木杆称为站人。纸架上部还有两个纸夹，若纸垛的纸少于10张，则其在搁板上就立不稳，需要用纸夹将其夹紧在纸架上部。此外，纸架右上方还有个挂钉，用于挂刷子。

（四）

石阡仡佬族皮纸的性能分析

所测石阡香树园村皮纸的相关性能参数见表9.1。

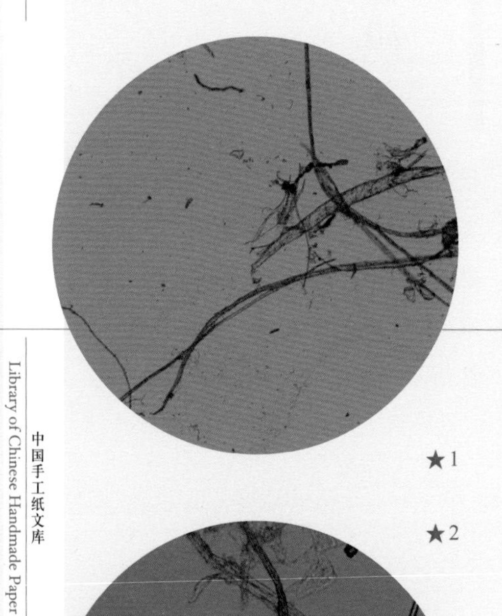

★1 香树园村皮纸纤维形态图(10×)
Fibers of bast paper in Xiangshuyuan Village (10× objective)

★2 香树园村皮纸纤维形态图(20×)
Fibers of bast paper in Xiangshuyuan Village (20× objective)

表9.1 香树园村皮纸的相关性能参数
Table 9.1 Performance parameters of bast paper in Xiangshuyuan Village

指标		单位	最大值	最小值	平均值
厚度		mm	0.094	0.046	0.074
定量		g/m^2	—	—	26.8
紧度		g/cm^3	—	—	0.362
抗张力	纵向	N	7.0	4.8	6.0
	横向	N	2.6	1.7	2.4
抗张强度		kN/m	—	—	0.280
白度		%	50.9	49.3	49.6
纤维长度		mm	5.97	0.60	1.83
纤维宽度		μm	22.2	0.4	5.5

由表9.1可知，所测香树园村皮纸最厚约是最薄的2.04倍，纸张厚薄差异较大，经计算，其相对标准偏差为0.26%。皮纸的平均定量为26.8 g/m^2。所测皮纸紧度为0.362 g/cm^3。

经计算，其抗张强度为0.280 kN/m，抗张强度值较小。

所测香树园村皮纸白度平均值为49.6%，白度较高，白度最大值约是最小值的1.03倍，相对标准偏差为0.20%，白度差异相对较小。

所测香树园村皮纸纤维长度：最长5.97 mm，最短0.60 mm，平均1.83 mm；纤维宽度：最宽22.2 μm，最窄0.4 μm，平均5.5 μm。所测皮纸在10倍、20倍物镜下观测的纤维形态分别见图★1、图★2。

五 石阡仡佬族皮纸的用途与销售情况

5 Uses and Sales of Bast Paper by the Gelo Ethnic Group in Shiqian County

(一) 石阡香树园村仡佬族皮纸的用途

1. 古时地方文书文献的抄录用纸

在没有机制纸的时代，当地的白皮纸是地方文书文献的重要载体。而随着机制纸的不断普及，白皮纸记录地方文书文献的功能逐渐减弱。

2. 日常生活书写用纸

当地百姓用白皮纸来记录一些资料，尤其是红白喜事时用的“礼尚往来”簿，这一则方便，不用外出去买纸；二则白皮纸适于写毛笔字，能长期保存。调查时仍有小孩用白皮纸来练字。

3. 裱糊用纸

一般当地宗教仪式、丧礼活动、乡村集庆时普遍使用白皮纸，如糊扎纸人、纸物、纸房，或用于剪成马、象、人、虎等图案。

4. 捆钞用纸

香树园捆钞纸为四夹纸，即晒纸时从纸垛上将4张纸一起揭下，贴于墙面晒干，白皮纸厚且具有很强的坚韧性，可谓手撕不破、水打不烂，近些年石阡当地一些银行专门到香树园村来订制捆钞纸。

⊙1

(二) 石阡香树园村仡佬族皮纸的销售状况

调查中了解的销售渠道信息，主要是造纸户拿到石阡县城批发给附近乡镇如龙塘、巨峰、大河坝等地的经营户，销售价格按照尺寸不同有0.2元/张、0.4元/张。被访谈的田儒德和熊德云夫妇，每年农闲时造纸4个月，可造30捆，按0.2元/张算，销售额为6 000元。若全年生产，则销售额为18 000元。由于造纸的经济效益不高且特别辛苦，目前在香树园村只有中老年人造纸。出生于20世纪70年代的田茂兵和田茂强兄弟，已弃造纸改去外地打工了。

⊙1
用石阡皮纸制成的旧日『礼尚往来』簿
Family gift money book with "Courtesy Demands Reciprocity" written on the cover, made of bast paper in Shiqian County

六
石阡仡佬族皮纸的相关民俗与文化事象

6
Folk Customs and Culture of Bast Paper by the Gelo Ethnic Group in Shiqian County

（一）丧葬习俗

仡佬族人过世后的当天晚上，丧家要请人唱孝歌，俗称“闹夜歌”，内容不固定，一般喜唱《三国演义》《杨家将》等，也可即兴编唱。歌前，需先烧香化纸。出柩前夜，要击鼓而歌，围尸跳跃，称“打绕棺”。其仪式由阴阳先生主持，棺上摆设东南西北中岳和十殿阎君牌位，下设香案、净茶礼品、明灯、文书。据访谈时村里老人回忆，以前写文书用的就是当地产的白皮纸。

出殡时，仡佬族无丢买路钱之俗。因仡佬族是本地先民，流传有“此山是我开，此树是我栽，我们死了人，还向谁买路”的民谚，代代相传。

（二）祭祀习俗

仡佬族在祭祀时，多处会用到手工纸。

从除夕到正月十五，仡佬族人都要到自己家的祖坟上去祭祀祖先。

祭秧苗土地的目的主要是祈求庄稼少遭病虫害，获丰收。一般在农历五月上旬由各家各户举行。备3束长线、一些纸钱、3~5个粽粑和少许酒菜在自家秧田里祭祀。烧化纸钱后，用竹竿将粽粑吊上插入田中。民间流传有“五月里来秧开排，秧苗土地要钱来，三束长线交与你，保佑禾苗长起来”的歌谣。

七
石阡仡佬族皮纸的保护现状与发展思考

7
Preservation and Development of Bast Paper by the Gelo Ethnic Group in Shiqian County

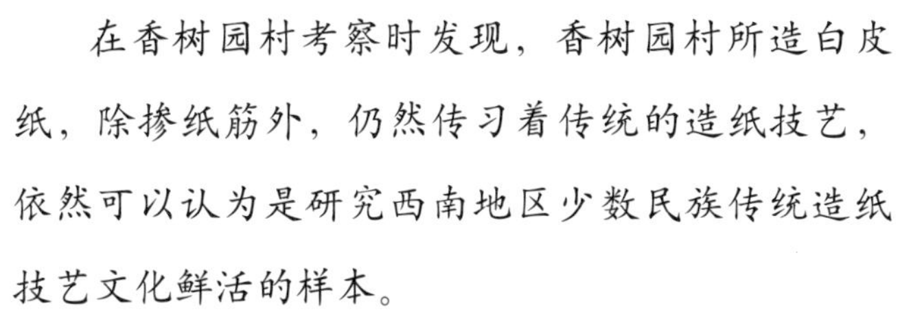

在香树园村考察时发现，香树园村所造白皮纸，除掺纸筋外，仍然传习着传统的造纸技艺，依然可以认为是研究西南地区少数民族传统造纸技艺文化鲜活的样本。

香树园村主要造纸工序多达30道，工艺十分复杂，多次重复蒸煮、踩料、舂料、洗料、选料等，充分体现了传统手工造纸工艺精益求精的特点。此外，香树园村白皮纸造纸工具名称极具地方特色，如踩皮料用的木板当地称为钢板等。此外，清明节上坟挂青、中元节封袱子等当地的祭祀民俗，也成为传统造纸文化的一部分，值得深入挖掘、整理和研究。

香树园村白皮纸多为年老者在农闲时操业制作，不计工时成本，以贴补家用。从目前看销路没有问题，但由于经济效益低、工艺复杂，鲜有年轻人加入，香树园村白皮纸的传承正面临着后继无人、逐渐消亡的态势。可以说香树园村白皮纸的传承与保护面临着较大压力。

针对目前的现状，调查组建议地方政府将香树园村仡佬族白皮纸列入非物质文化遗产保护规划，以引导资助的方式，促进造纸户更积极地对香树园村仡佬族手工造纸技艺进行传承与保护；对传统的不掺纸筋的香树园村白皮纸传统工艺进行深入挖掘、整理和宣传，发展其进入当代消费型的书画用纸、装饰用纸等中高端市场，使香树园村仡佬族白皮纸能推陈出新，从生产性传承的角度获得更大的发展空间。

⊙1

⊙1
访谈结束后的告别
Saying goodbye to local papermakers

⊙1

⊙1
造纸户的院墙外
Outside of a papermaker's courtyard

石阡仡佬族皮纸

Bast Paper by the Gelo Ethnic Group in Shiqian County

香树园村皮纸透光摄影图

A photo of bast paper in Xiangshuyuan Village seen through the light

第二节

石阡仡佬族竹纸

贵州省
Guizhou Province

铜仁市
Tongren City

石阡县
Shiqian County

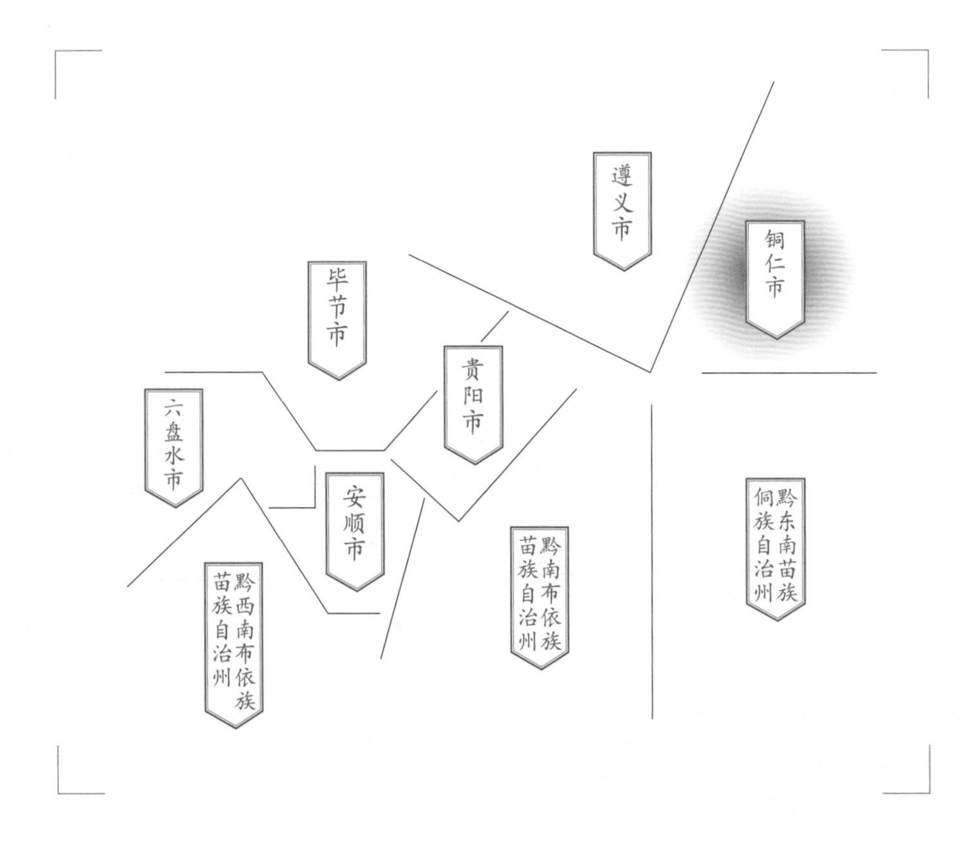

调查对象

坪地场仡佬族侗族乡下林坝行政村
大沙坝仡佬族侗族乡关刀土行政村
仡佬族竹纸

Section 2
Bamboo Paper
by the Gelo Ethnic Group
in Shiqian County

Subject

Bamboo Paper by the Gelo Ethnic Group in Guandaotu Administrative Village of Dashaba Gelo and Dong Town, Xialinba Administrative Village of Pingdichang Gelo and Dong Town

一

石阡仡佬族竹纸的基础信息及分布

1 Basic Information and Distribution of Bamboo Paper by the Gelo Ethnic Group in Shiqian County

在贵州古史中钩沉的学者发现了有关石阡竹纸的记载，即清代田雯所撰的《黔书》中赞云："石阡纸，极光厚，可临帖。"《黔书》成书于康熙二十九年（1690年），作者田雯在1687~1691年任贵州巡抚，治黔有方。《黔书》记录了他治理贵州的经验和对黔地山川地理、风土人情的谙熟，为后世了解贵州石阡县清初造纸情况亦留下了珍贵的一笔[1]。从简约的记述文字来看，当时石阡造纸工艺已经处于较高水平。用于书法临帖的纸，对毛笔润墨书写时的固色吸水等方面要求较高。例如，宋代人推崇用竹纸书画，因为竹纸的优点在于：滑、发墨色、宜笔锋；卷舒虽久，墨终不渝等[2]。此外，"光厚"指的是手工纸成纸的质量，即滑泽有光、厚重结实。古人讲究纸质匀薄，石阡纸偏厚，不算极品，但是光泽度好，显示出石阡县造纸工匠们精巧的用心。研究手工纸多年的学者刘仁庆指出，如要做到纸面光润无瑕，没有绒状纤维起毛的现象，那么踏竹麻要认真，剥竹麻要配合，焙纸的火墙要及时清理以便保持油面光滑[3]。田雯的这番记载透露出中国尊崇文字书写的历史岁月中石阡纸艺曾经的繁荣。

据民国《铜仁府志》记载，石阡县的手工纸有三种：以构树皮制成的纸叫皮纸，以竹纤维制成的纸叫竹纸，以竹子、杂草混合制成的纸叫草纸。三种纸的质量和功用不尽相同，皮纸、竹纸都适合书写，属于质量较好的文化用纸一类，草纸质地粗糙，主要为百姓祭祀等活动中焚烧冥镪（纸钱）之用。然而，随着手工纸技艺越来越强地受到现代机制纸的挑战与冲击，石阡纸早已失去往日的市场和知名度，由于没有

⊙1

⊙1 关刀土村的纸坊在榨纸
Pressing the paper in a papermaking mill in Guandaotu Village

[1] 朱兴泉.田雯与《黔书》[J].贵州档案,2000(4):37-38.

[2] 刘仁庆.中国古纸谱[M].北京：知识产权出版社,2009:111.

[3] 刘仁庆.造纸趣话妙读[M].北京：中国轻工业出版社,2008:156.

石阡仡佬族竹纸生产地分布示意图

Distribution map of the papermaking sites of bamboo paper by the Gelo Ethnic Group in Shiqian County

考察时间
2008年12月

Investigation Date
Dec. 2008

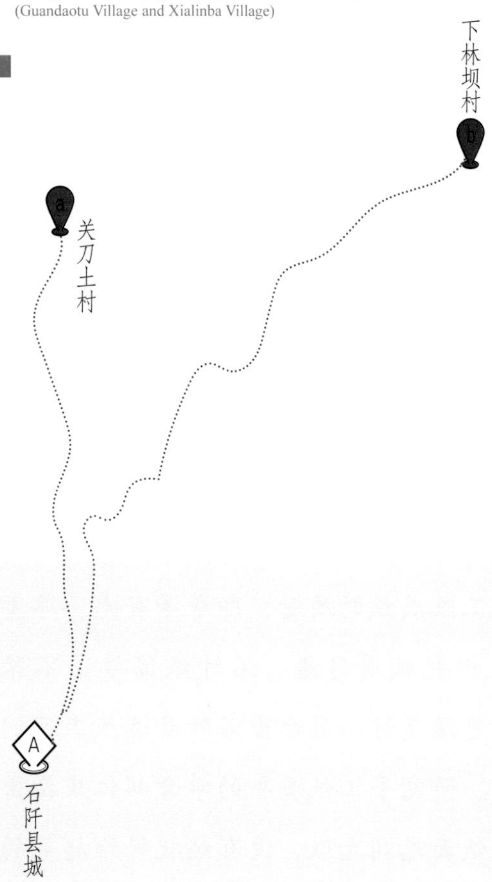

路线图
石阡县城
↓
关刀土村、下林坝村
Road map from Shiqian County centre to the papermaking sites (Guandaotu Village and Xialinba Village)

地域名称

A 石阡县 汤山镇
① 甘溪仡佬族侗族乡
② 白沙镇
③ 本庄镇
④ 坪山仡佬族侗族乡
⑤ 石固仡佬族侗族乡
⑥ 大沙坝仡佬族侗族乡 → a 关刀土村 造纸点
⑦ 坪地场仡佬族侗族乡 → b 下林坝村 造纸点

造纸点名称

位置分布

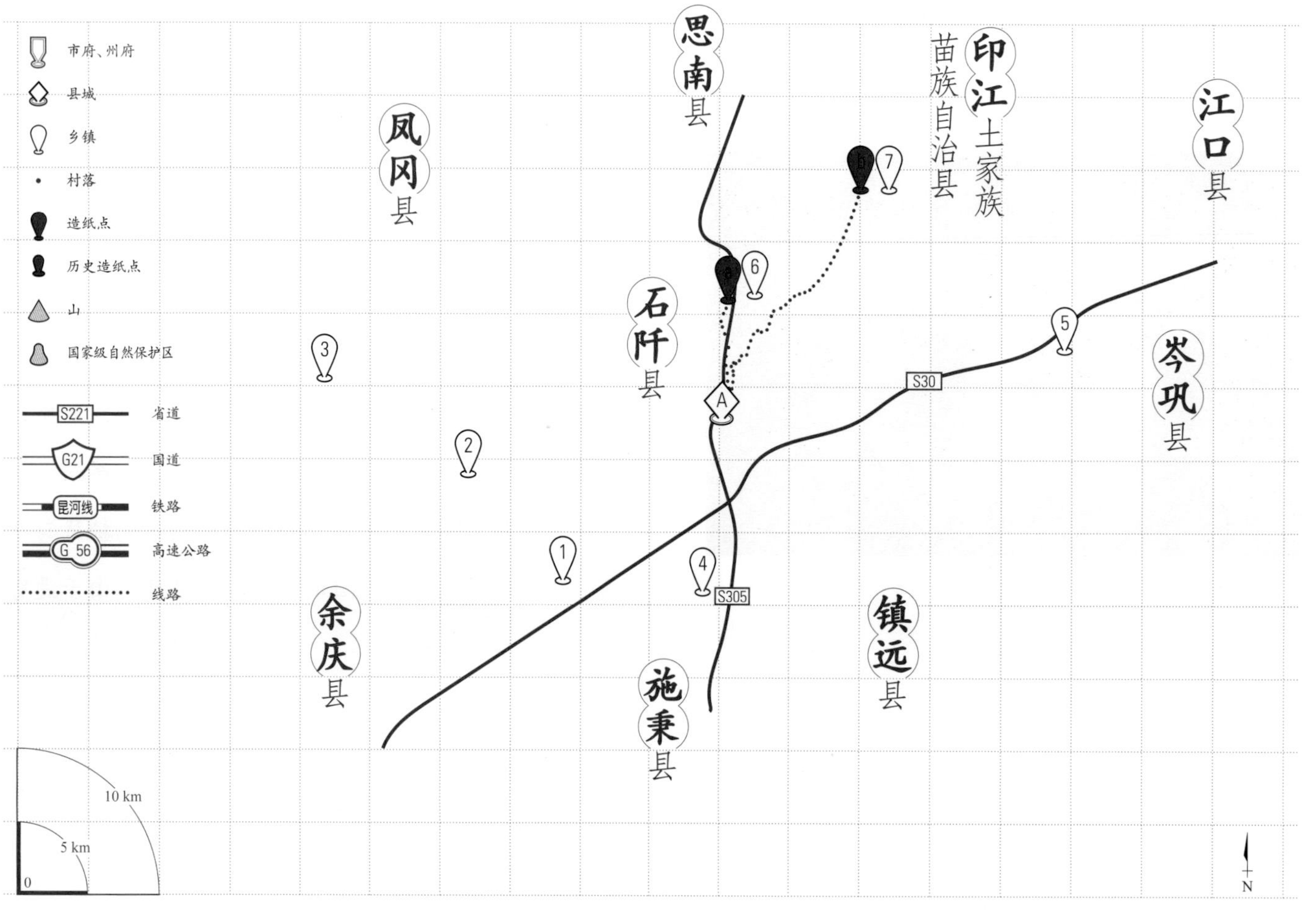

被纳入各级非物质文化遗产名录而失去获得政府保护的机会，古老的造纸工艺只能依靠自身的力量在简陋的造纸作坊里悄然维系。

实地田野调查中获得的信息显示，石阡手工竹纸制作的活态分布在坪地场仡佬族侗族乡下林坝村、大沙坝仡佬族侗族乡关刀土村等地。

二 石阡仡佬族竹纸生产的人文地理环境

2 The Cultural and Geographic Environment of Bamboo Paper by the Gelo Ethnic Group in Shiqian County

石阡县位于贵州省东北部、铜仁市西南部，地处东经107° 44′ 55″~108° 33′ 47″、北纬27° 17′ 5″~27° 42′ 50″，东临江口、岑巩县，南接镇远、施秉县，西连凤冈、余庆县，北靠印江、思南县，总面积2 173 km^2。

石阡古有“山国”之称，境内山峦起伏，沟谷纵横，岩溶地貌明显，最高海拔1 869.3 m，最低海拔388 m。属中亚热带季风湿润气候区，冬无严寒、夏无酷暑，平均气温16.8 ℃，雨量充沛。溪河纵横交错，大小河流117条，水资源丰富。地热资源遍布全县，出露密度居全省之首，达19处之多，平均水温34.7 ℃，富含锌、硒、氡、锶、偏硅酸等对人体有益的微量元素，符合国家饮用矿泉水标准，可饮可浴，2012年获得“中国温泉之乡”美誉。

⊙1

⊙2

石阡历史建置较早。秦始皇帝二十八年（公元前219年），置夜郎县于今县境西部。元至元年间，置石阡军民长官司于今治所。明永乐十一年（1413年），置石阡府。1911年，初沿袭清制，后调整行政区划，石阡建新县制并延续至今。

石阡旅游资源丰富，历史文化遗产、文物名胜众多。这里有国家级文物保护单位万寿宫等明代古建筑群，贵州省级文物保护单位府文庙、太虚洞和“红二六军团司令部”旧址。坪地场乡是石阡县所辖的18个乡镇之一，也是9个仡佬族侗族乡之一，面积31.69 km^2，东北与思南、印江两县相邻，有汉、仡佬、侗等族居民。

三 石阡仡佬族竹纸的历史与传承

3 History and Inheritance of Bamboo Paper by the Gelo Ethnic Group in Shiqian County

以竹造纸，曾是石阡县许多山区农民的衣食之源和谋生之本。1992年版的《石阡县志》记载了石阡县的造纸历史：“石阡盛产构皮、毛竹、梭草等富含纤维的植物，清同治八年（1869年），徐志文、杨志全由印江来石阡白沙田沟开创白皮纸作坊，纸销黔东南一带，水田沟村40多户人家就有一半设槽买帘，办造纸作坊。1949年全县有6处造纸，从业30多户，年产纸15 t左右。1952年产量达20 t。1953年农村各处造纸，组成造纸生产小组，商业部门与之订立生产合同。1954年土纸生产达33 t。1956年土纸生产小组和个体造纸手工业者组成的造纸生产合作社有三个，土纸产量达68 t。青阳造纸社1958年转为合作工厂，1959年下放为大公社工业。1958年12月，龙塘、白沙造纸厂生产合作社合并为北塔新建国营造纸

⊙1 石阡山区冬景
Winter view of the mountain area in Shiqian County

⊙2 下林坝村的侗族民居
Dong ethnic residence in Xialinba Village

厂，1959年4月建成投产，有职工82人。1962年转为集体性质，1964年12月下马。各公社和生产队共建成造纸厂（组）30多个，总产量达631.3 t，产品远销至湖南省、广西壮族自治区等地。1981年社队办厂13个，从业49人，产量达28.2 t。1985年有合作、联营、个体造纸厂15个，从业71人，造有‘烧纸、粗纸、白皮纸’，年产约15.71 t。”

据坪地场乡下林坝村仍在从事手工造纸的杨功纯老人介绍，他的手艺是从父亲杨宗斌那里继承来的。杨宗斌从20世纪50年代开始造纸，那时正是集体公社时期，国家提倡“自力更生，艰苦奋斗”的精神，鼓励人民自己动手，实现丰衣足食。当时村里人重拾数百年来老祖宗留下的造纸手艺，家家造纸，干得热火朝天。后来，造纸手艺在20世纪80年代的工商经济大潮面前逐渐萎缩。讲究“慢工出细活”的土纸生产无法在短期内带来较大的经济效益，因此从事的人越来越少，现在只剩寥寥几户人家了。

⊙3

⊙4

⊙3
坪地场山中的小溪
Stream in the mountain of Pingdichang Town

⊙4
龙塘坑造纸作坊的山路上十几年前曾是繁华市场
Country road to the Longtangkeng Papermaking Mill (used to be a prosperous market)

四 石阡仡佬族竹纸的生产工艺与技术分析

4 Papermaking Technique and Technical Analysis of Bamboo Paper by the Gelo Ethnic Group in Shiqian County

石阡县坪地场乡下林坝村的造纸作坊就建在被当地人叫作"龙塘坑"的地方，龙塘坑位于思南、印江两县交界处，依山傍水。造纸的原料和纸药用材来自附近的群山，造纸用水来自环绕群山的河水，整个造纸过程都在山中的作坊里完成，作坊里的种种设备也由山中的木、竹、石制作而成。从下林坝村头到造纸作坊，翻山越岭要步行40分钟才能到达。交通上的不便，是古老的造纸技艺无法发扬光大的因素之一。龙塘坑造纸作坊为下林坝村集体共有，村民们形成了一种约定俗成的规则：村里谁家有料，想造纸都到这个集体作坊来，然后造纸户们分工合作，彼此帮助生产，一般是五家合成一个互助组，每家出一个人（纸工）。

(一) 石阡仡佬族竹纸的生产原料与辅料

下林坝村用来造纸的竹子主要有阳山竹、金竹、斑竹、苦竹，这是凭借多年的手工造纸经验选取的竹类（见表9.2）。通常，生长1~3年的竹子正好适合造纸。不同竹子的生长期不同，贵州侗族有句谚语："金竹三月生嫩笋，苦竹六月不知春。"当地造纸户们每年在农历四、五、十一、十二月农闲时砍伐竹子。据调查组了解到的信息显示，石阡仡佬族竹纸的生产原料来源有两种：一种是自家人上山砍伐一些竹子作原料，但这远远不够造纸所需；另一种是依赖于从附近农户手中收购大部分原料，平均价格为0.16~0.2元/kg。

就造纸原料而言，纤维素含量越高，制浆获得率越高，造纸原料的纤维素含量应在40%以上

⊙1 下林坝村造纸作坊的水源 Water source of the local papermaking mills in Xialinba Village

表9.2　石阡县部分竹株的基本特征*
Table 9.2　Basic information of some varieties of bamboo in Shiqian County

竹种		苦竹	金竹
林分特征	多年生竹眉径（cm）		2.125
	1年生竹眉径（cm）	1.610	2.104
单竹特征	秆高（cm）	508.75	502.15
	地径（cm）	2.645	2.295
	平均节长（cm）	20.350	14.899
	平均节粗（cm）	1.384	1.329
单竹生物量	秆重（kg）	1.045	1.200
	全重（kg）	1.800	2.195

表9.3　石阡县部分竹种造纸性能指标*
Table 9.3　Papermaking performance parameters of some varieties of bamboo in Shiqian County

竹种	苦竹	金竹
1% NaOH抽出物（%）	19.03	19.91
纤维素（%）	42.27	41.77
SiO_2（%）	1.04	0.47
多戊糖（%）	16.86	15.24
木素（%）	26.67	29.93
灰分（%）	2.35	1.39

才比较经济；木素含量越少，制浆越容易，消耗的制浆辅料如石灰、烧碱等才会越少；多戊糖含量高，打浆较容易，纤维结合度好，成纸的透明性强；灰分和SiO_2含量高，碱回收困难，纸药消耗多，污染程度大；1% NaOH抽出物多，碱耗量大，经济效益低，污染程度大。以苦竹、金竹为例，从表9.3可以看出，两种竹子的纤维素都在40%之上，木素都在30%以下，多戊糖成分都大于15%，灰分和SiO_2含量不高，1% NaOH抽出物不多，与贵州其他的竹类相比，是最适宜的、经济的造纸原料。

下林坝村竹纸生产使用的辅料包括石灰、纯碱、“滑”和水。

石灰、纯碱是传统造纸工艺中必不可少的辅料。竹子不像麻、树皮、藤等造纸原料容易制浆，且竹茎结构紧密，纤维坚硬，打浆帚化难度大，因而需要选用嫩竹并经过沤泡、发酵后才能制浆抄纸。在嫩竹的沤泡发酵过程中，石灰、纯碱发挥着重要的作用——经过石灰和纯碱溶液高温处理过的竹原料，粘连在纤维之间的木素、果胶被除掉，纤维素纤维分散开来而成为纸浆[4]。下林坝造纸工序中，第一次蒸料前要过石灰水，添加纸药时也要加入一定比例的石灰，第二次蒸料时要加入纯碱。

当地造纸户称捞纸时添加的纸药为“滑”。下林坝村纸工们用从山上采集来的香叶、木姜子叶加水熬制成滑水。滑水不但制作成本低廉，而且其所具有的多种效能也无可替代。滑水能使浆料均匀、分散地悬浮于水中，滤水性能好，因此浆料上帘品质高，纸页容易分揭，提高了成纸率[5]。

水是所有造纸都离不开的重要辅料之一。过石灰水、煮料、洗料、加滑、捞纸等主要造纸工序中都需要加水处理。石阡水源丰富，水质优良，利于造纸。调查组实测下林坝村造纸作坊附近的河水含微量元素0.345‰，沟水含微量元素0.425‰，pH均为6.5。

* 表9.2、表9.3均根据张喜的研究整理而成，参见：张喜.贵州主要竹种的纤维及造纸性能的分析研究[J].竹子研究汇刊，1995(4)：14-30.

[4] 刘仁庆.中国古纸谱[M].北京：知识产权出版社，2009：110.

[5] 刘仁庆.中国古纸谱[M].北京：知识产权出版社，2009：111.

(二)

石阡仡佬族竹纸的生产工艺流程

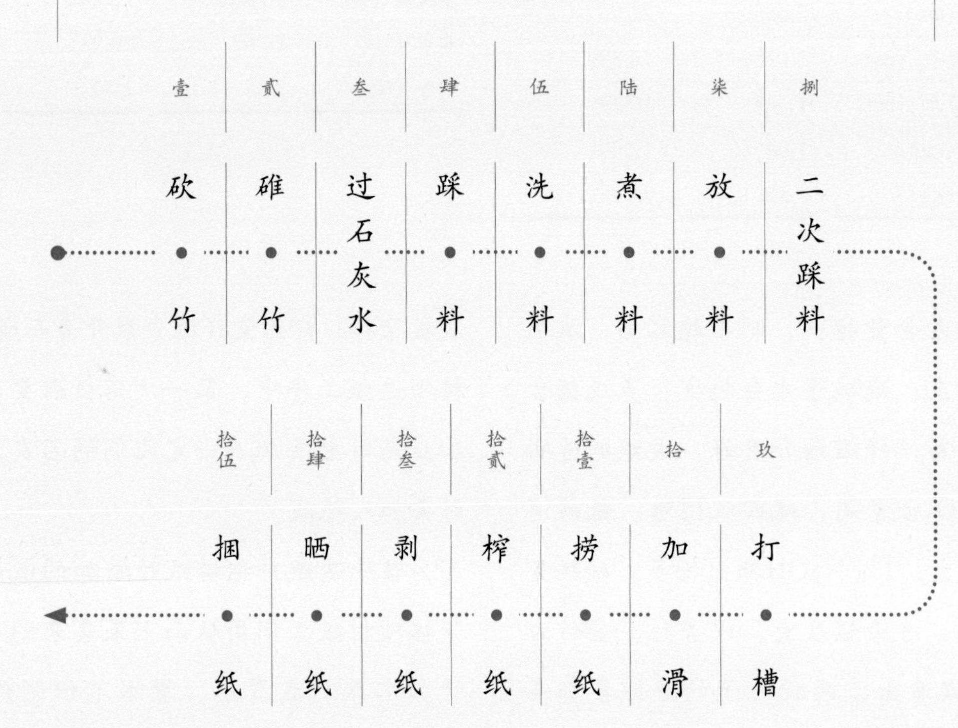

壹 砍竹 1

每年农历四、五、十一、十二月农闲时砍伐1~3年生的竹子。竹子被砍下后，需切成段，每段长2 m，然后捆成捆备用。此时的竹子原料当地称为“竹麻”。

贰 碓竹 2

用碓将竹麻粉碎。一个人一天可粉碎200 kg竹麻。

叁 过石灰水 3

将粉碎的竹麻一把把过石灰水，500 kg竹麻需要150 kg石灰，两三个人一天可过2 500 kg竹麻。

⊙1

肆 踩料

4 ⊙2

一人用捞钩将料子捞出来，再由四人穿上钉鞋用力踩料，然后由两人用锄头将竹料打烂。四个人一天可踩料4 000 kg。

⊙2

伍 洗料

5 ⊙3

将料子放入河水中，由两人在河里用手洗料，将料子中的石灰渣等杂质洗净。

⊙3

陆 煮料

6 ⊙4

料子洗净后放入窑子中，并往其中加入100 kg纯碱，封窑并用大火煮半个月。煮料时，窑子要加满水，中间不需再加水，共需用煤5 000 kg。

⊙4

柒 放料

7

将料子捞出来，放在料池中半个月，一料池可放500 kg料。

⊙1 盛石灰水的石槽 Stone trough holding limewater

⊙2 踩料用的槽 Trough for stamping the papermaking materials

⊙3 洗料的河水 River for cleaning the papermaking materials

⊙4 依山而建的窑子 Papermaking kiln built in the mountain area

捌 二次踩料

8 ⊙5⊙6

取出料子，放到踩槽中，由一个人赤着脚反复踩。踩槽底部并非水平，而有呈约45°的坡度，踩料的人在用力时需要拄着一根棍子以保持平衡。一次可放25 kg料子，需踩20分钟。

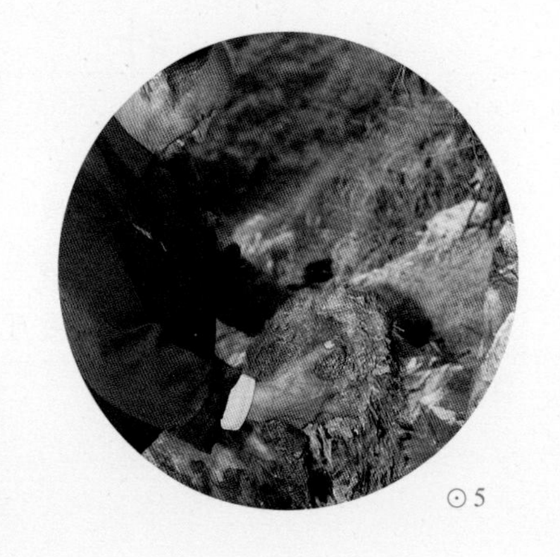

⊙5

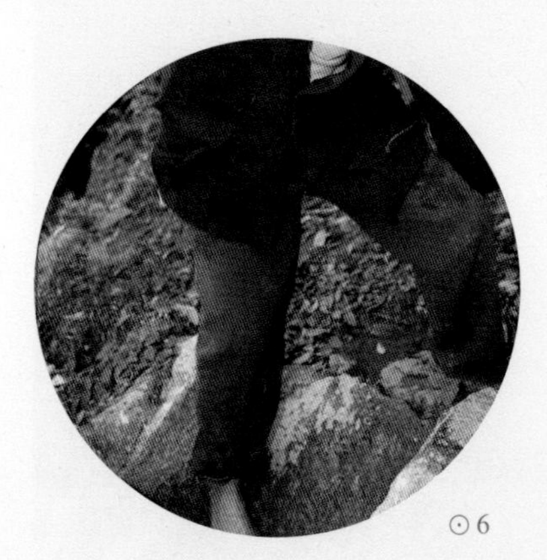

⊙6

玖 打槽

9 ⊙7⊙8

用簸箕将踩好的料子移至槽子里，然后用拱耙将料子拱散，再用捞筋棍将料子搅匀，并随时捞出挂在捞筋棍上的粗纤维。

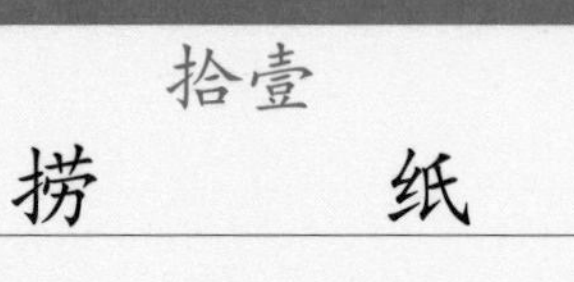

⊙7

⊙8

拾 加滑

10

将10 kg木姜子叶加2.5 kg石灰放入50 L（一挑）水中，煮一天后加250 L水稀释，即成“滑”。一槽要加2~3瓢“滑”。

拾壹 捞纸

11 ⊙9~⊙12

捞纸时，槽子左侧与内侧靠近边沿处搭着十字交叉地绑在一起的两根棍子，棍子正好搭在水面上，用于放置纸帘，掌握捞纸时摆动动作的平衡等，方便操作。把纸帘固定在帘架上，开始捞纸。分两次捞成，先由外向里捞，再由右向左捞。把捞好的帘架平放在十字交叉的木

⊙9

⊙5 取出料子 Lifting the papermaking materials

⊙6 在踩槽中踩料 Stamping the papermaking materials in the trough

⊙7 用拱耙打散料子 Stirring the paper pulp with a rake

⊙8 用捞筋棍搅匀 Stirring the paper pulp with a lifting stick

⊙9 捞纸 Scooping and lifting the papermaking screen out of water

棍上，卸下纸帘，并把纸帘翻扣在竹席上，用手均匀地抹平纸帘，再轻轻揭帘，湿纸便平整地扣在竹席上了。一槽可抄1 300帘（约4 000张），一年大约造纸300天，一天20分钟。

⊙10

⊙11

拾贰 榨纸

12

⊙13

当捞出的湿纸达到竹席旁立着的两根木桩的高度时，压上木板，再用木棍做杠加力压榨湿纸。一般情况下，10分钟左右即可压干。

⊙12

拾叁 剥纸

13

将榨干的纸垛运回家后，一张张地将纸剥开。

⊙13

拾肆 晒纸

14

剥开的纸以10张为一刀，一刀刀地挑晒在竹竿上，一般一天一夜即可晒干。冬天需加温，即将纸挑晒在室内，用柴火熏干。

拾伍 捆纸

15

把晒干的纸一刀刀地叠起来，每20刀为一捆，用结实的棕绳捆扎。纸的尺寸为16.5 cm × 36 cm。

⊙10 取下捞好湿纸的纸帘 Removing the papermaking screen from its supporting frame

⊙11 把纸帘上的湿纸翻扣在竹席上 Turning the papermaking screen upside down on the bamboo mat

⊙12 揭帘 Removing the papermaking screen

⊙13 压榨用的工具 Pressing device

（三）

石阡仡佬族竹纸生产使用的主要工具设备

手工造纸一直是石阡山区农副产业的一部分，所使用的生产工具及设备比较粗朴、简单，大多借助已有的农业用具。比如砍竹用的砍刀，碓竹用的石碓，晒纸用的竹竿，挑水用的水桶，装料用的簸箕，捆纸用的棕绳等。使用的材料也都是就地取材，多为木、石、竹、草、铁等。在下林坝村，调查组看到的传统造纸工具及设备主要有：

壹 窑子 1

用于煮料，类似于烧火的灶，直径约2.7 m，深约2.3 m，用石块垒砌而成，靠山而建，顶部支起煮料的锅。

⊙1

⊙2

⊙3

贰 踩槽 2

用于踩料，由四块石头就地砌成，槽底的石面呈斜坡状，有利于踩料时不断地沿着斜坡碾压料子。踩槽旁立有一根与拐杖一般高的棍子，给踩料人作支撑身体用。踩槽建在窑子与河岸之间。

⊙4

叁 料池 3

由大石块砌成的长方形池子，用于放料。

⊙5

⊙1 窑子 Papermaking kiln

⊙2 窑子入口 Entrance to the papermaking kiln

⊙3 窑子顶部支起煮料大锅的地方 Place for holding the boiling wok on the top of the papermaking kiln

⊙4 踩槽 Trough for stamping the papermaking materials

⊙5 料池 Pool for holding the papermaking materials

肆 槽子 4

又称“抄纸槽”，用于抄纸，外沿呈长方形，内侧挨着山体，上面用竹竿、木檩搭成棚子以遮蔽风雨。其对面是放置湿纸的竹席和木榨。

⊙6

伍 纸帘 5

竹质，用于抄纸，尺寸为(16.5×3+4.5) cm×(8+36+5.5) cm。

⊙7

陆 帘架 6

由木条、竹条作框，用以固定纸帘，在水中始终保持纸帘平行以使纤维均匀地分布在纸帘上。帘架内径尺寸约为58.5 cm×52 cm。

⊙8

柒 拱耙 7

木质，用于打槽，主要作用是把料子打散，让纤维在槽子中快速分散开来。

⊙9

捌 捞筋棍 8

细竹竿，捞纸前用其把料子搅匀，并捞出粗纤维，以便回收后再进行踩料加工。

⊙10

⊙6 木榨 Wooden presser

⊙7 纸帘 Papermaking screen

⊙8 帘架 Frame for supporting the papermaking screen

⊙9 拱耙 Stirring rake

⊙10 捞筋棍（放在槽子沿上） Lifting stick (put on the edge of the trough)

（四）

石阡仡佬族竹纸的性能分析

所测石阡关刀土村竹纸的相关性能参数见表9.4。

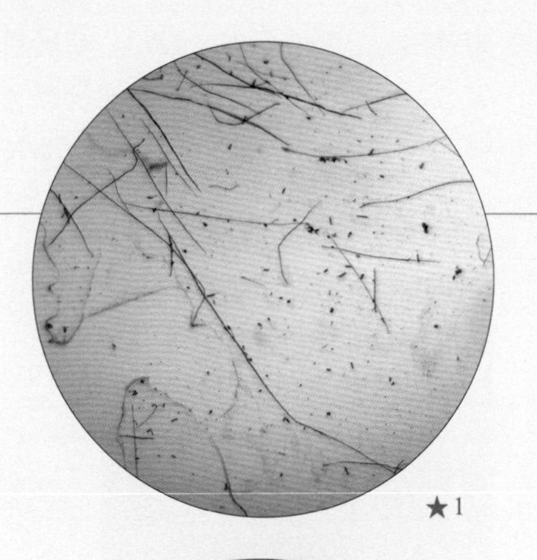

★1

表9.4　关刀土村竹纸的相关性能参数
Table 9.4　Performance parameters of bamboo paper in Guandaotu Village

指标		单位	最大值	最小值	平均值
厚度		mm	0.210	0.120	0.154
定量		g/m²	—	—	46.6
紧度		g/cm³	—	—	0.303
抗张力	纵向	N	4.5	2.1	3.3
	横向	N	3.0	1.8	2.3
抗张强度		kN/m	—	—	0.187
白度		%	20.6	19.7	20.0
纤维长度		mm	5.70	1.00	2.17
纤维宽度		μm	34.0	1.0	16.0

★2

性能分析

由表9.4可知，所测关刀土村竹纸最厚是最薄的1.75倍，相对标准偏差为2.80%，纸张厚薄较为一致。竹纸的平均定量为46.6 g/m^2。所测竹纸紧度为0.303 g/cm^3。

经计算，其抗张强度为0.187 kN/m，抗张强度值较小。

所测关刀土村竹纸白度平均值为20.0%，白度较低，白度最大值约是最小值的1.05倍，相对标准偏差为0.33%，白度差异相对较小。

所测关刀土村竹纸纤维长度：最长5.70 mm，最短1.00 mm，平均2.17 mm；纤维宽度：最宽34.0 μm，最窄1.0 μm，平均16.0 μm。所测竹纸在10倍、20倍物镜下观测的纤维形态分别见图★1、图★2。

★1
关刀土村竹纸纤维形态图(10×)
Fibers of bamboo paper in Guandaotu Village (10× objective)

★2
关刀土村竹纸纤维形态图(20×)
Fibers of bamboo paper in Guandaotu Village (20× objective)

五
石阡仡佬族竹纸的 用途与销售情况

5
Uses and Sales of Bamboo Paper by the Gelo Ethnic Group in Shiqian County

坪地场乡下林坝村竹纸的尺寸只有一种，用途也很有限，主要用于制作烧香拜佛时的烧纸，也有少量用于制作练习毛笔字的草纸。

这种竹纸的销售情况不是很理想。一般由造纸户拿到乡里去卖。每逢当地街日，即赶集的日子（每月农历初五、初十），附近有需要用纸的人都会去买纸，销售范围在花桥、大坝厂、杨柳、石阡县城等地，基本上属于自产自销的个体手工业类型。5年前每捆卖2元，现在每捆能卖5元，一个纸槽一天能生产20捆。考虑到造纸的成本，石灰0.3元/kg，煤450元/t，碱3元/kg，可以说下林坝村造纸户所操持的手工纸业目前基本上无利可图或略有微利。

六
石阡仡佬族竹纸的 相关民俗与文化事象

6
Folk Customs and Culture of Bamboo Paper by the Gelo Ethnic Group in Shiqian County

仡佬族葬礼的形式是击鼓而歌、焚烧纸钱。

石阡县仡佬族葬礼中，死了老人的人家（俗称“孝家”）在堂屋中设灵堂，院子里支起帐篷设“歌堂”。傍晚，歌堂四周挤满了围观者，歌堂内灯火通明，茶酒齐备，歌手们围坐于一面大鼓前。三通鼓响，歌师说唱起来，其他歌手随声应和。演唱分三步：开场、“接亡灵”、唱孝歌。伴有烧纸行为的是“接亡灵”，通过说唱、烧纸、哀哭等仪式，把“亡灵”从地狱里接回家。歌词大意是：

“接亡来，接亡来，亡在十二殿阎君南狱面前哭哀哀。

十二殿阎君当堂坐，抓住亡人要钱财。

要钱资，要钱财，天空降下钥匙来。

左手投，右手开，开笼箱，取钱财。

烧张纸，化张钱。火化钱财交于你，又把亡人接到十一殿阎君面前来！”

歌词不断重复地唱（除更换阎君名号外），将“亡灵”从十二殿一直接到一殿。每当唱到“烧张纸，化张钱”时，孝家安排专人在一旁烧化纸钱。此外，还以这种唱法讲述“接亡灵”时若碰到各种拦路鬼神，也要一一火化钱财打通关口，最后才能将“亡灵”接到灵堂归位。

⊙1

⊙1 石阡山里人走山路 Local residents walking on a country road

七 石阡仡佬族竹纸的保护现状与发展思考

7 Preservation and Development of Bamboo Paper by the Gelo Ethnic Group in Shiqian County

相对于那些已经获批为“非物质文化遗产”保护项目或传习基地的造纸地区而言，石阡的手工造纸工艺是一种完全处于原生态的文化存在形式。就文化调查者而言，石阡县下林坝以及关刀土等村落的造纸现状正好提供了一个少数民族乡土原生样本的参照。在石阡县下林坝村，调查组发现，传承技艺者给我们展示的是自己家造纸时朴素的技术过程，并没有那些经过设计后表演给游客或参观者观看的元素。

石阡下林坝造纸技艺之所以能够沿袭至今，依赖的是一种来自地方群体内部需求的动力。尽管下林坝造纸的品种十分单一，几乎仅供给民众祭祀祖先神佛用，然而民间信仰产生了有力量的精神需求，其外化为有较强坚持度的祭祀仪式中的行为。手工纸作坊在祭祀用纸的生产方面依旧有民族习俗文化空间，但其经济收益甚微。

因此，调查组建议：

（1）在普遍促进手工造纸非遗技艺旅游化的形势下，为像石阡竹纸这样少数民族乡土原生态的手工造纸技艺的生存提供较为宽松的政策环境和自由的发展空间，这样既不用将祭祀用纸视为宣传“迷信”的辅助物品而进行打压，又能结合乡土商贸帮助造纸户获得更广的销售渠道。

（2）结合贵州黔东北仡佬族文化的传习光大计划，重塑手工造纸技艺作为民族文化基因而受到尊重的身份和地位，培育和激发老纸工愿意传承、年轻人愿意学习造纸技艺的兴趣，使仡佬族造纸技艺获得薪火相传的内在驱动力。

⊙2
⊙3

⊙2 访谈中热情的造纸老人 Hospitable old papermakers

⊙3 下林坝村的造纸农家宅居 Papermakers' residence in Xialinba Village

石阡仡佬族竹纸

Bamboo Paper
by the Gelo Ethnic Group
in Shiqian County

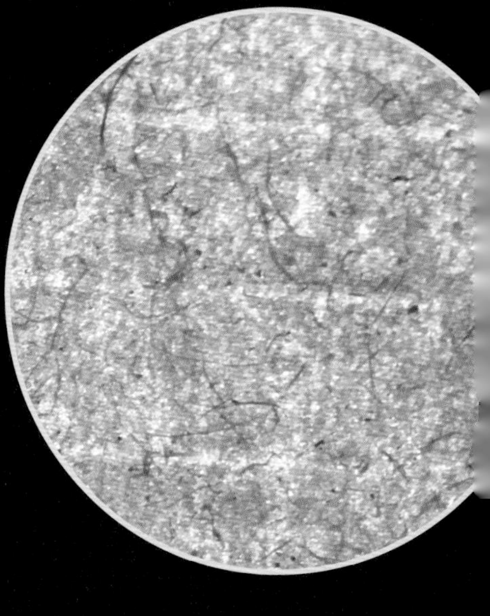

关刀土村竹纸透光摄影图
A photo of bamboo paper in Guandaotu Village
seen through the light

第三节

江口土家族竹纸

贵州省
Guizhou Province

铜仁市
Tongren City

江口县
Jiangkou County

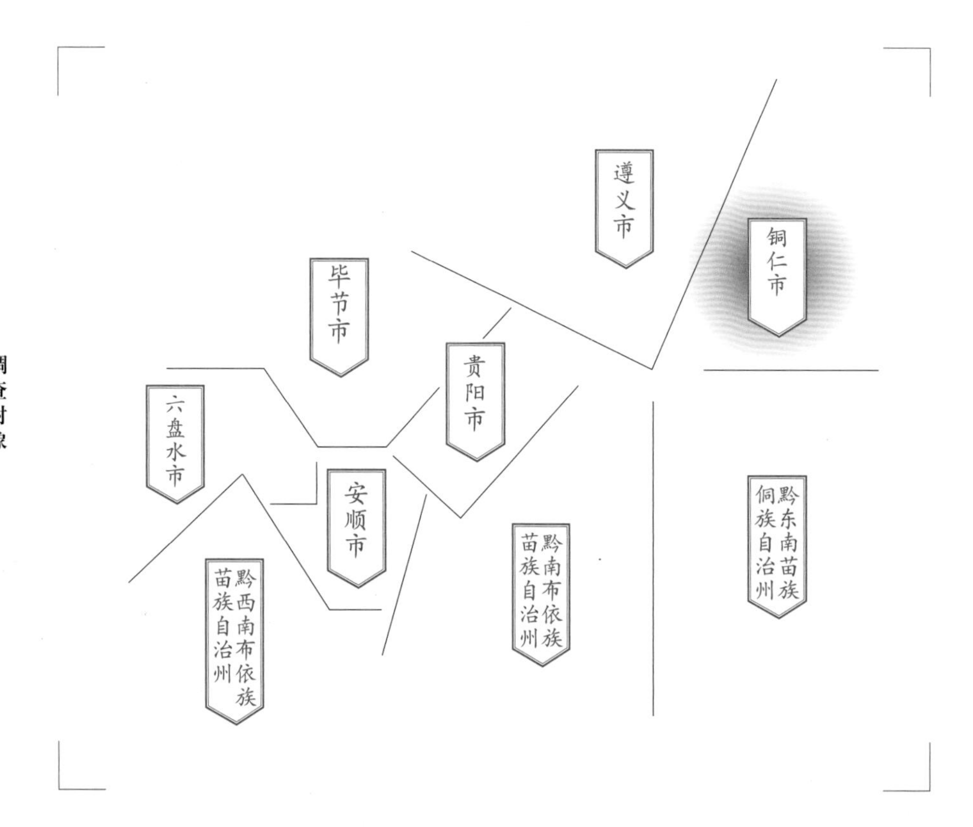

调查对象

太平土家族苗族乡云舍行政村
怒溪土家族苗族乡河口行政村
土家族竹纸

Section 3
Bamboo Paper by the Tujia Ethnic Group in Jiangkou County

Subject

Bomboo Paper by the Tujia Ethnic Group in Hekou Administrative Village of Nuxi Tujia and Miao Town, Yunshe Administrative Village of Taiping Tujia and Miao Town

一
江口土家族竹纸的基础信息及分布

1
Basic Information and Distribution of Bamboo Paper by the Tujia Ethnic Group in Jiangkou County

江口土家族竹纸是指江口境内以土家族为主的村民利用慈竹、山竹为原料通过手工操作造的纸。据1994年版《江口县志》记载，其主要产地集中在江口境内太平河两岸与举忙坡周围一带。所产纸除用于书写外，多作祭祀焚化之用。

2008年12月9~11日调查组入村调查时，江口县所造竹纸主要分布在怒溪土家族苗族村河口村、太平土家族苗族乡云舍村等地。河口乡位于江口县东北部，距县城约37 km；云舍村坐落在锦江之源太平河流域中段的河坎谷地，距江口县城5 km。两地均植被茂密，盛产竹子，为竹纸生产提供了充足的原料。两地造纸历史悠久，调查时怒溪乡还保存着上百个传统的造纸作坊，而在云舍村，造竹纸不但是一种传统手艺，而且至今仍是乡民一项重要的经济来源。

调查组入村时看到，云舍村的土家族百姓家家都有造纸作坊，户户都能造纸，已成为远近闻名的造纸村，是江口县土纸第一大生产基地。由于云舍村是贵州著名的土家文化旅游目的地，因此近年来土法造纸还成为游客们非常喜欢的一项参与性的旅游项目。

⊙1

⊙1
云舍村口农家生活小景
Rural view of Yunshe Village

路线图
江口县城
↓
河口村、云舍村
Road map from Jiangkou County centre to the papermaking sites (Hekou Village and Yunshe Village)

江口土家族竹纸生产地分布示意图

Distribution map of the papermaking sites of bamboo paper by the Tujia Ethnic Group in Jiangkou County

考察时间
2008年12月

Investigation Date
Dec. 2008

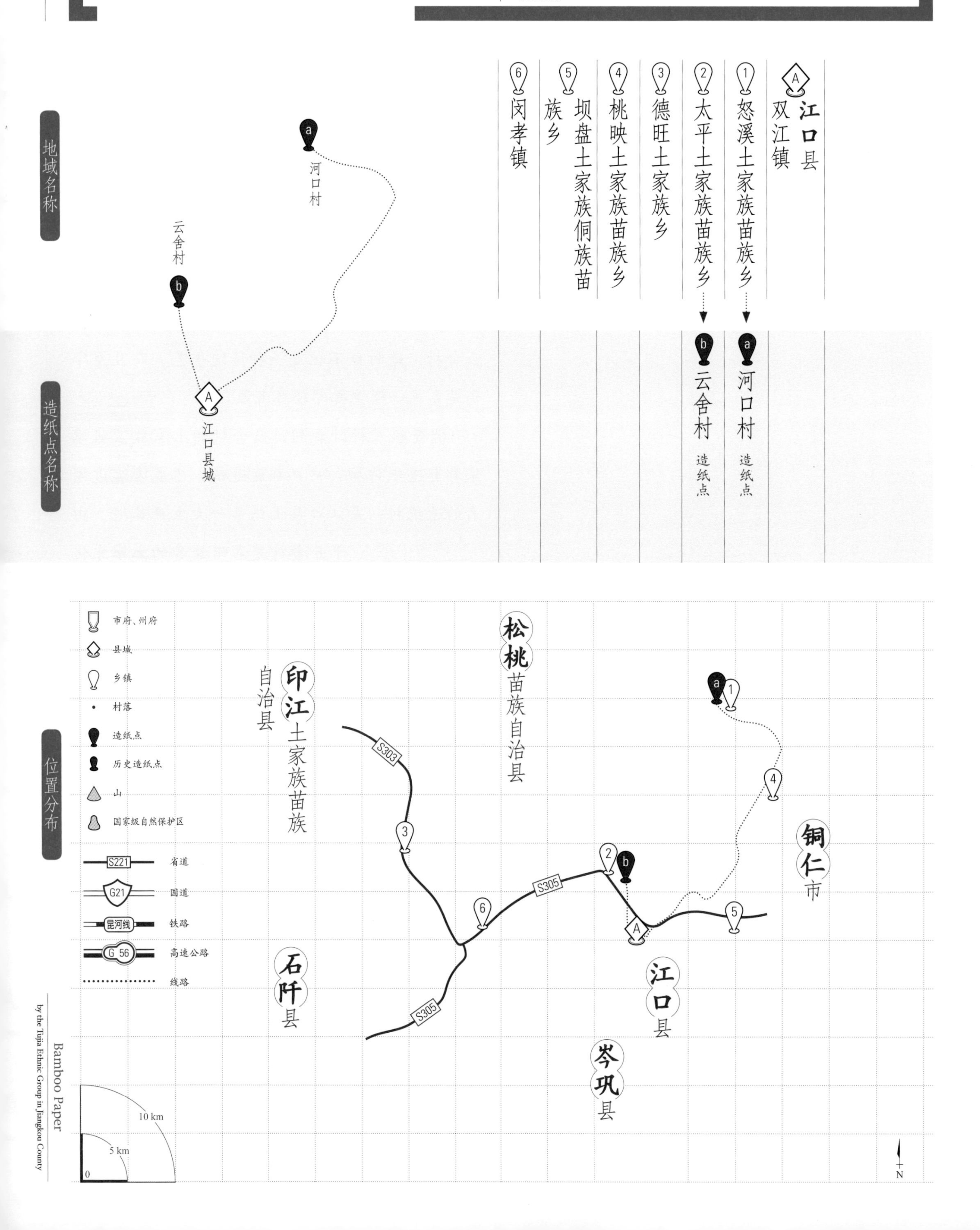

二
江口土家族竹纸生产的人文地理环境

2
The Cultural and Geographic Environment of Bamboo Paper by the Tujia Ethnic Group in Jiangkou County

江口县东西长52 km，南北宽55 km，总面积1 873 km^2，东邻铜仁市，南连岑巩县，西毗石阡县、印江土家族苗族自治县，北接松桃苗族自治县，处于贵州高原向湘西丘陵过渡的斜坡地带。全县地势由西、北、南三个方向向东倾斜，最高点在县西北凤凰山，海拔 2 579.5 m，最低点在溪口乡石埂，海拔275 m。境内河谷纵横，溪河密布，主要河流有太平河、闵孝河、桃映河、车坝河，均属长江流域沅江水系。全县丘陵地貌占总面积的90%以上。

其建制始于唐代，当时于此地置辰水县，明置铜仁县，1913年改为江口县。江口之名，以省水、提水两河（今闵孝河、太平河）于此汇流得名。1958年撤县。1961年复设江口县至今。

怒溪土家族苗族乡古称“漏旗”“漏溪”，位于江口县东北部，东接桃映乡，南邻坝盘乡，西接太平乡，北与松桃县寨英镇交界，距江口县城37 km，距铜仁市区45 km，距铜仁大兴机场58 km。

太平土家族苗族乡地处梵净山东麓，距江口县城12 km，境内有国家级自然保护区梵净山、

⊙1

⊙2

⊙3

⊙1 怒溪乡河口村的民宅 Local residence in Hekou Village of Nuxi Town

⊙2 怒溪乡的山景 Landscape of Nuxi Town

⊙3 云舍村边的小河 River alongside Yunshe Village

⊙1

⊙2
⊙3

贵州省级风景名胜区太平河与云舍土家民族文化村，辖1个居委会和7个行政村共146个村民组，少数民族人口占总人口的84.4%，其中以土家族、苗族、侗族为主，是一个典型的少数民族乡。

太平土家族苗族乡云舍村是一个土家民俗文化村，439户1 700余人中，98%的村民都是土家族的杨氏后裔，是江口县乡村的第一大寨。云舍土家族有着上千年的悠久历史，至今仍然保留着非常丰富多彩的民族风情习俗，被誉为“中国土家第一村”。调查组步入云舍土家民俗文化村时，首先映入眼帘的便是土家人的几十间造纸作坊。据当地造纸乡民的传说，土家人的祖先拜蔡伦为师，精学造纸，造纸技艺几百年来代代相传。

三
江口土家族竹纸的历史与传承

3
History and Inheritance of Bamboo Paper by the Tujia Ethnic Group in Jiangkou County

江口土家族竹纸的生产历史悠久，据2008年12月9~11日调查的造纸户们的乡土口传记忆，最早可以追溯到清康熙年间，距今已有300多年的历史。造纸工艺代代相传，保留了传统朴实的手工操作方法。鼎盛时期，全县有造纸户1 000余家。2009年调查时，怒溪乡的芦家寨、老屋基、张家湾、孟家屯、骆家屯、郑家屯、麻阳溪，坝盘乡的勤劳村以及太平乡的云舍村还采用土法造纸。

20世纪50年代末，江口县在怒溪乡巴古寨大湾一带的河口办起了全区最大的古法造纸作坊——怒溪纸厂。据《江口县志》记载，1953年底全县有97户286人造纸，年产纸135 t；1958年“大跃进”时期，全县有造纸厂19个，职工534人，年产纸535 t，总产值2 941万元，按当年的货币消费水平，已是相当大的产值。但随着经济形势的快速恶化，当年

⊙1 云舍村的造纸棚
Papermaking shed in Yunshe Village

⊙2 调查组成员访谈云舍村造纸人
A researcher interviewing a papermaker in Yunshe Village

⊙3 调查组成员访谈河口村造纸人
A researcher interviewing a papermaker in Hekou Village

年底大部分纸厂相继关闭，仅留怒溪纸厂、董家坡纸厂。1960年3月，董家坡纸厂关闭，62名工人全部下放农村。自1950年到20世纪80年代，地方政府先后多次开办手工纸厂，但效益均不佳，大多因亏损而倒闭。20世纪80年代中期后就变成了一家一户独立生产的业态。

据造纸户介绍，传统江口土家族造纸技艺都是通过师傅传徒弟、父传子等方式流传下来的。由于江口一带少数民族先民高度迷信鬼神，加上江口本地人掌握这种技术和造纸原料丰富，所以直到20世纪后期江口当地的造纸户依然较多，土纸销量大，销路也广。但21世纪初，随着机械造竹纸的发展和城镇化生活的移风易俗，手工竹纸的生产受到了较大冲击，从事土纸生产的作坊和人数都已明显减少。

⊙4

⊙5

四
江口云舍村土家族竹纸的生产工艺与技术分析

4
Papermaking Technique and Technical Analysis of Bamboo Paper by the Tujia Ethnic Group in Yunshe Village of Jiangkou County

(一)
江口云舍村土家族竹纸的生产原料与辅料

据造纸户介绍，云舍村所造竹纸的原料现为当地生长的山竹，以前使用过阳竹、慈竹。

云舍村造纸用的“滑”（纸药）有两种：滑叶汁和杨桃藤汁。

取滑叶树的叶子，装入袋子，用水煮开后，用手挤，滑叶汁就会从袋子里渗出来。滑叶树的叶子立夏时采一次，农历七月十五时又可再采一次。抄一天纸，需要0.5 kg滑叶。

⊙4 云舍村的连片纸坊 Papermaking mills in Yunshe Village

⊙5 进入云舍村的桥 Bridge leading to Yunshe Village

杨桃藤即猕猴桃藤。将新鲜的猕猴桃藤截断，敲破，放入桶中，也可用袋子装，加入或放入水中，浸泡出汁，即得杨桃藤汁。抄一天纸需要用0.5 kg杨桃藤汁，需要注意的是，老杨桃藤不能用。

(二)

江口云舍村土家族竹纸的生产工艺流程

根据调查组2008年12月10~11日的实地调查，总结江口云舍村土家族竹纸的生产工艺流程如下：

壹	贰	叁	肆	伍	陆	柒	捌	玖
砍竹	划竹	泡竹	洗涤	堆垛	二次泡竹	拌料	打槽	放水

拾	拾壹	拾贰	拾叁	拾肆	拾伍	拾陆	拾柒
加滑	抄纸	去纸边	压榨	二次压榨	揭纸	晒纸	捆纸

壹 砍竹 1

每年夏至前后一个多月，砍伐当地长至2~2.7 m高的山竹。由于山上的竹子生长得慢，根据地形高低，山下的山竹夏至前砍，山上的山竹夏至后砍。

贰 划竹 2

用柴刀将竹子对半划开。

叁 泡竹 3

根据料塘大小，把竹子截成段，每段长2 m左右，比料塘长度稍短些，此时的竹料称为竹麻。将竹料铺放在料塘之中，每铺一层竹麻撒一层石灰，待最上面一层竹麻铺好后，再撒一层石灰，并用石头压住竹麻，然后将料塘加满水泡竹，泡竹最少需一个月时间。泡竹时，每50 kg竹麻需用10~12.5 kg石灰。

肆

洗　涤

4

竹麻泡好后，将料塘的水放掉，加入清水洗竹麻，反复多次，直至洗干净为止。

伍

堆　垛

5

竹麻洗净后，从料塘取出，码放在干燥的地上堆成垛，发酵半个月。

陆

二　次　泡　竹

6

将料塘清洗干净，再将发酵后的竹麻放入料塘，并在其最上面覆盖稻草，用石头压好，加水浸泡一个月。

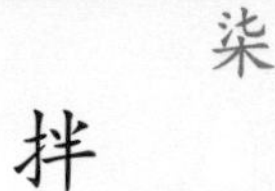

柒

拌　料

7　⊙1⊙2

二次泡竹后，将竹麻再次清洗干净，按一天抄纸所需竹麻的量将竹麻放入踩料池中，并用脚将竹麻踩融。踩料的时间长短根据竹子老嫩和先期泡竹时所放石灰多少而定，具体来说，若石灰放得多，踩一小时左右即可；若石灰放得少且竹子老，一般要踩一个半小时左右。一般一次可踩50~65 kg。据造纸户介绍，以前造纸用过水车、脚碓、棒子、牛碾来拌料，如果是老竹麻，必须要用水车来舂料。1978年后，上述拌料方式都较少使用，现在基本上用脚踩料。

⊙1

⊙2

捌

打　槽

8　⊙3

将踩融的竹料放入抄纸槽中，加水，先用打槽棍将料打匀，有力气的打几十下即可，没力气的需多打几下。然后用撸筋棍再次搅槽，并随时将附着在撸筋棍上的纸筋巴捞出。

⊙3

玖

放　水

9

打好槽后，放掉一半水，静置。其目的是使纸料能够较快沉淀。

⊙1/2
拌料
Stamping the papermaking materials

⊙3
用撸筋棍搅拌
Stirring the papermaking materials with a lifting stick

⊙4

拾 加滑

10 ⊙4

第二天将纸槽加满水，再次打槽，然后抄纸。待抄好第一张纸后，将制好的滑加入纸槽中，再次打槽。据造纸户介绍，抄第一张纸时不能加滑，如加滑，则因太滑而使纸浆上不了帘子，成不了纸。

拾壹 抄纸

11 ⊙5~⊙8

将纸帘放在帘架上，左手将纸帘与帘架握住，右手抓住帘架上的提柄，先由外往里将纸浆舀入纸帘，接着由右往左挖水，再由左往右送水，待水滤去，将纸帘从帘架上取下，转身将纸帘反盖于湿纸垛上，揭下纸帘，一张湿纸便留在纸垛上。一般一个人一天可抄11~12捆纸，年纪大的可抄7~8捆纸。据造纸户介绍，最近40年用的是一帘三纸的帘，以前用的是一帘二纸的。

拾贰 去纸边

12 ⊙9⊙10

一垛纸抄好后，用方木棒压一下纸垛的纸边，使之易于分开。随后用手将前后纸边的边角料去掉，以回收再利用。

⊙5

⊙6

⊙7

⊙8

⊙9

⊙10

⊙4 加滑 Adding in papermaking mucilage

⊙5 由外往里舀水 Scooping the papermaking screen from far to near

⊙6 由右往左挖水 Scooping the papermaking screen from right to left

⊙7 由左往右送水 Scooping the papermaking screen from left to right

⊙8 盖纸 Turning the papermaking screen upside down on the wet paper

⊙9/10 去纸边 Trimming the deckle edges

拾叁 压 榨

13 ⊙11~⊙13

去完纸边后，在湿纸垛上盖上一张旧纸帘，纸帘上再依次加上榨板、木方、码子等，先用手榨缓慢压榨一会，再将手榨换成大榨，用钢丝绳将大榨和滚筒连起来，将手榨插在滚筒的榨孔中，缓慢扳动手榨，直至将纸榨干。压榨快的需半小时，慢的则需2小时，甚至更长时间。

⊙11

⊙12

⊙13

拾肆 二 次 压 榨

14 ⊙14⊙15

松榨，将纸垛一分为三，随后再将三个纸垛叠在一起压榨。

⊙14

⊙15

拾伍 揭 纸

15

松榨后，将纸垛运回家。用手将纸一张张地揭开，每5张为一提。

拾陆 晒 纸

16

将揭好的纸一提提放在草坪、地面上晾晒，如遇阴雨天，则挑晾在屋内的竹竿上。阳光强烈时半天即可将纸晒干；一般天气，如天晴有风，需一天一夜；阴雨天时则需两三天，甚至更长时间。

拾柒 捆 纸

17

以100提为一担，10担为一捆，将纸数好，理齐，两边加竹片，用竹篾捆好。

⊙11 盖榨板 Putting on the pressing board

⊙12 手榨压榨 Pressing the paper with hand presser

⊙13 大榨压榨 Pressing the paper with large presser

⊙14 二次压榨 Pressing the paper for the second time

⊙15 纸垛 A pile of paper

(三)

江口云舍村土家族竹纸生产使用的主要工具设备

壹 纸帘 1

由细竹丝编成，其上刷漆，油光发亮，一般还会用红油漆在纸帘上写明主人的名字。纸帘长79.5 cm，宽48 cm。

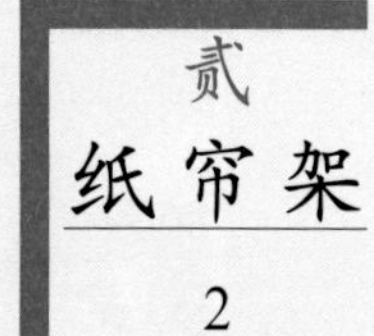

贰 纸帘架 2

木质，长84 cm，宽57 cm。手柄为竹质的。纸帘架中有很多竖排小木棍，用于承重并便于滤水；另有两根横排小木棍，上用塑料带缠绕，其位置和纸帘隔线相对应，便于将纸一分为三。

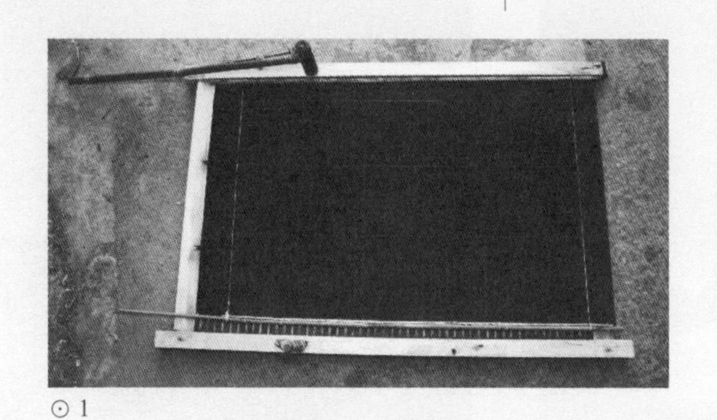

⊙ 1

⊙ 2

(四)

江口云舍村土家族竹纸的性能分析

所测江口云舍村竹纸的相关性能参数见表9.5。

表9.5 云舍村竹纸的相关性能参数
Table 9.5 Performance parameters of bamboo paper in Yunshe Village

指标		单位	最大值	最小值	平均值
厚度		mm	0.210	0.110	0.144
定量		g/m^2	—	—	29.2
紧度		g/cm^3	—	—	0.203
抗张力	纵向	N	9.4	6.4	7.2
	横向	N	3.9	1.2	2.7
抗张强度		kN/m	—	—	0.330

⊙ 1 纸帘及纸帘架 Papermaking screen and its supporting frame

⊙ 2 纸帘架 Frame for supporting the papermaking screen

续表

指标	单位	最大值	最小值	平均值
白度	%	21.0	20.5	20.8
纤维长度	mm	5.59	0.58	1.95
纤维宽度	μm	32.0	3.0	13.0

由表9.5可知，所测云舍村竹纸最厚约是最薄的1.91倍，相对标准偏差为3.50%，厚薄差异相对较小。竹纸的平均定量为29.2 g/m^2。所测竹纸紧度为0.203 g/cm^3。

经计算，其抗张强度为0.330 kN/m，抗张强度值较小。

所测云舍村竹纸白度平均值为20.8%，白度最大值约是最小值的1.02倍，相对标准偏差为0.17%，白度差异相对较小。

所测云舍村竹纸纤维长度：最长5.59 mm，最短0.58 mm，平均1.95 mm；纤维宽度：最宽32.0 μm，最窄3.0 μm，平均13.0 μm。所测竹纸在10倍、20倍物镜下观测的纤维形态分别见图★ 1、图★ 2。

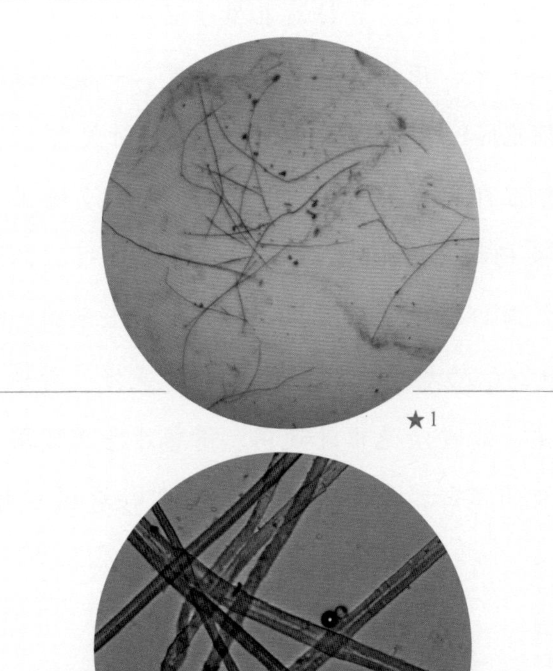

★1 云舍村竹纸纤维形态图（10×） Fibers of bamboo paper in Yunshe Village (10× objective)

★2 云舍村竹纸纤维形态图（20×） Fibers of bamboo paper in Yunshe Village (20× objective)

五 江口河口村土家族竹纸的生产工艺与技术分析

5 Papermaking Technique and Technical Analysis of Bamboo Paper by the Tujia Ethnic Group in Hekou Village of Jiangkou County

2008年12月9~10日，调查组实地考察了怒溪乡的河口村，了解了怒溪乡土家族竹纸的制作情况。河口村是以土家族、侗族、苗族为主的村落，辖22个村民组36个自然村寨，历史上一直是江口土家族竹纸具有代表性的生产地。

(一) 江口河口村土家族竹纸的生产原料与辅料

怒溪乡河口村生产竹纸的原料有慈竹和山竹。慈竹为当地产；山竹当地没有，多为购买。

据造纸户介绍，怒溪乡竹纸使用的纸药有野棉花汁、滑根汁、神香叶汁和滑精（PAM）。

野棉花汁：将野棉花连秆带花采回来以后，捶烂，装在袋子里，直接放在槽子里浸泡，用时把袋子拿起来，用手将滑药挤出即可。一般一袋可装野棉花1~1.5 kg，可用3~4天，只在冬天时使用。

滑根汁：滑根弯弯曲曲，生长在沙地里，有点像杨桃，但因其不能结果子，又叫"公杨桃"。河口当地没有滑根，须从市场上采购，价格为0.2~0.25元/kg，以5~6 cm粗为最好。滑根只用表皮，取皮浸泡出汁使用。滑根汁冬天可用，其他时间不行，原因不详。一般50 kg滑根可用2个月（以天天造纸计算）。

神香叶汁：每年4月神香叶树叶长到4~5 cm时，便可以采摘使用。叶子摘后又可长出新叶来，可以一直摘。所制滑药一年四季都可用。将采摘的叶子放入袋子中，加水，用柴火煮，一般煮20分钟即可将叶子煮烂，用时用手挤下袋子即可出汁。

滑精（PAM）：现代化工产品，粉末状。主要由四川人带来销售，价格为8.5~9元/kg。滑精一天只要两勺即可，大约25 g，0.5 kg滑精可用10~20天。自2005年开始使用滑精后，其他滑药基本不再使用。

（二）江口河口村土家族竹纸的生产工艺流程

根据调查组的实地调查访谈，总结河口村土家族竹纸的生产工艺流程如下：

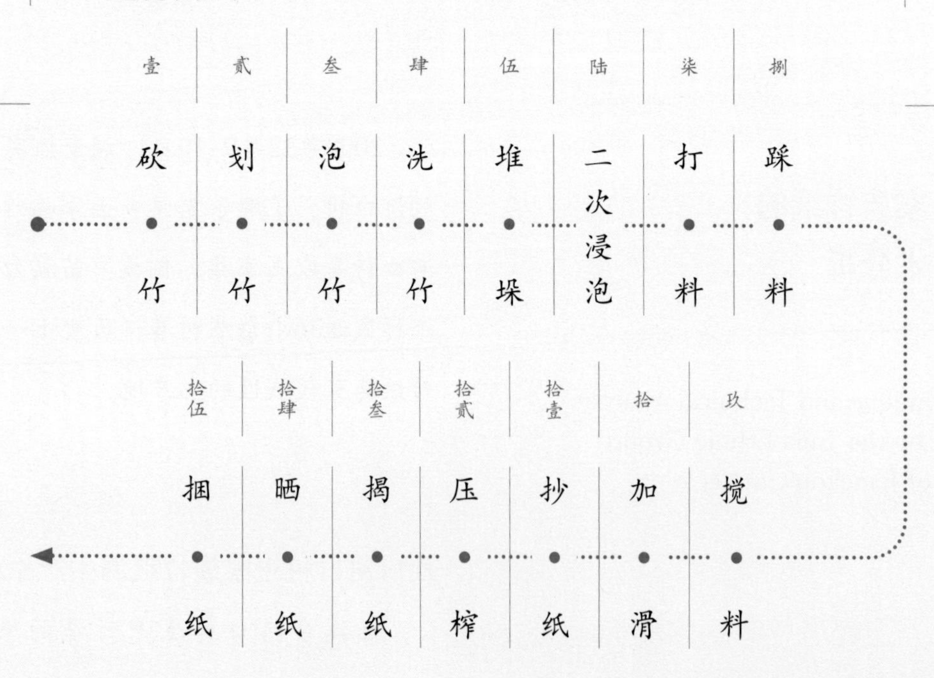

壹 砍竹

1

每年农历九月到次年农历二月是砍伐慈竹的时节，多数造纸户都是自己到山上砍，以砍伐当年生的竹子为最好。一年可砍2 500~3 000 kg，基本够用，如不够，也可以到市场上购买，一般50 kg竹子售价20~30元。

贰 划竹

2

用柴刀将竹子对半剖开，并砍成段，每段长1.3~1.7 m，此时的竹子称为竹麻。

叁 泡竹

3 ⊙1

竹麻大约20 kg为一把，先用竹篾捆起来，然后铺放在料塘中，放一层竹麻撒一层石灰，最下面是竹麻，最上面是石灰，再用石头压住，加水浸泡4个月，中间不用换水。

肆 洗竹

4

用锄头敲打泡好的竹麻使石灰脱落，并换水清洗。经过几次换水清洗并将竹麻洗净后，把水排掉，同时将料塘中的石灰也清理干净。

⊙1

伍 堆垛

5

将洗净后的竹麻放在洗净的料塘中堆置一个月。

陆 二次浸泡

6 ⊙2

一个月后，往料塘中加水，再次将竹麻浸泡半个月之后就可以取出使用。竹麻如不用则放在料塘中继续浸泡，时间再长也没有关系。

⊙2

⊙1 待泡的竹料 Bamboo materials to be fermented

⊙2 浸泡池 Soaking pool

柒

打料

7

据造纸户介绍，以前用脚将竹麻踩碎、踩融。一个人踩一天的量够一个人一天抄纸所用。1992年开始使用水车打料，一个人一天可打出够20~30个人使用一天的料。现在多使用打料机打料。原先打料机是用来打红薯的，后来才将它用于打竹料。一个人一天可打出够三四个人使用一天的料，打料机需打10次才能将竹料打得很碎。

捌

踩料

8

料子打碎后需要再放入踩槽中用脚踩融，一般踩十几分钟即可。

玖

搅料

9 ⊙3

将25 kg左右踩融后的料子放入抄纸槽，加水，并用槽棍将其搅匀，几分钟即可，然后将水放掉，使竹料沉淀。在抄纸前重新往抄纸槽内加水，然后用槽棍搅槽。此时搅槽并不是将沉淀的竹料全部搅起，而是凭经验，根据所抄纸的厚薄、个人抄纸习惯等，将沉淀的竹料部分搅起、搅匀。

⊙3

拾

加滑

10 ⊙4

搅料后，在抄纸槽中加入滑药，并用槽棍搅匀。

⊙4

拾壹

抄纸

11 ⊙5~⊙12

将纸帘置于帘架上，双手平端帘架并压住纸帘，将帘架由外往里插入纸浆，舀浆入帘，随后端起，再由右往左将帘架插入纸浆，再次舀浆入帘，随后平端帘架，将帘架平置，揭起纸帘，转身将湿纸反扣在纸垛上，揭帘离纸，这就完成了整个抄纸动作。此处由外往里舀浆入帘的工艺被称为“头道水”，由右往左舀浆入帘的工艺被称为“二道水”。抄出的纸左边比右边厚，这有利于后期数纸。数纸时从厚的一边开始数。

⊙5

⊙6

⊙3 搅料 Stirring the papermaking materials

⊙4 加滑 Adding in papermaking mucilage

⊙5/12 抄纸和扣纸 Scooping and lifting the papermaking screen out of water and turning it upside down on the board

⊙7 ⊙8 ⊙9 ⊙10

⊙11 ⊙12

拾贰 压榨

12 ⊙13~⊙15

抄纸时直接将湿纸置于木榨上，以免二次搬运。待纸抄到一定高度，就开始压榨，用木榨十几分钟即可将湿纸垛压干。压榨时，开始最为关键，要慢，不能过快，否则纸垛会被压爆，造成损失。如果抄纸过程中需去休息，可在湿纸垛上放纸帘、榨板、两木头压一下，目的是压去部分水，因为纸左边比右边厚，造成纸垛一边高一边低。如果不压，湿纸容易由高处滑到低处。如果一直抄纸，则不需要这样做。河口村使用的纸帘是一帘三纸的帘，亦可把中间的两根"砂"去掉，变成一帘一纸的大纸，大纸压榨时也是压十几分钟即可。

⊙13 ⊙14

⊙15

⊙13/15 榨纸 Pressing the paper

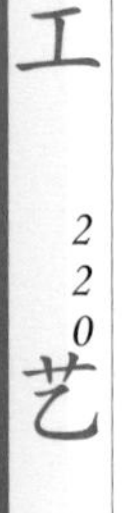

拾叁 揭纸

13 ⊙16⊙17

纸垛压干后，将纸一张张地揭下来，每2张略错开。

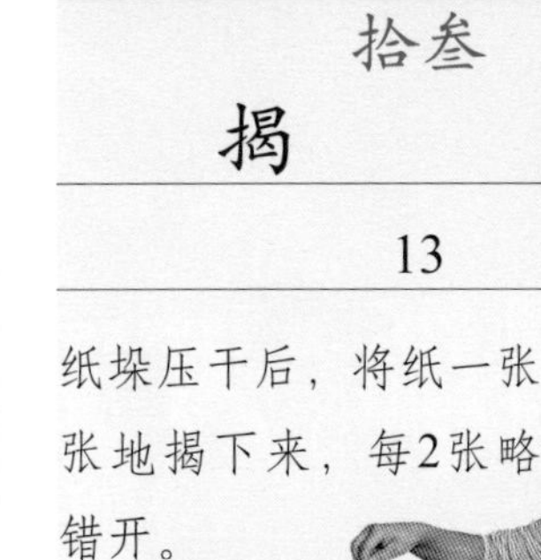

⊙16 ⊙17

拾肆 晒纸

14 ⊙18

揭纸后，将纸分叠晒干，有5张一叠的，也有3张一叠的。两种纸的尺寸分别为13.2 cm × 66.7 cm和12 cm × 60 cm。晒纸时，只要是能见到阳光、可放纸的地方都可以晒。夏天阳光充足，5张一叠的纸1小时即可晒干，3张一叠的纸半小时即可晒干；冬天晴天时5张一叠的纸2~3小时可晒干，3张一叠的纸一个多小时可晒干；如遇阴雨天，将纸放在屋里，或摆在地上，或挑在竹竿上晾干，5张一叠的纸需2~3天才能干，3张一叠的纸也需1~2天才能干。

⊙18

拾伍 捆纸

15 ⊙19⊙20

纸晾干后，收起，按叠理齐，然后将纸对折，两侧各放一个竹片，用竹篾捆一道。5张一叠的，400叠为一捆，价格为80元/捆；3张一叠的，500叠为一捆，价格为50元/捆。

⊙19

⊙20

六 江口土家族竹纸的用途与销售情况

6 Uses and Sales of Bamboo Paper by the Tujia Ethnic Group in Jiangkou County

江口当地所产竹纸主要用于各民族的民俗祭祀活动中，可用于制作焚烧的纸钱、挂青用纸，也可用于制作卫生纸、书写纸，但调查时这些用途都已经消失。

从销售渠道来看，竹纸的产量仍有一定规模，但其主要销售地域依然为江口本地，也有外销的情况，主要是集中在黔东南地区的近邻县乡。总的来说，江口竹纸的用途和销售区域都较小。

从销售方式来看，河口村的造纸村民在农历日期中尾数为二、七的日子，将纸运到怒溪乡的市场上去卖，偶尔也有人上门购买。2009年前后，竹纸零售价格可卖到80元/捆（5张一叠的规格）、50元/捆（3张一叠的规格）。

⊙16/17 揭纸 Peeling the paper down

⊙18 在家门外的地上晒纸 Drying the paper on the ground

⊙19 捆纸 Binding the paper

⊙20 可销售的成品竹纸 Bamboo paper ready for sale

七 江口土家族竹纸的相关民俗与文化事象

7 Folk Customs and Culture of Bamboo Paper by the Tujia Ethnic Group in Jiangkou County

1.“真钱”与“假钱”

调查中了解到，江口民众对于所用的纸钱有“真钱”“假钱”之说。他们把纯手工造的竹纸钱当作“真钱”，把半机械造的竹纸钱当作“假钱”。他们认为只有“真钱”老祖宗才能“收到”，祭祖才能产生效用。因此，虽然纯手工造的竹纸价格相对高一些，但大家仍愿意去买。民众的这一“买账”行为使得纯手工造纸仍有较大的生存空间。而“假钱”的存在，是由于周边一些县的民众不在乎“真钱”“假钱”的说法。因此，造纸户也可采用半机械化生产来提高造纸效率，进而获取更大的经济效益。

2. 节日用纸习俗

（1）春社。从立春起，数到第五个戊日，就是社日。这一天不仅要敬土神，为新坟挂社青，还要准备酒、肉、糖果等祭品，烧香、化纸钱、放鞭炮祭奠。

（2）清明节。江口人过清明节，要备办酒、肉、菜肴等祭品，为故去的亲人挂青扫墓。为坟墓培上新土后，要在坟顶插一束白纸，放鞭炮、化纸钱、烧香祭奠。

（3）七月半。即农历七月十五，又称鬼节。相传次日鬼怪横行，不宜出门。因此，江口民间过七月半时不许孩子乱跑，还要烧纸钱敬鬼神。祖宗也要回家看望子孙，家人要把房屋收拾干净，准备酒、肉、香烛，为祖宗烧包封（送给祖宗的“真钱”信封），为野鬼焚纸钱。土家人还相信，七月半时若看见蛇、大虫进屋，不能打，这是祖先回家来了，做子孙后辈的要烧纸、烧香祭祀。

⊙21

⊙21 河口村打纸钱的老人
An old papermaker making joss paper in HeKou Village

3. 丧葬用纸习俗

当老人落气瞑目时，举家痛哭，焚烧纸钱、鸣放鞭炮送亡灵“归天”。入殓时，要在遗体周围用土纸或死者的旧衣服塞稳当，再盖好寿被，最后盖棺。天黑以后，请道士给死者“开路”，吹响牛角或海螺，然后锣鼓齐鸣，孝子跪在灵前。道士一边打钹，一边唱“开路词”，其间要酌酒几次，烧化纸钱、包封两次。晚上10点后开始“打绕棺”。孝子们以长幼次序排列，在道士引导下绕着棺材走。绕棺时，道士说唱古今，从一月唱至十二月，每唱完一个月，就要烧香焚纸奠酒，唱到十二月的花开时，要“吹花”，即将纸钱点燃后朝棺材上吹，青年们互相吹火扔火烫人，称为“捞花”。出殡日凌晨4点许，举行闭殓仪式，每个晚辈剪下衣角放在死者手里。另有一个三角形白布袋，内装烧化的纸钱灰，系在死者胸前布扣上，放在右肋下。送葬时，一人在前丢“买路钱”，接着是“引魂幡”，长子背“望山钱”，其他孝子端灵牌、持孝幛以及花圈，棺材随后。从圆坟日起，每七天要烧一次包封，七七四十九天为“满七”。当地的仡佬族下葬时不丢“买路钱”，也不做“买山”，他们认为自己是本地的老户、拓荒者，无需向别人买地[6]。

4. 梵净山朝山拜佛用纸

江口云舍村位于梵净山脚下，山上佛教寺庙遍布，规模宏大，号称有48座寺院，是全国佛教文化旅游和生态旅游的重要目的地之一，每年前来的考察人员、游客、香客数以万计[7]。梵净山每年的香客对祭祀用纸有大量的需求，而云舍竹纸正是香客的重要消费品。

八 江口土家族竹纸的保护现状与发展思考

8 Preservation and Development of Bamboo Paper by the Tujia Ethnic Group in Jiangkou County

(一) 江口土家族竹纸的生产性传承现状

在传统农业社会里，受祭祀文化等的影响，土法造纸厂的市场宽广，因此从事生产的人也多；而到了工业化的今天，一方面农业社会型的消费市场急剧萎缩，另一方面手工造纸是一项工序繁多、劳动强度很大的工作，其经济利润很低，因此，从事手工纸生产的人越来越少，也鲜有年轻人愿意学习这些传统技艺。调查中，云舍村与河口村坚持祖业的造纸户大多只是在农闲季

[6] 江口县地方志编纂委员会. 江口县志[M]. 贵阳：贵州人民出版社.1994.134-136.
[7] 张凤科. 佛教圣地——梵净山[M]//政协江口县文教联谊委员会. 江口文史资料. 第5辑：梵净山专辑,1998:43-45.

节（主要在冬天）从事生产活动，以往从事造纸的大多数人员都外出务工或经商，这使得江口的手工造纸传统技艺面临无人或少人传承的局面。

与此同时，由于云舍村土法造纸需要大量的嫩竹，加之竹子生长缓慢，嫩竹被砍后，竹子面积越来越少，甚至需要外购原料。这既破坏了植被，也增加了造纸的成本。

由于尚未充分认识到手工纸制造的文化价值和历史价值，当地政府不但对此缺乏保护措施，反而在开发云舍土家民族文化村旅游的过程中拆除了相当一部分经营多年的手工竹纸作坊。

⊙ 1

⊙ 2

⊙ 3

⊙ 4

⊙ 1 河口村的山道 Country road in Hekou Village

⊙ 2 云舍村口的纸坊 Papermaking mill in Yunshe Village

⊙ 3 云舍村的旅游宣传牌 Tourism billboard in Yunshe Village

⊙ 4 调查组成员在通往云舍村的山道上 A researcher walking on the country road to Yunshe Village

(二) 江口土法造纸文化的发展与保护

作为一种具有鲜明民族特色的非物质文化遗产，江口手工竹纸制作工艺具有重要的文化传承意义和价值，应得到发展和保护。

云舍土法造纸作为一种文化遗产，无论是有形的，还是无形的，都应该确保其在不失技艺原真性的前提下，能进入市场并通过市场运作，完成积极的保护及潜能的开发工作。由于云舍村已是贵州著名的旅游景点，因此，应将尚存规模的手工造纸与当地的傩戏、饮食文化、特色建筑、自然风光互相统一，共同促进，形成旅游资源组合，打造更多元的旅游文化内容。

要实现这一新业态的转变，有下述问题值得关注和促进：

(1) 云舍土法造纸的价值在于保存了历史的记忆，因此，在开发的过程中必须坚持体现它的历史文化价值，但本真性并非是完全再现、僵固传递，而是在不扭曲历史文化的记忆下，不断融入新的元素，以新的方式，包括现代媒体方式，呈现于外来人面前。

(2) 虽然土法造纸不会对环境产生实质性的污染，但规模化的原料采集加工对山水环境的破坏还是很明显的，因此，在以旅游文化立足的云舍村，长期性地大量生产低端竹纸并不是很好的选择。可考虑将土法造纸由产纸卖钱向技艺展演和纸加工品消费的旅游经济转变。

(3) 挖掘纸文化，打造文化旅游，树立产业化的发展思路，对云舍村土法造纸重新进行科学的品种定位，制定合理的营销战略，集中力量培育文化品牌，将这种文化资源优势转化为品牌经济优势。

(4) 唤醒云舍村土家族对本民族土法造纸的情感，加强民众对土法造纸技能和土作坊群的保护和传习意识，使江口土家人明白保护土法造纸技术和土作坊群的民族文化基因价值。

江口土家族

竹纸

Bamboo Paper by the Tujia Ethnic Group in Jiangkou County

云舍村竹纸透光摄影图

A photo of bamboo paper in Yunshe Village seen through the light

第四节

印江合水镇皮纸

贵州省
Guizhou Province

铜仁市
Tongren City

印江土家族苗族自治县
Yinjiang Tujia and Miao Autonomous County

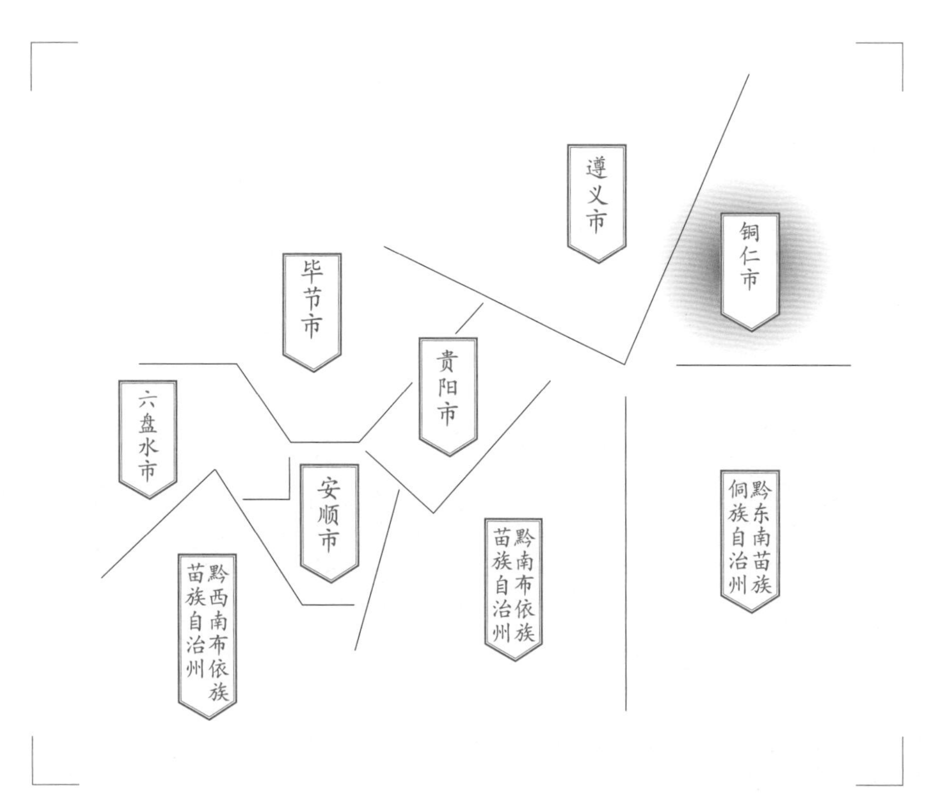

调查对象
合水镇兴旺行政村皮纸

Section 4
Bast Paper
in Heshui Town
of Yinjiang County

Subject
Bast Paper in Xingwang Administrative Village of Heshui Town

一
印江合水镇皮纸的基础信息及分布

1
Basic Information and Distribution of Bast Paper in Heshui Town of Yinjiang County

在黔东北印江土家族苗族自治县合水镇，清澈的印江河穿镇而过，沿河两岸从兴旺村至新场村之间7.5 km范围内的村寨中，分布着众多传统手工造纸作坊，其中以兴旺、香树坪、坪楼三个行政村区域的作坊最为集中。作坊区域总面积约7 000 m^2，传承着一个完整的传统造纸工艺体系，所留存和正在使用的造纸工具和设施体系，数量和规模都很可观。

2008年12月，调查组对兴旺村展开了重点考察以及重点访谈了当地造纸户。兴旺村距合水镇镇政府所在地0.8 km，下辖10个村民组，约1 840人，土家族、苗族聚居，以蔡伦古法造纸闻名，其手工造纸业态是印江手工造纸的典型代表。兴旺村依山而建，一走入兴旺村，"噼啪"的脚碓声便不时传入耳帘；屋外的墙壁上，一排排整齐晾晒的纸张时常可见；村舍旁水渠边的抄纸房中，造纸户正在抬帘抄纸。在离村不远处的山脚下，沿着蜿蜒的印江河畔，兴旺村的两侧各分布着一个规模较大的造纸作坊群。两处的建筑各有特点，一处以草棚为主，一处以瓦房为主。蒸料窑林立其间，河畔的水车正在转动着，步入其间，历史的沧桑和传统的厚重扑面而来。

⊙1

⊙2

兴旺村手工造纸以楮皮为主要原料，造出的皮纸当地习称为"白皮纸"，又称为"文章纸"。调查组入村时，兴旺村全村有100多户从事手工造纸，年产白皮纸300余吨。兴旺村所产纸的主要用途有书写，绘画，制作纸伞、斗笠、窗花、鞭炮引线及祭祀等。

⊙1 兴旺村以草棚为主的造纸作坊群 Thatched papermaking mills in Xingwang Village

⊙2 成排的蒸料窑 Row of steaming kilns

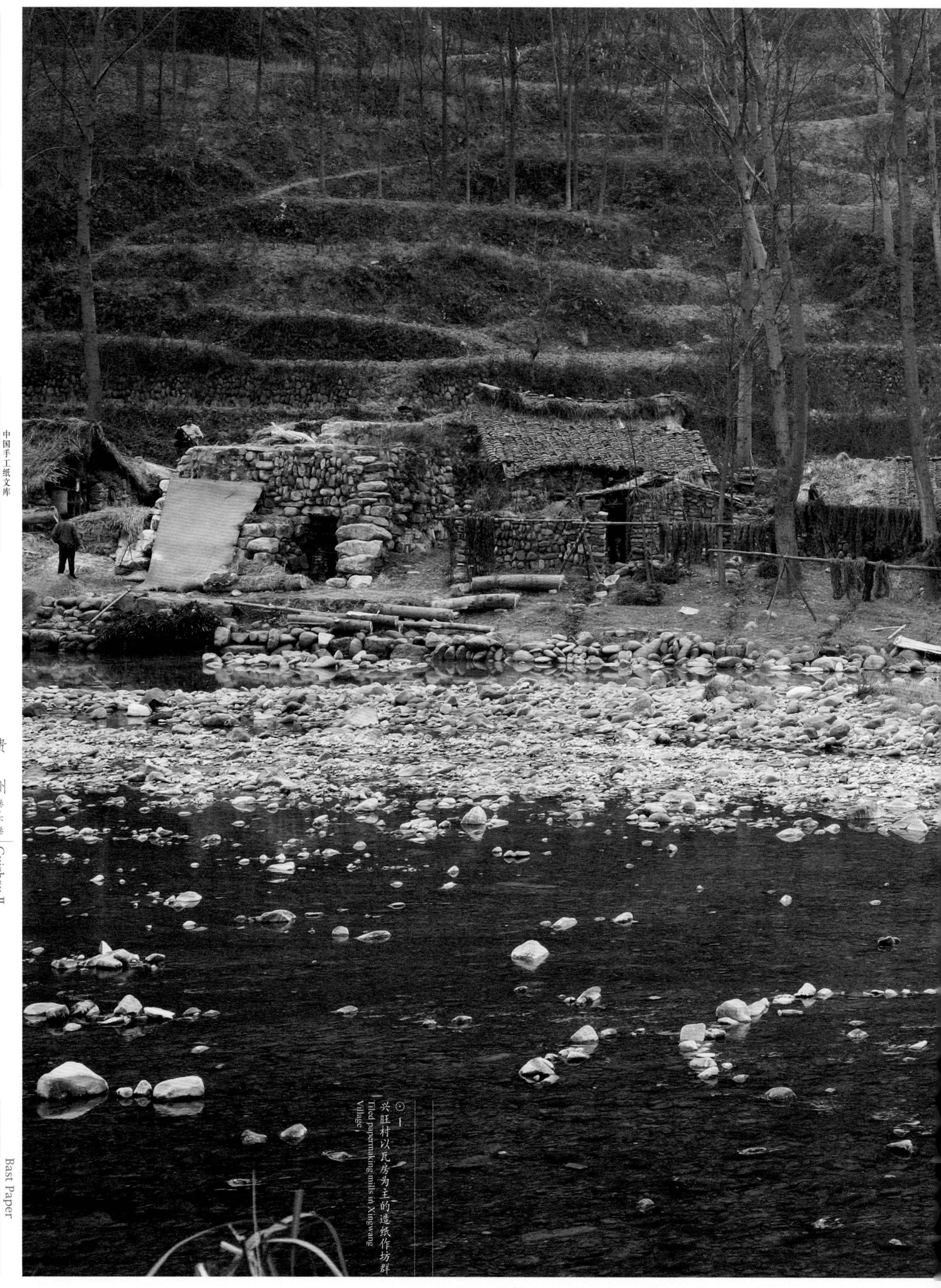

⊙1
兴旺村以瓦房为主的造纸作坊群
Tiled papermaking mills in Xingwang Village

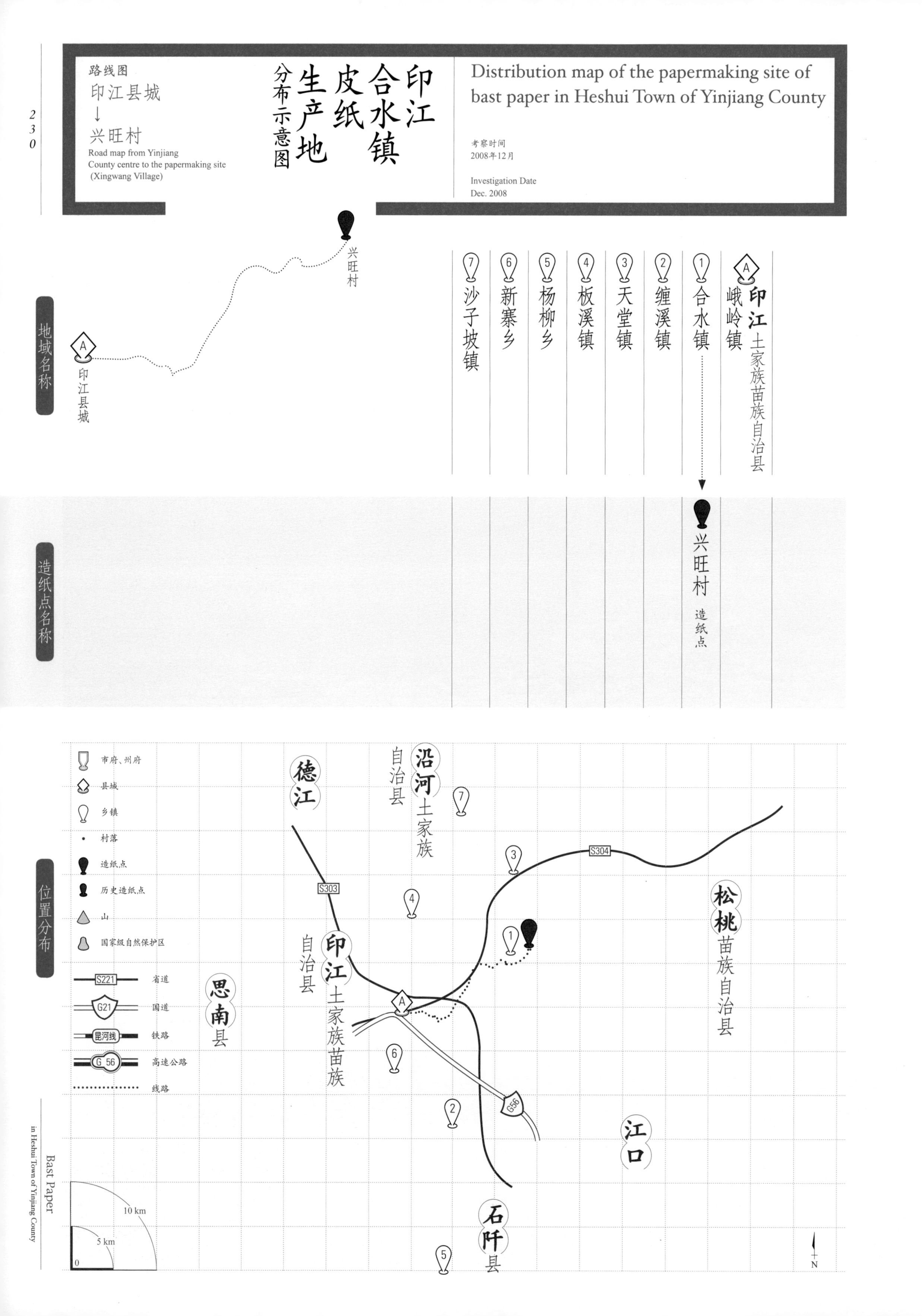
230
路线图
印江县城
↓
兴旺村
Road map from Yinjiang County centre to the papermaking site (Xingwang Village)
印江合水镇皮纸生产地分布示意图
Distribution map of the papermaking site of bast paper in Heshui Town of Yinjiang County
考察时间
2008年12月
Investigation Date
Dec. 2008
地域名称
兴旺村
印江县城
印江土家族苗族自治县
峨岭镇
合水镇
缠溪镇
天堂镇
板溪镇
杨柳乡
新寨乡
沙子坡镇
造纸点名称
兴旺村
造纸点
位置分布
市府、州府
县城
乡镇
村落
造纸点
历史造纸点
山
国家级自然保护区
S221
省道
G21
国道
昆河线
铁路
G 56
高速公路
线路
德江
沿河土家族自治县
S304
S303
松桃苗族自治县
印江土家族苗族自治县
思南县
G56
江口
石阡县
10 km
5 km
0
N
Bast Paper
in Heshui Town of Yinjiang County

二
印江合水镇皮纸生产的人文地理环境

2
The Cultural and Geographic Environment of Bast Paper in Heshui Town of Yinjiang County

印江古称邛江，唐置思邛县，宋改邛水县。明弘治七年（1494年）设印江县，“邛江”改为“印江”。乡土传说改名源自当时的一个典故：皇帝在读地方官员呈报的邛江政况时，一不留神，把“邛”字误读为“印”字，君无戏言，从此“印江”之名就出现在中国的行政版图中。1986年，中华人民共和国国务院正式批准建立“印江土家族苗族自治县”。

印江土家族苗族自治县地处贵州省东北部、铜仁市西部，总面积1 969 km²，“八山一水一分田”，为典型的山区农业县。境内有印江河、车家河、洋溪河等河流，均属乌江水系。全县总人口43.76万，土家族、苗族占70%以上，少数民族民间文化资源丰富，有印江傩堂戏、印江狮子灯、苗子祭鼓节、印江油纸伞、梵净山团龙供茶等。印江境内书画学习爱好者众多，有“书法之乡”的美誉。此外，印江的洁净空气和水土环境也孕育了高品质的自然条件，又有“中国长寿之乡”的称号。

⊙ 1

⊙ 2

⊙ 1
印江的山水
Mountains and rivers in Yinjiang County

⊙ 2
路边的大型印江文化宣传牌
Cultural billboard of Yinjiang County by the road

印江日照时间长，雨量充沛，年降水量1 100 mm左右，属中亚热带温暖湿润季风气候区。县境东部的国家级自然保护区——梵净山，得名于“梵天净土”，被誉为“贵州第一名山”，是中国五大佛教名山之一，设有我国唯一的弥勒佛道场。

印江县合水镇为1992年新建镇，距离梵净山约20 km，因源自梵净山的木黄河、永义河在此地交汇成印江而得名。印江河在合水镇过境20 km处，水力资源比较丰富，为白皮纸的生产提供了优质水源。

合水镇少数民族人口占总人口的92.5%，以土家族、苗族为主，有土家长号、花灯、金钱杆等乡土民俗和民间非遗文化。

合水镇所辖兴旺村，也是一个以土家族、苗族为主聚集的村落，以“蔡伦古法造纸”而闻名。兴旺村地少人多，制作白皮纸为村民主要的副业之一，造纸户以蔡、卢、田、帅四姓为主。2000年，当地政府将兴旺村列为“民族风情观光村”。以“中国蔡氏古法造纸艺术之乡”而誉称的印江，其核心地正是合水镇。

三
印江合水镇皮纸的历史与传承

3
History and Inheritance of Bast Paper in Heshui Town of Yinjiang County

关于印江合水白皮纸的起源，2008年底调查组入兴旺村调查时，造纸户讲述了两个不同的版本：(1) 当地蔡姓造纸户认为他们是蔡伦后裔，明洪武年间，精通古法造纸的蔡伦后裔的一支为求生计，辗转从江西来到合水，见这里山清水秀，造纸条件良好，便在此安家落户，制作白皮纸。正因为如此，该地至今每家都供有蔡伦祖师神位，并在每年农历三月十九蔡伦生辰时举行纪念活动。(2) 据合水卢姓造纸户相传，白皮纸的生产始于明末清初，造纸技艺由当地卢姓始祖卢依远从江西入黔时带入。卢依远一开始寻找了很多地方，最终因合水盛产构皮而在此定居造纸，于是，当地造纸业不断发展起来。

据《民国贵州通志·风土志》记载，印江白纸“莹洁如玉，似明代白棉纸，临帖印书并精绝”。在清代，印江纸还被指定为贵州科举考试专用纸之一。

从清末到民国，自合水镇的兴旺村至新场村，沿印江河两岸五六百户人家皆从事白皮纸的生产。抗日战争期间，合水白皮纸的生产达到鼎盛，因供不应求，在重庆经营手工纸的五大商号经理岳茂林、岳金廷、张山云、王德山等曾长期在合水街上“坐收”白皮纸。为了维护造纸户和消费者的利益，当时各造纸户共同集资成立了具有行业协会性质的“蔡伦会”，从会员中选举德高望重者为会首，操持会中日常事务。

⊙1

⊙2

⊙1 调查组成员进入兴旺村 Researchers arriving at Xingwang Village

⊙2 印江河畔的造纸现场 Papermaking site alongside Yinjiang River

四
印江合水镇皮纸的生产工艺与技术分析

4
Papermaking Technique and Technical Analysis of Bast Paper in Heshui Town of Yinjiang County

(一)
印江合水镇皮纸的生产原料与辅料

合水镇兴旺村传统白皮纸所用原料为构皮（楮皮），但自20世纪50年代起，造纸户为了生存以及提高竞争力，开始在构皮浆中加入现代造纸工业生产所用的由旧书、旧报等制成的再生纸浆，这一方式一直沿用至今。在合水当地除非订制，否则大多不会用纯构皮进行生产。

兴旺村使用的纸药为被当地人称作“松蒿”的植物。松蒿汁作为浆料的悬浮剂，能调节浆料的滤水性，改善纸浆的上帘性能，有利于纸的分揭，提高成纸率等。

辅料有石灰、纯碱等。构皮通过石灰浸渍后，再堆置、蒸煮，可除去其中所含部分果胶、色素等物质。纯碱用于蒸料的工序中，可脱除构皮中大量的木素。

⊙ 3
采下来的松蒿
Phtheirospermum japonicum used as papermaking mucilage

⊙ 4
调查组成员向造纸人认真核实工艺
A researcher confirming the papermaking technique with the local papermakers

(二)

印江合水镇皮纸的生产工艺流程

据调查组对兴旺村的实地调查和多轮访谈，归纳合水镇皮纸的生产工艺流程如下：

壹	贰	叁	肆	伍	陆	柒	捌	玖	拾	拾壹
砍构树	剥构皮	晒构皮	泡构皮	浆构皮	堆构皮	上窑	蒸皮	出窑	漂花皮	踩花皮

拾贰	拾叁	拾肆	拾伍	拾陆	拾柒	拾捌	拾玖	贰拾	贰拾壹	贰拾贰
晾料子	踩料子	进水	榨筒子	捧料子	蒸料子	漂料子	二次踩料子	拆料子	舂皮板	切皮板

贰拾叁	贰拾肆	贰拾伍	贰拾陆	贰拾柒	贰拾捌	贰拾玖	叁拾	叁拾壹	叁拾贰	叁拾叁
洗料	舂料粑子	搅槽	抽松蒿	坎垛	舀纸	压垛	晒纸	收纸	齐纸	捆纸

壹
砍构树
1

每年3~8月，选择2~3年生的构树枝条砍伐。在砍伐时，造纸户通过察看枝条表皮颜色的深浅来判断枝条的生长年限，老的构树枝条不选择，一年生的构树枝条因纤维少，出皮率低，一般也不选择。

贰
剥构皮
2

砍下构树枝条后，随即剥下其皮。

叁
晒构皮
3

将剥下的构皮运回家，摊放在地上晒，天气好时2天即可晒干，天阴时需4~5天。

肆
泡构皮
4 ⊙1

将晒干的构皮按每捆2~2.5 kg捆好，然后运到河里，一排排整齐地码放在水流平缓的河中，在清澈的河水中浸泡2~3天。

⊙1

⊙2

伍
浆构皮
5 ⊙2

将泡软后的构皮运到石灰池边，然后用捞钩一捆捆地钩起并浸入石灰浆中，使构皮沾满石灰水，随即钩出。一般一个人一天可浆构皮500 kg左右。

⊙3

⊙4

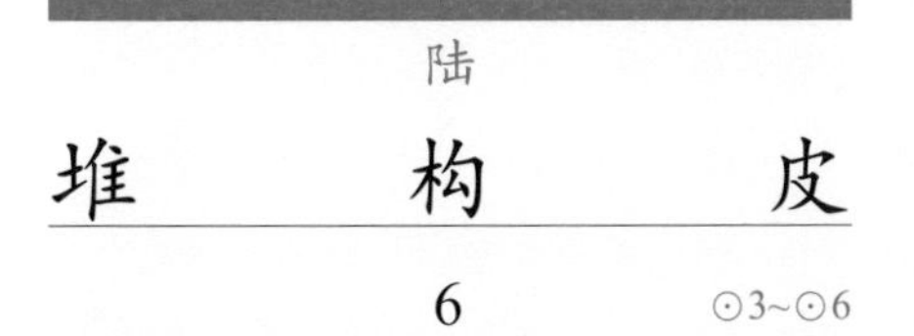

陆
堆构皮
6 ⊙3~⊙6

一般两人配合，一人浆皮，一人钩皮堆放。将浆好的构皮整齐地码放在石灰池边，码成一堆后，将一把把稻草整齐有序地铺盖在构皮堆上，最后搓一根粗草绳将稻草穿结连为一体，防止稻草被风吹走。铺盖稻草是为了避免将浆过石灰的构皮直接暴露在空气中，从而使构皮中的石灰晒干或被风吹干，同时可防止雨水将石灰冲走。一般需要捂盖堆置3~5天，每堆有1 000~1500 kg构皮。构皮一般堆放在石灰池与蒸料窑之间。

⊙5

⊙6

⊙1 泡构皮 Soaking paper mulberry bark

⊙2 浆构皮 Fermenting paper mulberry bark

⊙3/4 堆构皮 Piling paper mulberry bark

⊙5 盖稻草 Covering the papermaking materials with dried straw

⊙6 结草绳 Twisting straw robe

柒 上窑

7 ⊙7~⊙9

构皮堆放发酵后，拨开盖草，站在蒸料窑上，用捞钩将构皮一一钩入蒸料窑的横担上，并一圈圈一层层整齐地码放。当构皮与蒸料窑上口基本平行时，取料槌，用力捶打，将窑内堆放的构皮捶紧捶实，然后在其上面覆盖塑料薄膜，并用草把和石块将塑料薄膜四周压住。

⊙7

⊙8

⊙9

⊙ 10

⊙ 11

⊙ 12

捌 蒸皮

8 ⊙10~⊙12

构皮上窑后，往蒸料窑中部的蒸锅中加满水，然后在蒸料窑底部燃火蒸料。当蒸锅中的水沸腾后，随着蒸汽的上升、温度的增高，构皮会逐渐变软变粑，然后揭开塑料膜，取来料槌，再次用力捶打构皮，使窑中的构皮更加紧密，捶实后，窑内构皮的高度会变低些，然后用捞钩再钩放些构皮将窑填满，用料槌再次捶紧，盖上塑料膜，继续燃火蒸料。经过多次捶料、加料、蒸料，一般4~5天后，窑内构皮不再下沉。窑火工大（火烧得旺些）时，一天加一次构皮；窑火工小（火烧得小些）时，两天加一次构皮。构皮上窑蒸6~7天后熄火，将窑内构皮全部有序地钩出，然后颠倒次序将构皮再次码放入窑内，捶紧，盖窑，燃火，再继续蒸。从构皮上窑到翻窑继续蒸，一般总共要蒸10~16天。在蒸料时蒸料窑中蒸锅里的水会逐渐减少，每天需往蒸锅内加2~3次水，或在确保蒸锅不会干锅的情况下每天加满一次水即可。在蒸料窑的一侧有一个加水口，水是从这里加入蒸锅的，也可从这里观察蒸锅内水量的多少。一个蒸料窑一般可放1 500~2 000 kg构皮。以前蒸料时用的是柴火，而现在多用煤烧，每蒸50 kg构皮需用20 kg煤，煤的价格为0.52元/kg。

玖 出窑

9 ⊙13

构皮蒸好后，用料钩一一将其从蒸料窑中钩出。3~4个人一天可钩完1500 kg构皮。钩出构皮后，将其挑运到河边，一捆捆地在水中用手搓揉后，再次整齐地摆放在河里，用河水自然冲洗。此时蒸过的构皮斑点较多，当地称为花皮。

⊙ 13

⊙7/8 钩料上窑 Lifting the papermaking materials onto the kiln

⊙9 捶窑 Hammering the papermaking materials

⊙10 盖窑 Covering the kiln

⊙11 窑中蒸料 Steaming the papermaking materials in the kiln

⊙12 造纸人正在观察蒸锅里的水量 A papermaker checking water volume in the steaming wok

⊙13 流水冲洗皮料 Cleaning the papermaking materials in the running river

拾 漂花皮

10 ⊙14

花皮随水流自然漂洗2~3天后，用手一捆捆地将花皮依次在水中抖动、搓揉，然后将花皮捆提起翻身，再次浸入水中摆放整齐，让河水再冲洗1~2天。

⊙14

⊙15

拾壹 踩花皮

11 ⊙15

将漂好的花皮放入河里再次搓揉、清洗，然后运到岸边，一捆捆地放在钢板（石板）上用脚踩踏，使花皮变得更加松软。一个人踩3~4天可踩完1 500 kg花皮。

拾贰 晾料子

12 ⊙16~⊙20

踩好花皮后，将捆着的腰带解开（捆构皮，一般亦用构皮来捆，当地称之为腰带），整齐地搭挂在晒杆上自然晾晒，晾晒的同时将其中较宽的花皮拣出，用手撕成细条。阳光强或风大时，3~4天即可晒干，阴天则需一周左右。晾晒期间还需将花皮翻晒一次，使花皮晾晒得更均匀，否则有些花皮被压在下面不易晒干，可能会腐烂。晒干后的花皮叫作料子，因其很硬，且较为细长，故又称作铁线料子。一个人2~3天可完成1 500 kg花皮的搭晒。料子晒干后捆扎，储存备用。

⊙16

⊙17

⊙14 漂花皮 Cleaning the spotted papermaking materials

⊙15 踩花皮 Stamping the spotted papermaking materials

⊙16 晾料子 Drying the papermaking materials

⊙17 翻晒料子 Turning the papermaking materials for drying

⊙18

⊙19

⊙20

拾叁
踩料子

13 ⊙21

把铁线料子放到河里浸泡十几分钟，待其柔软后再放到钢板上用力踩，一个人5~6天可踩完1 500 kg料子（一般一个人一天可踩250~300 kg料子，包括清洗、榨筒子）。

⊙21

拾肆
进水

14

将踩好的料子运到河里，再次清洗，把料子上附着的皮壳等杂质洗净。

拾伍
榨筒子

15

将洗净的料子放到钢板上用脚踩干。

拾陆
捧料子

16

将料子捧松，然后加碱，一次捧多少料子，可根据生产需求而定，一般捧一次的料可用两周。1 500 kg构皮加纯碱30 kg，纯碱价格为4元/kg。

拾柒
蒸料子

17

在料子捧松并加入纯碱后，需上窑子蒸，原来在印江河边有专门的捧灰窑子，比蒸料窑小些，现在已不存在。生产规模小的造纸户，常在自家灶上蒸；生产规模大的造纸户就用蒸料窑蒸。一般一次最多可蒸500 kg构皮，最少可蒸100 kg，蒸两天两夜，每蒸500 kg构皮用煤150 kg。

⊙18 捆料子 Binding the papermaking materials

⊙19/20 挑料子 Carrying the papermaking materials

⊙21 踩料子 Stamping the papermaking materials

⊙ 22

拾捌 漂料子

18 ⊙22~⊙28

揭窑，将蒸好的料子从蒸料窑中扔到铺有塑料布的地上，然后挑运到河里，将料子一把把地在水中漂洗，然后将料子整齐地摆放在河里，让河水冲洗3~4小时，将料子中的碱液等杂质洗净。为防止料子被水冲走，通常每摆放一把料子就在其上面压上一块鹅卵石。一个人一天可漂200~250 kg构皮。

⊙ 23

⊙ 24

⊙ 25

⊙ 22 / 23
揭窑扔料
Lifting the papermaking materials from the kiln to the ground

⊙ 24 / 25
挑料至河中
Carrying the papermaking materials to the river

⊙26

⊙27

⊙28

拾玖
二次踩料子
19

将漂干净的料子运到钢板边，并一把把地放在钢板上用脚踩，将料子中水分挤出。一个人一天可踩250~300 kg料（只踩料），最多可踩500 kg。

贰拾
拆料子

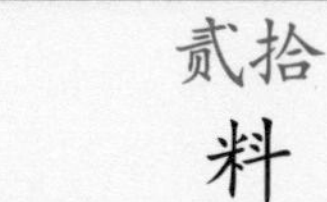

20 ⊙29

⊙29

用手将料子中夹杂的皮壳等杂质拣除干净，一天可拣250 kg左右料子。

贰拾壹
舂皮板
21 ⊙30⊙31

用水碓或脚碓将料子一把把地舂融。水碓一人翻料子即可，脚碓需两人配合，一人踩碓，一人翻料，随着碓头的上下碓打，不时地翻转、折叠料子，直至将料子打碎、打融，看不见构皮纤维束，结成一皮板状。若用水碓，一天可舂200~250 kg皮板；若用脚碓，一天可舂100~150 kg皮板。

⊙30

⊙31

兴旺村的水碓都设在印江河畔；为了生产、生活上的方便，造纸户都将脚碓设在自家房前屋后的空地上。

⊙26/28 漂料子 Cleaning the papermaking materials

⊙29 拆料子 Picking out the impurities

⊙30 脚碓 Foot pestle

⊙31 水碓 Hydraulic pestle

贰拾贰

切 皮 板

22 ⊙32

将舂好的皮板折成数叠，放在切皮凳上，再用一根铁丝圈套在皮板上，下面坠一木棍，取来切皮刀，造纸人骑跨在切皮凳上，右脚着地，左脚踩住木棍，带紧铁丝固定住皮板，双手持刀，由前往后将皮板切成薄薄的皮片。切皮板时，随着皮板长度的减少，不时调整铁丝圈的位置，固定皮板，直至将皮板切完。一个人一天可切150~250 kg皮板。

⊙32

贰拾叁

洗 料

23 ⊙33

皮料切好后装入编织袋中，封口，拎到河边，放入浅而清澈的河水中，用脚踩，并不时翻折袋子，将皮料中细小的杂质及颗粒从袋中挤出，洗净皮料，再将袋子放到高于水面的石板上继续踩，直至将袋中水分踩干为止。

⊙33

贰拾肆

舂 料 粑 子

24 ⊙34

把洗净的皮料放在碓窝舂，需用多少量就舂多少，一般一天需用10~15 kg皮料。如果在构皮纸浆中加入工业纸浆，则需事先将碾好的工业纸浆混入皮料中，一起舂捣，将纸料舂细舂融。混入工业纸浆的比例按市场的需求来定。舂好后的纸料称为料粑子。

贰拾伍

搅 槽

25 ⊙35⊙36

将料粑子放到捞纸槽里，加水，先用搅料棒搅槽十几分钟，然后用槽棍打槽将料粑子搅散、搅融，使料粑子纤维均匀分布在纸槽中，整个过程几分钟即可完成。

⊙34

⊙35

⊙36

⊙32 切皮板 Trimming the papermaking materials

⊙33 踩踏洗料 Stamping the papermaking materials for cleaning

⊙34 舂料粑子 Beating the papermaking materials

⊙35/36 搅槽 Stirring the papermaking materials

⊙37

贰拾陆

抽　松　蒿

26　⊙37~⊙42

将松蒿放入纸药池中加水浸泡，出汁。纸药池底部有一出口，将一个很长的蒿口袋（布袋）一头套住纸药池的出水口，一头放入捞纸槽中，利用纸药池与捞纸槽的高度差，将所需松蒿汁抽到捞纸槽中。蒿口袋有过滤的作用，可得到纯净的松蒿汁，阻挡住松蒿皮肉、纤维等流入纸槽。当松蒿汁不易自行从蒿口袋中流入捞纸槽时，需用手挤压蒿口袋，使松蒿汁迅速渗出。待纸浆中加入松蒿汁后，先用拱耙将纸料拱搅起来，然后将拱耙倒拿，一物二用，以柄为棍搅动纸浆，最后放下拱耙，再用双手在纸浆上层搅动，整个过程约3分钟，使松蒿汁与纸浆充分融合，纸浆悬浮，分布均匀。

⊙38

⊙39

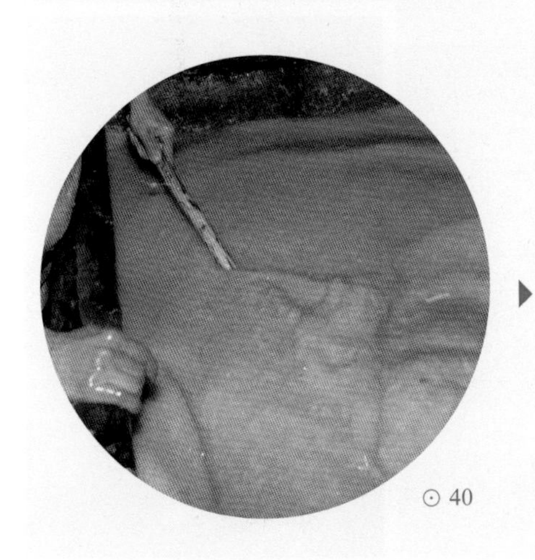

⊙ 40

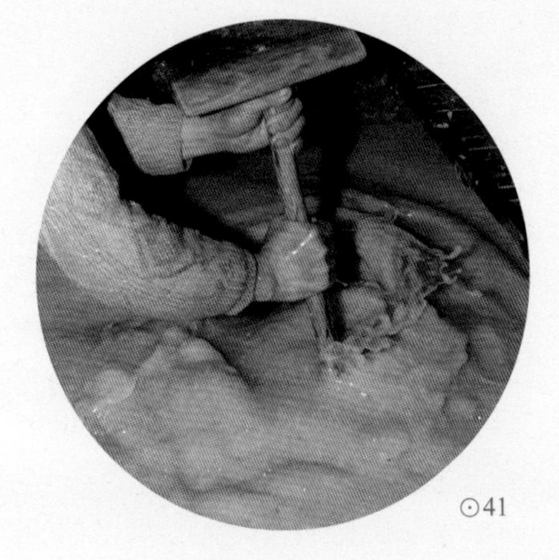

⊙41

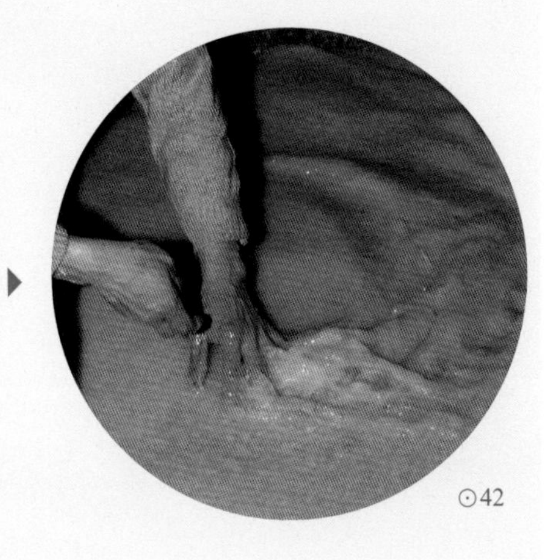

⊙42

⊙37/39 抽松蒿 Adding in *Phtheirospermum japonicum* as papermaking mucilage

⊙40/42 搅纸浆 Stirring the paper pulp

⊙43

⊙44

⊙45

⊙46

贰拾柒
坎垛

27 ⊙43~⊙45

多采用一人吊帘捞纸的方法捞纸。捞纸时捞的前10张纸叫坎垛，因为前10张捞的纸很厚。坎垛捞好后，停工一小时，让料下沉一些，再开始捞纸。当地称捞纸为舀纸。在停工期间，为保持坎垛一直有水分，先将纸帘、帘架洗干净，然后将纸帘铺在坎垛上面。

⊙47

⊙48

⊙43 坎垛 First ten pieces of paper

⊙44/45 洗纸帘及帘架 Cleaning the papermaking screen and its supporting frame

⊙46/48 舀纸 Scooping and lifting the papermaking screen out of water and turning it upside down on the board

贰拾捌
舀　纸

28　⊙46~⊙51

舀纸时，用两根“架尺”将纸帘与帘架固定，然后端起帘架，从“旁关”处将纸帘斜插入纸浆中，随即翘起帘架，往回带，顺势将帘架“插水”端没入纸浆中，翘起，再向前送，平端帘架，然后前后振荡几次。再将“旁关”端沉入纸浆中，以获得一边较厚的“纸头子”；立即抬起，从“插水”端沥去多余纸浆，钩回推杆，将帘架搭在推杆上。拿走“架尺”，揭起纸帘，转身将纸帘反扣在纸垛上，揭离纸帘，一张湿纸便留在纸垛上，随即取一瓷片在计数器上计数。一般一个人一天可舀纸800帘左右。舀完几十张纸后，纸浆会渐渐下沉，影响舀纸质量。为了得到厚薄均匀的纸张，需用槽棍将料再次搅起，然后支起帘架，在舀纸前再用“架尺”在纸帘两侧搅动纸浆后，继续舀纸。当纸槽里纸浆减少后，需要向纸槽内补充纸浆，之后，同样需进行搅槽后再开始舀纸。

⊙49

⊙50

⊙51

贰拾玖
压　垛

29　⊙52⊙53

一垛湿纸舀好后，用木榨压垛，经过几次扳榨、紧榨、松榨的过程，缓慢地将纸垛中的多余水分榨出，并紧榨静置到第二天早上，压干纸垛中的水分。现在有的造纸户改用千斤顶压垛。

⊙52

⊙53

⊙49/50 再次搅槽 Stirring the papermaking materials for the second time

⊙51 加纸浆 Adding in paper pulp

⊙52 压垛 Pressing the paper

⊙53 千斤顶压垛 Pressing the paper with a lifting jack

⊙54

叁拾
晒纸

30 ⊙54⊙55

将压干的纸垛搬靠在纸垛架上，从右上角开始，将湿纸一张张地从纸垛上揭下，就势刷贴于平整的墙面上，以前规定每5张为一贴，错落晾晒，现在5张、7张、10张为一贴的都有，根据场地大小、天气情况，每贴的数量各有不同。厚的夹皮纸（二合一的二层纸）最多3张为一贴。兴旺村晒纸，都是将纸垛压干后运回家，在室外平整的墙面上刷贴晾晒。

⊙55

叁拾壹
收纸

31

一张张、一贴贴地将晒干的纸从墙面上撕下来。

叁拾贰
齐纸

32

将撕下的纸四个角相对，理齐。

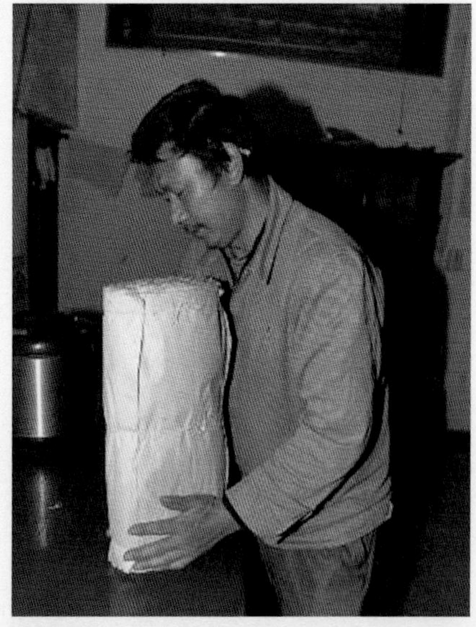
⊙56

叁拾叁
捆纸

33 ⊙56

每20贴为一刀，10刀为一捆，用一张厚夹纸将纸捆包好。现在有时用包装袋包装。一捆纸可卖150~160元。

⊙54 撕垛 Peeling the paper down

⊙55 刷贴 Pasting the paper on a wall for drying

⊙56 捆纸 Binding the paper

（三）

印江合水镇皮纸生产使用的主要工具设备

壹 甑子 1

又称“窑子”“蒸料窑”，用于蒸构皮，有上、中、下三层结构，最下层用于生火，中层为可加满水的大铁锅，在锅上横搭4~5根长约170 cm、直径14~15 cm的松木锅杆（圆木），锅杆之上为上层，为码放构皮处。实测甑子上口直径230 cm，上层高190 cm，可蒸1 500 kg构皮。

⊙57

⊙ 58

贰 灰池 2

又称“石灰池”，可放入石灰、水，用于浆构皮。多设在甑子边，在地上挖出方坑而得，旁边多有空地，用于堆放浆好的构皮。

⊙ 59

叁 料槌 3

木质，实测槌头长约70 cm，直径约13 cm；槌杆长约78 cm，直径约4 cm，质量约15 kg。

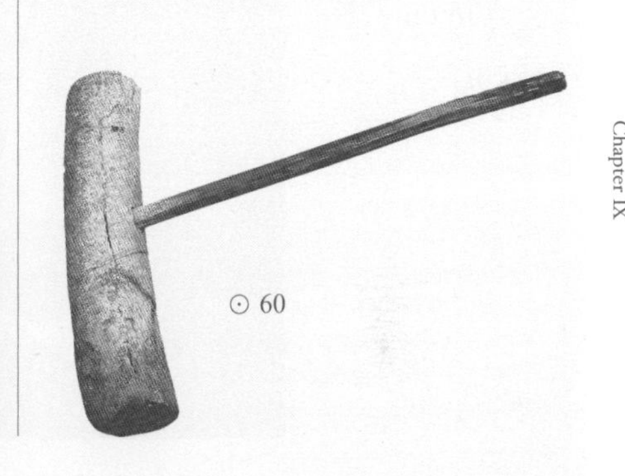

⊙ 60

肆 捞钩 4

用于钩捞构皮，实测总长约208 cm，其中钩头长约14.5 cm。

⊙ 61

⊙ 57/58 甑子 Papermaking kiln for steaming

⊙ 59 灰池 Fermenting pool

⊙ 60 料槌 Mallet for beating the papermaking materials

⊙ 61 捞钩 Lifting hook

伍
石碓与舂料棒
5

兴旺村多用其舂捣混合工业纸浆与构皮纸浆。实测石碓碓窝上口直径约36 cm，深约45 cm。

⊙ 62

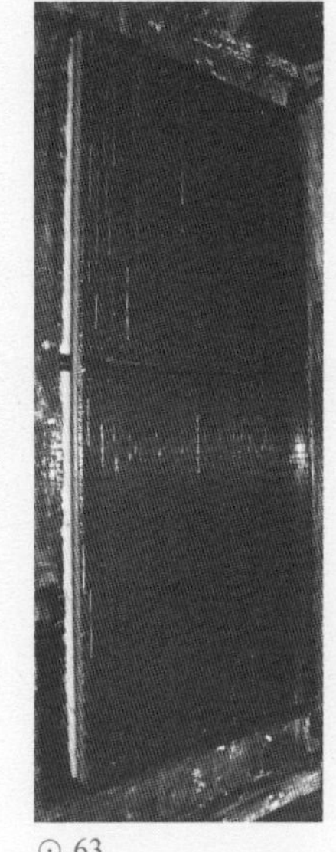
⊙ 63

陆
纸帘
6

兴旺村常用纸帘的尺寸为105.5 cm×43.5 cm，是一帘二纸型纸帘，纸帘中间有一粗线将纸帘分隔成左右两部分，所抄出来的纸，左右大小不等，长度分别约为45 cm和54 cm，这种一隔二分成大小不等两部分的纸帘是调查组至调查时为止见到的第一个，颇有特色。另测得兴旺村还有尺寸为67 cm×48.5 cm的纸帘。

柒
纸帘架
7

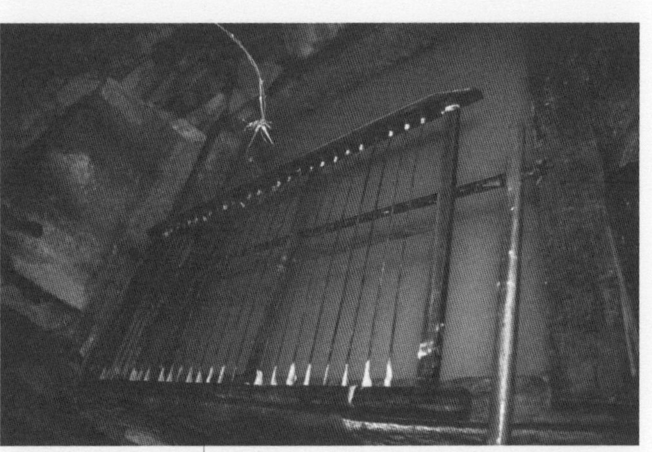
⊙ 64

实测纸帘架外部尺寸多为106 cm×55 cm和67 cm×60 cm（内部尺寸为62.5 cm×55.5 cm）两种。纸帘架左右两侧叫走水，舀纸时纸帘架靠近人的一边叫插水，远离人的一边叫旁关。舀出的纸靠近旁关一侧为纸头，靠近插水一侧为纸尾，纸头比纸尾厚。旁关：木质，长约106 cm，宽约10 cm；架尺：木质，长约46.5 cm，宽约2.2 cm，高约2.2 cm。

捌
拱耙
8

木质，T字形，用于搅槽。

⊙ 65

⊙ 62 石碓与舂料棒 Stone pestle and pestle stick

⊙ 63 纸帘 Papermaking screen

⊙ 64 纸帘架 Frame for supporting the papermaking screen

⊙ 65 拱耙 Papermaking rake

玖 捞纸槽 9

石质，用于盛放纸浆。实测长约135 cm，宽约116 cm，高约53 cm，底部石板厚约5.5 cm。

⊙ 66

拾壹 计数器 11

在桌上刻两组线条格，分别代表个位、十位。（大瓷片每片算10，小瓷片每片算1，大瓷片只有5个，只能算到50，50以上要自己脑记。图所示计数器上数字为162，即162=50×3+10×1+1×2。）

⊙ 68

拾 蒿槽 10

石质，用于盛放、浸泡纸药——松蒿。实测长约67 cm，宽约32 cm，高约38 cm，与捞纸槽相连，高于纸槽，底部留有出口。

⊙ 67

拾贰 石碾 12

由水车带动，用于制作工业纸浆，兴旺村工业纸浆的制作多集中在此处。

⊙ 69

⊙ 66 捞纸槽 Papermaking trough

⊙ 67 蒿槽 Trough for holding *Phtheirospermum japonicum*

⊙ 68 计数器 Counting apparatus

⊙ 69 水动力石碾 Hydraulic grinder

(四)

印江合水镇皮纸的性能分析

所测印江兴旺村皮纸的相关性能参数见表9.6。

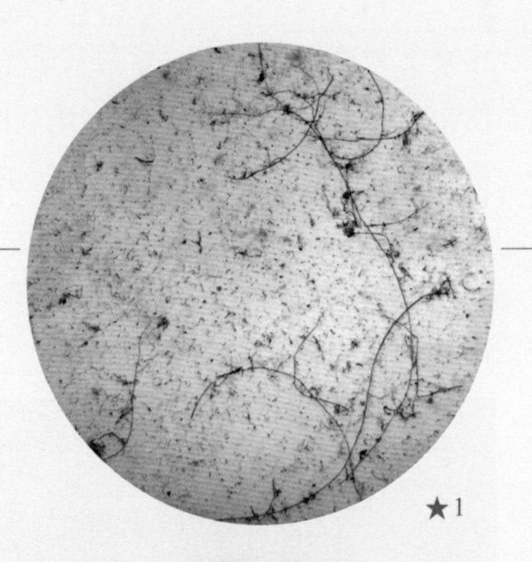

表9.6 兴旺村皮纸的相关性能参数
Table 9.6 Performance parameters of bast paper in Xingwang Village

指标		单位	最大值	最小值	平均值
厚度		mm	0.110	0.070	0.084
定量		g/m^2	—	—	15.9
紧度		g/cm^3	—	—	0.189
抗张力	纵向	N	11.9	8.5	10.3
	横向	N	3.2	1.2	2.2
抗张强度		kN/m	—	—	0.417
白度		%	35.2	34.6	35.0
纤维长度		mm	12.83	0.96	4.26
纤维宽度		μm	27.0	3.0	12.0

性能分析

由表9.6可知，所测兴旺村皮纸最厚约是最薄的1.57倍，相对标准偏差为1.0%，纸张厚薄较为一致。皮纸的平均定量为15.9 g/m^2。所测皮纸紧度为0.189 g/cm^3。

经计算，其抗张强度为0.417 kN/m，抗张强度值较小。

所测兴旺村皮纸白度平均值为35.0%，白度较低，白度最大值约是最小值的1.02倍，相对标准偏差为0.22%，白度差异相对较小，可能是因为印江合水皮纸加工时没有经过较强的漂白。

所测兴旺村皮纸纤维长度：最长12.83 mm，最短0.96 mm，平均4.26 mm；纤维宽度：最宽27.0 μm，最窄3.0 μm，平均12.0 μm。所测皮纸在10倍、20倍物镜下观测的纤维形态分别见图★1、图★2。

★1 兴旺村皮纸纤维形态图(10×)
Fibers of bast paper in Xingwang Village (10× objective)

★2 兴旺村皮纸纤维形态图(20×)
Fibers of bast paper in Xingwang Village (20× objective)

五 印江合水镇皮纸的用途与销售情况

5 Uses and Sales of Bast Paper in Heshui Town of Yinjiang County

（一）印江合水镇皮纸的用途

1. 传统用途广泛

在没有机制纸的年代，合水白皮纸广泛用于书写家谱、志书、公文、契约，绘画，制作斗笠、窗花、灯笼、纸扇、纸伞、鞭炮引线及祭祀等生活与文化、民俗活动中。当代，因机制纸的出现和现代文化的转型，其适用范围锐减。如今，书写公文、契约更多地使用机制纸，即便是祭祀用纸亦受机制纸的冲击，市场空间缩小。

2. 书写历史契约

在调查组实地考察过程中，偶然得遇一批使用合水白皮纸书写的契约，大多有准确的纪年标

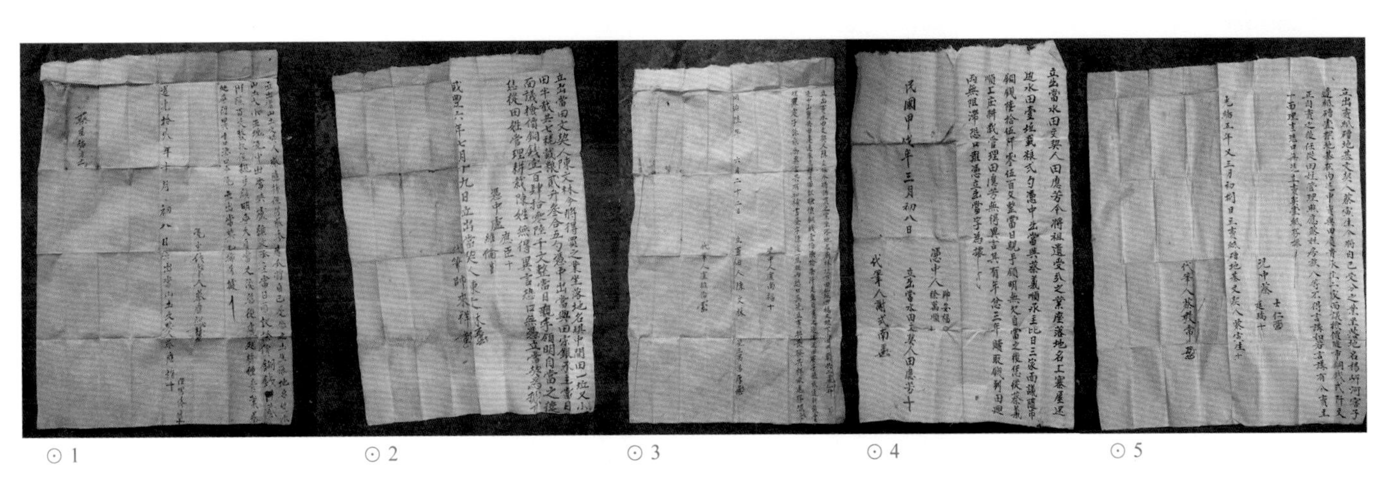

⊙1 ⊙2 ⊙3 ⊙4 ⊙5

记，历史跨度从清道光、咸丰、同治、光绪至民国时期，数量可观。从中我们寻见买卖捞纸槽的契约，这一历史资料的发现印证了合水白皮纸生产的历史悠久。从纸质本身来看，这批契约为纯构皮制作，有的薄如蝉翼，有的坚厚如皮，其制作工艺大多很精细，这从实物上佐证了合水白皮纸昔日的辉

⊙6

⊙1 光绪年间用白皮纸书写的卖纸槽地基文契 Contract for selling papermaking field written on white bast paper during Guangxu Reign of the Qing Dynasty

⊙2 道光年间用白皮纸书写的契约 Contract written on white bast paper during Daoguang Reign of the Qing Dynasty

⊙3 咸丰年间用白皮纸书写的契约 Contract written on white bast paper during Xianfeng Reign of the Qing Dynasty

⊙4 同治年间用白皮纸书写的卖山土文契 Contract for selling mountain field written on white bast paper during Tongzhi Reign of the Qing Dynasty

⊙5 民国时期用白皮纸书写的卖水田文契 Contract for selling paddy field written on white bast paper during the Republican Era of China

⊙6 薄如蝉翼的白皮纸 Thin and transparent white bast paper

煌与技艺的精湛。

3. 制作鞋样与鞋屉

合水白皮纸还用于传统手工制鞋过程中鞋样的剪制，这种用途在当地至今仍然保留。在考察中，调查组获赠两件已使用了20多年、用当地白皮纸制作的鞋屉，外形似公文包，通过纸张的黏结、折叠、展开后，呈现出一个个立体的、独立分隔的储物空间，设计精巧、独具匠心，主要用于储藏鞋样及针线、布条等杂物。这种用白皮纸做的鞋屉在其他地区至今未见过。

（二）

调查时的销售简况

合水镇皮纸目前产销量仍较大，销售基本上不成问题，但由于投入成本高、卖价低，故利润并不高。以前由供销社统一收购，调查时则多由私人老板上门收购，主要销往四川、重庆、湖南、湖北，以及贵州本地的印江周边县市。

⊙1

⊙2

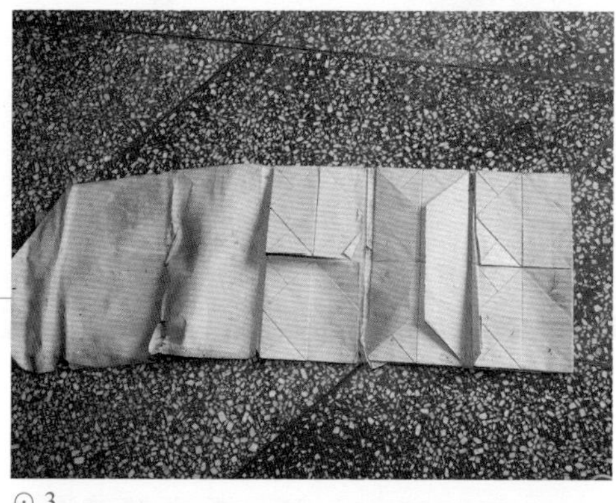

⊙3

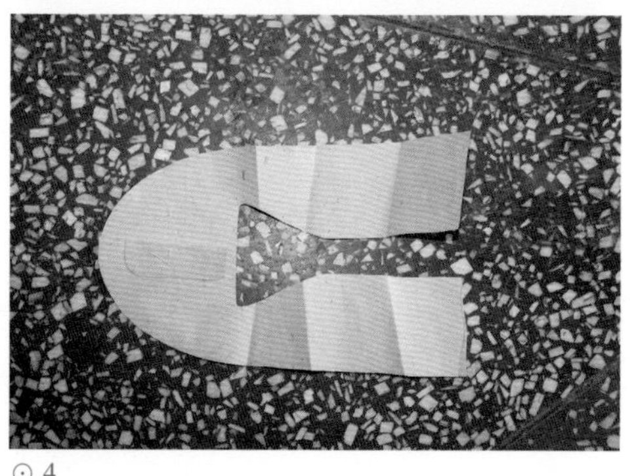

⊙4

⊙1 老太太向调查组成员赠送鞋屉
An old papermaker giving her shoemaking kit paper bag to a researcher

⊙2 鞋屉外观
Outside of the shoemaking kit paper bag

⊙3 鞋屉部分内部空间
Inside of the shoemaking kit paper bag

⊙4 用白皮纸制作的鞋样
Shoe sample made of white bast paper

六 印江合水镇皮纸的相关民俗与文化事象

6 Folk Customs and Culture of Bast Paper in Heshui Town of Yinjiang County

(一) 与蔡伦有关的习俗

1. 蔡伦舀纸

关于“舀纸”工序，当地有个生动的传说：蔡伦先师在试验“舀纸”这道工序时，历经数百次失败，仍想不出一个妙法。入夜，他精神疲倦而眠，梦见纸已舀成，醒来发觉原来是梦，但嘴边有黏糊糊的东西，一摸，原来是“梦口水”，顿然醒悟，认为是“神仙”指点，于是受其启发在纸浆里加入带黏性的松蒿。正是由于改进了这一“舀纸”工序，蔡伦造纸才得以成功。

2. 供奉蔡伦造纸先师

在合水镇造纸户家里房屋的中堂之上除供奉有天地君亲师及历代先辈的牌位外，大多供奉有蔡伦造纸先师的牌位。

⊙5

⊙6

⊙5/6 供奉蔡伦造纸先师 Worshipping Cai Lun, the originator of papermaking

(二) 祭祀习俗

1. 节日习俗

印江苗族视祖先为家神，十分敬重。每逢过年过节，都要在堂屋金承柱上的绊篼前点香焚烧用白皮纸制作的纸钱，祈求祖先保佑子孙平安。（绊篼，传说是苗族祖先逃难时用来装零碎物件的东西，后来成为祖先的象征物，代代相传，进行供奉。）

2. 丧葬习俗

白皮纸被广泛使用于印江土家族的丧葬习俗中，如用于制作葬礼中烧和丢撒的纸钱、亡人脸上的覆盖物及棺材底部的铺垫物等[8]。

[8]《印江土家族苗族自治县概况》编写组.印江土家族苗族自治县概况[M].贵阳:贵州人民出版社,1987:32.

(三)
造纸相关谚语

1.“造纸不轻松，七十二道工，道道须用工，外加一道口吹风。”[9]。

这句话简洁、平实地道出了造纸工艺的繁杂和艰辛，每一道工艺都需仔细、认真、用心去做，除“七十二道”工艺之外，“外加一道口吹风”准确地描述出揭纸时口吹纸角使其扬起的娴熟技艺。谚语的流传反映出造纸艺人朴实、勤恳的敬业精神。

兴旺村人民对造纸技术传统的描述和总结，是该地造纸传统技术文化中的重要组成部分，构建出造纸技艺的地方性技术知识体系。

2.“思南斗篷印江伞”

印江纸伞肇始于清末，盛极于民国。20世纪60年代仍在省内外畅销，当时有“思南斗篷印江伞”之说。旧时，印江苗族女孩出嫁时必用此伞。纸伞有花伞、普通伞两种。制伞原料为竹子、白皮纸、棉线、头发等。

白皮纸融入了印江人的生活，成为不可或缺的地方文化的组成部分。

(四)
书画用纸佳话

合水白皮纸作为书画用纸，在历史上乃至当代都被一些书画名家和书画爱好者所喜用，也产生了一段段画坛、文坛佳话和轶事。清末题写北京“颐和园”匾额、曾深受慈禧赏识的印江书法家严寅亮尤喜用印江合水白皮纸。1920年，梁启超曾托严寅亮购印江合水白皮纸，要求将纸的张幅改为107 cm × 53 cm，并加厚，按重量比照计价，故又名“称纸”。

1943年，徐悲鸿在四川、贵州期间创作的大量国画画作中，亦有不少是用合水白皮纸绘就的。徐悲鸿曾赞合水白皮纸曰：“白皮纸的吸墨力强，坚韧绵扎，细腻白泽，折不起皱纹，画与纸相得益彰。”1964年，著名画家李苦禅、郭味渠赞合水白皮纸性能好，适宜创作水墨画。类似这样的评价不绝于书。

1982年1月，贵州人民广播电台驻铜仁记者史可夫曾以《画家怀念印江皮纸》为题，报道了一些书画家对合水白皮纸的需要和感慨。同年12月4日，史可夫又以《“益使丹青添墨韵”——印江皮纸》为题，将李苦禅用印江合水纸绘的《卧鹭图》照片推荐登于《贵州日报》上，意在敦促印江合水生产绘画用的白皮纸。

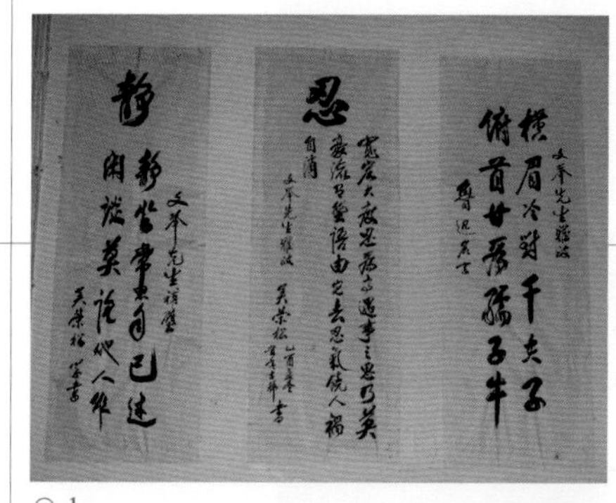

⊙ 1

⊙ 2

⊙ 1 用兴旺村白皮纸写的书法条幅 Calligraphy written on white bast paper in Xingwang Village

⊙ 2 造纸村民展示皮纸 Local papermakers showing bast paper

[9] 印江县政协文史资料委员会.印江白皮纸[M]//印江文史资料·第一集,1985:104.

七 印江合水镇皮纸的保护现状与发展思考

7 Preservation and Development of Bast Paper in Heshui Town of Yinjiang County

随着社会生活的演化，调查时，传统造纸业已不再作为合水镇的主要副业，很多从业人员改行或外出打工，诸多手工作坊也因此被遗弃或撤毁。而且，本已疲惫的传统造纸业又遭遇现代造纸工业的直接挑战，兴旺村古法造纸业显然也逃避不了这种冲击。传统造纸本身周期长、工艺复杂、经济效益低，从业者亦感非常辛苦。在现实的抉择中，合水镇的年轻人已很少有人愿做这一行，即使勉强参与进来，也无决心坚持，更别说献身其中了。合水皮纸的制作正逐步面临着后继无人的尴尬局面。

2009年8月，经过地方争取和多方的努力，合水蔡伦古法造纸技艺入选第二批铜仁地区级非物质文化遗产保护名录。2010年1月，印江皮纸制作技艺入选贵州省省级非物质文化遗产代表性项目名录。一系列的努力使一度濒于颓势的传统产业有了新的生机。2010年11月，印江县启动了“合水古法造纸作坊”维修工程，投入25万元资金，对合水镇兴旺村下寨、蔡家湾，木腊村桥头等地76个古法造纸作坊进行了维修或重建，其中维修作坊45间，重建作坊31间，涉及生产户231户，总面积为2 950.74 m^2。虽然传统造纸技艺危机仍然存在，但这一系列实实在在的努力有效地保护和传承了当地的古法造纸工艺。

合水镇兴旺村皮纸制作技艺精湛、工艺繁杂，有着完整的传统技艺传承体系。然而，由于后期加入了废旧纸边等工业纸浆，导致即使有传统精湛的技艺、很高的投入成本，也很难生产出昔日非常优质的白皮纸，很难创造出本身应有的经济效益和持续的文化影响，这令调查者倍感痛心。按现今生产的最终纸质来看，之前很多道引以为豪的传统技艺都可省去，多余的工序实则是一种资源的浪费，导致精湛的工艺并没有产生应有的经济效益和文化影响。设计和文化创意这些看似毫无必要的工序恰恰是未来印江合水镇皮纸重振辉煌的突破点。

⊙ 3

⊙ 4

⊙ 3 兴旺村连片的纸坊 Papermaking mills in Xingwang Village

⊙ 4 河边的老水车 Waterwheel by the river

印江合水

2 5 7

皮纸

Bast Paper
in Heshui Town
of Yinjiang County

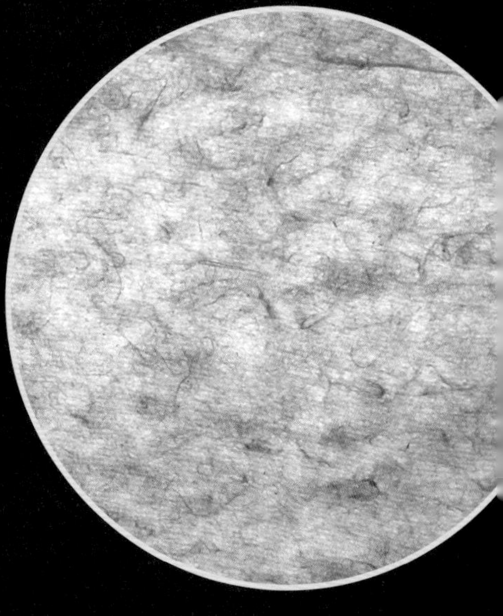

兴旺村皮纸透光摄影图
A photo of bast paper in Xingwang Village
seen through the light

第五节

印江土家族手工纸

贵州省
Guizhou Province

铜仁市
Tongren City

印江土家族苗族自治县
Yinjiang Tujia and Miao Autonomous County

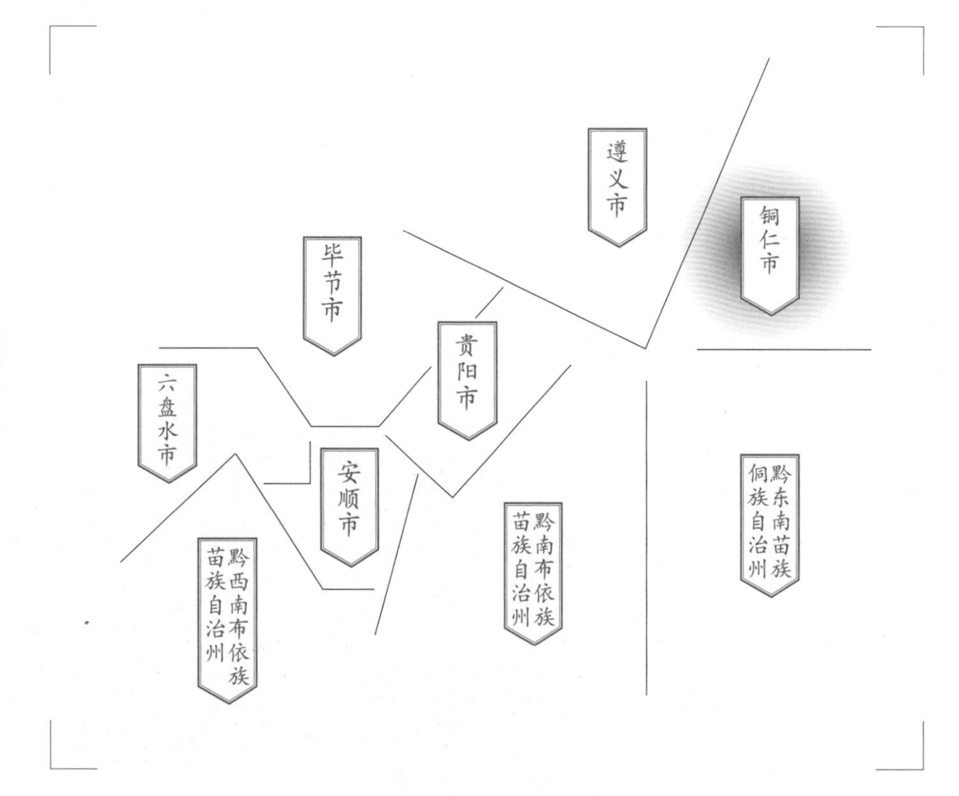

调查对象
沙子坡镇
六洞行政村
塘口行政村
土家族手工纸

Section 5
Handmade Paper by the Tujia Ethnic Group in Yinjiang County

Subject
Handmade Paper by the Tujia Ethnic Group in Tangkou and Liudong Administrative Villages of Shazipo Town

一 印江土家族手工纸的基础信息及分布

1 Basic Information and Distribution of Handmade Paper by the Tujia Ethnic Group in Yinjiang County

印江土家族苗族自治县沙子坡镇土家族手工纸主要分布在以土家族为主聚居的六洞行政村与塘口行政村，从事手工纸制作的多为土家族居民。虽然塘口村与六洞村相邻，但两村的手工纸生产却截然不同。六洞村以构皮为原料生产皮纸，制作的皮纸主要用于书写、祭祀、擦拭机器以及制作火炮引线、斗笠等；而塘口村则以慈竹为原料生产竹纸，制作的竹纸当地又称为“火纸”，主要用于祭祀、制作纸钱等。

2008年12月下旬，调查组沿着崎岖的山路，辗转找寻到位于武陵山深处的六洞村与塘口村时，发现曾经一度兴盛的手工纸制作在这两个村落中都已消失，只见到湮没在荒草中的捞纸槽及部分残旧的造纸工具。

六洞村于1998年停止了皮纸的生产，塘口村早在20世纪80年代初就已停止了竹纸的生产。然而一些造纸的老艺人并没有遗忘手工纸制作技艺，进行手工造纸的那段生涯成为他们一生中美好的回忆。

⊙1

⊙1 六洞村湮没在荒草中的捞纸槽 Abandoned papermaking trough in Liudong Village

路线图
印江县城
↓
塘口村、六洞村
Road map from Yinjiang County centre to the papermaking sites (Tangkou Village and Liudong Village)

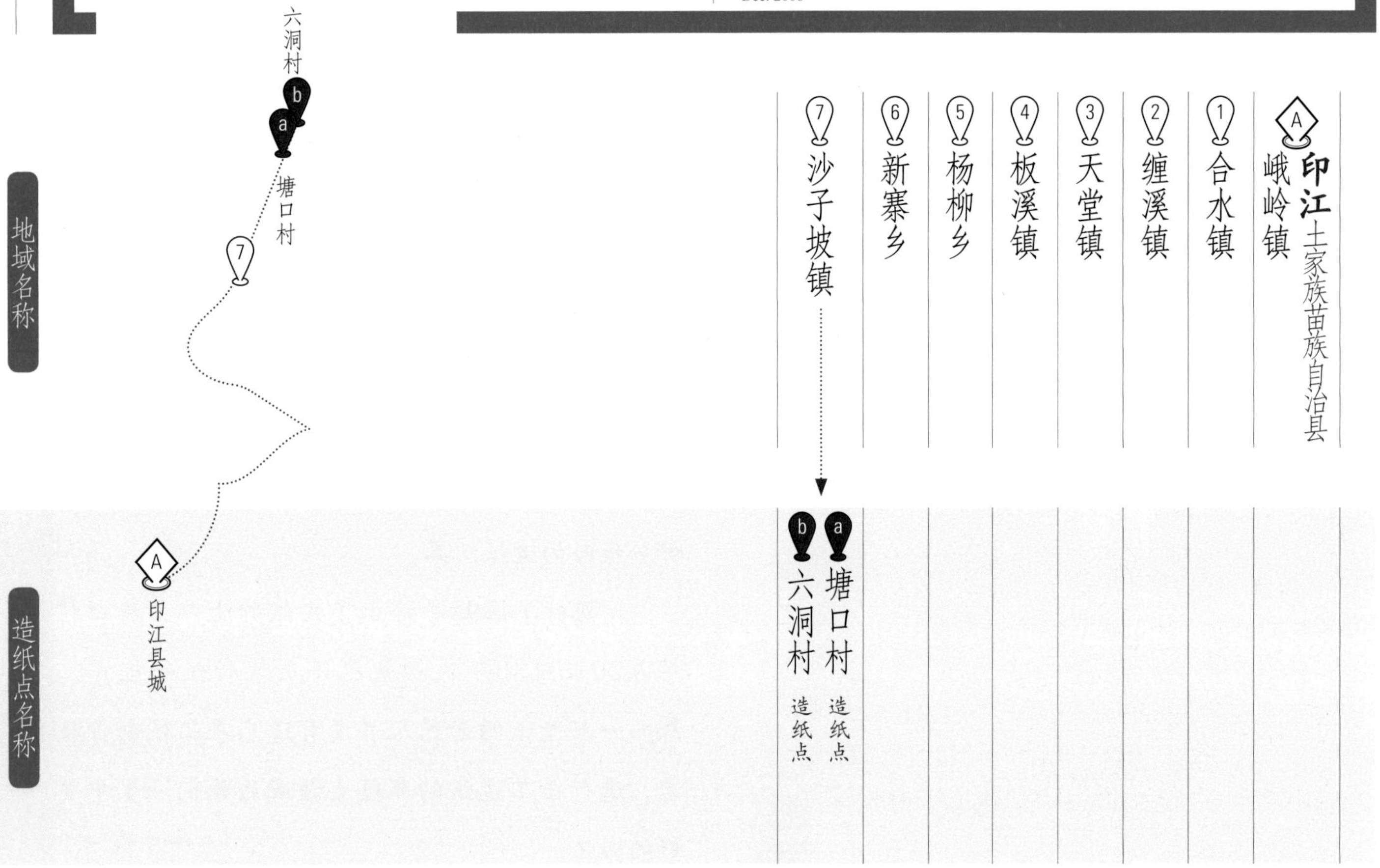

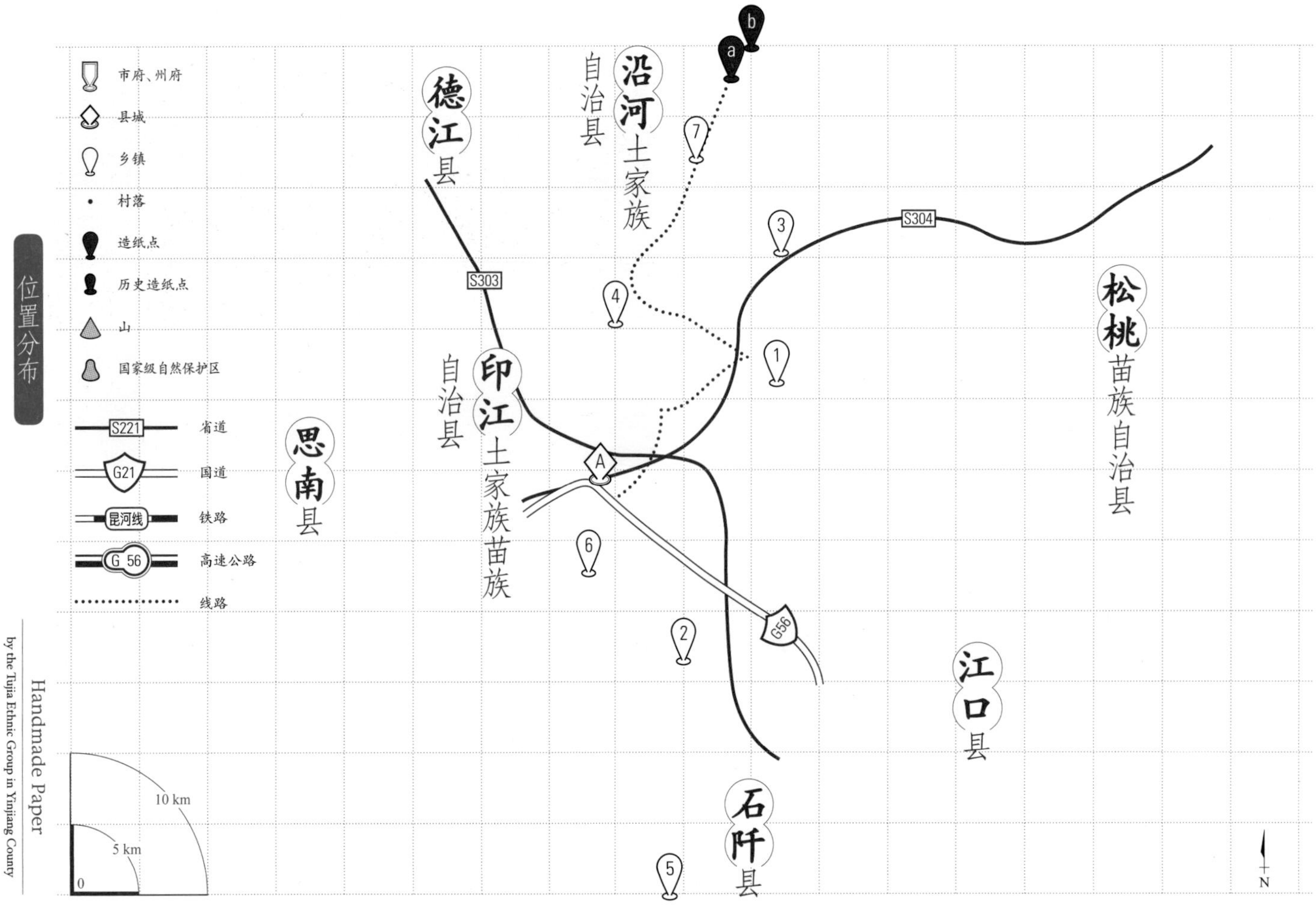

印江土家族手工纸生产地分布示意图

Distribution map of the papermaking sites of handmade paper by the Tujia Ethnic Group in Yinjiang County

考察时间
2008年12月

Investigation Date
Dec. 2008

地域名称

A 印江土家族苗族自治县 峨岭镇
1 合水镇
2 缠溪镇
3 天堂镇
4 板溪镇
5 杨柳乡
6 新寨乡
7 沙子坡镇

造纸点名称

a 塘口村造纸点
b 六洞村造纸点

位置分布

二
印江土家族手工纸生产的人文地理环境

2
The Cultural and Geographic Environment of Handmade Paper by the Tujia Ethnic Group in Yinjiang County

印江是土家族苗族自治县，民族、民间文化资源丰富。沙子坡镇位于县域北部，距离县城42 km，东、南两面与本县刀坝乡、天堂镇、板溪镇、杉树乡毗邻，西、北两面与德江县枫香溪镇、沿河县谯家镇交界，地处三县交界地段，乡土说法中有“印北金三角”的美称。

沙子坡镇设立于1992年，调查时，全镇总面积128 km^2，辖20个行政村204个村民组，6 449户，27 000多人。区域内气候温和、雨量充沛，属中亚热带季风湿润气候区，森林覆盖率达51.2%，年平均气温为15.1 ℃，年降水量1 000~1 100 mm，无霜期平均为277天，年日照时数约1 253小时。

⊙1

沙子坡镇居住地与耕地多在海拔700~900 m处，东、西两面有南北走向的三座平行山脉，夹车家河、六井溪河穿流其间，水资源较为丰富，有储水量达100多万立方米的消水坑水库和天生桥水库，有装机容量为6 000 kW的坨寨水电站以及龙塘窝水电站。境内矿产资源丰富，其中煤炭、铀矿、硫铁矿储量尤为丰富，有“黔东煤炭之乡”之称。

⊙1 高山上的六洞村 Liudong Village on a high mountain

六洞村地处沙子坡镇最北部、武陵山深处，

山高谷深，交通不便，经济落后，耕地面积0.1 km^2，辖5个村民组，调查时总人口457，是沙子坡镇最小的行政村。村寨依山而建，车家河从村边流过，捞纸槽沿河而建，离村寨很近。

塘口村与六洞村人文地理环境相近，地处高山环绕之中，捞纸槽也多建在车家河边，耕地面积0.46 km^2，辖14个村民组，调查时总人口1 700多，以土家族为主。塘口村竹纸的生产早在20世纪80年代初便已停止，现经济来源主要依靠种植烤烟及外出务工。塘口村文化生活丰富，农民常在劳动中放歌呐喊，内容多为劳作、爱情、伦理等，山歌音域达十三度，且八度大跳，音程特色凸显，当地称这种山歌为“高腔山歌”。

三 印江土家族手工纸的历史与传承

3 History and Inheritance of Handmade Paper by the Tujia Ethnic Group in Yinjiang County

印江沙子坡镇土家族手工纸的历史与传承在历史文献中没有记载，究竟源于何时，调查组尚未找到准确的文献资料与口述记忆的根据。

据六洞村造纸户介绍，六洞村造纸户多为杨姓土家族居民，造纸家族世代相传，至少已有百年历史。

调查组记录下六洞村3个家庭父子代代相传近百年的造纸传承，谱系如下：

表9.7 杨边畅户五代传承图
Table 9.7 Papermaker Yang Bianchang's genealogy of papermaking inheritors

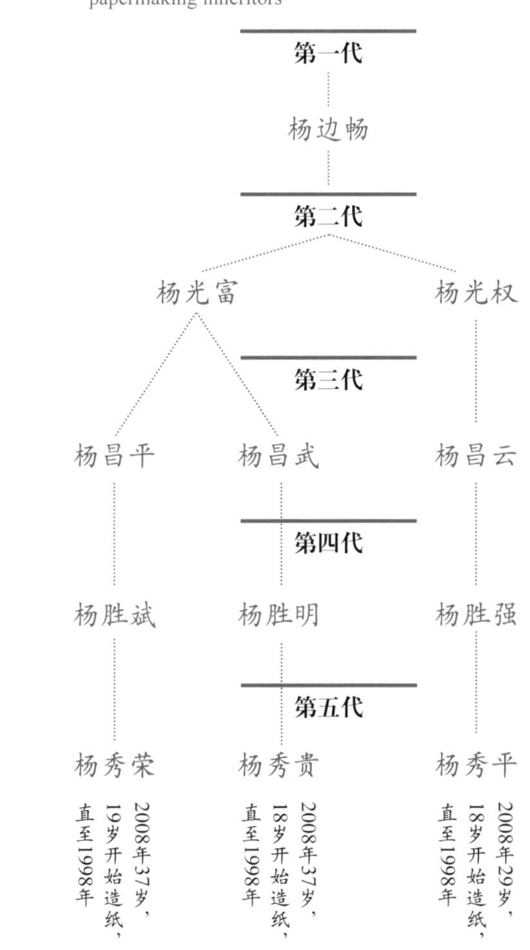

六洞村土家族皮纸传承到1998年，由于没有销路、利润低等原因皮纸的生产最终完全停止。

据造纸老艺人刘正坤讲述，至少在民国时期塘口村土家族就开始造竹纸了。

据《铜仁日报》刘大波所述，沙子坡镇有车家河、六井溪河两条河流顺山而过，水源看似丰富，却因地形成为缺水的乡镇。以前每逢天干，上万群众就要闹水荒；每逢下雨，家家户户都要拿出盆盆罐罐接“天水”，一盆水经过洗菜、洗脸、喂牛等多次使用后还舍不得倒掉，要是逢连续干旱，人畜饮水就更加困难了。自20世纪70年代起，沙子坡镇利用地形拦河、筑坝，兴修田间地头的小水池、小水窖，以积蓄水资源，并先后在塘口村旁的坝坨河上建起了一道道中小型水库、电站。由于耗水量大，手工造纸在水资源总体缺乏的乡镇势必会被淘汰。由于水位因水库、电站的修建而有所提升，因此早在1980年建在河边的造纸“窑子”就已被水淹掉，自那时起，塘口村就再也没人造“火纸”了。

⊙1

⊙2

⊙1
六洞村里的传承访谈
Interviewing the papermaking inheritors in Liudong Village

⊙2
调查组成员与昔日造纸人的合影
Researchers and former papermakers

四 印江六洞村土家族皮纸的生产工艺与技术分析

4
Papermaking Technique and Technical Analysis of Bast Paper by the Tujia Ethnic Group in Liudong Village of Yinjiang County

(一) 印江六洞村土家族皮纸的生产原料与辅料

原料：构树皮。构树为当地野生。

辅料：纸药（沙松树根）、石灰、碱（当地称地灰，即柴灰）。

(二) 印江六洞村土家族皮纸的生产工艺流程

调查组根据造纸户口述，记录六洞村土家族皮纸的生产工艺流程如下：

壹	砍构树
贰	刮构皮
叁	晒构皮
肆	泡构皮
伍	浆构皮
陆	沤构皮
柒	上窑
捌	蒸构皮
玖	洗料子
拾	泡料子
拾壹	踩料子
拾贰	搓料子
拾叁	晒料子
拾肆	二次泡料子
拾伍	二次踩料子
拾陆	二次搓料子
拾柒	加碱
拾捌	蒸料子
拾玖	捂料子
贰拾	二次洗料子
贰拾壹	碓舂
贰拾贰	切皮板
贰拾叁	淘料子
贰拾肆	下缸子
贰拾伍	打槽
贰拾陆	放滑水
贰拾柒	白纸
贰拾捌	压榨
贰拾玖	起纸
叁拾	晒纸
叁拾壹	收纸
叁拾贰	包装

壹
砍构树
1

造纸户在每年农历二至三月选取一年生的构树枝条砍伐。由于老构树皮不易煮烂，除少量砍伐两年生的构树枝条外，造纸户们一般都不会砍伐老的构树枝条。

贰
刮构皮
2

构树枝条砍伐回来后，将构皮从枝条上刮下来，一个人一天可刮50 kg左右的构皮。

叁
晒构皮
3

将刮好的构皮晒干或者晾干，一般需要4~5天。

肆
泡构皮
4

构皮晒干后储存备用，通常要放到每年的农历五至六月农闲时才开始蒸构皮。在蒸构皮前，需将晒干的构皮放入河水中浸泡，一般至少在河水里泡“一场”，即5天时间。

伍
浆构皮
5

将泡软的构皮分驮捆好，以2.5 kg干构皮的量为一驮。然后按每100 kg干构皮用50 kg石灰的比例，将构皮一驮驮地浸入配好的石灰浆中，随即取出。

陆
沤构皮
6

将浆好的构皮一驮驮地整齐码放成垛，沤置2~3天，同时沥去构皮中的大部分水分。

柒
上窑
7

将沤好的构皮运到窑子里码放整齐，码放时边放边用脚将构皮踩紧，装满窑子后再用“响子”在窑子顶部捶紧构皮，一般捶10分钟左右即可。窑子的大小决定了上窑构皮量的多少。当年生产时，六洞村最小的窑可放750 kg构皮，最大的窑可放1 500 kg构皮。

捌
蒸构皮
8

构皮上窑后，烧柴用大火蒸10天左右，如火力小则要蒸15天。蒸5天后要翻锅，即将窑子里的构皮上下翻动一次，然后再蒸5天。蒸750 kg构皮，需用柴（当地称为棒棒柴）30捆，约2 250 kg。

玖
洗料子
9

从窑子中取出蒸烂（软）后的构皮，并放入河水中，将料子上的石灰洗掉。两个人一天才能洗完一窑构皮（750 kg）。蒸烂后的构皮当地称为料子。

拾
泡 料 子
10

将洗好的料子放在河里继续浸泡3~5天。

拾壹
踩 料 子
11

将料子一驮驮地从河里取出，并放在木板上用脚踩，将料子外面的黑壳踩融。一个人五天才能踩完一窑（750 kg）料子。

拾贰
搓 料 子
12

将踩融后的料子再放入水中，用手搓揉，把料子搓散，以除去黑壳等杂质。

拾叁
晒 料 子
13

将料子再次晒干。阳光强烈时两天即可晒干，雨天可能要在家中晾10天，甚至半个月。

拾肆
二 次 泡 料 子
14

再次将晒干后的料子放在河水里浸泡，同时用手揉，泡几分钟即可。

拾伍
二 次 踩 料 子
15

再次将泡好后的料子放在木板上用脚踩。所踩料子的量视当天的用量而定，一天可踩100 kg料子，一窑料子要一周时间才能踩完。

拾陆
二 次 搓 料 子
16

再次在水中用手搓洗料子，并将料子搓散。

拾柒
加 碱
17

料子搓散后，按每50 kg构皮用25~30 kg地灰（即柴灰）的比例拌入地灰。

拾捌
蒸 料 子
18

将加入地灰的料子直接放入蒸子里蒸1.5小时。一蒸当地又称为一“搅”。蒸料的多少视蒸子的大小而定，有的用家里做饭的锅蒸，有的用窑子蒸。

拾玖
捂 料 子
19

料子蒸好，待蒸汽满后，就灭火，将料子在蒸子中捂一个晚上。

贰拾
二 次 洗 料 子
20

将捂好的料子从蒸子中取出，放入河里，让河水冲洗，并不时搓洗以洗净料子，除去地灰等杂质。

贰拾壹
碓 舂
21

将洗净后的料子用脚碓舂融，并舂成皮板状，一个皮板一般要舂1 600碓（次）。舂碓时，需两人配合，一人在碓头，清皮板，一人用脚踩碓。也有用水碓的，只需一人即可。一天可舂25 kg，舂料的多少视每天的用量而定，一个人一天要用10 kg左右的构皮。

贰拾贰 切皮板

22

用皮刀将皮板轧碎，10分钟切12个皮板，够一天用。12个皮板可造500张纸（73 cm×53 cm）。

贰拾叁 淘料子

23

将切碎的皮板装入布袋中，拿到河里淘洗，12个皮板几分钟就可洗净。

贰拾肆 下缸子

24

将淘洗干净的料子放入槽缸（捞纸槽）。

贰拾伍 打槽

25

槽缸加水后，用槽棍打槽1 000~2 000棒，将料子打细并分散开来。

贰拾陆 放滑水

26

打槽后，在槽缸中加入滑水（纸药）即杉松树根汁后，再次打槽，使皮料纤维均匀悬浮在水中。一天需用0.5 kg杉松树根。

贰拾柒 白纸

27

双手持两根夹持将纸帘压在帘架上，平端帘架，将纸帘由前往后舀入纸浆，提起，前后振荡几次，再由后往前将纸帘舀入纸浆，提起，前后振荡几次，将多余纸浆从纸帘上倾泻，随即将帘架放在支杆上，取下纸帘，转身将纸帘反扣在湿纸垛上，揭离纸帘，一张湿纸便留在纸垛上。

贰拾捌 压榨

28

待一垛纸舀好后，在湿纸垛上先放草平铺，再压上木板，最后用木榨将纸垛中多余的水分榨出。

贰拾玖 起纸

29

将纸垛搬回家，并用手将纸垛的四边扳翘起来，便于将湿纸一张张地撕下。

叁拾 晒纸

30

将撕下来的湿纸刷贴在纸焙上，用火烘焙几分钟即可。纸焙为两面，贴成上下两排，每排所贴纸张数视纸焙大小而定，一般从左到右可贴7~12张纸，73 cm的纸可贴6张。每2张纸面略错开，由上到下贴5张。纸焙一般长约5.3 m，高约2 m。

叁拾壹 收 纸

31

将晒干后的纸一张张地从纸焙上揭下来并理齐。

叁拾贰 包 装

32

将纸折成三小折，10张为一贴，100贴为一捆，上下各用一张双层的夹纸捆好，再用纸搓成的绳捆三道。

（三）印江六洞村土家族皮纸生产使用的主要工具设备

壹 纸帘与帘架

1

六洞村保留下来的纸帘及帘架已破损不堪。从造纸户那里了解到六洞村土家族皮纸的主要尺寸有73 cm × 53 cm、40 cm × 50 cm（一帘二纸）、26.7 cm × 40 cm（一帘三纸）三种。

⊙1

贰 捞纸槽

2

由青石板围制而成，用于盛放纸浆。

⊙2

叁 皮 刀

3

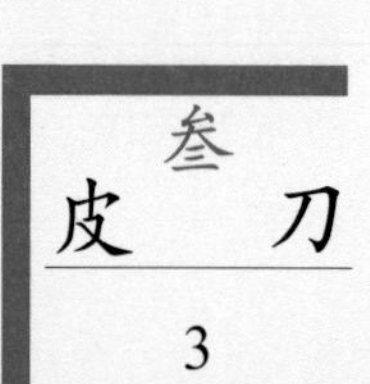

⊙3

铁质，用于切皮板，两端有柄，便于握刀。

⊙1 残留的纸帘与帘架 Broken papermaking screen and its supporting frame

⊙2 荒弃的捞纸槽 Abandoned papermaking trough

⊙3 皮刀 Iron knife for trimming the paper

五 印江塘口村土家族竹纸的生产工艺与技术分析

5 Papermaking Technique and Technical Analysis of Bamboo Paper by the Tujia Ethnic Group in Tangkou Village of Yinjiang County

经重点访谈造纸人游成贵和游天云，总结乌江村竹纸的生产工艺流程如下：

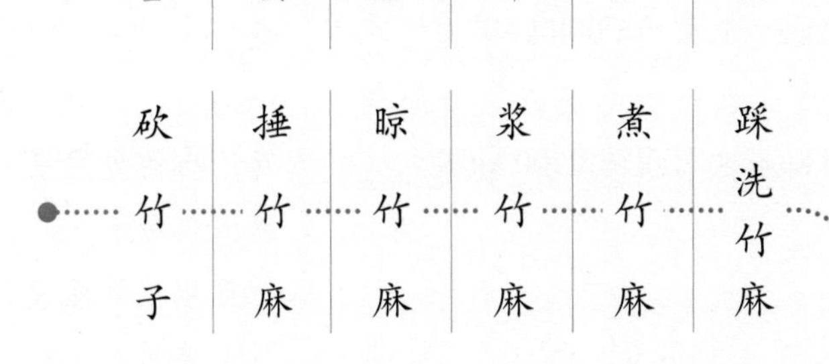

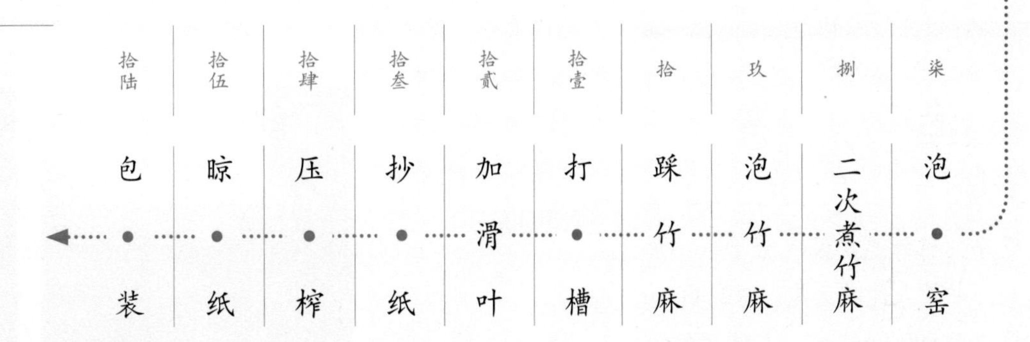

壹 砍竹子 1

每年农历十二月至次年农历二月都可砍一年生的慈竹。

贰 捶竹麻 2

用柴刀或锤子把竹子捶破，一个人一天可捶500 kg左右。

叁 晾竹麻 3

将打碎后的竹麻捆起，晾干，一般晾1~2个月。

肆 浆竹麻 4

竹麻晾干后，挑回家，放在池子里，加入水和石灰，开始浆竹麻。2 500 kg竹麻需用5 000 kg石灰。石灰多为造纸户自己烧制，每烧制5 000 kg石灰需用约2 500 kg煤。

伍 煮竹麻 5

把浆好的竹麻捆成把（一把质量1.5~2.5 kg），然后放到窑子里煮，一次煮竹麻的多少视窑子大小而定，一般一窑可煮2 500 kg竹麻，有的也可煮3 000 kg。将竹麻浸泡在清水里，煮一个月，随着窑子里水汽的蒸发，每隔4~5天需往窑子里加一次水，以使竹麻始终浸泡在水中。煮竹麻时，每天烧三道火，即早、中、晚各给窑子添一次煤，每次添煤约30 kg。

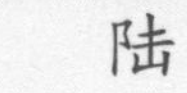

陆 踩洗竹麻 6

⊙1⊙2

等竹麻煮散、煮烂后，再观察竹麻在窑子中的高度是否下降。这时将竹麻从窑子中取出，穿上钉鞋踩竹麻，将竹麻踩软、踩散，然后将竹麻洗干净。一般由几家造纸户合作完成，每次大约有10个人同时踩竹麻，一天可踩250 kg竹麻，需四个人四天才能洗干净。

柒 泡窑 7

把洗净的竹麻再次装入窑子里，加满水，并放入纯碱浸泡。每2 500 kg竹麻需用100 kg纯碱。在使用纯碱浸泡之前多使用荞壳灰（荞麦壳烧成的灰）浸泡。

⊙1

⊙2

捌 二次煮竹麻 8

再次将竹麻放在燃窑中煮15天。

玖 泡竹麻 9

竹麻煮了15天后，将火熄灭，把水放掉，并把竹麻取出，洗净窑子，然后再将竹麻放入窑子，加满水浸泡，不煮，让竹麻腐烂，浸泡1~2个月。

拾 踩竹麻 10

竹麻泡好后，取出洗净，然后放入踩料池中用脚踩融。一槽10 kg竹麻1~2小时可踩融。

⊙1 钩竹麻示意
Showing how to pick out the bamboo materials with a hook

⊙2 踩竹麻示意
Showing how to stamp the bamboo materials

拾壹
打槽
11

竹麻踩融后放到纸槽里，加水，手握两根撸筋杆打槽，使竹麻纤维均匀分布在水中，然后静置几个小时，让竹麻纤维自然沉淀，再将纸槽中的水缓缓放掉。

拾贰
加滑叶
12

在纸槽里再次加入清水，然后加入滑药，用两根撸筋杆将竹麻纤维搅起，再次打槽，使其均匀悬浮在纸槽中。0.5 kg滑药可抄纸2~3天。

拾叁
抄纸
13 ⊙3~⊙6

将纸帘平放在帘架上，合上帘架上的提杆，右手抓住提杆，左手抓牢纸帘与帘架的左端，平端纸帘，先由右往左将纸帘插入纸浆中，端起，再由外往里将纸浆抄制在纸帘上，然后将帘架搭放，移开提杆，揭下纸帘，转身，将纸反扣在纸垛上。每抄一槽纸需用10 kg竹麻，一个人一天可抄完，10 kg竹麻可以抄20捆纸，每捆100张，一捆可卖1元钱。

⊙3 ⊙4 ⊙5 ⊙6

拾肆
压榨
14

抄好一垛（2 000张）纸后用木榨将其缓慢压干，压榨时需经过两次紧榨、松榨过程。如仅压榨一次，则压不干；如果将压力加大，急着将纸压干，则会将纸垛压爆。压榨过程需0.5~1小时。

拾伍
晾纸
15

纸垛压干后，每10张为一贴，揭起，搭在竹竿上晾干。搭晾时，一般生火烤，一天即可烤干。如搭晾在楼上，且不生火，则2~3天可晾干。

拾陆
包装
16

纸晾干后，收起，包装，待售。

⊙3/6 抄纸示意
Showing how to scoop and lift the papermaking screen out of water and turn it upside down on the board

(三)

印江塘口村土家族竹纸生产使用的主要工具设备

壹 纸帘 1

实测造纸户保存的旧纸帘尺寸为：58(26+26+6)cm×51(6.5+35.5+9)cm。塘口村使用的纸帘是一帘二纸的纸帘，纸帘由棕线与竹线分割，抄纸时，棕线与竹线处的纸浆较其他处薄，压榨后，用手可将纸张从隔线处分开，除去纸头、纸边，一张纸可变为两张小纸。塘口村竹纸尺寸多为32.5 cm×26 cm。棕线在当地又被称为"拦水线"，竹线被称为"隔沙"。

⊙1

⊙2

贰 纸帘架 2

实测造纸户保存的旧纸帘架内部尺寸为63.2 cm×46.5 cm，外部尺寸为68 cm×52 cm，用于盛放纸帘，上面安装有活动提杆。

叁 钉鞋 3

用于踩竹麻。鞋面、鞋底为牛皮刷桐油手工制作而成，鞋钉为铁质。钉鞋整体质地坚硬，为方便脚进入，鞋做得比较大，穿上时需用草塞紧，以防止鞋脱落。

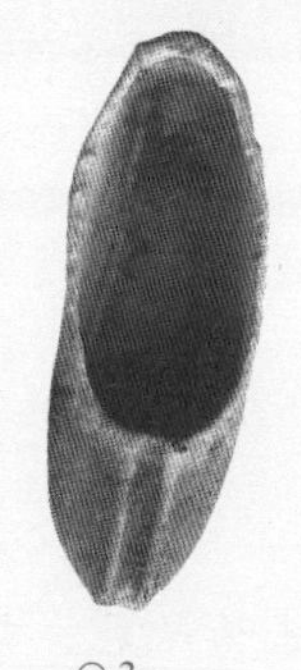

⊙3

肆 双捞钩 4

用于钩放竹麻。实测造纸户保存的双捞钩总长约207 cm，铁杆长约61 cm，前端插有两个铁钩，钩长约17 cm。塘口村有的造纸户使用单捞钩。

⊙4

⊙1 纸帘 Papermaking screen

⊙2 纸帘架 Frame for supporting the papermaking screen

⊙3 踩竹料用的钉鞋 Spiked shoes for stamping the papermaking materials

⊙4 双捞钩 Stick with double lifting hooks

六 印江土家族手工纸的用途与销售情况

6 Uses and Sales of Handmade Paper by the Tujia Ethnic Group in Yinjiang County

印江沙子坡镇六洞村土家族皮纸主要用于书写、擦拭机器、祭祀以及制作火炮引线、斗笠等。调查中，造纸人杨秀平取出自己珍藏的、用自己制作的皮纸线装的斋事礼簿与新婚礼簿。

印江沙子坡镇六洞村土家族皮纸主要是拿到附近集市去卖，如本县的沙子坡、铅场坎、小井、刀坎、龙溪、桥家、甲石以及沿河县，四川省的南届、龙池等地。六洞村土家族皮纸的生产以家庭为单位，主要于农闲时制作，以3个人的家庭生产单位为例，一年可生产30捆纸，按1998年每捆售价65元来计算，销售额为1 950元。六洞村土家族皮纸制作造纸户每年需按产值缴纳10%的税，扣除税收外，连本带利造纸户实际所得为1 755元。

印江沙子坡镇塘口村竹纸主要用于祭祀、制作烧纸钱，为周边村寨所使用。由于已有20余年没有造纸，因此现在保留下来的竹纸少之又少，调查时造纸老艺人刘正坤将他仅剩的几张竹纸赠予了调查组。

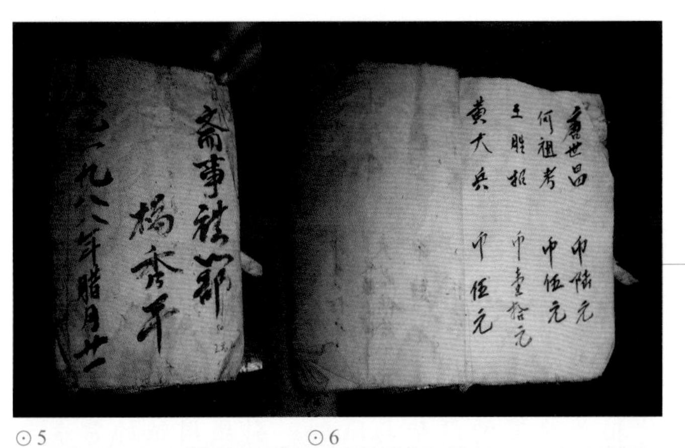

⊙5 ⊙6

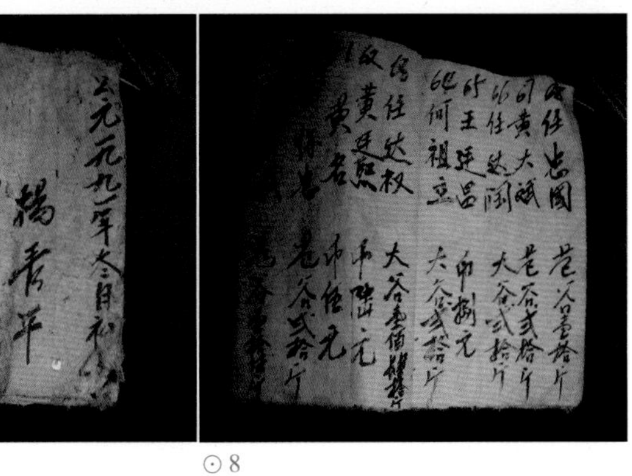

⊙7 ⊙8 ⊙9

⊙5/6
六洞村土家族皮纸制作、书写的斋事礼簿
Praying signature book made of bast paper by the Tujia Ethnic Group in Liudong Village

⊙7/8
六洞村土家族皮纸制作、书写的新婚礼簿
Wedding signature book made of bast paper by the Tujia Ethnic Group in Liudong Village

⊙9
仅存的竹纸
Last pieces of bamboo paper kept by an old papermaker

七
印江土家族手工纸的相关民俗与文化事象

7
Folk Customs and Culture of Handmade Paper by the Tujia Ethnic Group in Yinjiang County

印江沙子坡镇造纸户以特殊的方式纪念着曾经辉煌过的时代。调查时发现，虽已停产多年，但年老的造纸户都珍藏了一些自己亲手制作的手工纸。老人们将其视为珍宝，不肯轻易示人。老人们用大半生时间操持着手工造纸业，手工纸承载着他们的喜怒哀乐，伴随着他们走过人生的酸甜苦辣，保留一点自己亲手制作的手工纸以备百年之后能与自己相伴，也算是预先为自己死后带入另一个世界储备的冥钱。

八
印江土家族手工纸的保护现状与发展思考

8
Preservation and Development of Handmade Paper by the Tujia Ethnic Group in Yinjiang County

印江沙子坡镇六洞村土家族皮纸最大的用途是制作“火炮”引线。随着有关“火炮”制作相关政策的调整与限制，引线用纸的需求市场消失，皮纸销售受阻。随着社会的发展与进步，交通、通信状况的改善，山外的繁荣与精彩深深吸引着地处大山深处的人们，加之皮纸制作的利润低，为了家人有更好的生活条件，造纸户纷纷选择放弃世代相传的造纸手艺，走向山外另谋他业。自1998年起，六洞村土家族造纸户停止了皮纸生产活动。

2008年调查时，六洞村土家族皮纸技艺连同往日的辉煌都已成为回忆，纸槽、窑子早已淹没在杂草中，这给手工纸的传承与保护带来了思考与警示。任何手工技艺的经济形式最终都会反映在其所表现的目标市场上。手艺人若想依靠手艺

⊙1

纸需水量大，生产中产生的废液又会对水源造成一定的污染。沙子坡镇作为一个缺水严重的乡镇，出于对水利设施的修建和环境保护的考虑，塘口村造纸的“窑子”最终淹没在水中，纸工们纷纷改行，“火纸”的生产退出了历史的舞台。塘口村火纸仅作为一种记忆停留在塘口村人们的脑海中、言语边，当年最年轻的纸工现也已近中年。破旧的纸帘、纸架，长长的捞钩，奇异结实的钉鞋无不向人们讲述着那段破竹为纸的历史。塘口村“火纸”的历史生动地展现了生产生活需求与环境保护相对立的案例，为了保护环境与合理分配水资源，造纸户失去了个人的权益空间。

塘口村“火纸”虽已退出历史舞台，但其制作技艺作为我国传统手工造纸的一部分，亦是一份宝贵的非物质文化遗产。其造纸工具——钉鞋是自本项目调查以来调查组首次见到的实物，有其独特的研究、收藏价值。虽造纸艺人还在，工具也还在，技艺的火种还没有完全熄灭，塘口村的“火纸”却已不适用于非物质文化遗产的恢复性、生产性保护，只可作为地方历史文化发展的一部分做展陈式保护，留住记忆，留住文化。

塘口村“火纸”的历史给我们留下了深深的思考。非物质文化遗产的保护应根据实际情况进行具体分析，不能一概而论，哪怕是同一种技艺，也应根据不同的特点（所处环境、区域、时间等）制定不同的保护模式。

实现自己的价值，满足自身及家庭的生活需求，并保护好手工纸，首先就要保护好其赖以生存的目标市场。只有在目标市场上实现经济价值，手艺人才能得以生存，手艺才能得以发展，手工纸才能得以永续传承。

在调查中，有的造纸户表示如现有销路，他们还可恢复皮纸的生产。

在手工造纸中，水源是一个重要的条件。造

⊙2

⊙1
六洞村外小河边的废纸槽
Abandoned papermaking trough by a river in Liudong Village

⊙2
在造纸老人家中夜谈
Interviewing an old papermaker at night

印江土家族皮纸

Bast Paper by the Tujia Ethnic Group in Yinjiang County

六洞村皮纸透光摄影图
A photo of bast paper in Liudong Village seen through the light

第十章
黔东南苗族侗族自治州

Chapter X
Qiandongnan Miao and Dong Autonomous Prefecture

第一节

岑巩侗族竹纸

贵州省
Guizhou Province

黔东南苗族侗族自治州
Qiandongnan Miao and Dong Autonomous Prefecture

岑巩县
Cengong County

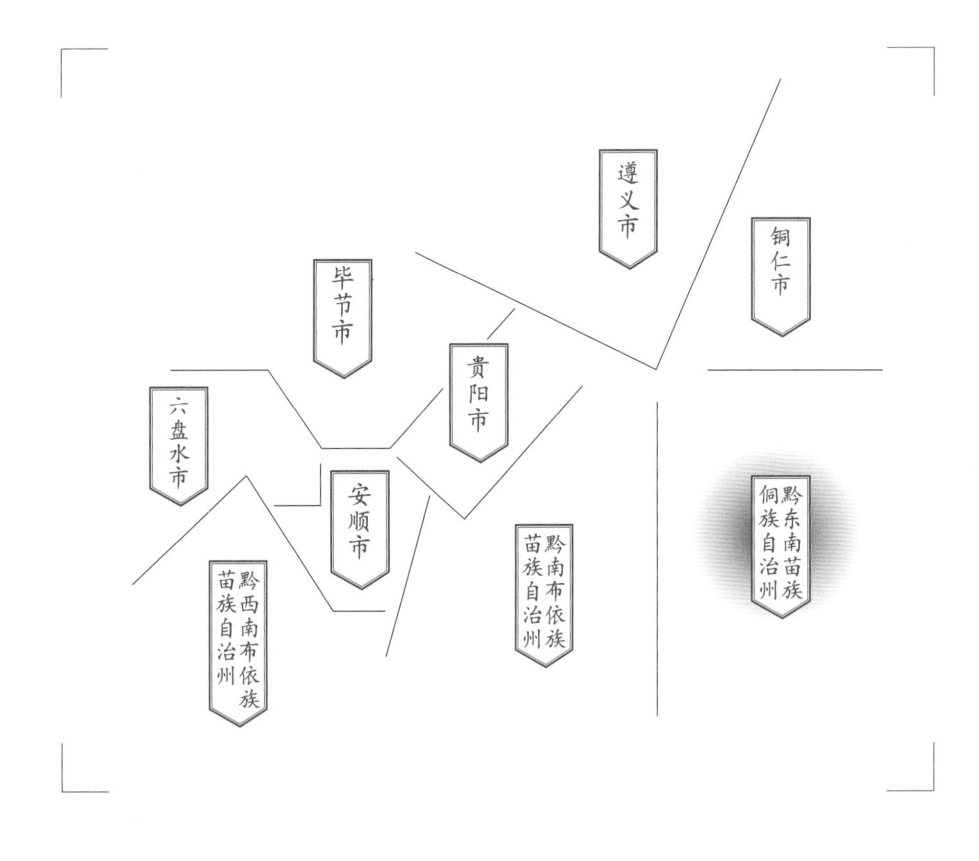

调查对象

羊桥土家族乡龙统行政村
水尾镇白水行政村和腊岩行政村
侗族竹纸

Section 1
Bamboo Paper by the Dong Ethnic Group in Cengong County

Subject

Bamboo Paper by the Dong Ethnic Group in Layan and Baishui Administrative Villages of Shuiwei Town, Longtong Administrative Village of Yangqiao Tujia Town

一
岑巩侗族竹纸的基础信息及分布

1
Basic Information and Distribution of Bamboo Paper by the Dong Ethnic Group in Cengong County

岑巩竹纸的当代主产地在岑巩县水尾镇的白水行政村以及腊岩行政村，其制作者以侗族村民为主，工艺上采用对竹料进行蒸煮的“熟料法”制作；另一相对集中的产地在县辖羊桥土家族乡的龙统行政村，采用石灰直接发酵而不经蒸煮的“生料法”制作。其中，白水行政村境内的龙鳌河沿岸村落为岑巩侗族竹纸的代表性生产地。

白水村竹纸采用的原料为当地所产的“阳山竹”（即植物学中的绵竹），它具有纤维品质匀细、成浆率高等特征。贵州省非物质文化遗产项目公布的材料显示，白水村竹纸技艺传承系400余年前由黄姓侗族先人从湖南省靖州苗族侗族自治县迁徙而传入岑巩县水尾镇的白水村一带，至今已在黄氏宗族中传习了21代。

2011年6月调查组入村调查所收集的信息显示，白水行政村约有80户人仍在从事竹纸生产，传习状况及生产规模维系仍属良性。

岑巩侗族竹纸在工艺上自称有72道工序，所产竹纸具有绵白、薄韧的品质特点，除在本区域销售以外，湖南新晃侗族自治县与贵州玉屏侗族自治县也是旺销区域。历史上及当代岑巩侗族竹纸的主要用途是制作祭奠焚烧用纸，因而也被称为“岑巩火纸”或“思州火纸”。

岑巩侗族竹纸2007年被列入第二批贵州省级非物质文化遗产名录。

⊙1

⊙2

⊙1
进入白水村的乡村公路
Country road leading to Baishui Village

⊙2
调查组成员与村民在白水村纸坊中交流
A researcher communicating with villagers in a papermaking mill in Baishui Village

岑巩侗族竹纸生产地分布示意图

Distribution map of the papermaking sites of bamboo paper by the Dong Ethnic Group in Cengong County

考察时间
2011年6月

Investigation Date
June 2011

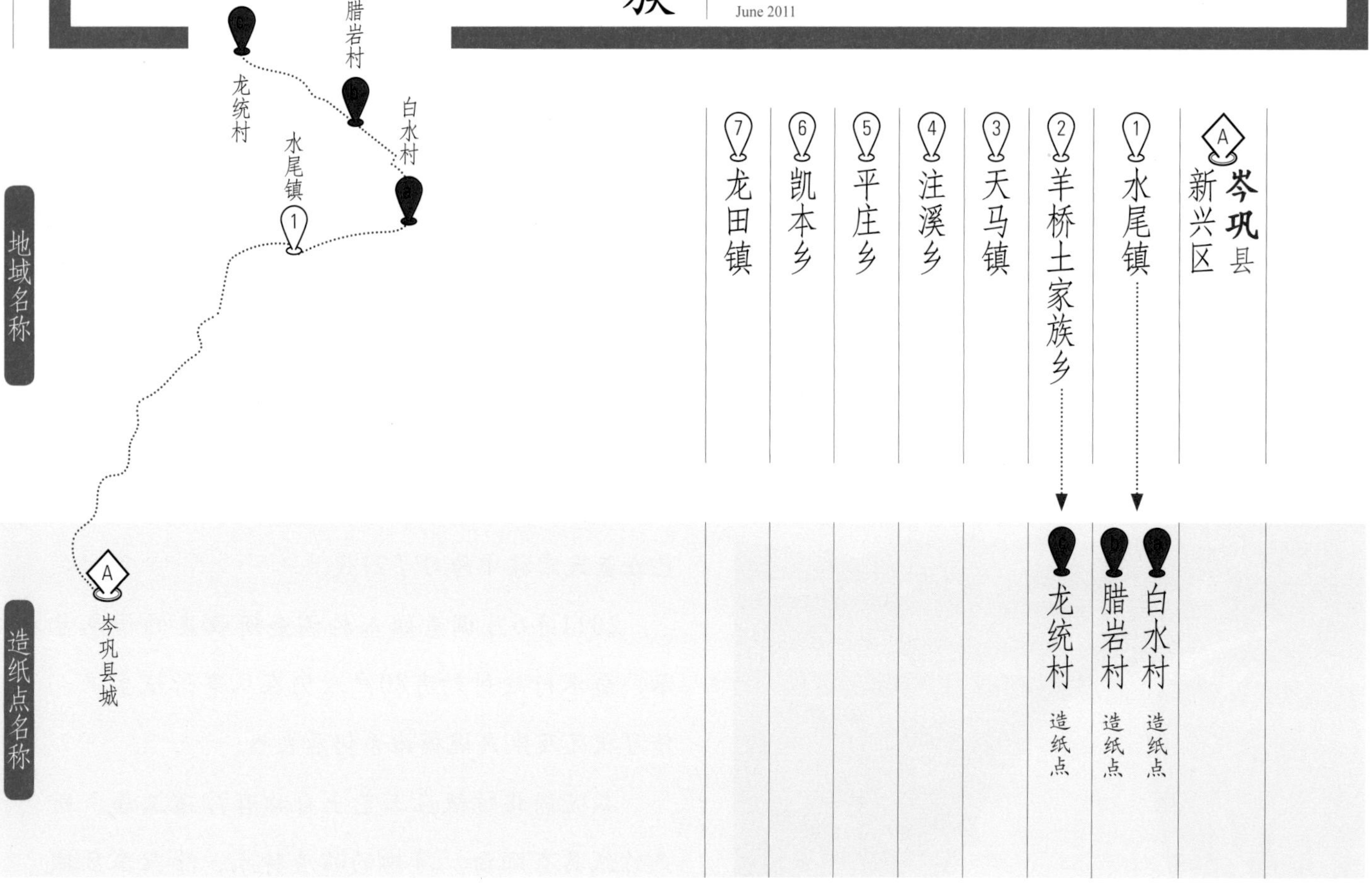

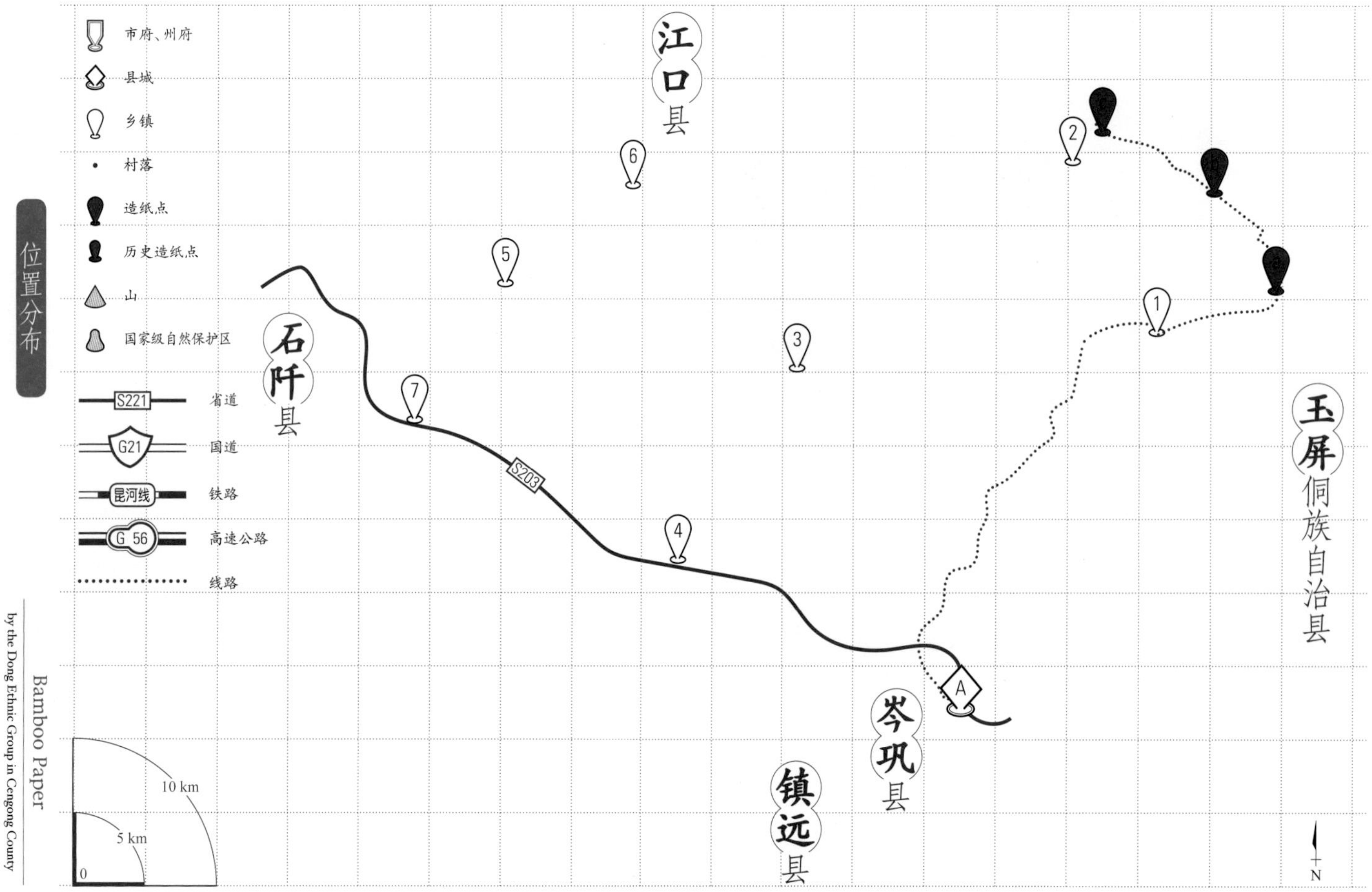

路线图
岑巩县城
↓
白水村、腊岩村、龙统村

Road map from Cengong County centre to the papermaking sites (Baishui Village, Layan Village and Longtong Village)

地域名称

A 岑巩县 新兴区
1 水尾镇
2 羊桥土家族乡
3 天马镇
4 注溪乡
5 平庄乡
6 凯本乡
7 龙田镇

造纸点名称

白水村造纸点
腊岩村造纸点
龙统村造纸点

位置分布

二
岑巩侗族竹纸生产的人文地理环境

2
The Cultural and Geographic Environment of Bamboo Paper by the Dong Ethnic Group in Cengong County

岑巩县位于贵州省东部，隶属于黔东南苗族侗族自治州，东邻铜仁市玉屏侗族自治县，南与西南接壤本州镇远县，西、北两面与铜仁市的石阡县、江口县相界，地处东经108° 20′ ~109° 3′、北纬27° 9′ ~27° 32′，县域面积1 486 km²。

岑巩县属中亚热带季风气候区，四季分明，雨热同季，通常从4月上旬即进入雨季。岑巩湿润气候明显，年平均降水量1 188 mm。流域面积超过10 km²的溪流有33条，分属于舞阳河、龙鳌河、龙江河三大河系，可资利用的水能资源相当丰富。年平均气温16 ℃，年日照时数在1 220小时以上。森林覆盖率达50.35%，有林地约533 km²。

岑巩县境内多为喀斯特地貌，属中山丘陵寒武熔岩地貌。已发现5处溶岩洞群、200余处溶洞景观、3个万米以上的溶洞（即万佛长廊洞、云门洞、将军洞），其中地处县内平庄乡包东村的万佛长廊洞以洞内多达万余个形似佛仙的钟乳奇石而得名，全长16 130 m，分天厅、上宫、中楼、地殿、水道5层，呈现天然壮丽的“地下宫殿”景观。

岑巩县域交通便利，柏油公路已通达所有乡镇，湘黔铁路、“320”国道、G65高速公路（上海—瑞丽）均穿城而过，正在建设中的上海至昆明的高铁干线将通过岑巩。航空方面，县城距铜仁大兴机场100 km，距贵阳龙洞堡国际机场220 km。公路方面，到达黔东南州首府凯里市及铁路枢纽湖南省怀化市均只需一小时。

⊙1

⊙1 流经水尾镇的龙鳌河 Long'ao River flowing through Shuiwei Town

岑巩宜农宜林，特色性物产丰富，思州绿茶为历史名茶，曾长期被列为朝廷贡茶，当代品牌有岑峰牌思州剑雪茶、天仙毛峰茶等。思州牌文旦柚品质出众，曾荣获2007年度中国国际林业产业博览会金奖及2008年度北京奥运会“奥运推荐果品”大奖。岑巩县的油桐产量及品质均有显示度，是国家级重点油桐生产基地县，境内车坝河两岸有著名的“十里桐林”景观。每年4月，上万亩桐林银花怒放，可谓“河面漂霜随人走，路旁落花似堆银”。岑巩县为贵州省优质稻米基地县，盛产优质大米，以岑巩米制作的思州米粉、思州年糕、思州汤圆均为地方知名食品。

岑巩古称思州，唐以前历属西南夷、夜郎国、五溪蛮辖地，自唐武德元年（618年）设思州、思州军、思州司府以来，宋、元、明、清一直归思州建制统辖，且一度成为领68个府、州、县及蛮夷长官司的军民安抚司的中心地，既是荆楚夜郎文化的融合地，也是贵州制度文化的重要发祥地之一，地方史学界因此有“先有思州，后有贵州”之说。民国二年（1913年）废思州府而设思县，民国十九年（1930年）改名为岑巩县至今。

调查时，岑巩县辖7镇4乡1管理委员会，境内有汉、苗、侗、仡佬、土家、布依等19个民族杂居。2009年岑巩全县总人口22.9万，其中少数民族人口11.5万，占总人口的50.2%。岑巩多民族文化与民俗特色鲜明，遗存或流传的代表样式品位高、经典性强。

思州白崖悬棺葬位于古思州城东3 km处的思阳镇（今城关镇）桐木村一带的白崖岩上。白崖岩高72 m，宽350 m，临龙江河北岸。悬崖绝壁上今存悬棺洞穴18处，其中4处为纯人工凿成，呈“口”字形，棺木尸骨均存。据考古专家的研究，其葬式为仡佬族祖先葬俗，历史已逾千年。

中木召古城旧址位于大有乡木召村中木召寨，距县城23 km。古城建筑宏伟浩大，古木成林，院内全为青石和六棱砖铺地，城中通往8个方向的道路则全为鹅卵石铺设，每条路均长3 000~5 000 m。古城建有药院、盐库、蜡库、练兵场等系列设施，占地面积约34 700 m^2，建筑讲求高度对称，且以12的倍数为吉祥数布局，具有典型的宫殿建筑文化特征。关于古城的来历，考古专家进行研究后形成4种推测：古夜郎国的一处都城、古土司王世居地、古苗王都城、古名宦庄园。整个庄园的建造若以4 000个劳力来测算需15年方能完工。

陈圆圆墓。据《岑巩县旅游文史资料》所记乡土传说，清康熙十七年（1678年），吴三桂称帝后在湖南衡州病逝，清军欲灭吴家九族。“陈圆圆闻讯后，在吴三桂生前军师马宝的护送下，带领吴三桂的儿子吴启华、孙子吴仁杰等人在黑夜乘舟横穿洞庭湖，逆沅江而上，沿贵州舞阳河支流——龙鳌河，直扑思州龙鳌里（今岑巩县境内）……在思州鳌山寺隐藏一段时间后，于康熙十八年（1679年）削发为尼，改佛名‘寂静’，字‘玉庵’。后又在距鳌山寺10余千米的思州天庵寺出家。”康熙二十四年（1685年），陈圆圆安排吴启华定居在原始森林中的狮子山，并将寨名取为马家寨，以铭记马宝冒死护送的恩情。陈圆圆葬于今水尾镇马家寨的狮子山，今存雍正六

⊙1

⊙1 传说中的陈圆圆墓
Tomb of Chen Yuanyuan according to the local legend

年（1728年）所立墓碑："故先妣吴门聂氏之墓位席。孝男吴启华，媳涂氏立。孝孙男吴仕龙、杰……立。"碑文所称"聂"字时为创新字（恰与后代简化字相同），无人能懂，但吴氏家族内一直称之为始祖陈老太婆墓，而不称聂氏墓。今马家寨居民1 100余人均为吴姓。

思州石砚。砚材产于思阳镇星石潭，因属思州古地而得名。又因石含天然金星微粒微片，而称为金星石砚。其优质砚石通体墨黑发亮，金星闪烁，细腻似婴肤，发墨如油，属历史名砚，康熙皇帝曾特选思州石砚为"御砚"入藏。1979年1月1日，国家轻工业部将"思州石砚"列为全国传统工艺美术品"刻砚"类的八大名砚之一。

思州拕锣。岑巩土家族、侗族、苗族村寨的独特民俗，流行于水尾、天星、羊桥等乡镇。"拕锣"源于宋代，其时土家等族先民僻居深山，以燃爆竹筒、打响棍、敲鼎罐防兽防袭。后演化为用具有民族特色的铜锣鼓驱兽，并逐渐发展成为每年庆丰收与重大祭典时的打击乐舞。

思州拕锣以铜锣鼓等打击乐器构成独具风格的打击交响乐，并形成了类似曲牌的丰富的锣鼓词，如《鸡拍翅》《牛擦痒》《青蛙跳水》《马蜂过坳》，诙谐逼真的仿生调极富特色，被誉为"中国少数民族打击乐历史的活化石"。

当代"拕锣赛"通常从农历正月初三开始，以自然村寨为单位的"拕锣队"在20支以上，每队少则10人，多则数百人，现场观众则有万人以上，他们全都会集到水尾镇大树林和羊桥乡杨柳沟，场面蔚为大观。

思州傩。据《岑巩县旅游文史资料》记载，思州傩始创于东汉永寿元年（155年），至今已有1 800多年的历史。表演分法事表演、傩技表演、正戏表演三大部分，时间安排在深夜，成为标准的"神鬼戏"。供奉的神灵"上至天庭玉皇，中至人间人皇，下至地府阎王"，共计200余位，表演程序上先祭后戏，祭戏交替。思州傩表演时有120余种不同的人脸面具，均为形象怪异的立体浮雕，大小近似真人脸。

思州傩的一大特色是特技表演，几乎都是令人匪夷所思的绝技。如《上刀山》是手握刀叶，脚踩刀刃，登上12把钢刀与红、黑、白三色布组成的刀梯，并且在刀梯上跳舞；《脚踩红犁》是赤脚在烧红至1 000 ℃左右高温的钢上随意踩跳，而且助手还不时往钢上喷油或酒，令人胆战心惊。其他诸如《手捞油锅》《口衔红铁》《口吐红火》《喊竹》等表演让人叹为观止。

1995年8月，香港凤凰卫视向全世界30多个国家转播了思州傩技傩戏专题片，形成了积极的国际传播。

龙鳌河风景名胜区。又名"车坝河"，发源于石阡、江口、岑巩三县交界的朝阳坡，因河水绕着龙鳌山东南流而得名。该河流经岑巩县的凯本、羊桥和水尾3个乡镇，全长39.7 km。龙鳌河沿岸峰峦叠嶂，涧幽水清，景观奇秀。如潭生幽雾、细雨漫天的龙王潭飞瀑景区，飞瀑倒转回旋

⊙2

⊙3

⊙2 思州砚作坊 Ink slab mill in Sizhou Prefecture (ancient name of Cengong County)

⊙3 神奇的思州傩特技 Magical Nuo Show in Sizhou Prefecture (ancient name of Cengong County)

如喷水银龙的龙鳌闹海景区，孤崖绝壁砌有古代石墙石屋和悬葬葬门的雀儿垄垦景区，洞悬钟乳灿若云锦，恍如“千年雨”洒向行船的大明洞景区，绿色崖山百门千窗、色彩瑰丽的一百单八洞溶洞景区……奇绝胜景令人目不暇接。此外，车坝与下茂马村寨的思州傩技艺、车坝沿河两岸的十里油桐林，以及龙鳌河沿岸以白水村为代表的手工竹纸制作也是极具特色的景观。1995年，龙鳌河风光被公布为贵州省级风景名胜区。

⊙1

⊙1
龙鳌河沿岸竹纸作坊
Bamboo papermaking mill along the Long'ao Riverside

三
岑巩侗族竹纸的历史与传承

3
History and Inheritance of Bamboo Paper by the Dong Ethnic Group in Cengong County

岑巩侗族竹纸的主产地在县辖水尾镇白水行政村与腊岩行政村以及羊桥土家族乡龙统行政村。水尾镇与羊桥乡均位于岑巩县的最东端，龙鳌河由羊桥乡流经水尾镇延向玉屏侗族自治县与湖南省新晃侗族自治县。在公路未通的历史时期，这条水上航道是湘西与黔东南物资运输的重要通道，岑巩侗族竹纸的历史运销也一直以龙鳌河水运为主。同时，在龙鳌河沿岸采竹料十分便利且可就地利用水源造纸，因此造纸的侗族村落都分布在龙鳌河沿岸。历史上，以家家户户造竹纸闻名的烂褥河村即散布在河两岸，2 km长的沿河作坊群曾经是相当壮观的百年手工造纸一大景观，调查时烂褥河村已归属白水行政村辖区。

有关岑巩侗族竹纸技艺的起源年代及类型，根据调查组2011年6月进入白水行政村与多位村民的访谈，以及岑巩县申报贵州省省级非物质文化遗产项目——“民间火纸制作技艺”的正式材料，认定它是在400余年前的明代晚期由湖南靖县（今湖南省靖州苗族侗族自治县）传入的，属于输入型技艺起源，而技艺输入的主体正是当年由靖县迁徙到岑巩水尾镇白水村的黄姓侗族祖先。由于侗族无文字，加之汉族文字的地方历史文献对相关民间竹纸缺乏历史记述，因而有关岑巩侗族竹纸的历史与传承谱系基本上来自白水村黄姓宗族的口述记忆以及当代人的整理归纳。

据“民间火纸制作技艺”贵州省非物质文化遗产项目公布的传承谱系可知，黄姓侗族竹纸已传承21代，有姓名记忆的谱系辈分排行为：万—上—之—世—大—懋—锌—久—贵—俊—秀—忠。而列入传承谱系的代表性传承人名单为：黄万润，黄上贵，黄之忠、黄之国，黄世平，黄大喜，黄懋明，黄锌海（又名黄海生），黄久金、黄久炳，黄贵金、黄贵华，黄俊松、黄俊财、黄俊章、黄俊明、黄俊春，黄秀木、黄秀明、黄秀线、黄秀龙、黄秀炳，黄忠勇。

⊙2

⊙3

⊙2 水稻田中的造纸作坊 Papermaking mill in a paddy field

⊙3 纸坊中的黄姓纸工 Local papermaker Mr. Huang in a papermaking mill

显然，白水村竹纸生产规模较广，上述传承谱系只是择黄姓造纸宗族脉络记忆较清晰的一支进行的申报归纳，实际的传承状况应当是更为普遍和多元化的。

2011年6月下旬，调查组进入白水行政村的多个村民组进行调查访谈。据调查所获得的数据，白水行政村现有12个村民组，其中11个村民组约有80户造纸农户仍传习祖业。而据时任白水村村长黄俊平及村支书黄治文的介绍，2011年白水村全村约有360户，其中从事造纸的户数要视口径与波动情况而定。例如黄村长与黄支书家中也“舀纸”，他们虽然自身都掌握造纸的手艺，但由于担任基层行政职务则雇人“舀纸”，不过他们认为自己也可以算作造纸人。此外，白水村造纸户有长年生产的，也有间断生产的，有一定的变化区间。在田野调查时有一个较特别的发现：水尾镇侗族竹纸生产地的传统提法是烂褥河村，当代提法却是白水村及腊岩村，但同属水尾镇的另一个行政村于河村也有较兴旺的竹纸生产，该村与白水村隔河相望，调查组也曾渡河实地探访，推测于河村应是原中心区烂褥河村当年的分布遗存。另据2013年12月25日《西部开发报》刊登的采访于河村老支书、造纸传人黄俊龙的文章介绍，水尾镇的白水、腊岩、于河、长坪4个行政村均有从事竹纸生产的历史与当代业态。

⊙1

综合访谈调查得知，水尾镇以白水村为中心的侗族竹纸制作规模虽然较历史高峰期家家户户均造纸已有明显的萎缩，但其仍然保持着较合理的手工技艺分布形态与良好的生产性传承。

⊙2

⊙1
白水村小山坡上的煮料窑
Boiling kiln on a hill in Baishui Village

⊙2
岩山脚村的一处造纸作坊群
Papermaking mills in Yanshanjiao Village

四
岑巩侗族竹纸的生产工艺与技术分析

4
Papermaking Technique and Technical Analysis of Bamboo Paper by the Dong Ethnic Group in Cengong County

2011年6月23日，调查组重点访谈了白水行政村岩山脚村民组的造纸人黄贵春、黄俊有、黄云祥，以及补充访谈了中寨村民组的黄贵堂、黄贵培，他们均为侗族村民。黄贵春1961年出生，1985年当兵复员回村后即从事造纸，已有26年“舀纸”经历，其妻子姚菊莲1964年出生于湖南新晃侗族自治县；黄俊有1966年出生，14岁即在村中“舀纸”，属于以造纸为主业的造纸人，已有31年“舀纸”经历；黄云祥1945年出生，已有近40年“舀纸”经历，其妻子姚火青及3个女儿都会“舀纸”。综合访谈与现场考察，调查组对白水村侗族竹纸的工艺流程形成了一定的认知。

⊙3

(一)
岑巩侗族竹纸的生产原料与辅料

白水村竹纸生产使用的主要原料为当地产的绵竹，植物学上属于禾本科，主要分布于中国西南部云南、四川、贵州等地，其种类也有蛮竹、凤尾竹、芒竹等别称，岑巩当地又称为“阳山竹”。造纸技工们认为，岑巩本地生产的阳山竹除了竹茎肉厚外，其纤维均匀细密、出浆料率高，属优质制造竹纸原料。据造纸户们介绍，与岑巩县相邻的玉屏侗族自治县与湖南新晃侗族自治县也生产竹纸，其所用原料就是从水尾镇一带采购的。

白水村竹纸生产使用的辅料包括纸药、纯碱与石灰。最常用的一种纸药是被当地造纸户称为“神仙树”的树叶，它形似桂花叶，浸泡成汁液后，作为“舀纸”时分张用，通常在天气温热时使用，黄贵春说立夏后即用。该树叶一般从湖南的新晃侗族自治县采购。另一种纸药当地造纸户

⊙4

⊙3 阳山竹 Local Yangshan bamboo (*Bambusa intermedia* Hsueh et Yi)

⊙4 纸坊中的姚菊莲 Local papermaker Yao Julian in the papermaking mill

称为“野棉花”，因其花骨朵类似棉花而得名，用其藤茎浸泡成汁液使用。关于第一种纸药的学名，造纸户与调查组一时均未能确定。

据刘仁庆（纸史研究专家）的相关研究，“野棉花”应是流行纸药植物黄蜀葵。“黄蜀葵属锦葵科，又名棉花滑，系一年生或多年生直立草本植物，茎高0.9~2.7 m。适宜生长在山间潮湿肥沃的沙质土壤中。黄蜀葵在广东、广西、贵州、云南、湖北、江西等地皆有分布。造纸户一般在秋末冬初进山采集、挖葵根。”[1]。

据造纸人黄贵春介绍，冬天及气温低时，会使用“野棉花”及杨桃藤浸汁作为纸药，而杨桃藤的主要来源地也是新晃侗族自治县。

纯碱主要用于蒸煮竹料，这源于水尾镇竹纸采用“熟料法”制作。据造纸人黄云祥回忆，白水村的传统方式是用桐子灰（当地亦称为桐壳灰）煮料，而龙鳌河一带正是油桐的集中产地，有著名的“十里桐林”景观，其桐子灰原料充足且获取便利。桐子灰一直使用到1982年，后来改用纯碱。

(二)
岑巩侗族竹纸的生产工艺流程

岑巩侗族竹纸的相关研究资料及申报贵州省级“非物质文化遗产”的材料均记述白水村侗族竹纸有72道制作工序，白水村当地造纸户们也确认了这一说法，但具体工序却未见于任何文献或研究文章中。这可能是因为中国手工造纸文化记忆习俗中经常有72道工序或108道工序之“吉数”说法，以及将一个标准工序拆分成多道小工序的习惯做法。

调查组通过立足跟踪重点造纸户的流程调研，结合相关文献与造纸户的共性描述，将白水村竹纸的生产工艺流程概述如下：

壹 砍竹 → 贰 破竹 → 叁 泡竹 → 肆 沤竹 → 伍 煮竹 → 陆 踩竹 → 柒 洗麻 → 捌 煮麻 → 玖 泡麻 → 拾 制浆 → 拾壹 入槽 → 拾贰 加纸药 → 拾叁 舀纸 → 拾肆 榨纸 → 拾伍 分纸 → 拾陆 晒纸 → 拾柒 捆纸

[1] 刘仁庆. 中国书画纸[M]. 北京：中国水利水电出版社，2007:30.

⊙1

壹
砍竹
1 ⊙1

每年的农历七至九月为砍竹最理想的时节，选择当年生枝叶将发或新发的竹子整根砍下后截成160~230 cm长的竹段，具体长度一般根据造纸户沤竹的石灰池与煮竹的土窑的大小确定。由于阳山竹在白水村一带分布很多，因此一般造纸户自家的竹子即可满足原料需求。

贰
破竹
2

将砍下的新鲜竹段用铁锤锤破，然后置于外面晒干，再将晒干的干料捆成小捆，每小捆2 kg左右，一般用竹篾捆扎。据黄贵春介绍，在20世纪90年代以前，主要是用脚踩木碓破竹的，调查时仍有造纸户使用木碓破竹。

叁
泡竹
3

将捆扎后的干竹放入河水中浸泡，一般浸泡10天左右，以浸泡的竹茎变软变粑为度。调查时发现也有在水池中泡竹的，但由于造纸作坊均沿龙鳌河分布，所以多数造纸户还是选择在河水中浸泡。

肆
沤竹
4 ⊙2

将泡软的竹料放入石灰水池，池中为生石灰加水形成的热石灰液，其配比为100 kg竹料投放约40 kg生石灰。竹料在池中蘸满石灰浆液后即拿出，码成一堆堆沤发酵，一般需10天左右。

伍
煮竹
5

将堆沤发酵好的竹料放入蒸煮窑里，然后注满清水，表面覆盖塑料布，密封后用大火煮7天左右。

⊙2

陆
踩竹
6

将煮好的竹料从窑里取出，赤脚或穿胶皮长筒鞋踩竹，使竹料纤维分丝帚化。据黄贵春等造纸人介绍，以往是穿稻草鞋来踩竹料的，但时间长了脚容易磨烂，后来就改为赤脚踩，现在经济条件好转后则常穿胶鞋踩。在白水村造纸工序中，经过蒸煮踩料后的竹料称为竹麻。

⊙1 堆放的竹麻原料
Piled raw material of bamboo materials

⊙2 料塘中沤竹
Fermenting the bamboo in a pool

柒

洗　麻

7

将用脚踩后的竹麻拿到河水或池水里清洗，并将竹麻里的石灰浆液和附着的杂质在清水里洗干净。

捌

煮　麻

8　⊙3

将洗干净的竹麻放入蒸煮窑里，加入纯碱煮3~4天，添加纯碱的比例为每100 kg竹麻加10 kg纯碱。

⊙3

玖

泡　麻

9

循环放清水入窑以洗净纯碱水液，然后将竹麻放在窑内的清水里浸泡约30天，如果夏天很热则浸泡15天左右。其中，在浸泡的第一天要将窑内水烧热，称为“沤火”。

⊙4

⊙5

拾

制　浆

10　⊙4⊙5

取出泡好的竹麻置于干净之地或清水池中备用。需要“舀纸”时，则取竹麻用木碓击打舂融，一般一团料要舂打15~20分钟，直至形成纸浆料团或料饼。

拾壹

入　槽

11

将舂融并已完全帚化的纸浆料放入“舀纸”的纸槽中，用搅拌棍在清水里搅打，使纸浆料均匀地散布在水中。

拾贰

加　纸　药

12

加入已浸泡好的药液，冬天用杨桃藤与野棉花汁液，夏天则用神仙树叶的汁液。然后再次用搅拌棍搅打，使纸药液与纸浆水充分混融。

⊙3
煮竹麻
Boiling the bamboo materials

⊙4/5
打碓
Beating the papermaking materials

拾叁
舀　　纸

13 ⊙6~⊙9

实际上即抄纸工序，岑巩当地习称“舀纸”。用抄纸竹帘在浆料池里往前插入舀水，再回荡入水，提起滤水后便形成一张湿纸膜。然后将帘架搁在槽边上，从帘架里提起抄纸竹帘，转身倒扣在待榨的木板上，轻轻揭起竹帘，一张湿纸便留在榨纸板上。通常一个熟练纸工一天能抄约3 000张纸。

⊙6

⊙7

⊙8

⊙9

拾肆
榨　　纸

14 ⊙10

⊙10

待抄出的湿纸在榨纸板上一张张重叠至约40 cm厚时，即可用木质的纸榨进行压榨。通常是用轳辘绞紧钢丝绳，并利用杠杆原理将湿纸的水分挤去，一般榨至原湿纸高度的1/3即可。

拾伍
分　　纸

15 ⊙11

将压榨好的半湿纸块从榨床上取下，放在凳子或桌子上，用手将纸一张张地轻轻揭开，按相邻两张纸边缘相距约2 cm依次叠放，每10张为一贴。

⊙11

⊙6/9 舀纸 Scooping and lifting the papermaking screen out of water and turning it upside down on the board

⊙10 榨纸 Pressing the paper

⊙11 分纸 Splitting the paper layers

⊙12

拾陆
晒　纸
16　⊙12⊙13

将分好的纸以贴为单位铺在屋内或室外的干净地面上晾晒，待干后即可包装。

⊙13

⊙14

拾柒
捆　纸
17　⊙14

将晾晒干的纸以10贴为一刀，10刀为一捆叠放，并用细竹篾捆扎，这就是可销售的成品纸的流通规格了。

(三)
岑巩侗族竹纸生产使用的主要工具设备

壹
纸　窑
1

主要用于蒸煮阳山竹料以及竹麻。在小山丘或高坡上砌建的土窑或砖石窑，直径一般约1.2 m，深度约0.6 m，窑下部放一口大铁锅，锅沿留有放水用的小孔，锅四周用鹅卵石加水泥或纯水泥涂壁密封，蒸煮竹料时便在锅内或锅中放一排竹木支屉，如同蒸笼的笼屉，将竹料放在屉上煮。坡下另开一坑洞为火膛，煮料时即在火膛内点火。调查时建窑的成本约2 000元。

⊙15

贰
纸　碓
2

木质，用于将鲜竹段捶破的工具，以及将竹麻捶烂、使竹麻变成纸浆团饼的制浆工具。白水村比较流行的是脚踏碓。

叁
纸　槽
3

盛放纸浆液及纸药与纸浆混合液的水池，用于“舀纸”及“打槽”工序，呈长方体形，有石质槽、木质槽或石木混合槽。白水村当代则以水泥槽为主。

⊙17

⊙16

⊙12/13 晒纸 Drying the paper

⊙14 捆纸 Binding the paper

⊙15 于河村的纸窑 Papermaking kiln in Yuhe Village

⊙16 中寨村的纸碓 Beating pestle in Zhongzhai Village

⊙17 岩山脚村的纸槽 Papermaking trough in Yanshanjiao Village

肆 滤药袋 4

盛放纸药原料的布袋，用于将在水中浸泡后的纸药渗滤成纸药汁液。

伍 纸帘 5

竹质，是舀纸浆液成为湿纸膜的必备工具。外为帘架，上有提手，抄纸竹帘置放在帘架上。竹帘由细竹丝编成，每根竹丝之间的间隔为0.2 cm，呈经纬线交错结构。实测黄贵春家竹帘尺寸为54 cm×48 cm。另据黄贵春介绍，细帘通常从四川采购，订货后，卖方会寄到中寨村民组的一位造纸人家中，2010年一副竹帘的价格约200元。

⊙18

陆 纸榨 6

木质，利用杠杆原理制成的传统榨纸工具，用于挤出湿纸膜中的水分。调查时，白水村造纸户常用的还是木榨，现代金属榨纸设备尚未流行。

⊙19

(四) 岑巩侗族竹纸的性能分析

所测岑巩白水村竹纸的相关性能参数见表10.1。

表10.1 白水村竹纸的相关性能参数

Table 10.1 Performance parameters of bamboo paper in Baishui Village

指标		单位	最大值	最小值	平均值
厚度		mm	0.073	0.036	0.052
定量		g/m^2	—	—	18.1
紧度		g/cm^3	—	—	0.348
抗张力	纵向	N	9.8	6.2	7.5
	横向	N	7.5	4.1	6.1
抗张强度		kN/m	—	—	0.453

⊙18 纸帘 Papermaking screen

⊙19 岩山脚村的纸榨 Pressing device in Yanshanjiao Village

续表

指标	单位	最大值	最小值	平均值
白度	%	30.5	30.3	30.4
纤维长度	mm	11.79	0.84	2.71
纤维宽度	μm	24.0	3.0	12.0

由表10.1可知，所测白水村竹纸最厚约是最薄的2.03倍，相对标准偏差为1.30%，纸张厚薄不一。竹纸的平均定量为18.1 g/m^2。所测竹纸紧度为0.348 g/cm^3。

经计算，其抗张强度为0.453 kN/m，抗张强度值较小。

所测白水村竹纸白度平均值为30.4%，白度较低，白度最大值约是最小值的1.01倍，相对标准偏差为0.13%，白度差异相对较小。

所测白水村竹纸纤维长度：最长11.79 mm，最短0.84 mm，平均2.71 mm；纤维宽度：最宽24.0 μm，最窄3.0 μm，平均12.0 μm。所测竹纸在10倍、20倍物镜下观测的纤维形态分别见图★1 、图★2。

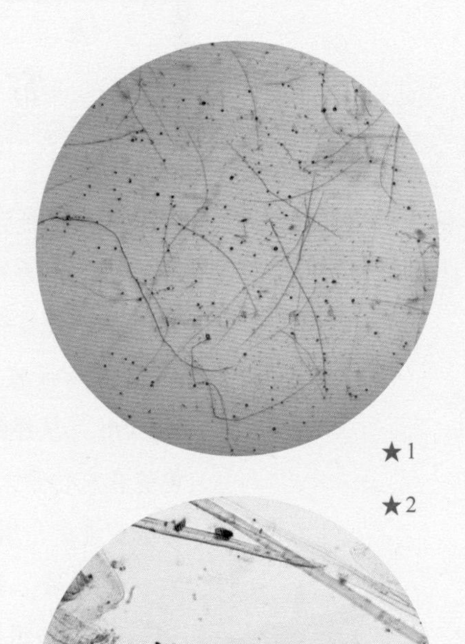

★1

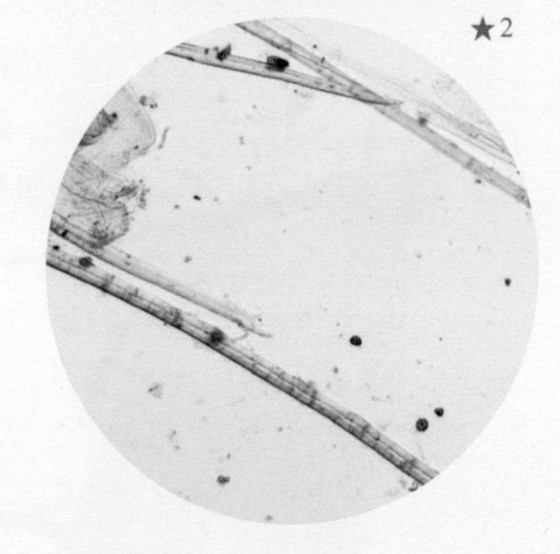

★2

★1
白水村竹纸纤维形态图(10×)
Fibers of bamboo paper in Baishui Village (10× objective)

★2
白水村竹纸纤维形态图(20×)
Fibers of bamboo paper in Baishui Village (20× objective)

五 岑巩侗族竹纸的用途与销售情况

5 Uses and Sales of Bamboo Paper by the Dong Ethnic Group in Cengong County

(一) 岑巩侗族竹纸的用途

调查组入村调查时，造纸户还在坚持用非常传统的制作工艺生产岑巩侗族竹纸：原材料的加工采用自然的方式，如晾晒；蒸煮过程中未加任何化学材料；破碎、压榨也以传统人力及工具为主。因而他们生产出的竹纸保持了纯粹的手工技艺，纸质柔软轻薄，纤维韧性强，光洁度与白度在中国西南地区民间竹纸里尚属上品。

1. 祭奠

岑巩竹纸历史上又名“思州火纸”，可见从过去到当代祭奠都是其最主要的用途，而且当代似乎只剩下这一种用途了。综合调查组对造纸村民的访谈记录及地方文献的记述，祭奠用途又可分为3种类型：

其一是村民日常生活中祭祀祖先亲人时的焚烧用纸。如清明与七月半（当地有“月半纸清明香”之说）、冬至以及办丧事仪式上，均要烧纸。这类烧纸的方式是烧纸钱送给祖先丧亲，即在竹纸上用錾子錾打成纸钱后焚烧，表示送钱给先人用，通常还会用竹纸包一个信封一并焚化。另外，也有村民直接烧纸及制作挂青用的钱串。

其二是地方民俗节日的文娱活动中都会频繁用到竹纸。如“思州拖锣”和傩戏表演中许愿、还愿等都涉及“烧纸祭”的仪式。

其三是若干的婚庆嫁娶中也会用到竹纸。据白水村村长黄俊平及造纸人黄俊有介绍，嫁娶进屋时节水尾镇乡间有焚竹纸的习俗。

2. 书写

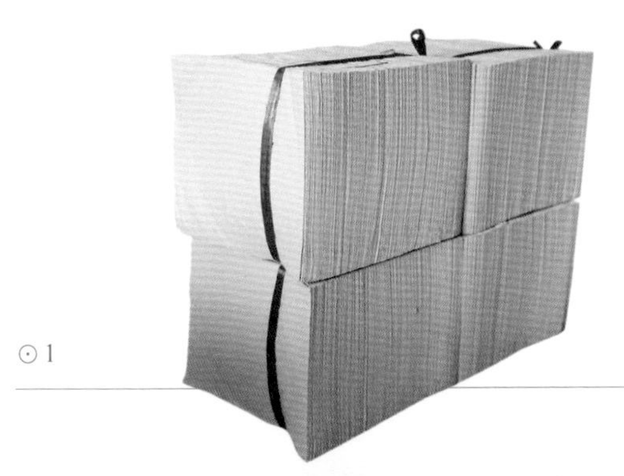
⊙1

⊙2

⊙1 优质的白水村竹纸 Quality bamboo paper in Baishui Village

⊙2 纸钱 Joss paper

据白水村造纸人黄贵春及黄云祥的记忆，20世纪中叶机制书写纸还不流行时，乡间流行用岑巩竹纸作为一般的书写纸，如小学生练习毛笔字、家中或村庄里记事簿、地契房契以及手抄傩戏唱本等的用纸等。黄贵春自述早年还制作过用于书写的竹纸。不过，调查时这一用途已基本消失。另有清代与民国前期著名的手写“思州文书”，今存约20 000件，最早的为清康熙三十七年（1698年）的存世品，但早期存品主要为绵纸材料，尚不清楚古思州绵纸是岑巩本地产纸还是周边输入的纸。

3. 印谱

虽然侗族本身无文字，但在汉文化影响下逐步接受了汉字的使用，同时也开始用汉字记述家族与宗族谱系，而这类谱基本上会用当地生产的竹纸印刷或抄写。今存民国三十六年（1947年）水尾镇白水村《黄氏族谱》的载体即为岑巩竹纸，调查时白水村岩山脚村民组的造纸人黄俊有也说他们家族里有《黄氏族谱》存世。

(二)

岑巩侗族竹纸的销售状况

岑巩侗族竹纸除在本县及邻近的玉屏侗族自治县，以及江口县、石阡县的少数民族乡镇销售外，主要通过与湖南省新晃侗族自治县产销的紧密联动流通到湖南省怀化市所辖湘西若干地区。这是一个有历史渊源的侗族竹纸产供销地区。早在400余年前，造纸技艺就随黄姓先民从靖州苗族侗族县带入岑巩，部分原料（如制作纸药的植物）来自新晃侗族县，上门采购的纸商来自新晃侗族县，造出的纸挑到新晃与玉屏侗族县去销售，这基本上是在一个经典的侗族文化区内形成的乡村交易市场。

岑巩水尾镇销售竹纸的基准单位是刀，每10张为一贴，每10贴为一刀。从售价来看，2008年前后，湖南新晃纸商上门收购的竹纸价格约4元/刀，而2011年白水村竹纸的最低价为7元/刀，纸价涨幅明显，说明销售很旺。以重点调查的造纸人黄俊有家的造纸收益为例：黄家共四口人，只有约2 000 m^2地，两个孩子目前都在读书，放假在家时也帮忙干分纸、晒纸等活。由于田少，黄俊有夫妻两人以造纸为主业，每年有5~6个月时间从事造纸工作，其余时间种水稻及干农杂活。黄俊有2010年造纸4 000余刀，按7元/刀计算，年度毛收入约28 000元。而另一重点调查对象黄贵春与其妻子姚菊莲两人每年造纸4~5个月，每年的产量为4 000~4 200刀。因为自己砍阳山竹，其他纸药、纯碱、石灰成本很低，因而净收益在乡村中还是相当不错的。

当然，黄俊有与黄贵春两户是以造纸为主业的，而一般造纸户是农田、山场、经济作物与造纸兼顾的，据黄俊有访谈时的说法，造纸年收入10 000~20 000元属正常情况，可见白水村手工竹纸业的收益普遍较为良好。

在访谈黄俊有与黄贵春时获知传统烂𬓇河一带竹纸的销售模式是：民国年间与20世纪50年代，烂𬓇河村几乎家家户户都造竹纸，销售则以造纸户自己挑运到附近集市（包括玉屏及湖南新晃的乡镇集市）上交易为主。1958年人民公社成立后，则由生产队（相当于自然村）将造纸户们组织起来生产，原料的供给和成纸的销售则由供销社统一收购，造纸户们只管造纸，按每天的“舀纸”定额计工分。20世纪70年代中后期，则出现专门抽几个人代表生产队销售竹纸的模式，类似于业务承包。黄贵春说自己也被安排过专职卖纸。

从调查时掌握的信息来看，2000~2011年间，水尾镇使用熟料法制作的竹纸供不应求。白水村造纸人黄俊有在访谈时表示，这几年都是新晃一带的纸商上门收纸，他自己已经不再外出卖纸，缺纸的时候，纸商会在湿纸边等着下单订货，甚至有的纸商自己捆扎刚晒干的纸，还有一些纸商互相竞价买纸。虽然其说法的普遍性尚不能确定，但足以说明销售情况确实良好。在机制竹纸已流行的大环境下，手工竹纸依然旺销，除了与岑巩侗族竹纸品质较优有关外，还与黔东南与湘西少数民族百姓认为用机制竹纸制作的纸钱是“假钱”，祖宗不收，用手工造纸制作的纸钱才是“真钱”的乡土认知有关。

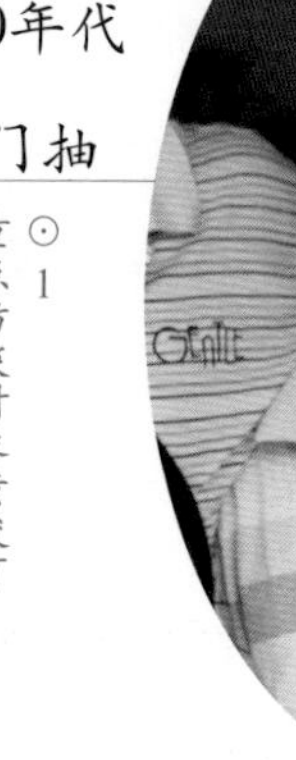

⊙1
重点访谈对象黄俊有
Local papermaker Huang Junyou, an important interviewee of the research

六
岑巩侗族竹纸的相关民俗与文化事象

6
Folk Customs and Culture of Bamboo Paper by the Dong Ethnic Group in Cengong County

1. 龙鳌祭

龙鳌祭分布于县境内龙鳌河流域水尾镇、羊桥土家族乡及凯本乡的苗寨。最具代表性的祭地是水尾镇新场、大树林等红苗村寨。祭祀时间是每年中秋节的深夜，祭祀地点则在村寨内的宗祠里。

祭祀方式：在祠堂里面贴上龙神画像，目的是祈求龙鳌河的龙神保佑来年风调雨顺、五谷丰登。祭仪由苗人寨老主持。第一步是“献生”，即摆放供品后由寨民依次向龙神敬香，然后燃放鞭炮与火铳；第二步是“念祭文”，即半夜鸡叫时，由祭师带领全体寨民面向龙神像三拜九叩行大礼，祭师念祭文毕则烧纸钱并燃香烛；第三步是“上熟”，即将牛、羊等供品的肉割下并放到大锅里煮熟，用盘子盛上熟肉置于供桌上再供半小时，然后全寨人一起享用。

2. 开窑祭

岑巩水尾镇的造纸乡民对竹纸制作技艺保持着敬畏与自豪之情，传统的做法是每年准备开窑煮竹料前，都会在窑前焚烧竹纸、长线（纸钱）以及燃香进行祭祀，以表示对神灵及造纸祖先的感谢，同时也祈望获得保护而使造纸顺利。

3. 造纸民谚“七十二道半，还有打烧不能算”

这是岑巩造纸人对竹纸工序之多之细的一种形象表述，即本身造纸工序就已有七十二道半之多，这还没有算上打纸钱的多道工序、烧香祭窑的仪式，以及焚纸钱的祭祀仪式。这句民谚反映了造纸人对岑巩侗族竹纸技艺的自豪之情。

4. 造纸顺口溜

造纸小调或顺口溜：“舀纸郎，舀纸郎，清早起来进槽房，问你一天舀纸多少张？舀纸就像牛喝水，榨纸就像猴搬砖。”意思是说，天蒙蒙亮即到纸坊，整天弯脚弓背，像猴子和牛一样，劳累一生。这段话描述了造纸的艰辛，感叹造纸的不易。

5. 寄纸钱习俗

这是当地独特的焚烧纸钱给祖先亲人的民

⊙2

⊙2 中寨村打纸钱的老人
An old man making joss paper in Zhongzhai Village

⊙1

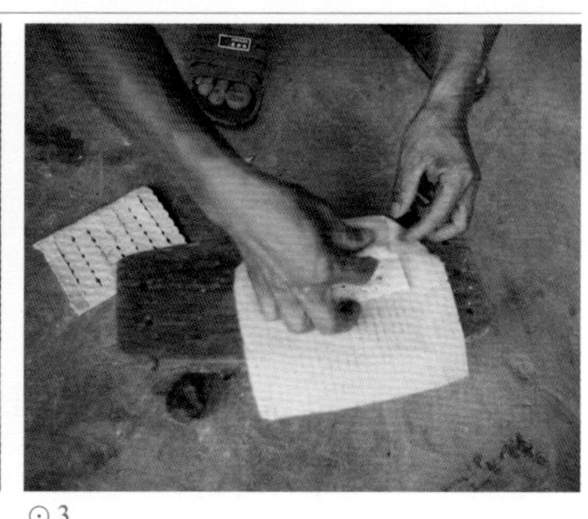

⊙2 ⊙3

俗。岑巩侗族竹纸的最大用途是制作作为祭奠用的纸钱，即用錾子在竹纸上錾打出数目不等的圆形钱串。特殊之处在于：当地通常在使用纸钱时，预先用同样的竹纸包在纸钱外成信封状，然后在信封表面写上收钱人的姓名、送钱的时间、送钱人的姓名、所送钱的数量。

结合前述黔东南与湘西侗族、土家族、苗族民众对用手工造纸制作的钱才是“真钱”的强烈认知倾向，可见该区域少数民族民众对祭祖先的真挚情感。

七 岑巩侗族竹纸的保护现状与发展思考

7 Preservation and Development of Bamboo Paper by the Dong Ethnic Group in Cengong County

（一） 岑巩侗族竹纸的传承保护现状

通过对水尾镇造纸村落和地方政府相关部门的调研，调查组认为岑巩手工竹纸技艺与产业的生产性传承和保护仍处于相对正常的状态，这在机制竹纸流行、乡村农民职业选择多样化的当代已属大不易。在中国南方包括贵州省竹纸制作的很多区域，手工竹纸都因为产品低端和用途单一而陷入经营困境或生产性传承中断的危境，那么岑巩侗族竹纸传承和保护现状特色及内涵要素的

⊙1 打纸钱 Making joss paper

⊙2/3 包纸钱寄给祖先示意 Showing how to wrap joss paper as sacrificial offerings to ancestors

表达该如何认识呢？

1.

岑巩侗族竹纸生产的模式一直以家庭为制作单元，表现为千家万户分布聚合的业态。从造纸的户数来看，调查时的2011年水尾镇中心造纸区各村落已比最兴盛时的20世纪50年代家家户户造纸有了大幅萎缩，如有360余户的白水行政村，只有约80户造纸，有40户左右的白水村斜井村民组，只有4~5户造纸。相邻的于河行政村、腊岩行政村及羊桥乡的龙统行政村未能获得比较数据，但推测应该不会比核心产地白水村的传习性强。

⊙4

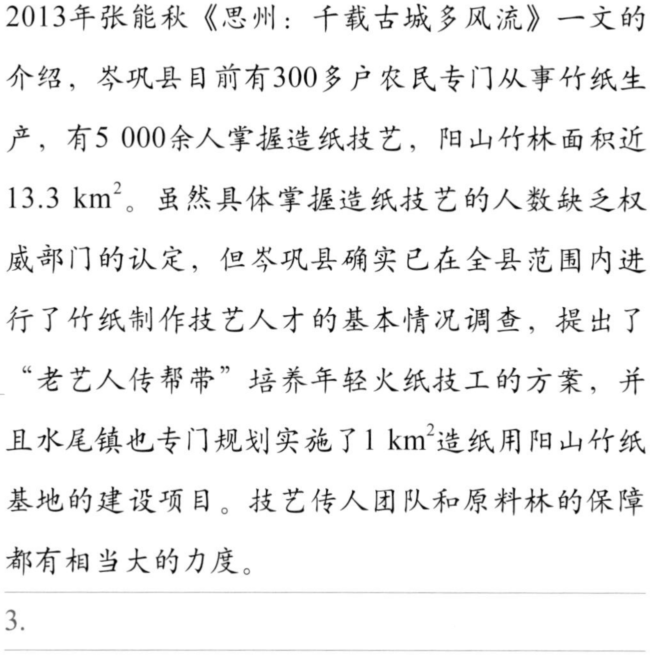

2.

岑巩侗族竹纸的当代技艺传承脉络清晰。据2013年张能秋《思州：千载古城多风流》一文的介绍，岑巩县目前有300多户农民专门从事竹纸生产，有5 000余人掌握造纸技艺，阳山竹林面积近13.3 km^2。虽然具体掌握造纸技艺的人数缺乏权威部门的认定，但岑巩县确实已在全县范围内进行了竹纸制作技艺人才的基本情况调查，提出了“老艺人传帮带”培养年轻火纸技工的方案，并且水尾镇也专门规划实施了1 km^{2}造纸用阳山竹纸基地的建设项目。技艺传人团队和原料林的保障都有相当大的力度。

3.

岑巩侗族竹纸相当难能可贵的是在当代仍整体维系了纯自然、纯手工的造纸工艺，而且是在百家千户分散造纸，多乡、镇、村独立分布的格局下实现的。从非物质文化遗产的传承保护目标来判断，其规模化的原生性业态之纯粹令人惊喜。至于拒绝化学材料和机械加工方式进入传统工艺目标的原因，调查组考察后的理解是：

其一，黔东南、湘西地区岑巩竹纸主消费市场区域对手工造“火纸”为“真纸”、机制“火纸”为“假纸”的普遍信念与心理，以及该区域侗、土家、苗等民族对祭祖祭亲等的高度重

⊙5

⊙4
送调查组成员去于河村的小船
The boat taking the researchers to Yuhe Village

⊙5
岩山脚村的连排纸坊
A row of papermaking mills in Yanshanjiao Village

视，可以视为一种强烈的小区域乡愿文化信念的支持要素。实际上，据黄云祥、黄贵春等造纸人叙述，白水村的纸坊也试过掺锯木粉、水竹等杂料生产半机制竹纸，但很快就因缺乏认同而放弃了。

其二，岑巩竹纸的销售价格与市场需求仍有优势。造纸户数缩减明显、市场需求旺盛使售价强劲上升，如前述在白水村调查时获知2008年售价约4元/刀，2011年调查组入村时已达7元/刀。岩山脚村民组造纸人黄俊有家一年中约有6个月造纸，毛收入近28 000元。2013年11月21日《贵州民族报》调查白水村岩山脚村民组造纸村民黄锦发、黄成发、黄贵春时，他们自述每月造纸收入均能达到5 000元，每户年收入能达50 000元，并表示比外出打工强多了。手工竹纸既然销售如此兴旺，而且收入也相当不错，当然就不需冒消费者认同风险改变材料工艺或工具了，这应该视为经济融合消费文化的支持要素。

其三，鉴于岑巩侗族竹纸已成为贵州省的非物质文化遗产保护对象，按照中国国家文化部“非遗保护项目”的分级要求，属地各级政府主管部门应有若干规定内的促进政策。调查组调研了岑巩县委宣传部和水尾镇政府，了解到相关主管部门已开展对侗族竹纸制作技艺、传承谱系等多方面资料的搜集整理，并正在推动代表性传承人体系的建立，同时也明确提出“保存完整的传统火纸制作工艺”，在火纸制作中尽量避免使用现代化设备和化学制剂。“立项建设‘思州火纸工艺园’，包括火纸工艺古法生产作坊。”这可以视为来自政府政策资源的支持与引导要素。

综合以上叙述的保护发展现状，可以认为至调查组入镇、村的年段，岑巩侗族竹纸的资料型（博物馆式）保护与生产型保护均具有良好的示范价值。

(二)
岑巩侗族竹纸的发展思考

根据调查组的现场调研与当代文献分析，认为该地域竹纸发展面临的挑战或隐忧主要表

⊙1

⊙2

⊙1
于河村正在烧窑的造纸人
A papermaker boiling the papermaking materials

⊙2
满池待洗的竹麻料
Bamboo materials to be cleaned

现为：

1. 用途过于单一

当代岑巩侗族竹纸的用途过于单一，历史上曾经流行过的书写用纸等用途全部消失，一旦祭祀用纸的乡土文化习惯或市场发生变化，则竹纸的生产立即就会受到无法调整的刚性冲击。从中国竹纸纸品的整体来看，四川省夹江市、浙江省富阳市等地的高端竹纸消费空间相当多样，而岑巩竹纸材料正宗、工艺纯正、加工细致、纸质优良，其本身具备提升拓展的品质。

究其原因，一是缺少与竹纸中高端制作系统的交流学习，因而看不见更多元的空间与技艺，只停留在祖传的、收缩的传承系统里；二是当代市场重新兴旺后，户数已大大减少的纸坊生产供不应求，陷入业务饱满、创新压力与动力暂时变为隐性的状态。

2. 依旧延续小农经济型的百户千家独立产销业态

岑巩侗族竹纸的生产延续着小农经济型的百户千家独立产销业态，一直没有构建传统工艺产业化、集成化的现代产销业态。其带来的挑战是在中国手工（包括半手工）造纸行业里竞争力难以提升，而且一旦传统的产销模式破裂，则竹纸的生存环境立刻就会非常严峻。岑巩地方政府也已意识到问题的严重性，在2008年即提出配合龙鳌河省级风景名胜区旅游带建设“思州火纸工艺园”的计划，这是实现“思州火纸”由传统农村家庭业态向现代乡村产业链业态转型的重要探索，但至2014年初，这一计划仍处于尚未启动的状态。

3. 面向年轻一代缺乏吸引力

虽然当代从事竹纸生产在岑巩乡村中属于收入相当不错的工作，但长时间的手工造纸是相当辛苦的体力活，就像水尾镇造纸民谚描述的“牛喝水”“猴搬砖”那样，而且会被束缚在很小的乡村空间里，因而当地愿意继承造纸祖业的年轻人少，主流纸工的年龄在40~60岁。调查中，调查组与造纸户、村镇干部数次讨论过这一后继乏人的隐忧，但反馈更多的是，年轻人不仅嫌造纸辛苦，还认为造纸太单调寂寞，不如走出偏远的乡村到城里谋求发展空间。从社会学角度探究，岑巩侗族竹纸面向年轻一代缺乏吸引力的这一隐忧具有当代中国社会转型的共性内涵，需要结合共性与区域业态发展个性探索积极的传习机制与发展模式。

⊙3

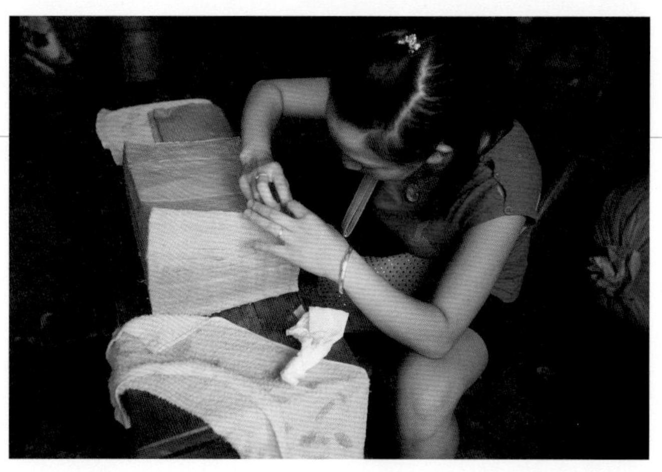

⊙4

⊙3
山脚下的料塘与竹料加工现场
Soaking pool and the locale for processing the bamboo materials at the foot of a mountain

⊙4
水尾镇陪同调研的年轻女孩正在学习揭纸
A young girl from Shuiwei Town learning how to peel the paper down

岑巩侗族竹纸

305

Bamboo Paper by the Dong Ethnic Group in Cengong County

白水村竹纸透光摄影图

A photo of bamboo paper in Baishui Village seen through the light

第二节

三穗侗族竹纸

贵州省
Guizhou Province

黔东南苗族侗族自治州
Qiandongnan Miao and Dong Autonomous Prefecture

三穗县
Sansui County

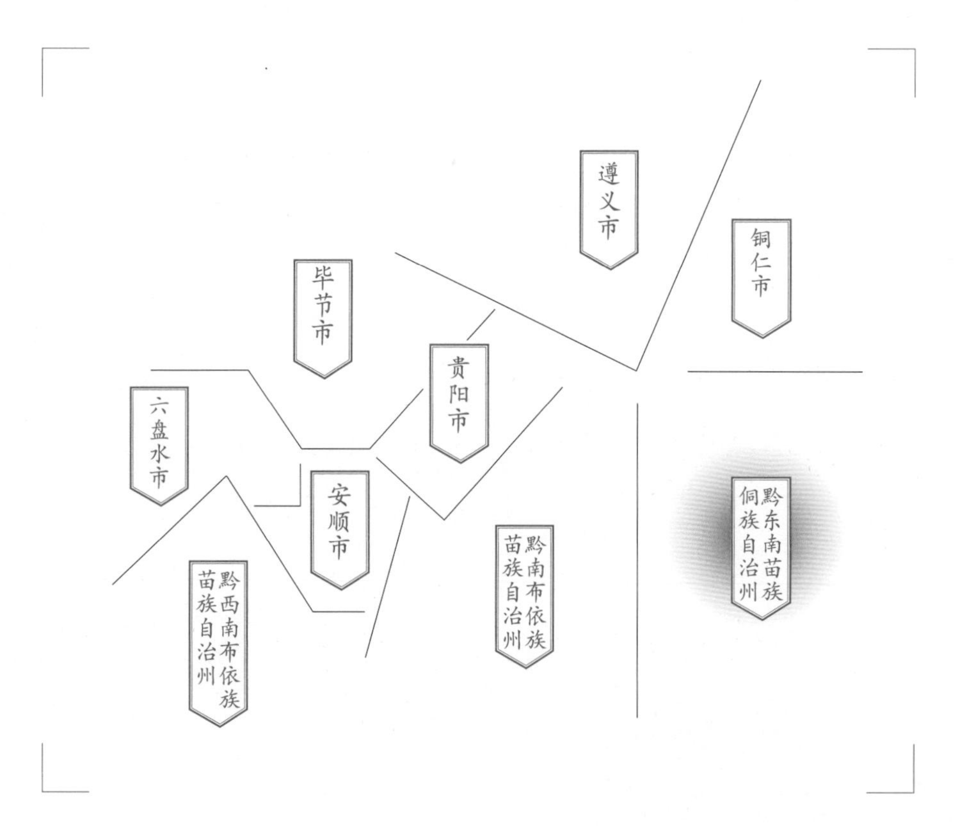

调查对象

八弓镇
贵洞行政村
侗族竹纸

Section 2
Bamboo Paper by the Dong Ethnic Group in Sansui County

Subject

Bamboo Paper by the Dong Ethnic Group in Guidong Administrative Village of Bagong Town

一 三穗侗族竹纸的基础信息及分布

1 Basic Information and Distribution of Bamboo Paper by the Dong Ethnic Group in Sansui County

三穗侗族竹纸的历史十分悠久，1994年版《三穗县志》载："三穗县民间素有造纸传统，采用土法制造绵纸、土纸，产地主要集中于泥山乡，已有两百多年历史。"但通过田野调查发现这种观点似乎偏于保守，调查组根据相关造纸家族的家谱记载发现，三穗造纸的历史可以上溯到600多年前的明代初年。

三穗县贵洞村竹纸以嫩白竹为原料制成，呈土黄色，纸色光亮，细腻、蓬松，有较好的韧性，纸面较为平整，厚薄较为均匀，当地人形容其纸质"绵、韧、软、薄"。历史上曾经大量作为包装用纸，调查时则多为乡民祭祀时使用。2006年6月贵洞竹纸传统手工技艺被选为三穗县首批县级非物质文化遗产保护项目，2008年以贵洞为代表的三穗竹纸手工技艺入选贵州省首批省级非物质文化遗产保护项目，2009 年4月下旬调查组入县时了解到地方政府正在筹备申报国家级非物质文化遗产保护项目。

贵洞等造纸村庄原属尼山乡，现已划归县城所在地的八弓镇。目前的三穗竹纸，其工艺集中传承地主要在八弓镇贵洞村。调查时三穗县内的界牌、蜜蜂、贵根及泥山诸村亦有技艺文化的活态保存，但因贵洞村以侗族造纸户为主，所产竹纸是三穗竹纸手工技艺传承的主要代表，调查组田野工作选取的展开点以贵洞村为中心。

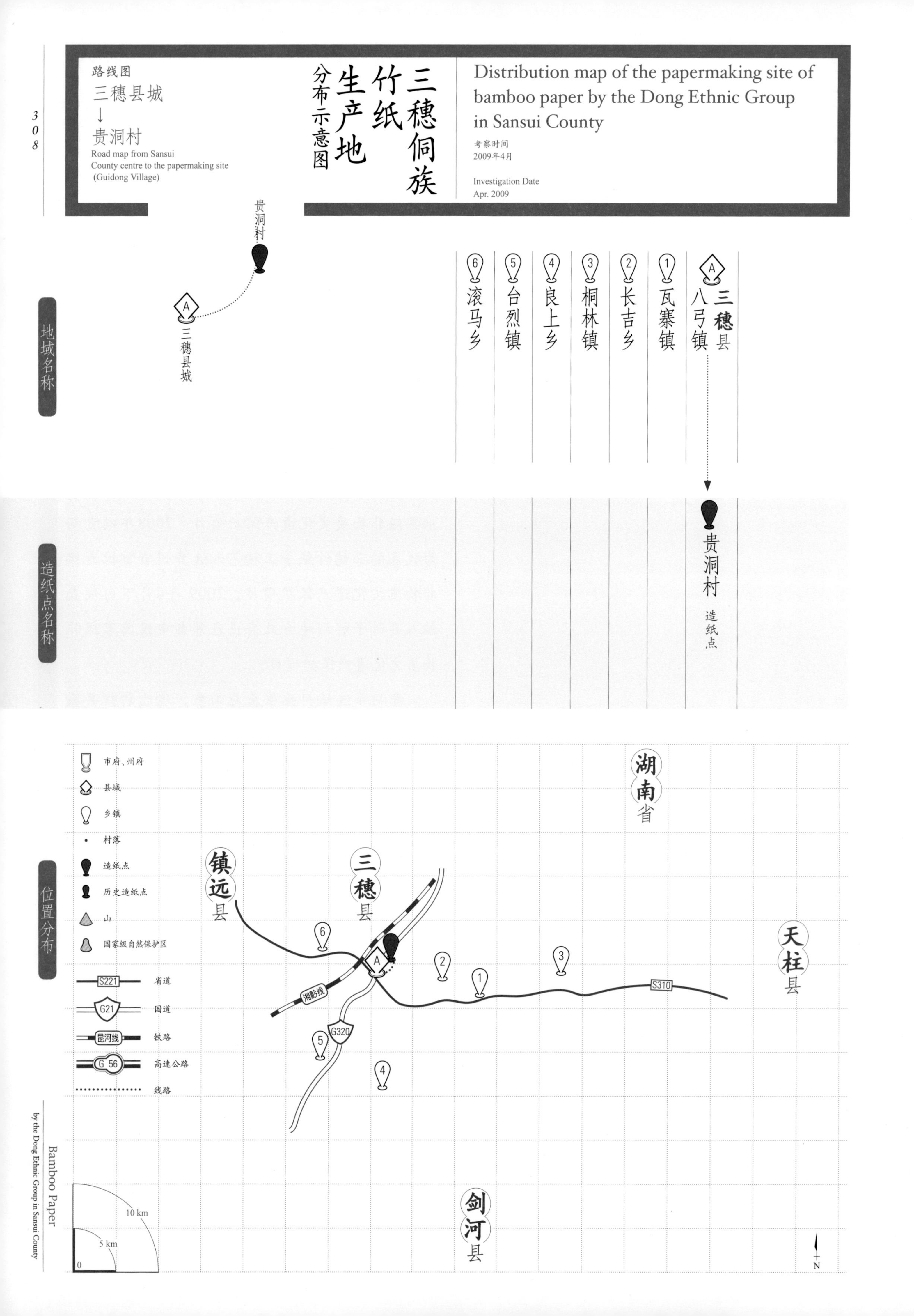
三穗侗族竹纸生产地分布示意图
Distribution map of the papermaking site of bamboo paper by the Dong Ethnic Group in Sansui County
考察时间
2009年4月
Investigation Date
Apr. 2009
路线图
三穗县城
↓
贵洞村
Road map from Sansui County centre to the papermaking site (Guidong Village)
贵洞村
三穗县城
地域名称
三穗县
八弓镇
瓦寨镇
长吉乡
桐林镇
良上乡
台烈镇
滚马乡
造纸点名称
贵洞村
造纸点
位置分布
市府、州府
县城
乡镇
村落
造纸点
历史造纸点
山
国家级自然保护区
S221
省道
G21
国道
昆河线
铁路
G 56
高速公路
线路
湖南省
镇远县
三穗县
天柱县
剑河县
湘黔线
G320
S310
10 km
5 km
0
N

二
三穗侗族竹纸生产的人文地理环境

2
The Cultural and Geographic Environment of Bamboo Paper by the Dong Ethnic Group in Sansui County

三穗县位于黔东南苗族侗族自治州东北部，东北与湖南省新晃侗族自治县毗邻，东南、西南与州内的天柱、剑河两县接壤，西北与镇远县相连，素有“黔东要塞”和“千里苗疆门户”之称。三穗县城驻地在八弓镇，距州府凯里市120 km。

三穗县全县总面积1 035.8 km^2，下辖5个镇4个乡159个行政村，境内居住着侗、苗、汉等16个民族，总人口22万，其中少数民族人口占75%，属北亚热带温和湿润季风气候区。

三穗县原名邛水县，北宋大观元年（1107年）始置，民国十五年（1926年）改名灵山县，民国十六年（1927年）因“秋收丰稔，一禾三穗”而于次年4月改称三穗县，此后一直沿用至今。

三穗人文荟萃，不仅有香火鼎盛持续数百年的佛教圣地永灵山、甘霖寺等古迹，以及雄伟挺拔的明代文武笔塔，还有历经沧桑的木界风雨桥、钉耙塘古战场遗址等，这些均是不可多得的历史人文景观。除了手工造纸之外，三穗竹编艺术也有400余年的历史，其作为贵州重要的非物质文化遗产蜚声海内外。2008年11月，三穗被国家文化部命名为“中国民间竹编文化艺术之乡”。

贵洞村位于三穗县东北部，距县城八弓镇政府驻地5 km，因洞得名。“贵洞”原名“鬼洞”，鬼洞即为以洞为居的苗族田氏自治的“竖眼大田溪洞”（历史名城镇远当时的建置名）下辖七十二洞之一，由田氏驻管（现“田家湾”“洞掌田”等古地仍在）。此洞全长4 km左右，宽5~50 m，为喀斯特地貌所独有的溶洞。村庄坐落在山谷之间，但山势并不陡峭，上游有泡木塘水库，得以灌溉谷中良田。贵洞虽然拥有河谷良田，但是人均土地面积有限，农业收入不足以满足生存所需，大部分村民还要依靠造纸补贴。而贵洞村相比其他村庄的独特之处，就是村旁的山上长满拇指粗的竹子，而不是树木，在三穗县只有贵洞村拥有这样大片的竹林。

⊙2

⊙3

贵洞村年平均气温14~15 ℃，温热的气候条件和酸性土壤适宜竹类生长，成片白竹面积达3.6 km^2，这为土纸制作和竹编生产提供了大量原料。

沿着贵洞村旁的小溪逆流而上，密布着传统的造纸作坊，这就是贵洞村造纸作坊群，目前有60%的纸坊近乎荒废，但从这些作坊群仍可以想象出贵洞当年手工造纸繁荣时的盛况。

⊙1

⊙1 贵洞村的乡土建筑 Local residences in Guidong Village

⊙2 贵洞村寨风景图 View of Guidong Village

⊙3 贵洞村造纸作坊群 Papermaking mills in Guidong Village

三
三穗侗族竹纸的历史与传承

3
History and Inheritance of Bamboo Paper by the Dong Ethnic Group in Sansui County

三穗竹纸历史大致可追溯到明洪武年间，相关文献资料以及调查组的民间走访资料都表明该县的造纸已有600年左右的历史。据当地杨姓造纸户家传《杨氏族谱》记载，该家族于明洪武年间迁入贵洞，以造纸为业，至今已传近30代。

2009年4月底造纸村民杨再祥和杨政权据家谱向调查组成员转述：该地有两杨姓造纸户，元泰定甲子年（1324年）老杨姓住入。“杨氏有八人由于战乱于明洪武年间从江西迁出，四人入湖南，四人入贵州，一人住雪洞（死后埋于雪洞），一人住三穗中团（死后埋于长坡），一人住新寨，其二子迁入鬼洞，一子住杨家湾（又名石万沟）（死后埋于老坟山），一子住鬼洞。当时造的纸叫思州纸，后因纸形如裹脚布，改名为裹脚纸。丙寅年天旱，竹林死完，杨氏从镇远大菜园卖绵竹麻，从河船到青溪姚湾上岸，挑回现煮造。小新寨秀通公有五子，二子杨再钦居鬼洞。”（杨政权口述）杨氏十三世祖新寨杨胜元之孙杨再清（现贵洞杨姓之迁入始祖）见纸业有钱可赚，于明洪武甲子年（1384年）迁入贵洞习艺，与老杨姓共享。清代中期老杨姓因故迁出后，再清公后裔将此纸艺改三合一为整张并发扬光大传于周边，上寨杨九千还亲赴四川学习雷氏竹帘制作技艺（近已失传），前后600余年，共历30余代，土纸制作一直在这一带传承（因两杨姓均为甲子年迁入，故外人称此处杨姓为二甲杨）。

三穗竹纸生产在近代曾经有过一个兴盛的时期，据黄维善主编的《三穗县地名志》记载，民国十九年（1930年），泥山土纸年产2.3万捆，有造纸户34户，从业人员50人。民国三十四年（1945年），仅泥山的纸槽就有126个。1950年，土纸年产量达10万千克，1953年则增至15.2万千克。

到1956年，地方政府在全国复兴手工业背景下大力发展当地手工造纸业，县政府利用贵洞的造纸技术和原材料，投资开办县造纸厂，纸厂分设三个车间，进行一条龙生产，年产红、绿、黄、白各种毛边纸13.3万千克，由县供销社统筹销往省内及湖南地区，成为当时县财政支柱产业之一，1959年创下22.7万千克的高峰产值。后因交通限制、生产技术落后、原材料缺乏等缘故于1962年4月停产，其古法造纸技艺又重回到了民间，由贵洞、界牌、蜜蜂、贵根四地民众保存，仍以家庭作坊形式延续至今。1987年全县造纸作坊一度达到479个，这可谓是现代三穗造纸最辉煌的时期。

⊙1

在1949年后的一段时间内，三穗地方政府也曾经兴办过半现代型的机器造纸工业，但终告失败，不过也正因为如此，该地传统造纸技艺才有了生存空间，能够以家庭作坊形式在民间延续至今。三穗手工造纸技艺一般以父传子、子传孙的家族传承式为主要传承方式。目前三穗土纸技艺传承和保护的核心问题是年轻人不愿继承祖业，其传习状况堪忧，有后继乏人之患，全村已有60%的纸坊近乎荒废。三穗地方政府为持续推动竹纸的保护工作而做出努力，如在上报省和国家的非物质文化遗产保护材料中，贵洞村已确定的三穗竹纸技艺传承人就有杨再祥、杨政权、杨秀芳、杨秀明、王怀炳、杨再金等。

⊙2

四
三穗侗族竹纸的生产工艺与技术分析

4
Papermaking Technique and Technical Analysis of Bamboo Paper by the Dong Ethnic Group in Sansui County

(一)
三穗侗族竹纸的生产原料与辅料

1. 竹料

三穗贵洞村竹纸的生产原料是当地造纸户所称的白竹，每年农历四月底至五月中旬砍，这时白竹刚长出叶子，当地叫“鸡毛叶”，质量比较好，至多砍到六月初十。但据杨再祥的补充说明，其实用这样的原料造出来的纸质量更好，可是所需石灰多，且较难碾烂，费时费力，增加成本，所以一般选嫩竹。旧时村民分的竹山较多，

⊙1 《贵州省三穗县地名志》书影
The Annals of Place Names in Sansui County of Guizhou Province

⊙2 调查组成员采访『非遗』传承人杨再祥（中）
Researchers interviewing local intangible cultural heritage inheritor, Yang Zaixiang (middle)

可请人砍下来放在山上，10天内再去运回来，若时间久了则不易断节，很难处理。

2. 纸药

三穗贵洞村竹纸生产使用以下几种纸药。

（1）猕猴桃根。

用斧头或铁锤将采集回来的猕猴桃根敲碎，用纱布袋装起来，使用时挤出即可。一般准备2 kg，可用2~3天。据杨再祥介绍，选取的猕猴桃根必须空心，空心的效果比较好。

（2）蝴蝶花。

一次采集0.75 kg干的蝴蝶花，可用4天。

（3）聚丙烯酰胺。

这是流行的化学纸药，但据造纸户反映效果一般，易回潮，造出的纸不是很干。一般来说，蓬松的干纸好卖些。

⊙1

（二）三穗侗族竹纸的生产工艺流程

据调查组2009年4月底对以杨再祥为主的村里造纸户的实地调查，总结三穗贵洞村竹纸的生产工艺流程如下：

壹	贰	叁	肆	伍	陆	柒	捌	玖
砍竹麻	破竹麻	捆纸	下塘	泡麻	洗麻	放水	下麻	沤麻

⊙1 敲猕猴桃根 Beating Chinese gooseberry root to make mucilage

拾	拾壹	拾贰	拾叁	拾肆	拾伍	拾陆	拾柒	拾捌	拾玖	贰拾	贰拾壹	贰拾贰	贰拾叁
二次泡麻	碾麻	下槽	打槽	捞筋	放水	关水	拱麻	加玄	抄纸	榨纸	抹垛	刮纸边	揭纸

贰拾肆	贰拾伍	贰拾陆
摇纸	晾纸	收纸

壹 砍竹麻

1

用镰刀将竹子砍成段，每段长1.5~1.7 m，打捆后拉回家，一般一捆的质量约30 kg，一个人一天可砍250 kg。

贰 破竹麻

2

也叫划竹，用镰刀从竹根往竹尖方向将竹子划成两半，效率高的，一个人一天可划500 kg左右。

叁 捆纸

3

把一大捆竹麻分成两小捆，便于下塘。

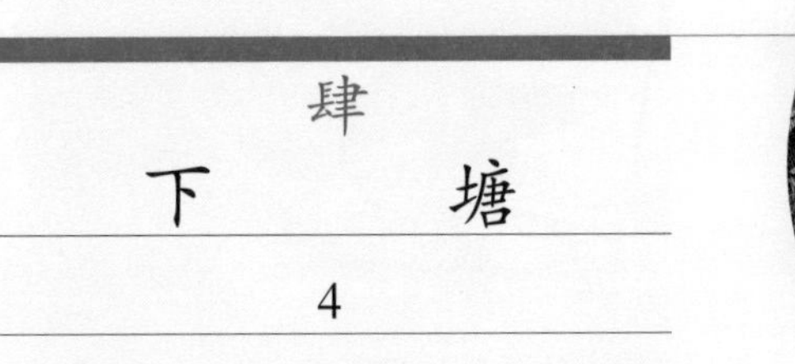

肆 下塘

4

将竹麻直接放到麻塘里，一层层累积堆放，每堆一层竹子，便立即加一次水，待水淹没过竹麻0.1 m左右后按50 kg竹麻加15 kg生石灰的比例撒石灰，依次下去，一般最多不能超过5层，太高了不好捞。两个人一天可累放2 000 kg竹子，一般两天完成下塘，大约能累放5 000 kg竹子，一个麻塘最多可放6 000 kg竹子，最少的只能放1 000 kg。下塘时，加石灰很有讲究，如果撒干石灰，则捞出来的竹子不好洗，因为石灰会沾在上面。

⊙2 泡麻
Soaking the papermaking materials

伍 泡麻

5 ⊙2

竹麻下塘后，泡两个月即可用，如果能持续泡三个月会更好，泡的时间越长质量越好，但最多不能超过一年，时间太长，质量虽好，但青皮层都是筋，就浪费了。

陆

洗　麻

6

用水将竹麻上的石灰冲洗干净，一个人两天可洗一大池（6 000 kg）。

柒

放　水

7

将麻塘里的水放干。不用清除麻塘里的石灰，沉淀的石灰可用于粉刷麻塘四周，使之不漏水且表面光滑。

捌

下　麻

8

先放4~5根木枕于麻塘底部，再垫上竹子，后放竹麻。放好竹麻后，将捆竹麻的篾条砍断，以便后期取麻。先在麻上面盖一层草，再用木枕或竹条压实，最后在其上面放石头。一般一个人两天就可全部下完。麻上盖草是为了保温，山草、谷草都行，如不盖，则表面的麻因日晒雨淋而浪费。

玖

沤　麻

9　⊙3

将竹麻放在麻塘里沤一个月，当地叫“发烧”，沤后的竹麻会变脆。沤麻很有讲究，如沤的时间太长，则竹麻容易霉坏，即容易变得脆且发黑，当地造纸户称之为“把竹麻烧坏了”，造出的纸质量就不好。

⊙3

拾

二　次　泡　麻

10

往麻塘里加水泡麻，如果天气好需泡20~30天，天气不好泡的时间会更长，泡到拿起来一踩就烂即可。

拾壹

碾　麻

11　⊙4

将泡好的麻取出，用石碾碾压，一次能碾30 kg，碾一次即够一天抄纸所用的量。将麻放在碾上，要碾到麻变粑为止，所以一般最少要碾半小时，若碾一小时，则能使竹料更细，质量更好。现在有些造纸户改用机器打浆，一次打30 kg，只需十几分钟。

⊙4

拾贰

下　槽

12　⊙5

先往纸槽里加水，然后将碾好的竹料全部放入纸槽。

⊙5

拾叁

打　槽

13

用槽棍将料打匀，据当地造纸户介绍，要用槽棍打300下，才能达到效果。现在也有造纸户用抽水机打槽，这样既节省了人力，也提高了速率，效果也不错。

⊙3 沤麻 Fermenting the papermaking materials

⊙4 碾麻示意 Showing how to grind the papermaking materials

⊙5 碾好的竹料 Processed bamboo materials

拾肆

捞　　筋

14　⊙6⊙7

用捞筋筷在纸槽里以“8”字形将竹筋捞出来，捞到筋少为止，这需要靠个人经验把握。如竹麻碾得碎，捞2~3次即可；若碾得不碎，捞20次也不行。

⊙6

⊙7

拾伍

放　　水

15

把纸槽里的水放干。

拾陆

关　　水

16

第二天抄纸前加水到距纸槽的最高边0.15 m左右。

拾柒

拱　　麻

17　⊙8

用拱耙将料拱起来一小部分，再用捞筋筷搅匀并将较为粗糙的竹筋捞出来扔掉。

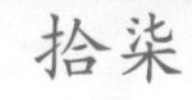

⊙8

⊙9

拾捌

加　　玄

18　⊙9

将装在袋子里的玄挤出来，拱一次麻则挤一次玄，再用捞筋筷搅匀。如玄太多，则多拱些麻。一般一天需加10余次玄。玄起到冲洗、光滑、分离的作用。

⊙6/7 捞筋 Picking out the impurities

⊙8 拱麻 Lifting the papermaking materials

⊙9 加玄 Adding in papermaking mucilage

拾玖
抄　纸

19　　⊙10~⊙12

抄纸时，手持帘架，首先由内往外推，趁水掀起波浪再由外往内拉，随后由右往左起头帘水；接着由右往左将水舀起来，这是二帘水。最后将湿纸反扣在纸垛上。头帘水固尾子，即纸帘尾巴上必须有麻；二帘水固头子，即把麻固定在头子上，头子厚些，这样造出来的纸才能揭开。抄了一段时间纸后，用刮板轻轻刮纸，使之去掉部分水。一个纸工一天可抄20刀（2 000张）纸。

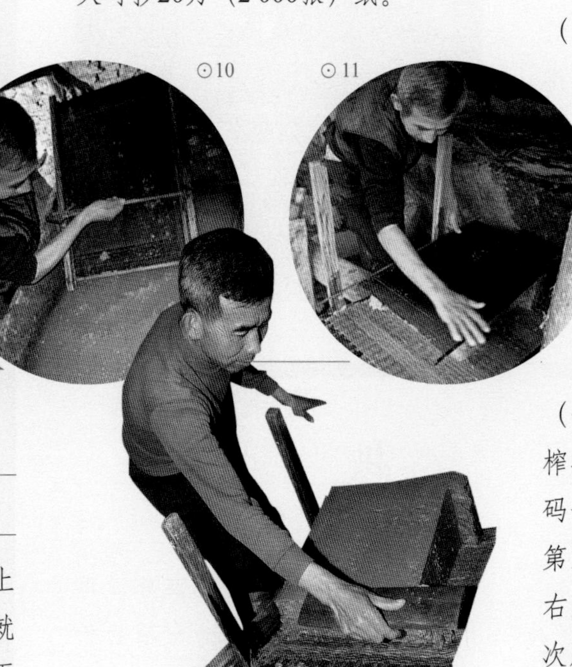

⊙10　⊙11　⊙12

贰拾
榨　纸

20　　⊙13⊙14

抄完纸后，在湿纸垛上盖一印帘（即废帘子），再压上盖板、扯码、码子、咬口。先将手杆置于咬口上，慢慢打杆，片刻后换上大杆子，再用钢丝绳将大杆子和碾子捆起来。接着将手杆置于碾子空洞中，缓慢压榨，大约半小时后，纸垛下降，松榨，加枕头板（一边厚一边薄，易于揭纸），装榨再压；大约压20分钟，再加一次码子，如果压得不好还没干则要加第三次，一般压榨总共需一小时左右。压榨时很有讲究，尤其是第一次压榨，时间要长些，如果动作快了则纸会断开。压榨到最后人跳上去都榨不动，则表明干了。

⊙13

⊙14

贰拾壹
抹　垛

21

榨纸结束后，用麻袋将纸垛四边上的水抹掉，抹一两次即可，随后就放榨。以前用棕丝抹垛，效果更好，这是由于棕丝滤水性更佳。

贰拾贰
刮　纸　边

22

用刮子将纸垛四边由下往上刮，使纸边上翘，便于揭纸。也可以由上往下刮，然后将纸垛翻转过来。

贰拾叁
揭　纸

23

用手将纸头打蓬松，第一张纸揭起17~20 cm后对折，后面每两张纸间隔1 cm左右，10张为一叠。第一张纸揭起来的长短有讲究，如揭得太短，则后面的纸间隔太小，容易粘在一起，导致揭不开；如揭得太长，则浪费时间。

⊙10/11 抄纸
Scooping and lifting the papermaking screen out of water and turning it upside down on the board

⊙12 刮纸
Scraping the paper

⊙13 压盖板
Pressing the paper with a cover board

⊙14 榨纸
Pressing the paper

贰拾肆 摇　纸

24

用摇纸筷将纸对分，再用摇纸板将纸刮平，避免纸卷起来。

贰拾伍 晾　纸

25

将纸晾在屋内竹竿上，使之阴干。天气好时一天就可以晾干，一般需要2~3天，如下雨则要一周左右。造纸户认为阴干的纸颜色不会发生变化，仍为金黄色且有亮度，这样的纸质量好，也比较好卖；而太阳晒干的纸较为粗糙，颜色不佳，不好卖。

贰拾陆 收　纸

26

将晾干的纸收下来，把纸尾子往中间折，再用纸头子把纸尾子盖起，折成三重。折纸时也有讲究，头子距边上4 cm左右，不能漏出头子，否则容易把头子弄烂。

(三)
三穗侗族竹纸
生产使用的主要工具设备

壹 石　碾

1

用于碾麻，实测石碾的碾盘内径约100 cm，外径约130 cm，因碾盘较小，故外加15 cm厚石头围一圈，便于使用。

贰 碾子（磙子）

2

高约30 cm，内边直径约67 cm，外边基本一样宽。

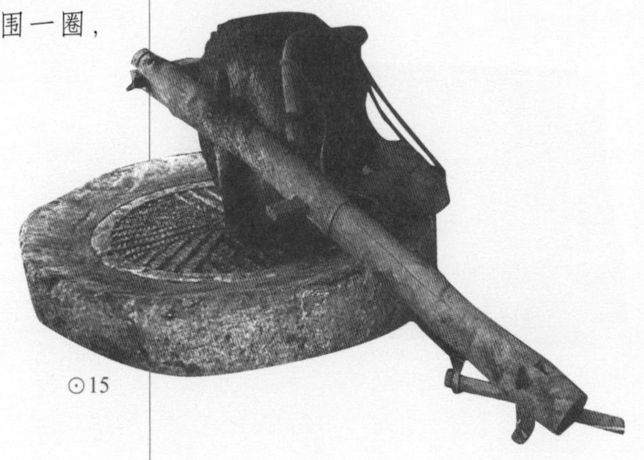

⊙15

⊙15 石碾 Stone grinder

叁 碾麻杆

3

实测碾麻杆长约180 cm。

肆 喂麻刀

4

将碾到外面的料弄平的工具。

伍
纸 槽
5

上宽下窄，实测杨再祥家纸槽尺寸为：最上面长约150 cm，宽约115 cm，高约65 cm。老的纸槽用石头砌成，新的纸槽主要用水泥板砌成。纸槽底部的四个角原用石灰和泥巴，现用水泥敷圆，敷圆的目的是便于安放槽席。面边倾斜约30°，方便人站立，若斜度太大，则帘架易碰到面边；若斜度太小，则不好站人，易碰到腰部。远离人的一边叫作背边，左右两边叫作挡边。

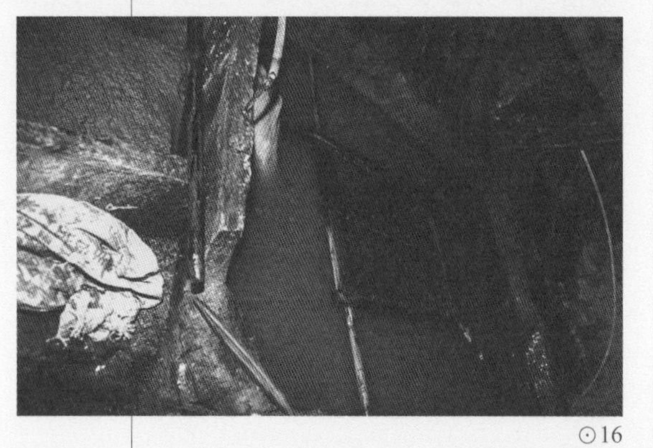
⊙16

陆
踩 槽
6

实测杨再祥家踩槽尺寸：长约60 cm，宽约65 cm，高约30 cm。

⊙17

柒
纸 帘
7

实测杨再祥家纸帘尺寸：长约58 cm，宽约43 cm。

⊙18

捌
帘 架
8

实测杨再祥家帘架尺寸：长约64 cm，宽约48 cm。

⊙19

⊙16 纸槽 Papermaking trough

⊙17 踩槽 Trough for stamping the papermaking materials

⊙18 纸帘 Papermaking screen

⊙19 帘架 Frame for supporting the papermaking screen

(四)

三穗侗族竹纸的性能分析

所测三穗贵洞村竹纸的相关性能参数见表10.2。

表10.2 贵洞村竹纸的相关性能参数
Table 10.2 Performance parameters of bamboo paper in Guidong Village

指标		单位	最大值	最小值	平均值
厚度		mm	0.090	0.010	0.021
定量		g/m²	—	—	23.9
紧度		g/cm³	—	—	1.138
抗张力	纵向	N	7.7	4.8	6.0
	横向	N	5.4	3.5	4.2
抗张强度		kN/m	—	—	0.340
白度		%	20.9	19.6	20.2
纤维长度		mm	3.49	0.66	1.74
纤维宽度		μm	27.0	4.0	12.0

由表10.2可知，所测贵洞村竹纸最厚是最薄的9倍，相对标准偏差为2.0%，纸张厚薄不均。竹纸的平均定量为23.9 g/m²。所测竹纸紧度为1.138 g/cm³。

经计算，其抗张强度为0.340 kN/m，抗张强度值较小，可能是三穗竹纸厚薄不一所致。

所测贵洞村竹纸白度平均值为20.2%，白度较低，白度最大值约是最小值的1.07倍，相对标准偏差为0.39%，白度差异相对较小。测试小组认为可能是因为竹纸未经过较强漂白，加上表面比较粗糙，所以纸张白度较低。

所测贵洞村竹纸纤维长度：最长3.49 mm，最短0.66 mm，平均1.74 mm；纤维宽度：最宽27.0 μm，最窄4.0 μm，平均12.0 μm。所测竹纸在10倍、20倍物镜下观测的纤维形态分别见图★1 、图★2。

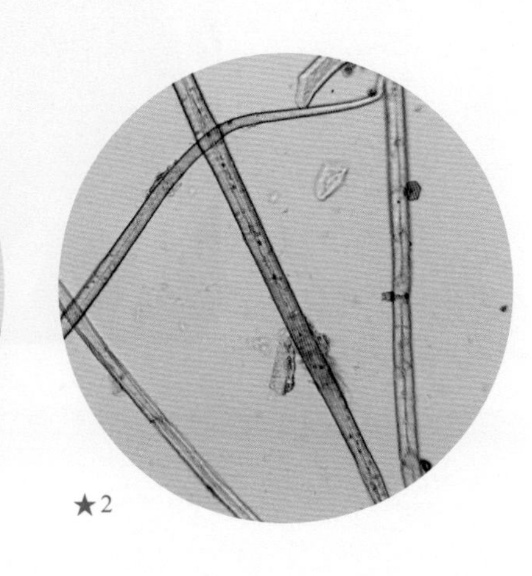

★1 ★2

★1 贵洞村竹纸纤维形态图(10×) Fibers of bamboo paper in Guidong Village (10× objective)

★2 贵洞村竹纸纤维形态图(20×) Fibers of bamboo paper in Guidong Village (20× objective)

五

三穗侗族竹纸的用途与销售情况

5

Uses and Sales of Bamboo Paper by the Dong Ethnic Group in Sansui County

历史上，三穗竹纸曾经应用很广，最大用途是包装，调查时则收窄很多，主要作为当地民众日常民俗用纸，用于民间红白喜事和祭祀活动。用法一般是：将纸剪成纸串，用钱錾打上钱眼，每一串钱均用红纸缠上，在红白喜事当天烧给祖宗。调查组入村调查时，贵洞村刚好有人娶亲，新媳妇进门前，需烧纸钱告知祖宗。

据调查时获知的信息，三穗竹纸的销售主要靠三穗本地的用户上门购买，以前还有贵州盘县、六枝、水城等地的用户过来购买，现在基本上没有了。村里也有部分造纸户将纸拉到周边市场去卖，如八弓镇，销售积极的造纸户一年可在附近市场卖掉2 000捆左右。销售价格因市场而波动，2009年4月底调查组前去调查时，质量最好的纸可卖7.5元/刀，差的也能卖到7元/刀。一般

⊙1

⊙2

一个抄纸工一天可抄20刀，即一天抄纸销售额有140~150元。因为当地民俗消费用纸的传统保存较完整，所以手工造纸收入在贵洞村比较理想。

六 三穗侗族竹纸的相关民俗与文化事象

6 Folk Customs and Culture of Bamboo Paper by the Dong Ethnic Group in Sansui County

(一) 节日与婚礼用纸习俗

苗族吃新节。苗语叫作“垄模”，即吃新米饭的意思。农历六月后逢“辛卯”或“己卯”日，家家煮糯米饭，杀鸡宰鸭，烧香焚纸，祭祀祖先和土地菩萨，企盼来年五谷丰登。

侗族七月半节。农历七月十五，当地侗族过七月半节。这天，身穿节日盛装的青年男女云集盛德山赶歌场，老人则带着香纸上山拜求菩萨保佑。

土家族腊月二十三杀年猪。是日，村民用稻草索把猪捆住，意在方便祖宗拖走食用，并烧香化纸祭祀祖先。这样做的意思是请老祖宗放心，子孙们过年有猪肉食用。

三穗当地新媳妇入男方家时要烧纸钱告知祖

⊙1 亲戚们帮忙将嫁妆送至新郎家
Relatives carrying the dowery to bridegroom's house

⊙2 烧纸钱
Burning the joss paper

先，让祖先知道小辈喜结良缘，家里人丁兴旺有保证。

(二)

丧葬用纸习俗

三穗县汉族丧葬礼俗中用纸之处颇多。老人去世停尸时，白纸盖脸，头枕“金鸡枕”，手握纸钱。灵堂设在堂屋，在灵柩前置香案、扎牌坊、贴孝联。祭奠时用纸更多，其祭品多数由白纸扎制而成。发丧路上，一人先行，撒“买路纸钱”。

(三)

谚语

“三月打笋笋不生，四月打笋笋登林。”即三月出笋，四月才能挖笋，不能太早。

⊙3

⊙3 告诉祖先小辈喜结良缘 Informing the ancestors of the offspring's wedding

七 三穗侗族竹纸的保护现状与发展思考

7 Preservation and Development of Bamboo Paper by the Dong Ethnic Group in Sansui County

(一)

三穗侗族竹纸的保护现状

长期以来，土纸制作在三穗县贵洞、界牌、蜜蜂、泥山、贵根等村的经济生活中起到了重要作用，解决了农闲时相当数量的农村剩余劳动力的出路问题。造纸原料取自农林副产品，促进了农业增产、农民增收。由于白竹资源丰富，因此三穗竹纸传统产业的基础相当稳固。

随着人们对非物质文化遗产的关注，当地政府也加强了对传统造纸工艺的传承与保护。从

⊙1

⊙1 贵洞村边的造纸作坊群
Papermaking mills alongside the Guidong Village

1998年开始，各级政府和旅游、文化等部门就开始拨付资金，用于民间手工技艺的宣传、引导、扶持和保护。2005年8月9日，三穗县成立了非物质文化遗产普查领导小组，把土法造纸列为保护对象。2006年6月30日，贵洞竹纸传统手工技艺入选三穗县首批县级非物质文化遗产保护项目。2008年以贵洞为代表的三穗竹纸手工技艺入选贵州省首批省级非物质文化遗产保护项目，为三穗传统造纸工艺的保护与发展奠定了良好的基础。

调查时，三穗已建立有专家指导的、以人民政府县长为组长的保护领导小组，制定了《三穗县非物质文化遗产普查申报工作实施方案》，成立了领导机构和工作机构，且有了一定的资金保障。最重要的是该地成立了以技艺精深的老手工艺人为主体的土纸传统工艺顾问小组，在宣传研究所下设立了土纸传统制作技艺研究保护室。一系列以政府为主导、乡土专家为顾问的举措，为三穗手工纸生产营造了较好的文化环境。

(二)
三穗侗族竹纸面临的压力和发展思考

然而，调查中也发现，在良好的形势下，三穗竹纸也面临着一系列问题：

1. 后继无人

由于土纸制作周期较长，原料加工、成品制作等环节都较为辛苦和枯燥，因此年轻人多选择外出打工，而不愿习此技艺，造成后继乏人的现象，这是手工技艺传承普遍面临的问题，也是最严峻的传承载体缺乏的问题。

⊙2

2. 出现毁林卖竹趋势

出于经济利益考虑，农户毁林卖竹趋势出现，使造纸原料骤减，这对竹纸产业打击不小。

3. 性价比不高

三穗竹纸主要以家庭型手工作坊为单位生产，效率低、成本高、收益小。性价比不高是60%的纸坊近年荒废的一个可能性因素。

4. 价格不易提高

三穗竹纸对质量要求较低，且主要用于红白喜事和祭祀，用途单一，这意味着其价格也不容易提高。在此现状下，其传承与保护存在较大困难。

⊙3

针对上述问题，调查组提出以下关于三穗侗族竹纸发展的思考：

（1）加强统筹和整合管理，将传承和保护落到实处，可考虑由政府出面成立民族传统文化园，将手工纸技艺与文化园中的其他技艺整合在一起，做出一个综合的地方民族文化展示平台，对地方旅游经济的发展起到聚合推动的作用，可以形成一股传承、保护、再生的合力。

（2）政府成立传习所，可以考虑与文化园相结合，这样可以打破父传子的单一传承模式，鼓励更多的年轻人投身手工纸技艺的传承中，可给予传承人适当的经济补贴，以解除其后顾之忧。

（3）加强对传统造纸所需竹林资源的保护，每年砍伐的数量和新生的数量要匹配，不能涸泽而渔，考虑借鉴“轮作制”，对竹林实行轮流砍伐，杜绝乱砍滥伐。

（4）政府鼓励和支持当地手工纸艺人不断创新，并引进和整合国内外手工纸研究学者的力量，提高手工纸的技艺，探索手工纸的新用途，摆脱对祭祀用纸的单纯依赖，不断满足新的消费空间需求，唯有如此，三穗手工纸的传承才能够真正活化、光大。

⊙2 废弃的纸槽房 Abandoned papermaking mill

⊙3 还在使用的碾坊 Grinding workshop still in use

竹纸

Bamboo Paper
by the Dong Ethnic Group
in Sansui County

贵洞村竹纸透光摄影图
A photo of bamboo paper in Guidong Village
seen through the light

第三节

黄平苗族竹纸

贵州省
Guizhou Province

黔东南苗族侗族自治州
Qiandongnan Miao and Dong Autonomous Prefecture

黄平县
Huangping County

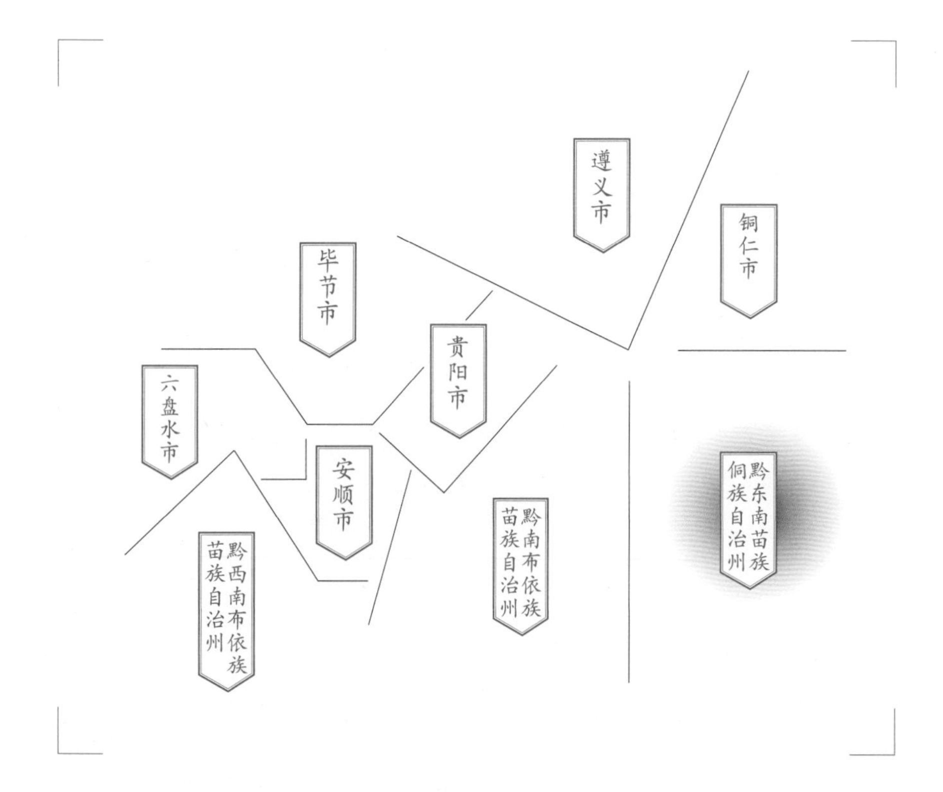

调查对象
翁坪乡满溪行政村苗族竹纸

Section 3
Bamboo Paper by the Miao Ethnic Group in Huangping County

Subject
Bamboo Paper by the Miao Ethnic Group in Manxi Administrative Village of Wengping Town

一
黄平苗族竹纸的基础信息及分布

1
Basic Information and Distribution of Bamboo Paper by the Miao Ethnic Group in Huangping County

2011年8月、2013年8月及10月，调查组先后3次实地入村考察黄平县翁坪乡满溪村的手工竹纸生产，黄平县仍在生产、使用的手工祭祀用纸为竹纸，当地也称为“凿纸”。另据调查中的访谈信息，黄平县以前还生产过草纸，也主要分布于翁坪乡。历史上，黄平手工纸在当地生活中曾有较为广泛的作用，除了主要用于祭祀外，还曾用于包装点心、糖等，以及用来做鞭炮等。造纸人主要为当地苗族百姓，调查时仍有十六七户在从事竹纸生产。

二
黄平苗族竹纸生产的人文地理环境

2
The Cultural and Geographic Environment of Bamboo Paper by the Miao Ethnic Group in Huangping County

黄平县隶属黔东南苗族侗族自治州，位于州域的西北部，县以地平（地名）“撅土为黄”而得名，东接施秉县，南邻台江县与州府凯里市，西连黔南苗族布依族自治州的瓮安县与福泉市，北与遵义市的余庆县接壤。黄平县城距州府凯里市49 km。

黄平是一个历史文化悠久的地域，春秋战国至汉代，分属牂牁国、黔中郡、且兰国管辖。

南宋宝祐六年（1258年）始修筑黄平城，黄平之名首见于史册；景定元年（1260年）置黄平元帅府，隶播州。

元至元二十八年（1291年）置黄平府，仍归播州，隶属元王朝的四川行省。

明洪武七年（1374年）置黄平安抚司，设州级建制并沿用到清末。

路线图
黄平县城
↓
满溪村
Road map from Huangping County centre to the papermaking site (Manxi Village)

黄平苗族竹纸生产地分布示意图

Distribution map of the papermaking site of bamboo paper by the Miao Ethnic Group in Huangping County

考察时间
2011年8月/2013年8月/2013年10月

Investigation Date
Aug. 2011/Aug. 2013/Oct. 2013

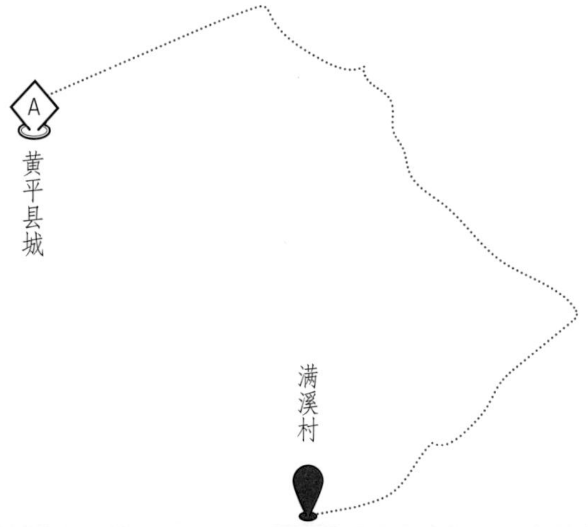

地域名称

黄平县
新州镇
① 翁坪乡
② 黄飘乡
③ 旧州镇
④ 谷陇镇
⑤ 野洞河乡
⑥ 上塘乡

造纸点名称

满溪村 造纸点

位置分布

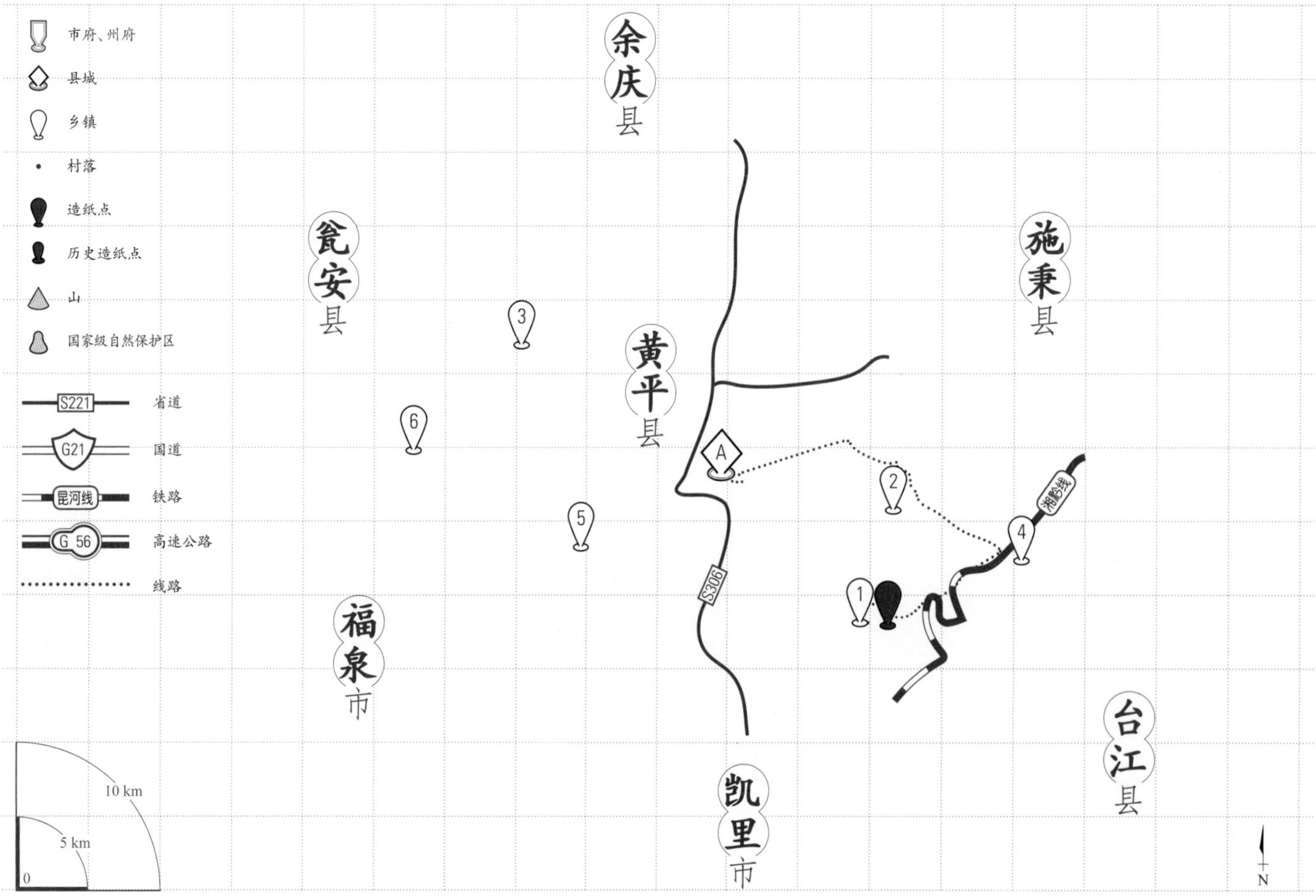

⊙1

⊙2

民国三年（1914年）废黄平州，改名为黄平县并沿用到当代。

据调查时的数据，截至2011年，黄平县县域面积为1 668 km²，下辖5镇9乡，包括县城所在的新州镇、县名来源的旧州镇以及产纸的翁坪乡和名称与纸相关的纸房乡等；户籍登记在册人口为38.5万，县境内共生活有20个民族，其中苗、布依、革家等少数民族人口占总人口的64%。

黄平县地处东经107°35′~108°12′、北纬26°43′~27°14′，呈由黔中丘陵平原向黔东低山丘陵过渡的地貌特征，海拔600~1 200 m，最高海拔1 367 m，最低海拔519 m。黄平县境内河流属长江水系，有重安江、阳河等百余条河溪。黄平县属亚热带季风气候区，气候四季分明，较为温和，年均气温15.1 ℃，最热的7月平均温度也较宜人，为24.7 ℃，雨量充沛，年均降水量达1 307.9 mm。

黄平县有着浓郁的少数民族文化风情，例如：

黄平苗族大歌。这是一组以讴歌历史英雄群体为中心的谣诗，共分为12路歌，表达天地日月生成、英雄神鬼事迹以及人间生活故事，有王宝、勾雄、勾当等英雄人物。

革家蜡染。革家人中流行手工印染技艺，其工艺为用蜡刀蘸上蜡液后，在土纺的白布上绘几何纹案及花鸟虫鱼形象，然后放到盛满靛蓝染料的缸中浸染，最后再用水将蜡煮脱融化即形成有图案的蓝花布。

哈戎节。黄平县革家人的一种宗族祭祖盛典，有10余年举办一次的，也有50年举办一次的。宗内每一房都会有自己的哈戎坪（空间场地），一般会杀猪宰牛，祭典通常持续三天三夜。

翁坪乡位于黄平县城东南52 km处，境内平均海拔890 m，最高点为牛岛坡石堆峰，海拔1 344.1 m，最低点为满溪河边，海拔584 m。

翁坪乡是一个典型的少数民族聚居乡，主要有苗族、革家等少数民族居住，苗族、革家民族风情浓郁，民族节日丰富多彩，主要有吹芦笙、斗牛、唱苗歌、斗鸟、赛马、民族服饰展等多种活动。

革家人在中国民族史上非常特殊，是没有被列入56个民族的一个少数民族族群，主要分布在贵州省的黄平县与关岭县、麻江县等地。当地统计的人口数为5万多，黄平县有2.1万左右，是革家人最集中的聚集地，其中黄平县的枫香寨有3 700余人，为革家第一大寨。革家人有语言但无文字，属汉藏语系苗瑶语族的分支，但与苗瑶语族之间均难交流。

革家人自称是上古传说中射日英雄后羿的后人，但民族来源识别未有明确结论。中华人民共和国公安部“公治2003118号”文件明确革家人在身份证民族栏上填为“革家人”。

红条石是翁坪乡特有的资源，省内其他地方

⊙1 黄平境内的山地河流
Mountains and rivers in Huangping County

⊙2 黄平县的苗族大歌表演
Miao ethnic polyphonic music show in Huangping County

⊙1

很少见到，其品位高，主要用于市政建设，也可作房屋装饰等用。犀牛河位于翁坪乡东部，其两岸山形地貌多姿、奇特、秀美，水质清澈，水流平缓，是度假、休闲的好去处。此外，湘黔铁路、株六复线贯穿白洗、党约、满溪三村，虎谷公路横穿全乡，全乡的交通通达性较好。

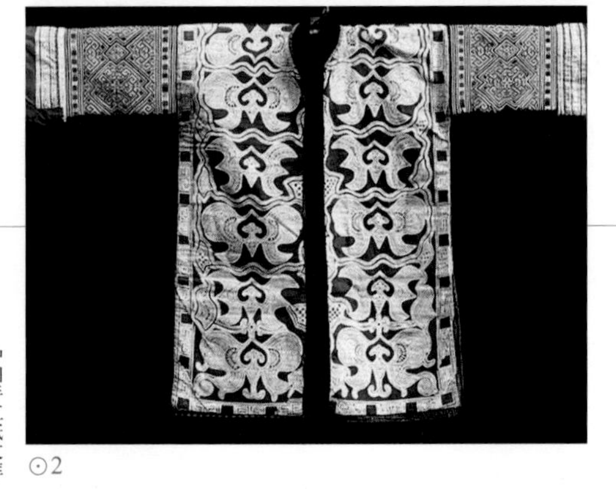

⊙2

⊙3

三 黄平苗族竹纸的历史与传承

3 History and Inheritance of Bamboo Paper by the Miao Ethnic Group in Huangping County

满溪村目前是黄平苗族竹纸的活态生产地，2013年调查组入村时仍有十六七户村民在生产竹纸，调查组重点采访了苗族造纸人王启光（1945年出生）、杨正美（1947年出生）夫妇。

据王启光夫妇口述，满溪村造竹纸的技艺起源是输入型的，是在很早的时候由本县重安镇纱砂村的师傅到满溪村传授的造纸技术，但具体什么年代已完全记不清了。据王启光介绍，他家的造纸历史到他时已是第七代，能够回忆起的传习谱系是：王垢龙—王垢九—王垢两—王启光，王启光的儿子以前读书，现在去外地打工，没有学过造纸。王启光是1982年分田到户后，跟村里造纸老人学会造纸的。据王启光说，当时也没有正式的拜师仪式，在现场看老一辈造纸人做，不懂就问，他们也热心教，学会以后就自己种竹子和买竹子造竹纸了。

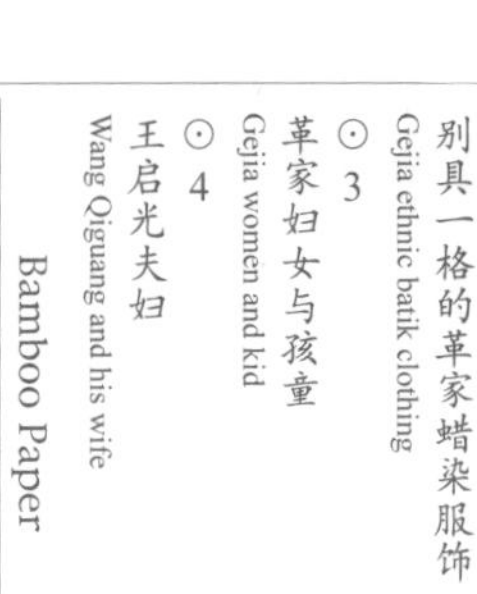

⊙4

⊙1 热闹非凡的革家哈戎节
Harong Festival celebrated by the Gejia people

⊙2 别具一格的革家蜡染服饰
Gejia ethnic batik clothing

⊙3 革家妇女与孩童
Gejia women and kid

⊙4 王启光夫妇
Wang Qiguang and his wife

四

黄平苗族竹纸的生产工艺与技术分析

4 Papermaking Technique and Technical Analysis of Bamboo Paper by the Miao Ethnic Group in Huangping County

(一)

黄平苗族竹纸的生产原料与辅料

黄平竹纸的造纸主原料为当地苗语称作“豆究”的一种竹子，这种竹子于农历五月底六月初生长，次年农历正月至二月砍，大约半年生，不能用太嫩的竹子。调查时，造纸户强调，太嫩的竹子不方便抄纸。竹子一般由造纸户自己在山场和屋边种，若不够用再买，如王启光家一年约用竹料1 000 kg，基本够用；也有少数造纸户从周边村民处购买的，2013年调查时，竹价为0.2元/kg，但造纸户需要自己去砍并拉到造纸现场。

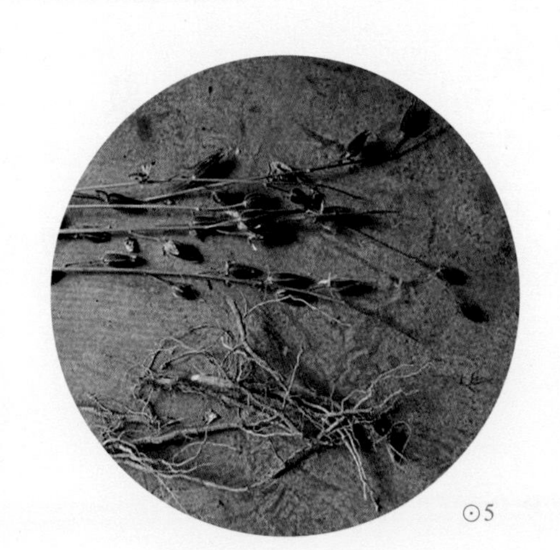

⊙5

黄平造纸用的纸药，在当地称为药水，由当地纸药树的树根、树皮、叶子等制成。一般农历四月二十左右种纸药树，农历六月时大概长到30 cm高，可用其叶子制纸药，如挖根用则要等到农历九月。造纸户通常会将纸药树的叶子、树根割回家，用手揉搓后，再放到簸箕里，置于抄纸槽中，抄一槽要用2~2.5 kg。天热时纸药只能用一天，否则会烂，失去药效；天冷时可放四五天。药水不可加太多，否则槽中纸浆水太滑反而影响抄纸。从实物来看，当地造纸户使用的纸药似为学名叫黄蜀葵的植物。

(二)

黄平苗族竹纸的生产工艺流程

黄平满溪村苗族造纸户以农业种植为主，只有农闲时才会造纸，因此当地农历九、十月造纸较普遍。

⊙5 纸药树原料 Raw material of papermaking mucilage

经深入调查，将黄平苗族竹纸的生产工艺流程概述如下：

壹	贰	叁	肆	伍	陆	柒
砍竹	敲竹	晒竹	泡竹	洗竹	泡清水	捞竹

拾肆	拾叁	拾贰	拾壹	拾	玖	捌
放水	压纸	捞纸	揉药	搅纸浆	碾竹	破竹

拾伍	拾陆	拾柒	拾捌	拾玖	贰拾
松榨	打纸	撕纸	晾纸	收纸	叠纸

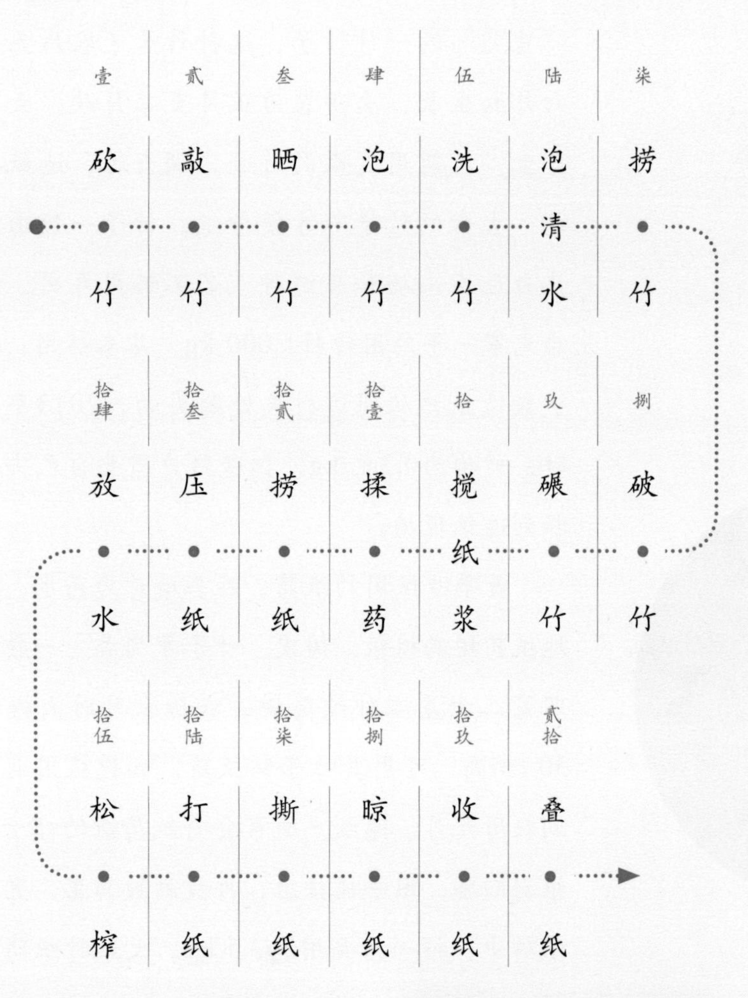

壹 砍竹

1

造纸户每年农历正月和二月用柴刀砍周边的竹子，一天可砍大的竹子250~300 kg，小的竹子100~150 kg，竹尖长约70 cm。将旁边枝条去掉后砍成截，每截长2~2.3 m，再拉回造纸地点。

贰 敲竹

2

用木槌将竹段敲碎，一个人一天可敲350~400 kg。

叁 晒竹

3 ⊙1

将敲碎的竹片捆成捆，每捆直径约25 cm，质量约15 kg。将捆成捆的竹子置于室外太阳下晒干，根据阳光的强弱，可晒20~30天。虽然造纸户说最好在下雨前将竹料收回来，否则竹料被雨淋后会变黑，从而对纸的品质产生影响，但因晒竹花的时间多且还需找地方晾，故就算下雨造纸户一般也不收回来。100 kg生竹子晒干后约为55 kg。

⊙1 晒竹 Drying the bamboo

肆 泡竹

4 ⊙2

将晒干的竹片直接置于泡竹塘，底层不用垫，两层竹子一层石灰，100 kg生竹子需用略多于10 kg的生石灰。为了不让竹片浮起来，顶层用石头压着，然后加水泡两个月左右。

⊙2

伍 洗竹 5

用手将泡过的竹片在泡竹塘里洗干净。

陆 泡清水 6

将洗干净的竹片再放入泡竹塘，加清水，再泡一个月。

柒 捞竹 7

从竹塘里捞出够一天抄纸用的竹子出来，不用洗。

捌 破竹 8

直接用柴刀将捞出的竹子破成段，每段长约3 cm，一个人一天可破150~200 kg生竹片。

玖 碾竹 9

⊙3⊙4

将约50 kg的生竹片放到碾子里碾成“面面”，即碾碎，碾得快需3小时，碾得慢需约7小时。碾料时，需要有人在旁边将碾出来的竹片扫进去，碾后的竹子称为“料”。

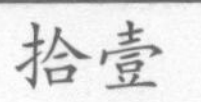

⊙3

拾 搅纸浆 10

将碾好后的料放到纸槽里，加水，用“耙耕”打散，然后大约搅拌一小时后，把水放掉，重新加清水。

拾壹 揉药 11

⊙5

加好水后，将纸药树的叶子、树皮、树根放到簸箕里，置于抄纸槽中，用手揉出适量的药水，接着用搅纸棍从上到下斜以“一”字形搅半小时左右。

⊙5

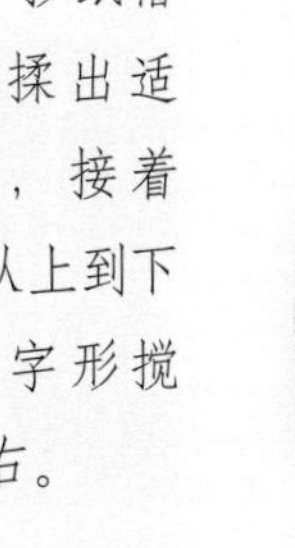

⊙4

⊙2 泡竹塘 Pool for soaking the bamboo

⊙3 正在碾竹料的碾子 Grinder for grinding the bamboo materials

⊙4 初步碾碎的竹料 Processed bamboo materials

⊙5 放到簸箕里的纸药原料 Raw material of the papermaking mucilage in a bamboo basket

拾贰 捞纸

12

⊙6~⊙10

捞纸需经过两道水，将纸帘置于纸帘架上，手持纸帘架，先由外往里舀水，接着由右往左舀水，水从左往右倒。将竹篾片置于压板的底板上，其上再放编织袋，然后将捞出的第一张湿纸置于编织袋上，此后则一张张叠加成湿纸垛。

⊙6 ⊙7 ⊙8 ⊙9 ⊙10

⊙11

拾叁 压纸

13

⊙11

捞完纸后，在纸桩上依次加废旧纸帘、压板，可在其上加石头先压，然后去掉石头，换上板矮页、垫码、大杆，套上钢索，小杆插在滚筒上压，待不能继续压时松榨，然后加一个垫码再压，一般共需加五六个垫码才能将纸压到所需要的干湿度，整个过程需要40~60分钟。

⊙12

⊙13

拾肆 放水

14

⊙12⊙13

压纸过程中，打开纸槽出水口，将水放掉，剩余的纸浆会沉淀在竹篾片上。第二天早上捞纸前再加清水。

⊙6/10 捞纸 Scooping and lifting the papermaking screen out of water and turning it upside down on the board

⊙11 石头压纸 Pressing the paper with a stone

⊙12 放水 Opening sluice to let water out

⊙13 纸浆沉淀 Precipitate of the paper pulp

拾伍 松榨

15

松榨，将压至半干的纸垛抬回家中。

拾陆 打纸

16

用打纸棒将纸面打松，轻轻打一次，可撕三提。

拾柒 撕纸

17

第一张纸撕到大约一半位置，接着撕第二张，后每两张间隔大约1.5 cm，一般十五六张为一提，少的有12张，多的有18张。稍稍折起后，整体揭起放在一边。

拾捌 晾纸

18 ⊙14

到一定厚度后，直接用手将纸一提提晾在屋内竹竿上，天气好时要晾一个月，天气不好时大约要晾50天。一般不着急收下来，待使用或晾纸场地不够时再收下来。黄平满溪村的习惯是将纸都晾在屋内，调查时造纸户认为如放在外面晒，万一刮风下雨，会将纸吹断淋坏。

拾玖 收纸

19 ⊙15

晾干后，用手将纸一提提收下来。据访谈时了解到的信息，一天捞的纸晒干后的质量约15 kg。

⊙14

⊙15

贰拾 叠纸

20 ⊙16⊙17

将纸一张张撕开，三张为一小叠，将一头理齐，然后对折。

⊙16

⊙17

这样就完成了整个造纸过程。

黄平满溪村竹纸一般不直接卖，而是先加工成纸钱，纸钱的主要制作过程如下：

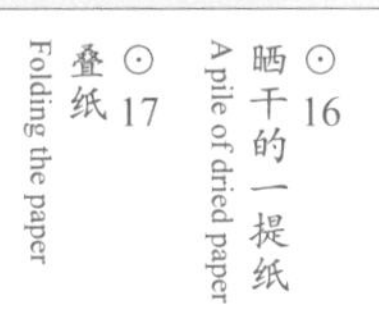

⊙14 晾纸 Drying the paper

⊙15 收纸 Collecting the paper

⊙16 晒干的一提纸 A pile of dried paper

⊙17 叠纸 Folding the paper

壹 凿纸

*1 ⊙18~⊙20

将若干叠纸整齐地摆放在凿纸桩上，其上用凿纸担压紧、固定，两侧各放一块砖头压着。用凿纸锤在纸上凿出一定数目的钱眼。

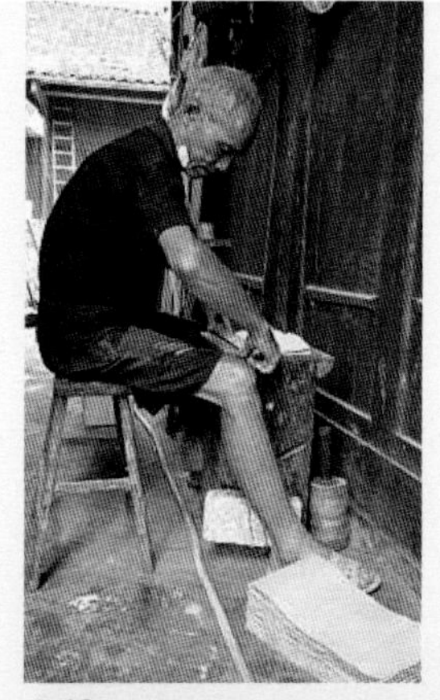

⊙18

⊙19

⊙20

贰 砍纸

*2 ⊙21⊙22

用砍纸卡确定纸的大小，再用砍纸刀将纸砍成合适大小的纸钱。

⊙21

⊙22

(三)

黄平苗族竹纸生产使用的主要工具设备

壹 搅纸树

1

即搅纸棍，实测王启光家所用的搅纸树长约114 cm，由直径约3.5 cm的竹子制成。

贰 打纸棒

2

实测王启光家所用的打纸棒直径：粗端约4.5 cm，细端约2.5 cm，总长约42 cm，手柄长约16 cm，大小不太一致。

⊙23

叁 纸帘

3

用苦竹编成，其上涂桐油。实测王启光家所用的纸帘中间有一隔线，为一帘二纸型，捞一次可得两张纸，纸帘内径约75.5 cm，宽约35 cm，一张纸的实际尺寸为75.5 cm × 17.5 cm。

⊙24

⊙18 用凿纸担压纸 Pressing the paper with a carrying pole

⊙19 / 20 凿纸 Perforating the paper to make joss paper

⊙21 / 22 砍纸 Trimming the paper

⊙23 打纸棒 Beating stick

⊙24 纸帘 Papermaking screen

肆

凿纸桩

4

木质，实测王启光家所用的砍纸木桩高50~55 cm，不太规则。旁边有一小竹筒，上放棉花和菜籽油，用以涂在刀上，使刀光滑不生锈。

⊙25

伍

砍纸刀

5

实测王启光家所用的砍纸刀长约24 cm，含柄宽约11.5 cm，不含柄宽约5 cm。

陆

砍纸卡

6

竹质，用于确定纸的大小，实测王启光家所用的砍纸卡总长约19 cm，卡槽长约11.8 cm。

⊙26

柒

凿　　纸

7

即纸钱，尺寸为16.5 cm × 11.8 cm。

捌

凿纸槌

8

木质，直径11.3 cm，实测王启光家所用的凿纸槌总长约30.6 cm，细柄长约13.5 cm。

⊙27

玖

凿纸担

9

木质，用于压紧、固定纸，长约52.3 cm，凿纸时两侧各吊一砖头压案。

⊙28

⊙25 凿纸桩 Trimming stake

⊙26 砍纸刀、砍纸卡 Trimming knife and trimming ruler

⊙27 凿纸槌 Mallet for perforating the paper

⊙28 凿纸担 Carrying pole for fixing the paper

(四)
黄平苗族竹纸的性能分析

所测黄平满溪村竹纸的相关性能参数见表10.3。

表10.3 满溪村竹纸的相关性能参数
Table 10.3 Performance parameters of bamboo paper in Manxi Village

指标		单位	最大值	最小值	平均值
厚度		mm	0.530	0.430	0.480
定量		g/m²	—	—	131.5
紧度		g/cm³	—	—	0.274
抗张力	纵向	N	26.1	20.3	23.4
	横向	N	13.6	7.7	10.4
抗张强度		kN/m	—	—	1.127
白度		%	17.7	16.8	17.3
纤维长度		mm	4.93	0.61	1.87
纤维宽度		μm	34.0	4.0	12.0

由表10.3可知，所测满溪村竹纸最厚约是最薄的1.23倍，相对标准偏差为0.32%，纸张厚薄差异较小。竹纸的平均定量为131.5 g/m²。所测竹纸紧度为0.274 g/cm³。

经计算，其抗张强度为1.127 kN/m，抗张强度值较大。

所测满溪村竹纸白度平均值为17.3%，白度较低，白度最大值约是最小值的1.05倍，相对标准偏差为0.33%，白度差异相对较小。

所测黄平竹纸纤维长度：最长4.93 mm，最短0.61 mm，平均1.87 mm；纤维宽度：最宽34.0 μm，最窄4.0 μm，平均12.0 μm。所测竹纸在10倍、20倍物镜下观测的纤维形态分别见图★1、图★2。

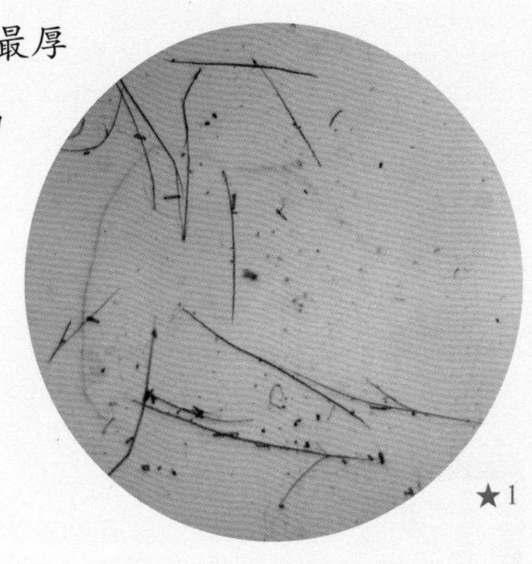
★1

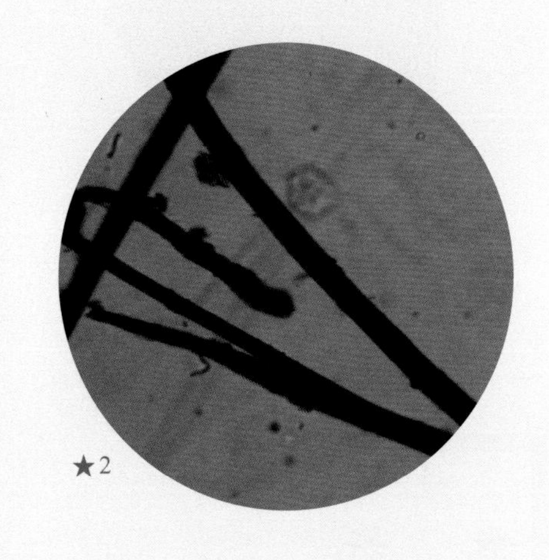
★2

★1 满溪村竹纸纤维形态图(10×)
Fibers of bamboo paper in Manxi Village (10× objective)

★2 满溪村竹纸纤维形态图(20×)
Fibers of bamboo paper in Manxi Village (20× objective)

五
黄平苗族竹纸的用途与销售情况

5
Uses and Sales of Bamboo Paper by the Miao Ethnic Group in Huangping County

黄平竹纸纸色较黄，较厚，纸面粗糙，有明显的纸筋。调查时了解到黄平竹纸的唯一用途是民间祭祀，当地红白喜事都用，但丧事类的“白喜事”用得更多。

黄平历史上也生产过相对精细的草纸，主要用于包装，如包糖、点心等，但当代已完全被塑料类包装材料替代，“草纸”业态也早已消失。

黄平竹纸销售的基本单位是“把”，分“大把”与“小把”。6张为一贴，8贴为一小把。调查时，一小把竹纸最低卖4元，最高可卖10元。两小把为一大把，质量约0.35 kg。春节前最后一次赶场时售价最高，清明节和春节时用量大。当地将清明节挂青时插在坟上的竹纸叫“挂白”，且每座坟上都会焚烧竹纸；农历七月十五中元节时售价最低，且用量小。

据王启光介绍，夏天时一个造纸户一天最多可抄20 kg干纸，冬天则为15 kg。王启光家最近一年约用1 000 kg生竹子，造出约300 kg纸。不计损耗，300 kg纸可加工成约860大把纸钱，调查时一大把纸钱的售价为8~20元，若以14元计算，销售额约1.2万元。

造纸主要支出为买竹子和石灰。竹子售价为0.2元/kg，石灰售价为0.6元/kg，1 000 kg生竹子需100 kg石灰。按王启光家最近一年的造纸量计算，买竹子需花费200元，买石灰需花费60元。相比销售收入来说，支出所占比例较小。

黄平满溪村竹纸主要是自产自销，造纸户将所造的纸制成纸钱后，赶场时拉到黄平县的翁坪、旁海、谷隆，凯里市的湾水等乡镇去卖，一次拉两担，一担为42把。

⊙1

⊙1 打纸钱 Making joss paper

六 黄平苗族竹纸的相关民俗与文化事象

6 Folk Customs and Culture of Bamboo Paper by the Miao Ethnic Group in Huangping County

1. 祭祀碾、槽

调查中了解到，满溪村一带较讲究的造纸户每年第一次用碾、槽前要烧香和纸，缘由是碾、槽可帮人赚钱，故新年第一次用时要烧香、纸来祭祀一下，以示感谢。王启光等造纸人并不知造纸始祖是谁，只是祭祀。曾经这是传统的乡俗，如今有些不讲究的造纸户已不烧祭了。

2. “造纸七十二道手序”

王启光听老人讲过一句俗语“造纸七十二道手序”，这说明当地造纸工序多且复杂，虽然现在的烧纸工序已没有那么讲究，但他认为最少也有四五十道工序。

3. 纸钱有讲究

调查中了解到，当地对纸钱的制作与使用颇有讲究：大纸，也称四路钱，每张有4列，每列有

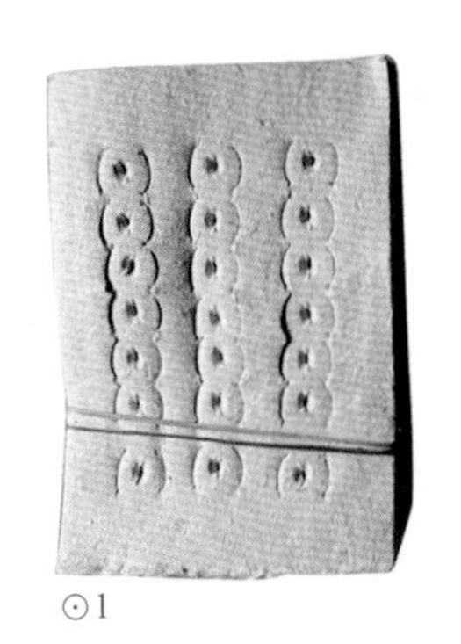
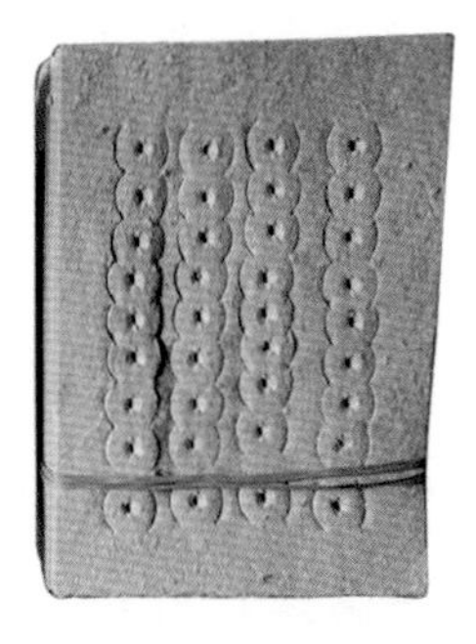
⊙1

9个眼，主要用于农历七月十五敬菩萨，据说历史上苗族等少数民族居民经常使用；小纸，也称三路钱，每张有3列，每列有7个眼，据说当地汉族居民经常使用。

虽然两种纸的面积大小相同，但大纸钱眼多些，代表钱更多，据王启光口述，以前主要用大纸，现在改用小纸。

七
黄平苗族竹纸的保护现状与发展思考

7
Preservation and Development of Bamboo Paper by the Miao Ethnic Group in Huangping County

黄平苗族竹纸的造纸原料充足，纸药树虽然需种植，但量也足够。2013年调查时，满溪村有十六七户仍在造纸，但全都是中老年人，年龄最小的也已经50多岁。一个造纸户一年最多造350 kg干纸，少的则不到100 kg。年轻人普遍觉得造纸不但累，而且赚钱不多，于是都不愿继承祖业。就连王启光也说，自己老了，下不了广（下广，即去广东打工，现泛指出去打工），如能下广，他也不造纸。

通过综合判断，调查组认为黄平竹纸的传承与发展目前面临着以下一些问题：

首先，从用途来说，黄平竹纸的唯一用途是民间祭祀，用途单一且销售范围窄，这使得黄平竹纸的发展空间有很大的局限性。

其次，从需求量来说，因为黄平竹纸只用于

⊙1
成品纸钱
Final product of joss paper

⊙2

民间祭祀，没有其他用途，所以整体来说需求量不大。同时由于竹纸的用途低端，加上成本等因素，也使得其往往只能在周边集市销售，不太容易销售到远地。满溪村造纸户每年农闲时造的纸主要满足附近村镇乡民在清明节、中元节、春节时的需求，其整体需求量不大。另外，现代祭祀用品如各种冥府银行钱币、机制竹纸在当地的出现和使用，使得传统的祭祀用品也有一定的改变，这显然抑制了手工祭祀用纸的发展空间。

最后，从经济效益来说，黄平竹纸用途低端，使得其价格不高，需求量不大，产量也不会太高，虽然相比销售额来说，生产竹纸所需成本较小，但其经济效益较一般。王启光也多次提到，纸不是太好卖。虽然春节前的价格比中元节时要高不少，但这主要与纸的总量不多有直接关系。调查中发现，由于手工造纸经济效益较低，满溪村的年轻人都选择出去打工而无意留在家中造纸。访谈中获知的信息是，造纸人中年龄最小的也已年过半百，整体上呈现后继无人的状态。

在一定时间内，由于传统祭祀的需求，以及老年人在家兼业赚钱，黄平造纸会基本保持现状，但不会有太大的发展，同时随着目前造纸人的年纪越来越大，黄平手工竹纸很可能会在不太长的时间内面临衰亡的局面。因此，实施抢救性的技艺影像保护已是迫切的文化任务。

在尽快实施博物馆式影像保护性传承的前提下，如果要使黄平手工纸业态可以稳定、持续地发展，很重要的途径之一是将手工纸和少数民族文化、生态旅游相结合，通过少数民族生态旅游带动苗族造纸文化，同时推动手工造纸为生态旅游添彩。翁坪乡是一个典型的少数民族聚居乡，主要居住有苗族、革家等少数民族，苗族、革家民族风情浓郁，民族节日丰富多彩，有吹芦笙、斗牛、唱苗歌、斗鸟、赛马、民族服饰展等多种活动。此外，犀牛河两岸拥有独具魅力的自然旅游资源。可考虑发展有特色的少数民族生态旅游，将苗族造纸及其他手工技艺、文化、民俗等纳入少数民族生态旅游的范畴，相辅相成。这不仅会给当地村民带来经济效益，同时也会使得民间意识里对传统技艺、文化、民俗等有新的认识和理解，自觉地保护包括手工纸在内的苗族传统技艺。

当然，由于目前有关黄平苗族手工造纸的文献研究和田野研究的积累非常少，因而较难深入完整地描述该地手工造纸的历史与文化，这需要进行后续的专项探究工作。

⊙3

⊙4

⊙2 满溪村外景 Landscape of Manxi Village

⊙3 翁坪乡的革家寨 Gejia ethnic residence in Wengping Town

⊙4 翁坪乡的犀牛河 Xiniu River in Wengping Town

黄平苗族

竹纸

Bamboo Paper
by the Miao Ethnic Group
in Huangping County

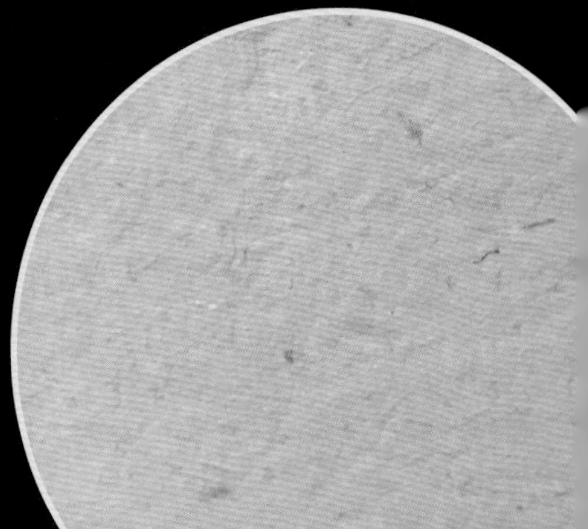

满溪村竹纸透光摄影图
A photo of bamboo paper in Manxi Village
seen through the light

第四节

凯里苗族竹纸

贵州省
Guizhou Province

黔东南苗族侗族自治州
Qiandongnan Miao and Dong Autonomous Prefecture

凯里市
Kaili City

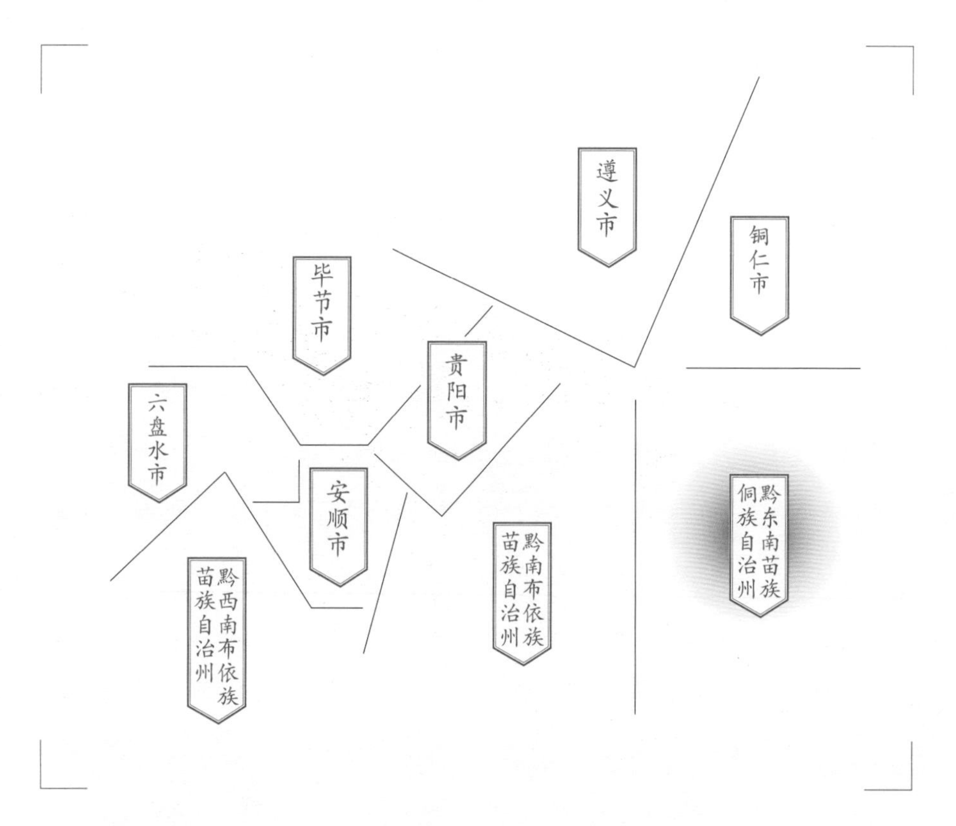

调查对象
湾水镇
苗族竹纸

Section 4
Bamboo Paper by the Miao Ethnic Group in Kaili City

Subject
Bamboo Paper by the Miao Ethnic Group in Wanshui Town

一
凯里苗族竹纸的基础信息及分布

1
Basic Information and Distribution of Bamboo Paper by the Miao Ethnic Group in Kaili City

凯里竹纸的当代生产分布于黔东南苗族侗族自治州州府所在地凯里市的湾水镇，主要由当地苗族村民传习制作。湾水镇的手工土纸在历史上曾具有较广泛的用途，2009年9月至2013年10月，调查组先后4次进入湾水镇考察，所了解到的信息是，湾水竹纸除了目前仍在广泛用于祭祀外，旧日还用于包装点心、糖、盐等乡土生活用品，以及制作鞭炮芯纸等，曾经广泛使用于地方百姓的日常生活中。

二
凯里苗族竹纸生产的人文地理环境

2
The Cultural and Geographic Environment of Bamboo Paper by the Miao Ethnic Group in Kaili City

凯里市是黔东南苗族侗族自治州州府驻地，贵州省东部中心城市。凯里地域历史悠久，战国至南北朝属且兰故地，元朝属麻峡县，明弘治年间置清平县，归属于都匀府，民国二年（1913年）改炉山县，1959年1月设凯里县，1984年1月改为凯里市。

据2011年的人口统计数据，凯里市常住人口为47.86万，有苗、汉、侗、布依、畲、水、仡佬等27个已定民族，以及革家、西家等待定民族。少数民族人口占总人口的75.4%，其中苗族人口占总人口的63.3%。

凯里有“中国百节之乡”和“苗岭明珠”之称，有世界上最长、最宽的风雨桥——凯里风雨桥，有世界上最大的苗寨——西江千户苗寨。“凯里”系苗语的音译，意为“木佬人的田”，苗语称木佬人为“凯”，称田为“里”。

凯里苗族竹纸生产地分布示意图

Distribution map of the papermaking site of bamboo paper by the Miao Ethnic Group in Kaili City

考察时间
2009年9月/2011年8月/2013年8月/2013年10月

Investigation Date
Sep. 2009/Aug. 2011/Aug. 2013/Oct. 2013

路线图
凯里市区
↓
湾水镇
Road map from Kaili City centre to the papermaking site (Wanshui Town)

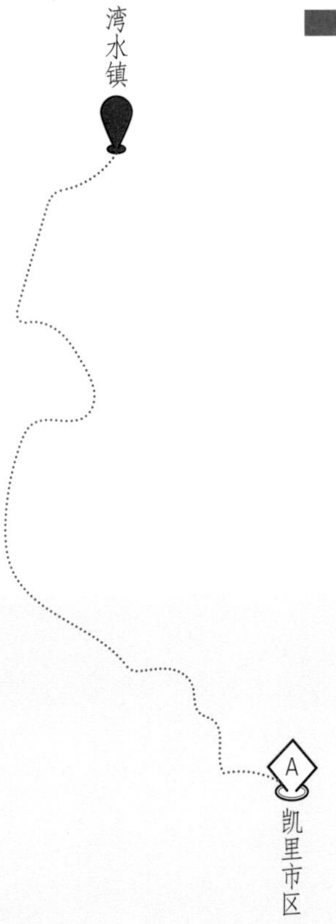

	A	1	2	3	4	5	6	7
地域名称	凯里市	湾水镇	龙场镇	舟溪镇	炉山镇	大风洞乡	三棵树镇	旁海镇
造纸点名称		湾水镇 造纸点						

位置分布

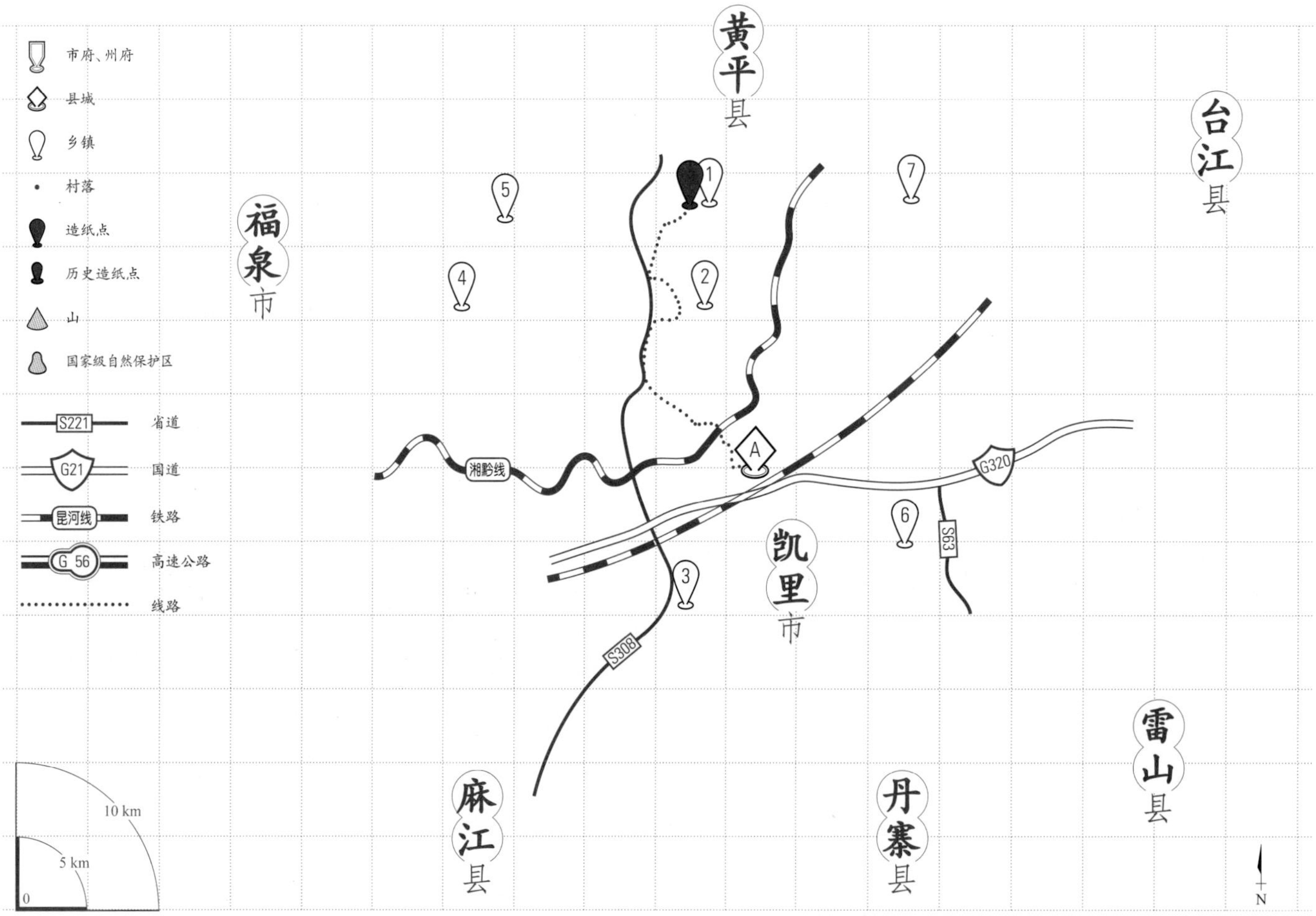

⊙1

⊙2

湾水镇是1992年经贵州省人民政府批准建立的新镇，地跨重安江两岸，位于凯里市北面，镇政府驻地距市中心37 km，东邻旁海镇，南接湾溪街道，西抵大风洞乡，北界黄平县。全镇总面积78.54 km^2，共辖 17个行政村，1个居委会，共5 580户24 991人，其中苗族人占98.7%以上，是苗族人聚居的乡镇。贯穿全境的重安江长22 km，两岸翠柳修竹，江面木船穿梭；巍巍苗岭树木郁葱，鸟语花香，苗疆的山水使人心旷神怡。镇内地域广阔，土壤肥沃，气候温暖湿润，日照时间长，雨量充沛，年无霜期315天以上，是造纸原料竹子优良的生长地，为湾水竹纸的生产提供了充足的原料。

三 凯里苗族竹纸的历史与传承

3 History and Inheritance of Bamboo Paper by the Miao Ethnic Group in Kaili City

2009年9月、2011年8月、2013年8月和10月，调查组4次到凯里市湾水镇进行竹纸制作的实地调研，先后采访了姜明亮（1939~2013）、姜文安（1936~）、姜明春（1964~）等苗族造纸师傅。据姜文安、姜明亮等人回忆，姜文安家最早从事造纸，但有关造纸技艺从何处、由何人传入已记

⊙3

⊙4

⊙1 流经凯里的重安江 Chong'an River flowing through Kaili City

⊙2 进入湾水造纸村落的吊桥 Suspension bridge leading to Wanshui Papermaking Village

⊙3/4 造纸人姜文安夫妇和儿子姜明春 Papermakers Jiang Wen'an with his wife and son Jiang Mingchun

⊙1

不清楚。姜文安的爷爷姜勾龙、父亲姜通甫、儿子姜明春都造纸，他们家至少有4代人造纸。姜文安17岁时开始造纸，27岁就以造纸为主业，直到78岁高龄才被迫放弃造纸。而姜明亮则是跟其父亲姜礼秀学会造纸的，其儿子没有造纸，他们家至少有3代人造纸。按调查组对姜文安和姜明亮两户的推算，湾水镇至少有百年的造纸历史。

四
凯里苗族竹纸的生产工艺与技术分析

4
Papermaking Technique and Technical Analysis of Bamboo Paper by the Miao Ethnic Group in Kaili City

⊙2

(一)
凯里苗族竹纸的生产原料与辅料

凯里湾水镇造纸原料当地称为洋竹，据调查组的现场考察知其实际上是慈竹类。据造纸户的说法，一般用2~3年生的老竹子，1年生的竹子也可以用，但水分多、纤维少、出料率低，因而用得较少。

湾水镇造竹纸需用纸药，当地称为药，用称为药树的植物的叶子、秆、根的汁做成纸药，平时用药树的叶子、秆，冬天则用药树的根。

⊙1 调查组成员与姜明亮合影
Researchers and the local papermaker Jiang Mingliang

⊙2 药树
Raw material of papermaking mucilage

⊙3

(二) 凯里苗族竹纸的生产工艺流程

虽然原则上只要有空就可以造纸，但湾水镇主要是农闲时造纸，造纸户们还是以农业种植为主。据调查中了解到的信息，一般每年有3个时间段造纸：第一个时间段是农历二月，一般是白天挖纸，晚上加工，造的纸主要在清明节时使用；第二个时间段是农历四月至六月底，这段时间都挖纸，农历六月加工，造的纸主要在中元节时使用；第三个时间段是农历十月至十二月，这段时间都挖纸，农历十一月至十二月加工，造的纸主要在春节时使用。由于天冷，第三个时间段所造纸的量相对前两个时间段会少一些。

由于入村调查的时段均未能赶上竹纸制作的时间，因此若干工艺流程没有留下现场的图文记录，经与姜文安一家及姜明亮的深入访谈交流，总结凯里湾水镇竹纸的制作工艺流程如下：

壹	贰	叁	肆	伍	陆	柒	捌	玖
砍竹子	捶竹子	捆竹子	晒竹子	泡竹麻	捞竹麻	砍竹麻	碾竹麻	踩纸浆

拾	拾壹	拾贰	拾叁	拾肆	拾伍	拾陆	拾柒
搅纸浆	加药水	挖纸	榨纸	分纸	撕纸	晒纸	揭纸

⊙3 采访造纸村民
Interviewing papermakers

壹

砍竹子

1 ⊙1

造纸户在农历正月下旬或八月上旬上山砍竹子，并将砍下的竹子抬回家。一天可砍竹子数量不定，多的可砍500 kg。一次一般砍1 000~1 500 kg，数量根据竹麻塘大小而定。

⊙1

贰

捶竹子

2 ⊙2

将竹子砍成段，每段长约2.3 m，用刀将较粗的竹段对半破开，然后用刀背或木槌将其捶破，较细的竹段直接用刀背或木槌捶破。

⊙2

叁

捆竹子

3

用竹篾将捶破的竹片捆成把，每把直径约20 cm，质量为7.5~10 kg。

肆

晒竹子

4

将成捆的竹片堆成堆，置于太阳底下晒，大约晒半个月时间。

伍

泡竹麻

5

待竹片晒成半干后，在竹麻塘内先撒一层石灰，再放两排竹片，后洒一层石灰浆，再垂直放两排竹片，后洒一层石灰浆。如此放4~5层，共约10排竹片，最上面洒大量石灰浆，再用几块石头压住，并放满水，泡3个月，泡好后的竹片叫作竹麻。1 000 kg半干的竹片需400 kg石灰。

陆

捞竹麻

6

用钉耙将竹麻捞出来，并在竹麻塘里将石灰洗干净。捞竹麻时，需要多少就捞多少，一般一次捞6把，一小时即可洗干净。

柒

砍竹麻

7

将竹麻抬回家，用柴刀或斧头将竹麻砍成段，每段长约7 cm，一个人砍一天才能砍完6把竹麻。

捌

碾竹麻

8

将砍断的竹麻一次性放入碾槽里，打开水闸，水流带动碾子碾料。一般需碾4~5小时；如果竹麻太硬，则需碾5~6小时，碾后的竹麻称为纸浆。碾竹麻时，需要一个人用钉耙将碾槽边上的竹麻扫进碾槽里，一般十几分钟扫一次，碾好后关水闸，并将纸浆装入簸箕里。

玖

踩纸浆

9

将纸浆抬到纸塘，分两次踩。踩纸浆时，边放水边踩，水量不能太多，漫过纸浆即可。第一道踩完后把下面的排水口打开，待污水放完后，关上排水口，再放入清水，如此反复三次，当地称为踩三道，每道大约需半小时。接着把踩好的纸浆放到纸塘前面的水泥板上。

⊙1 砍竹 Lopping the bamboo

⊙2 捶竹 Hammering the bamboo

拾 搅纸浆

10

将踩好的纸浆全部放到纸塘里，亦可分两次放，接着放满水后关水，用耙耙（即拱耙）将纸浆拱上来，几分钟后用竹竿划圈搅，将纸浆搅匀。整个过程需十几分钟。

拾壹 加药水

11

也称加滑药，将药放在簸箕里，用手揉，药水通过簸箕流下去，一天约需1.3 kg药。

拾贰 挖纸

12 ⊙3⊙4

用帘子架将滑水搅均匀，接着手持帘子架挖纸。挖纸时分两道，头道由外往里挖，后由右往左抬起来；二道由右往左挖。当地造纸户认为，挖头道水，竹两端纤维（毛毛）往外，挖二道后，毛毛往右偏，纸好揭，撕了不破，左边厚些，称为纸头。接着将帘子架置于竹竿上，手持帘子将其翻盖到坐板竹篾片或旧帘子上。多的一天可挖1 200帘左右，6把料挖得的纸干后质量约30 kg。

⊙3

⊙4

⊙3 头道挖纸示意 Showing how to scoop the papermaking screen for the first time

⊙4 二道挖纸示意 Showing how to scoop the papermaking screen for the second time

拾叁 榨纸

13

挖完纸后，在纸垛上直接加压板，依次加4~5个木桩、肩膀桩、榨杆，用钢丝绳将榨杆和搅车捆起来，把小杆插入搅车的洞里，手扳动小杆，榨20分钟左右，松榨，加一个木桩，如上再榨两次，至手扳不动后，人踩在榨杆上榨纸，直到没有水流下来为止，且用拇指按纸垛，若按不进去即算榨干了，总共约需一小时。

拾肆 分纸

14

如用一帘三纸的帘子，则需分纸；如用一帘一纸的帘子，则没有该步骤。分纸前，先松榨，接着将纸垛抬离坐板，放一木桩于坐板上，再将纸垛中间置于木桩上，两手把纸垛往两边扳，可得到三个小纸垛。

拾伍 撕纸

15

用捶纸棒捶纸面，并用手将纸头盘起来，当地称为盘纸头，后用手一张张从纸头往纸尾撕，每两张间隔约0.5 cm，5张为一排，天气好时可6~7张为一排。

拾陆 晒纸

16

用手将一排纸整体撕开，挂在屋内竹竿上，天气好时2~3天即可阴干，下雨或冬天则需一周才可阴干。虽称晒纸，实际上只是阴干。当地造纸户认为阴干的纸颜色黄，好卖；太阳晒的纸白些，不好卖。故只能阴干，不能晒干。

拾柒 揭纸

17

将阴干的纸取下来，一把把堆起来。

这样就完成了整个造纸过程。

据姜文安口述，以前竹子晒干后要在石灰里浆一下，然后置于石甑的大铁锅中，水浸过竹子，由钢管出气，煮一周。一次最多煮2 000~2 500 kg，最少也可煮1 000 kg，约耗200 kg煤。竹子煮后可以提高造纸效率，但耗煤多，因此1948年前后就不再煮竹了。

以前用钱錾打上钱印，再用砍纸刀切成合适大小即可得到纸钱。现在的制作工序大体相似，其过程如下：

把竹纸置于切纸凳上，上放纸钱架，再用三根活动木条压于其上，捆好，使之固定。

接着用钱錾按顺序打上钱眼，打完后将纸钱架去掉。

将打好钱眼的纸置于切板上，其上用一块木板压住，木板的边缘与切板的缝平行。左脚压木板，手持砍纸刀沿着木板将竹纸切开。其后用上述方式，按尺寸将纸切好后，即可得到纸钱。

⊙5

⊙6

调查中还发现，如竹子晒的时间过长，就会成为陈竹麻。此时，需将陈竹麻泡过后，洗干净晒干，再泡半个月就要赶紧用，否则泡的时间过长，竹麻容易烂。陈竹麻易碾，好挖，但挖出的纸颜色没有那么黄。

⊙7

据姜明亮介绍，湾水镇以前有一户人家生产过草纸，主要供给黄平县重安镇的糖厂，1958年后不再生产。草纸可用于包装糖、点心等产品。

造草纸的原料为竹和稻草，竹与稻草的比例为2：1。所用的稻草很有讲究，只能用籼米稻草，不能用糯米稻草。

据姜明亮介绍，稻草原料的大致制作工艺流程如下：

浸泡稻草－洗稻草－晒稻草－捆稻草－泡稻草－碾稻草。

用泡竹麻的石灰水将稻草泡一个月后，再将

⊙8

⊙5 装架 Fixing the shelf

⊙6 打钱眼 Perforating the paper

⊙7 打好钱眼的纸 Paper with holes on it

⊙8 切纸 Trimming the paper

⊙9

稻草洗干净，晒干，捆好备用。碾稻草前需先将稻草在清水里浸泡一晚，碾稻草时需待竹麻快碾好时才能放稻草下去同时碾。

竹麻和稻草一起碾好后，就可以抄纸，后续工序则和生产竹纸工序相似。

(三)
黄平苗族竹纸生产使用的主要工具设备

壹 碾子 1

石质，以碾槽外侧算，碾子直径约289 cm；碌子直径约93 cm，中心处厚约19 cm。

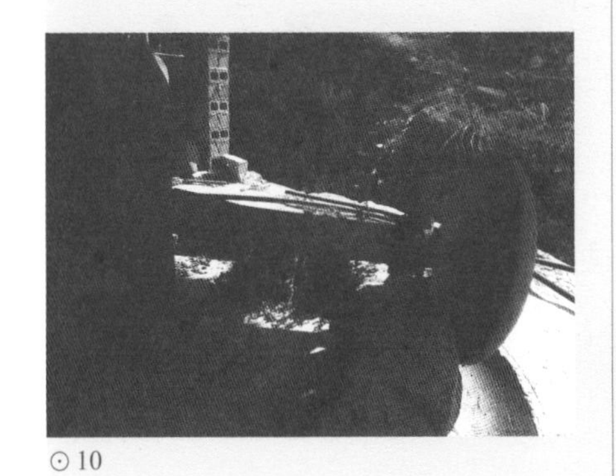

⊙10

贰 纸塘 2

上宽下窄，调查时所测纸塘上部长约190 cm，宽约100 cm；下部长约184 cm，宽约74 cm。纸塘前方正中央往里凹，挖纸时人站在凹处；其上粘有一块圆木头，挖纸时人腹部时不时会碰到最上方，圆木头可保护腹部不被水泥刮伤。

⊙11

⊙12

⊙9 姜明亮及其家人 Jiang Mingliang and his family members
⊙10 碾子 Stone grinder
⊙11 水车 Waterwheel
⊙12 纸塘 Papermaking trough

叁 帘子 3

由细苦竹丝编成，四周用木条固定。现有的帘子中间用两根小木棍将纸帘一分为三，所抄纸为一帘三纸，亦有一帘一纸的帘子。实测帘子长约89 cm、宽约49 cm，所造出的纸长约85 cm、宽约16 cm。亦有长约78 cm、宽约36 cm和长约69 cm、宽约40 cm等不同尺寸的帘子。

⊙13

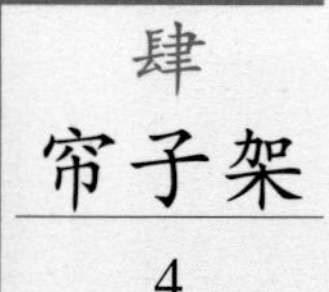

⊙14

以前的帘子上还有造帘子之人和帘子主人的姓名，如“马玉龙造，姜文安办”。

肆 帘子架 4

用枫香树（柏树）制成，密度较大。左边木条上有小凹槽，用于固定纸帘。

⊙15

⊙16

2011年，姜明春设计出带提手的帘子架，冬天抄纸时手就可以不用浸到水里，据他自述这是他自己想出来的，没见人用过。

伍 木榨 5

不同的木榨尺寸有所差异，实测木榨的架桩高约240 cm，方肩高约170 cm；坐板尺寸约为97 cm×48 cm，上可放两柱；搅车（即榨磙）长约85 cm，榨杆长约240 cm。

⊙17

陆 砍纸刀 6

铁质，长约23 cm，宽约7 cm。

柒 钱錾 7

长约9 cm，宽1.5~2 cm。

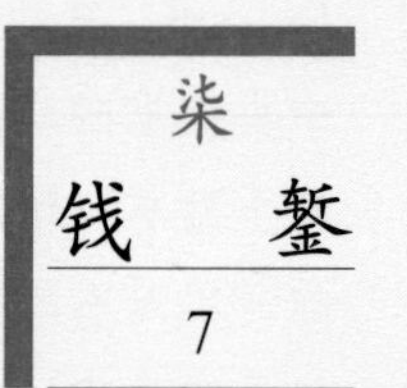

⊙18

⊙13 一帘三纸的帘子 Papermaking screen that can make three pieces of paper simultaneously

⊙14 有制帘人和造纸人姓名的帘子 Papermaking screen with the names of screen-maker and papermaker on it

⊙15 帘子架 Frame for supporting the papermaking screen

⊙16 带提手的帘子架 Papermaking frame with handles

⊙17 木榨 Wooden presser

⊙18 砍纸刀和钱錾 Trimming knife and chisel

捌 切纸凳 8

木质，所测切纸凳长约74 cm，宽约39 cm，高约60 cm。为了稳定，在距地高约30 cm处的木条上放一木板，其上再用石头压住。切纸凳上端木板上铺有麻布、塑料布，一侧边有三根两端带绳子的活动木条，切纸时可起到固定作用。

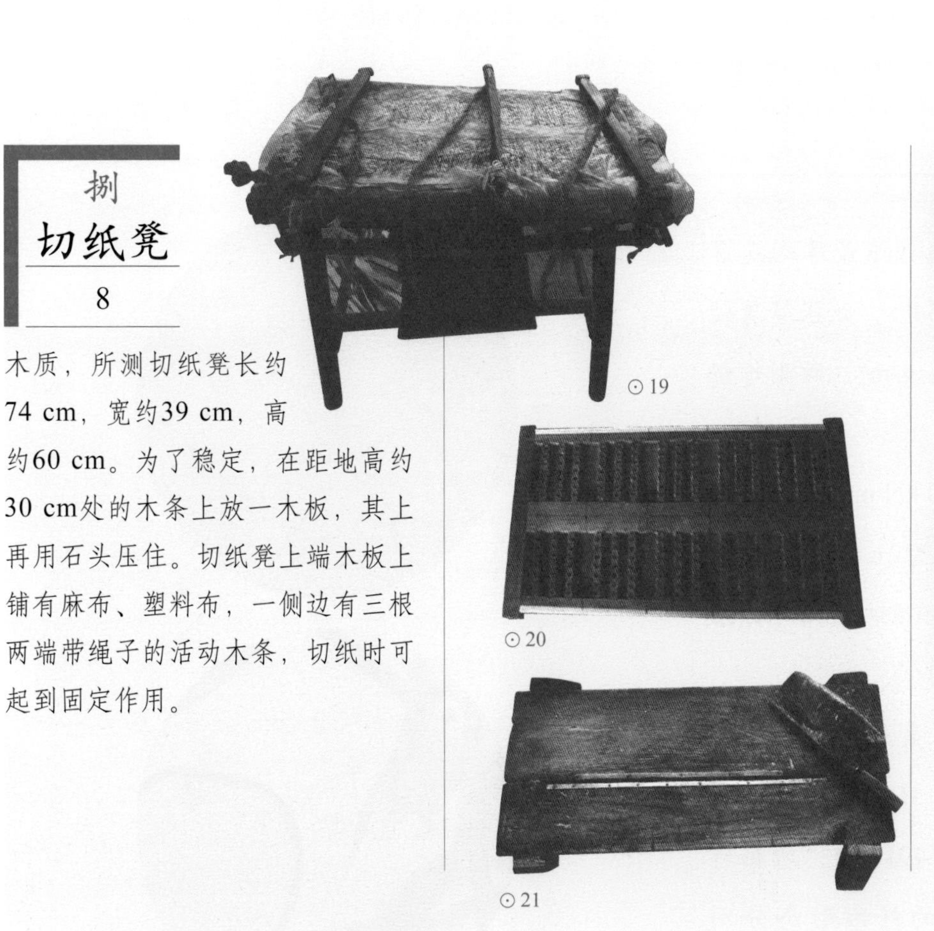

⊙19 切纸凳 Trimming table

⊙20 纸钱架 Frame for perforating the paper

⊙21 切板与切刀 Trimming board and trimming knife

玖 纸钱架 9

木质，用砍纸刀将竹纸按尺寸切好后，置于纸钱架下，用钱錾按顺序打上钱眼，即可得到纸钱。

拾 切　板 10

打好纸钱后，用裁刀将其裁小，实测切板长约85 cm，宽约36 cm，其中缝长约62 cm，宽约1.2 cm。

(四) 凯里苗族竹纸的性能分析

所测凯里湾水镇竹纸的相关性能参数见表10.4。

表10.4 湾水镇竹纸的相关性能参数
Table 10.4 Performance parameters of bamboo paper in Wanshui Town

指标		单位	最大值	最小值	平均值
厚度		mm	0.815	0.650	0.714
定量		g/m^2	—	—	10.8
紧度		g/cm^3	—	—	0.015
抗张力	纵向	N	13.6	9.5	11.1
	横向	N	7.7	3.8	6.0

续表

指标	单位	最大值	最小值	平均值
抗张强度	kN/m	—	—	0.570
白度	%	21.4	20.4	21.1
纤维长度	mm	33.00	1.00	2.96
纤维宽度	μm	19.8	0.7	14.0

由表10.4可知，所测湾水镇竹纸最厚约是最薄的1.25倍，相对标准偏差为0.54%，纸张厚薄较为均匀。竹纸的平均定量为10.8 g/m^2。所测竹纸紧度为0.015 g/cm^3。

经计算，其抗张强度为0.570 kN/m。

所测湾水镇竹纸白度平均值为21.1%，白度较低，白度最大值约是最小值的1.05倍，相对标准偏差为0.28%，白度差异相对较小。

所测湾水镇竹纸纤维长度：最长33.00 mm，最短1.00 mm，平均2.96 mm；纤维宽度：最宽19.8 μm，最窄0.7 μm，平均14.0 μm。所测竹纸在10倍、20倍物镜下观测的纤维形态分别见图★1、图★2。

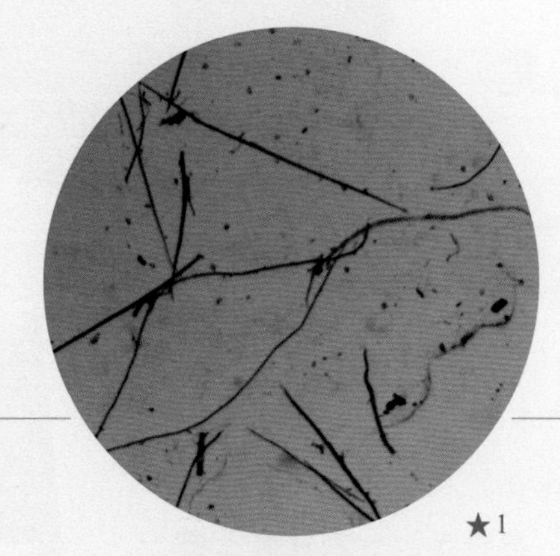

★1

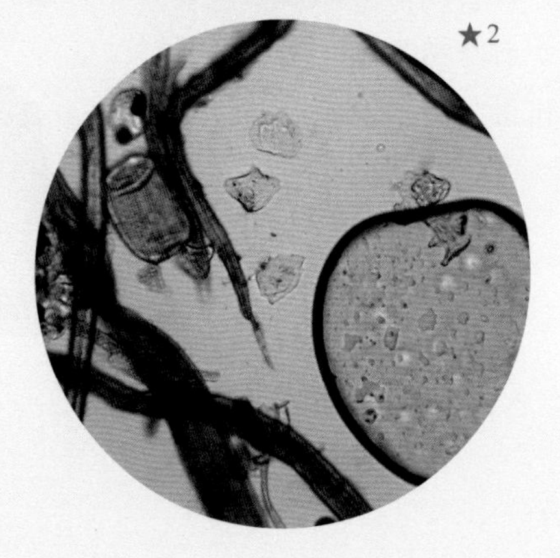

★2

★1 湾水镇竹纸纤维形态图(10×) Fibers of bamboo paper in Wanshui Town (10× objective)

★2 湾水镇竹纸纤维形态图(20×) Fibers of bamboo paper in Wanshui Town (20× objective)

五 凯里苗族竹纸的用途与销售情况

5 Uses and Sales of Bamboo Paper by the Miao Ethnic Group in Kaili City

(一) 凯里苗族竹纸的新旧用途

1.

湾水竹纸目前主要用于民间祭祀，红白喜事都用，“白喜事”用得更多些，如老人过世，身下要垫构皮纸，烧三行竹纸，用量较大。

2.

据访谈中造纸老人姜文安口述，数十年前制作卫生纸也是湾水竹纸的一个重要用途，甚至集体化管理的纸厂曾经有每个月给工人发1刀纸的福利，工人回家后通常将这1刀纸裁作卫生纸。据老人回忆，厂里规定，不管厚薄，每刀纸48张。

3.

凯里湾水镇以前以竹与稻为原料生产的草纸主要用于包装糖、点心等产品。

（二）

凯里苗族竹纸的销售与价格情况

1949年前，一块大洋可买3捆竹纸（一捆46刀），可见当年竹纸价格较高。而1978年后，烧封包的量少了，虽然用纸量随之减少，但造出来的纸都能销完，只是价格不再像以前那么高。调查组于2009年9月实地调查时，竹纸价格约8元/kg，如果是老板上门收，收购价约5.6元/kg。

⊙1

竹纸价格不高，但若加工成纸钱，则价格会高不少。过去纸钱8张为一贴，一捆为46贴（小张）。调查时则按驮来卖，纸钱尺寸不同，价格也不同。所测大的纸钱，长约18 cm，宽约14.5 cm，5元/驮；小的纸钱，长约17 cm，宽约12 cm，4元/驮。大的纸钱约0.22 kg/驮，小的约0.17 kg/驮。

如果一个人有半年时间在造纸，则约需1 300 kg生竹子，可生产出约400 kg纸钱，销售收入为9 000多元。

造纸主要的支出为购买竹子和石灰。按半年造纸，约需1 300 kg竹子计算，卖家砍好，和买家一起拖回来，价格为0.2元/kg，则购买竹子需260元；1 300 kg竹子晒成半干的竹片后，约430 kg，约需170 kg石灰，石灰价格为0.3元/kg，则购买石灰需51元。购买竹子和石灰共需311元。相比销售收入来说，支出较小。

⊙2

凯里湾水竹纸主要靠自产自销。据姜文安等人介绍，造纸户通常是将纸钱拿到湾水镇、凯里市、黄平县重安镇去卖，也有拿至六盘水市盘县的集市去售卖的。时常会有老板上门收购，但由于价格比自销低不少，造纸户一般不太愿意直接卖给转手获利的老板。

⊙1 湾水镇集市上的纸钱交易
Joss paper trade in a market in Wanshui Town

⊙2 湾水镇集市上销售纸钱场景
Scene of joss paper trade in a market in Wanshui Town

六 凯里苗族竹纸的相关民俗与文化事象

6 Folk Customs and Culture of Bamboo Paper by the Miao Ethnic Group in Kaili City

1. 秦叔宝与纸钱

姜文安口述了这样一则民间故事：据说秦叔宝考上状元后，天天得病。秦叔宝感觉夜里有鬼到家，于是请了3个人守门。为了犒赏守门人，秦叔宝每天花7个钱请他们吃夜宵，另给他们14个钱，即算起来每人一天工钱为7个钱，共21个钱。有了守门人后，秦叔宝的病很快就好了，因此秦叔宝继续请他们守下去。秦叔宝过世后，那3个人给他送纸钱——每张3排、每排7个眼的三行钱。

2. 纸钱有讲究

凯里湾水镇竹纸制成纸钱时非常有讲究，先用砍纸刀将纸切成小张，将4张纸对叠，然后盖上模子，用砍纸刀将其切成合适大小。凯里湾水镇纸钱有两种，小的是3排，每排7个眼；大的是4排，每排9个眼，1个眼代表1个铜钱。只有这两种纸钱才能用。

小的纸钱（3排，每排7个眼），当地称为主人钱，也叫作三行钱、真钱，主要是逢年过节时烧给老祖宗，家家都烧，多少不限。

大的纸钱（4排，每排9个眼），当地称为菩萨钱，也叫作四行钱、四路钱，主要用于敬菩萨，以前敬菩萨才能烧，后来和尚做庙会也烧四路钱。

⊙1

⊙2

⊙1 烧纸钱 Burning the joss paper

⊙2 三行钱与四行钱 Joss paper with different numbers of holes (3×7, 4×9)

七 黄平苗族竹纸的保护现状与发展思考

7 Preservation and Development of Bamboo Paper by the Miao Ethnic Group in Huangping County

调查组认为，凯里湾水镇苗族竹纸目前面临着以下问题：

首先，从用途来说，目前竹纸只用于民间祭祀，用途十分单一且属低端消费，这使得凯里湾水镇苗族竹纸不太容易拓展出新的发展空间。

其次，从需求量来说，因为只是用于民间祭祀，整体需求量不大，且竹纸售价低，考虑到成本等因素，往往只在周边集市自产自销，这从造纸户普遍不愿让跑乡串户的纸贩低价收购、经销的心态可见一斑。从每年三个造纸时间段主要分别满足清明节、中元节、春节的需求，即可看出以民俗固定消费定产的收缩模式。另外，各种机制纸钱及相关焚烧祭祀用品的流通和使用，使得传统的祭祀方式也有一定的改变，也使得手工竹纸的需求量逐渐减少。

最后，从经济效益来说，凯里湾水镇苗族竹纸所带来的经济效益较低。虽然相比销售额来说，生产竹纸所需成本较小，但其低端且缺乏规模业态的现状使得经济效益仍较低。调查中的现状是，21世纪初，由于造纸的效益较低，湾水镇很多出身造纸世家的年轻人都选择出去打工而不愿意继承造纸祖业。以湾水镇江口村乐安村民组为例，这里曾经家家户户造纸，男的负责碾料、抄纸、打纸钱等工序，女的负责撕纸、晒纸、揭纸等工序，分工协作，井井有条；2000年时，该村民组共104户，仍有50多户在造纸，而到2010年前后，仅剩下2户在村里造纸，而且已有1户陷入青黄不接的困境。

在较短时间内，由于传统祭祀的需求，以及老年人在家希望能兼业赚点钱补贴家用，湾水镇造纸或许会基本维持现状。但随着造纸人年纪越来越大，如不能迅速解决后继无人的瓶颈，凯里湾水镇苗族竹纸的活态无疑会走向衰亡。

针对凯里苗族竹纸的保护现状，其保护与发展重要的途径可能是将手工纸和苗族文化、生态旅游

⊙1

⊙1
江口村的旧纸坊
Old papermaking mill in Jiangkou Village

相结合，通过苗族生态旅游带动苗族人造纸，同时围绕民族生态旅游做造纸生产经营方式的调整。湾水镇苗族居民占98.7%以上，是苗族聚居度极高的乡镇，苗族文化风俗保留得相当完整。加之湾水镇山清水秀，发展有特色的苗族文化生态旅游，将苗族造纸及其他手工技艺、文化民俗等纳入苗族生态旅游的范畴，相辅相成，相得益彰。这不仅可以给当地村民带来一定的经济效益，同时也会积极地促进乡民对本民族传习久远的传统技艺文化有新的情感和理解，增强自觉地保护包括手工纸在内的苗族传统技艺的意识。

⊙2
湾水镇苗族村民的酒席
Miao people's feast in Wanshui Town

凯里苗族竹纸

Bamboo Paper by the Miao Ethnic Group in Kaili City

湾水镇竹纸透光摄影图

A photo of bamboo paper in Wanshui Town seen through the light

第五节

丹寨苗族皮纸

贵州省
Guizhou Province

黔东南苗族侗族自治州
Qiandongnan Miao and Dong Autonomous Prefecture

丹寨县
Danzhai County

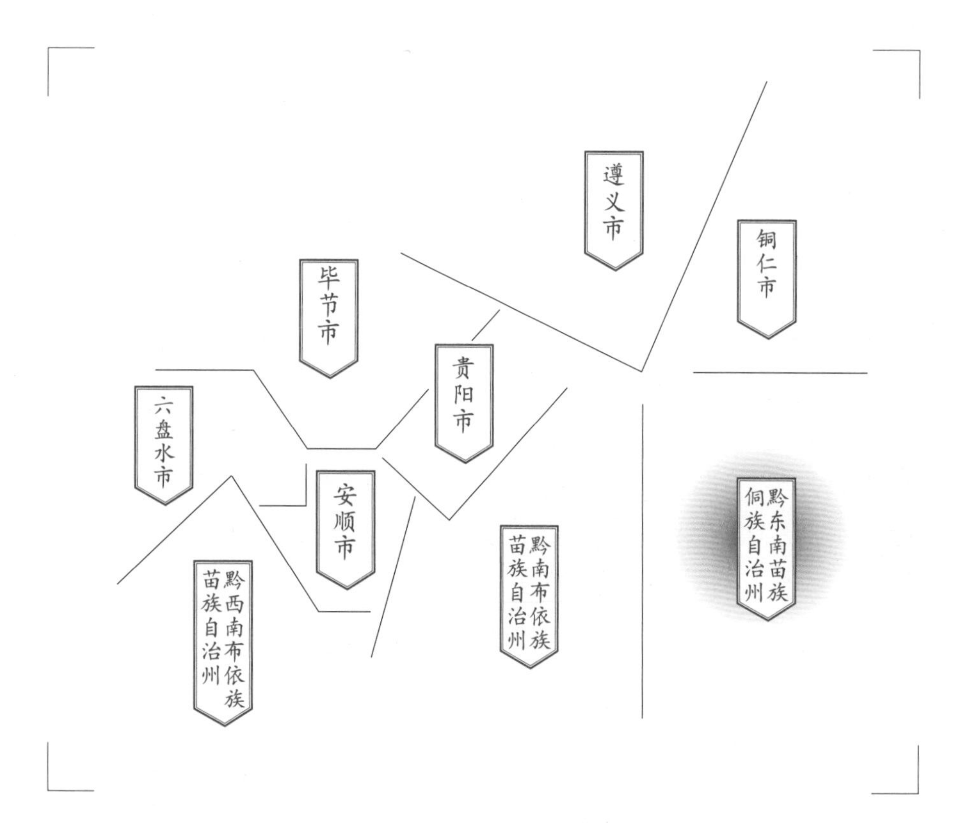

调查对象

南皋乡石桥行政村苗族皮纸

Section 5
Bast Paper by the Miao Ethnic Group in Danzhai County

Subject

Bast Paper by the Miao Ethnic Group in Shiqiao Administrative Village of Nangao Town

一 丹寨苗族皮纸的基础信息及分布

1 Basic Information and Distribution of Bast Paper by the Miao Ethnic Group in Danzhai County

丹寨县丰富的造纸原料——构树皮和优质的水源，为手工造纸提供了很好的条件。据石桥皮纸研究者文朝汉等人的研究，20世纪20年代末，都匀的罗培清等三名汉族造纸师傅在石桥村天然石桥对面的大岩脚石壁下设槽造纸，打开销路后，又去都匀请了一些造纸师傅来扩大生产。从此，石桥村的手工造纸业快速发展起来。各造纸户都招收一些学徒或帮工，其中大多数是苗族人，部分苗族人学会造纸后，开始自己建槽生产白皮纸。1949年，生产白皮纸的苗族村民已占半数以上，且其比例持续增大[2]。

丹寨县境内手工造纸的品种有书画纸、捆钞纸、竹料冥纸等，其中以构树皮制作的皮纸最为著名，主要产于石桥村。石桥皮纸的特点是洁白、细腻、折叠性好、拉力大，为民间广泛使用，近代军工业上也有它的用处。石桥白皮纸厂曾于1978年派人到安徽泾县学习宣纸生产技术，以构皮麻掺龙须草为原料，试图制作国画用纸，后经贵州省工艺美术公司鉴定，产品质量合格，石桥白皮纸厂成功生产出了适宜书画用途的国画纸。1982年，石桥的苗族造纸工人杨大文被推荐随中国传统技术展览团出国，在加拿大表演抄纸技艺，受到国际友人的欢迎和好评[3]。2006年，丹寨石桥的古法造纸技艺入选中国第一批非物质文化遗产保护名录。

⊙1

⊙2

⊙1 石桥山崖下的造纸作坊 Papermaking mill at the foot of mountain in Shiqiao Village

⊙2 杨大文于加拿大抄纸展演旧照 Photo of papermaking show performed by Yang Dawen in Canada

[2] 文朝汉．石桥白皮纸的民族工艺文化[M]// 潘光华，龙从汉．贵州民间工艺研究．北京：中国民族摄影艺术出版社，1991:344.

[3] 文朝汉．石桥白皮纸的民族工艺文化[M]// 潘光华，龙从汉．贵州民间工艺研究．北京：中国民族摄影艺术出版社，1991:347.

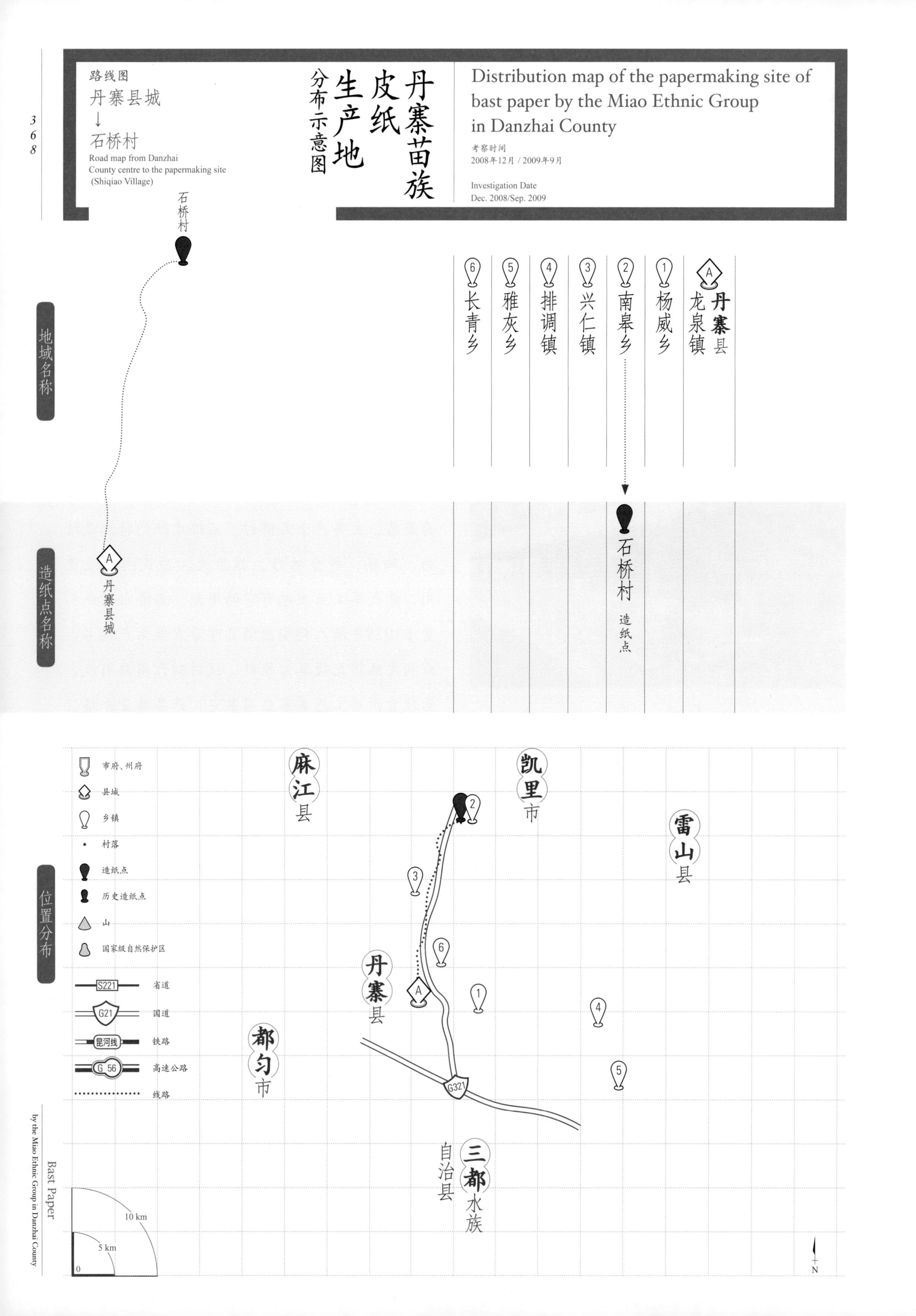

路线图
丹寨县城
↓
石桥村
Road map from Danzhai County centre to the papermaking site (Shiqiao Village)
丹寨苗族皮纸生产地分布示意图
Distribution map of the papermaking site of bast paper by the Miao Ethnic Group in Danzhai County
考察时间
2008年12月 / 2009年9月
Investigation Date
Dec. 2008/Sep. 2009
地域名称
造纸点名称
位置分布
石桥村
丹寨县城
A 丹寨县 龙泉镇
1 杨威乡
2 南皋乡
3 兴仁镇
4 排调镇
5 雅灰乡
6 长青乡
石桥村 造纸点
市府、州府
县城
乡镇
村落
造纸点
历史造纸点
山
国家级自然保护区
S221 省道
G21 国道
昆河线 铁路
G56 高速公路
线路
麻江县
凯里市
雷山县
丹寨县
都匀市
G321
三都水族自治县
10 km
5 km
0
N

二
丹寨苗族皮纸生产的人文地理环境

2
The Cultural and Geographic Environment of Bast Paper by the Miao Ethnic Group in Danzhai County

丹寨县为贵州省黔东南苗族侗族自治州辖县，地处贵州省东南部，东与雷山县接壤，南靠三都水族自治县，西与都匀市、麻江县交界，北抵凯里市。丹寨县城距黔桂铁路都匀站50 km，距湘黔铁路凯里站70 km，距西南出海通道贵新高等级公路54 km，距省会贵阳市210 km。

其地域隋朝属牂牁郡，唐朝属黔中道应州地，宋朝为夔州路绍庆府所领56羁縻州南部东段地带，元朝始有正式建置。清康熙十一年（1672年），改隶都匀府，雍正七年十二月（1730年1月），朝廷准建八寨厅，厅治在今老八寨。八寨厅建制一直沿至民国初年。1949年12月，建立丹寨县人民政府，隶属于独山专区，1952年11月改称都匀专区。

⊙1

⊙1 丹寨县的苗族斗牛盛会 Miao Ethnic Bullfighting Pageant in Danzhai County

调查时丹寨县辖3个镇、4个乡、1个国有农场，2006年12月1日，全县总人口为16.01万，县境内多民族聚居，有苗、汉、水、布依等18个民族，其中苗族人口占总人口的85.6%，是典型的苗族聚居县。

南皋乡位于丹寨县北部，地处清水江上游，东接雷山县，南邻本县兴仁镇，西与麻江县宣威镇隔清水江相望，北靠凯里市，乡政府驻地南皋村距丹寨县城37 km，距州府凯里36 km。全乡辖12个行政村，45个自然村寨，88个村民组，共2 949 户12 224 人，其中少数民族占全乡总人口的97%，是一个以苗族为主的少数民族聚居区。

南皋乡旅游资源丰富，乡村旅游发展方兴未艾，境内有乌高河和南皋河两条河流，两河在南皋村交汇，沿公路蜿蜒而行，流入清水江。公路沿线奇山秀水，风景秀丽，其中的穿洞全长约4 km，洞口宽约15 m、高约13 m，俨然一个大客厅，洞内有千姿百态的钟乳石及长年水流不断的暗河，大自然的鬼斧神工让这个天然溶洞充满了神奇的魅力。穿洞附近还有河流穿梭而过的天然石桥，石桥村正因此而得名。具有神话传说的大簸箕银子洞及其边上的壁画、年代久远的石桥诗刻、石桥皮纸手工作坊等都极具观赏和研究价值。同时，南皋浓郁的民族风情和纯手工制作且雍容华贵的苗族服饰，古老的芦笙舞及热闹非凡的望会节和翻鼓节等，吸引了众多的外来游客。南皋乡也因石桥皮纸而成为“中国古法造纸艺术之乡”。

石桥村位于丹寨县北部，距县城35 km，沅江上游的一条支流南皋河流经村旁，一座巨大的天然石桥横跨在河上，天然石桥对面的大岩脚石壁下，是石桥皮纸作坊旧址。

⊙1

⊙2

⊙1 天然石桥 Natural stone bridge

⊙2 石桥村入口介绍牌 Introduction to Shiqiao Village

三
丹寨苗族皮纸的历史与传承

3
History and Inheritance of Bast Paper by the Miao Ethnic Group in Danzhai City

近年来的一种流行说法是，丹寨手工纸是一种唐朝工艺，有着千年历史。这广泛反映在相关文章、网页及近年的申遗材料上。

其依据主要有：

(1) 水车形制和唐朝类似；

(2) 工艺和《天工开物》一书中描述的内容类似。

调查组通过先后数次进入石桥村进行考察所获得的信息及相关文献研究，认为丹寨手工造皮纸的历史可能并不长。

据文朝汉的研究观点，丹寨皮纸的生产始于20世纪20年代末。辛亥革命后，原本偏远的丹寨一带经济、文化获得一定发展，白皮纸用量日益增多，而当地市场上的白皮纸供不应求，人们只能去都匀等地买，有时还不一定能买到。当时家住石桥堡的汉族人杨仿文任八寨（丹寨）县第四区（南皋）区长，常常为买不到纸影响工作而烦恼。一次他本人亲自到都匀买纸时，看到生产白皮纸所用原料是构皮麻和岩杉树根（杉松树根），回来后组织人调查发现，南皋及周边山里有丰富的野生构树和岩杉树。因此，杨仿文便到都匀请来罗培清等三名汉族造纸师傅，在天然石桥对面的大岩脚石壁下建槽造纸，造纸获得成功后，所产白皮纸供不应求，于是又去都匀请师傅，扩大再生产[4]。

⊙3

⊙3 造纸胜地——大岩脚石壁 Papermaking scenic spot named Dayanjiao Cliff

调查组2009年前到当地进行过调查，并与相关专家做了进一步交流，对上述观点也有一定佐证。如贵州省博物馆梁太鹤先生20世纪80年代曾到石桥调查白皮纸，当时了解的情况也是白皮纸的生产始于20世纪20年代末。原都匀皮纸厂副厂

[4] 文朝汉.石桥白皮纸的民族工艺文化[M]//潘光华,龙从汉.贵州民间工艺研究.北京:中国民族摄影艺术出版社,1991:344.

长肖明远在调查组调查都匀皮纸历史时，也提到民国年间都匀人曾到丹寨去造白皮纸。

此外，成书于1931年的《八寨县志稿》记载：“楮，亦名榖，俗呼构皮麻，以其皮如麻，故名。山谷、郊野、园林皆有。本邑三、五两区尤广，商人收买运匀换皮纸，每年不下数千担，惜无人提倡造纸，听利外溢。树高丈余，叶似桑而燥，花雌雄异株，实如弹，丸入药，捣其皮为白纸，最细致而绵。”[5]可见旧县志明确认为八寨县，即现丹寨县原无造纸记述，虽然县志编撰者在介绍楮（构皮麻）时，提到“捣其皮为白纸，最细致而绵”，也就是说他们知道构皮麻可用于造纸，且知道“山谷、郊野、园林皆有”，甚至“本邑三、五两区尤广”，但遗憾的是“商人收买运匀换皮纸，每年不下数千担，惜无人提倡造纸，听利外溢”。“匀”在此是指都匀，这与都匀当时皮纸质量好、影响大是相吻合的，且都匀和丹寨相距不远。

丹寨皮纸和都匀皮纸之间的源流关系清楚地表明在《八寨县志稿》撰写时，县里还没有造纸。如完全按照《八寨县志稿》记载，不考虑资料的滞后效应等因素，则丹寨县造纸历史或许不会早于1931年；考虑到编撰县志有一个过程，资料可能会有一点滞后，有可能在《八寨县志稿》出版前一小段时间，即20世纪20年代末，丹寨已经有皮纸生产，这和文朝汉的研究结论是吻合的。

21世纪初，随着石桥皮纸制作技艺被列入第一批国家级非物质文化遗产保护名录，王兴武入选国家级项目代表性传承人，以及2006年亚太旅游协会援助中国贵州丹寨县石桥村“古法造纸”项目实施等一系列加强促进因素的落地，丹寨皮纸的业态在短时间内快速兴旺起来。造纸乡民在地方政府助推下成立了若干造纸专业合作组织，如“丹寨县黔山古法造纸专业合作社”“易兴古法造纸合作社”；国内外的宣传和专项旅游资源也开始流向原先僻远的苗族村寨，如电影《云上太阳》曾获第十七届美国赛多纳国际电影节最佳外语片、最佳摄影、最佳影片提名三项大奖，影片的播映显著提升了石桥造纸技艺的国际影响力。

⊙1
亚太旅游协会援助项目碑
Monument in memory of Pacific Asia Travel Association's assistance to local papermaking program

⊙2
黔山古法造纸合作社
Papermaking cooperative employing Qianshan traditional papermaking techniques

⊙3
易兴古法造纸合作社
Papermaking cooperative employing Yixing traditional papermaking techniques

[5] 王世鑫，等.八寨县志稿[M].民国二十年(1931年)刊本.台北:成文出版社,1968:323.

四

丹寨苗族皮纸的生产工艺与技术分析

4

Papermaking Technique and Technical Analysis of Bast Paper by the Miao Ethnic Group in Danzhai County

调查组曾于2008年12月、2009年9月两次前往丹寨石桥村调查。

(一)

丹寨苗族皮纸的生产原料与辅料

丹寨石桥苗族造皮纸所用原料为构树皮，造纸户认为，嫩构皮尤其是1~2年生的最好；3~5年生的老构皮，皮厚实，产量不高。最早以前用石灰、草木灰作碱，20世纪中叶后改用烧碱。调查时已流行用漂白粉漂白。

丹寨石桥苗族造皮纸需用纸药，以前用杉根，调查时流行用聚丙烯酰胺。

(二)

丹寨苗族皮纸的生产工艺流程

经现场考察与同造纸户的多轮访谈，总结丹寨石桥苗族造皮纸的工艺流程如下：

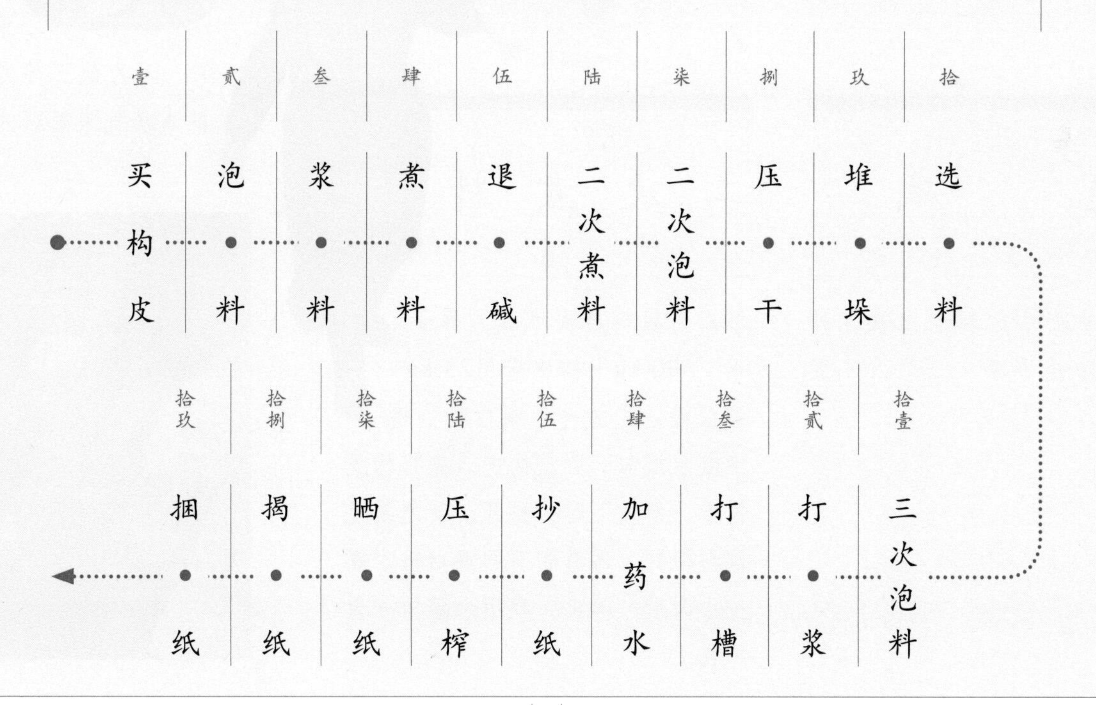

壹 买构皮

1

造纸户到附近乡镇买构皮，老嫩均可，但嫩构皮更好。

贰 泡料

2　⊙1⊙2

将买来的构皮泡在河里，夏天泡2~3天，冬天泡5~7天。泡之前先将构皮捆成捆，每捆大约4 kg。

⊙1

⊙2

叁 浆料

3　⊙3

将泡好的构皮取出来，沥干后，用烧碱水浆料，每140 kg干构皮需10 kg烧碱。

⊙3

肆 煮料

4　⊙4⊙5

用煤生火，将浆好的料整齐地堆放在甑子里，铺满整个甑子后，用脚踩实，踩时需注意不要让水喷出来伤到人，踩实后用塑料布盖紧，煮一天一夜。虽然说是煮料，但严格来说其实是蒸料，因为此工序是用甑桥下的水蒸气来蒸料的。

⊙4

伍 退碱

5

用料钩将煮好后的料取出来，并放到河里冲洗干净，三个人一天即可洗完。

陆 二次煮料

6

第二天将料再次堆放到甑子里，500 kg干构皮需用25 kg漂白粉。先放一半漂白粉，将半甑水搅匀，后放构皮（先将构皮拧干，但放下时要松开，使之易吸漂白粉），再放剩下的漂白粉，煮一小时后，停火，利用余温捂一天一夜。

⊙5

⊙1/2 河水中泡料
Soaking the papermaking materials in a river

⊙3 浆料
Fermenting the papermaking materials

⊙4 挑煤
Carrying the coal to heat papermaking materials

⊙5 煮料
Boiling the papermaking materials

柒 二次泡料
7

将料再放到河里泡一天，将漂白粉等冲洗干净。

捌 压干
8

将料捞起来，挤干并抬回家。

玖 堆垛
9

将压干后的料堆放在铺满鹅卵石的地上，这样料里残存的少量的水可以通过鹅卵石的缝隙流走。抄纸时，用多少料就取多少料。

拾 选料
10 ⊙6

把竹料上的结、黑斑点等杂质去掉。

⊙6

拾壹 三次泡料
11 ⊙7

将选好的料再拿到河里去泡，一般泡一晚即可。

⊙7

拾贰 打浆
12 ⊙8⊙9

将料放到打浆机里打浆，一次打半小时。

⊙8

拾叁 打槽
13

将挤干的料放在槽里，用槽棍将料打匀，几分钟即可，一槽一般能抄100张纸左右。现在有些造纸户用电钻来打。

⊙9

⊙6 选料 Picking out the impurities

⊙7 泡料 Soaking the papermaking materials

⊙8/9 打浆 Beating the papermaking materials with a machine

拾肆

加 药 水

14

凭个人经验，加一定量的化学药水，加一次料后就加一次药水。药水量要恰到好处，这样抄出来的纸光滑细腻且没有气泡。否则，若药水加多了，由于太滑腻，纤维结合不紧，提不起来，不能成纸；若药水加少了，纤维太多太重，在纸帘上分布不均匀，容易起棉花点。

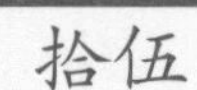

拾伍

抄 纸

15 ⊙10⊙11

现在都用吊帘，将帘架杆推开，前后振荡2~3次，即得一张纸。随后将纸帘从帘架上取下来，转身盖于湿纸垛上。

⊙10

⊙11

⊙12

⊙13

拾陆

压 榨

16 ⊙12~⊙14

抄完纸后，先在湿纸垛上盖一张废旧纸帘，再盖盖板、码子，用木榨缓慢压榨1~2小时，静置一夜，让水缓慢流完，第二天早上松榨。如压榨太快，则纸垛受力不均匀，容易爆掉。

⊙15

⊙16

⊙14

拾柒

晒 纸

17 ⊙15~⊙17

将纸贴在纸焙上，一排贴10张，贴5排，每两排间错开1 cm左右，即一面墙一次可晒50张，一个纸焙一次可晒100张。一般晒2小时左右即干。如果温度高或在晚上，可多贴几排。

⊙17

⊙10/11 抄纸 Scooping and lifting the papermaking screen out of water and turning it upside down on the board

⊙12/14 压榨 Pressing the paper

⊙15 揭纸角 Peeling the paper down from the upper corner

⊙16 撕纸 Peeling the paper down

⊙17 晒纸 Drying the paper

拾捌
揭　纸

18　⊙18⊙19

从左上角往右下角揭，也可从右上角开始揭，揭完一面墙50张，恰好为一刀。

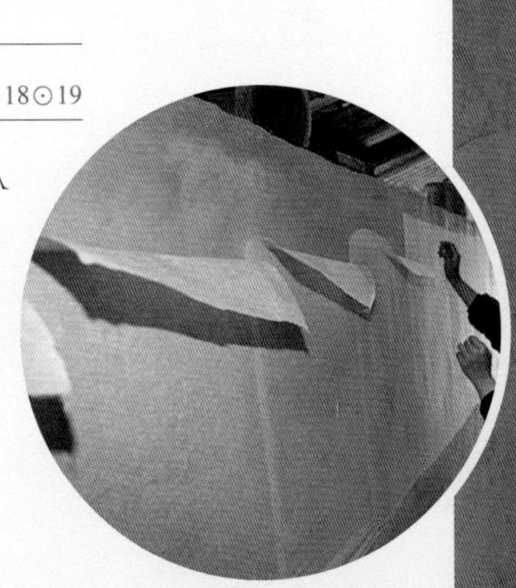

⊙18　⊙19

拾玖
捆　纸

19　⊙20⊙21

10刀为一捆，先用一张“刮草料”包好，再用纸绳捆绑。

⊙ 20

⊙21

调查时发现，现代工艺相比传统手工皮纸工艺已经做了一些改变，其区别主要如下：

壹
浆　料

1

传统工艺为将泡好的构皮取出来，待水滤干后浆石灰，100 kg构皮需40 kg石灰。

贰
煮　料

2

传统工艺为将料整齐堆放在甑子里，铺满一层，立即用脚踩实，接着再铺，铺满整个甑子后用稻草盖紧，纸厂成立后开始用塑料薄膜盖。一锅可放约150 kg构皮，用柴烧，煮了两天后，翻过来，再煮两天。

叁
退　碱

3

传统工艺是先将料放河里泡1~2天，再将石灰冲洗掉。

⊙18/19 揭纸 Peeling the paper down

⊙20/21 捆纸 Binding the paper

肆
二次煮料
4

传统工艺是先将料过一遍草木灰水，再放到甑子里，用同样方式煮一天一夜。

伍
选料
5

传统工艺需选料，把结、黑斑点等杂质去掉。现代工艺不需选料。

陆
舂料
6

传统工艺需用脚碓舂料，一人舂，一人转，一般舂一个晚上可满足一人一天抄纸所需，大约500张。现代工艺改成机器打浆。

柒
冲洗料
7

传统工艺在舂料后，还需冲洗料，即将料装在布袋里，用顶部为圆盘的棒不断上下捅，直到将里面的污物清洗干净，一天最多可洗8个料粑（即8袋），一个料粑可抄100张纸，洗干净后，把水挤干，即可倒入纸槽。现代工艺不需冲洗料。

捌
打槽
8

传统工艺用槽棍打槽，大约需半小时。现代工艺也有用槽棍打槽的，但只需几分钟，也有用电钻打槽的。

玖
加药水
9

传统工艺都加杉根泡的药水，现代工艺则用化学药水聚丙烯酰胺。

拾
抄纸
10

传统工艺采用双人抬帘的抄纸方式，师傅抬帘，徒弟帮帘。抄纸时，往左抄一次，往右抄一次，即得一张纸。左右方向的先后顺序根据个人而定。现代工艺采用单人吊帘。

拾壹
晒纸
11

传统工艺都用纸焙，用火烤。现代工艺则利用天气好时中午阳光强烈的特点直接晒，省钱。

此外，访谈时据原石桥纸厂三孔桥分厂的潘登海口述，当时黔东南州轻工业局派毛兴荣到安徽考察宣纸制作工艺，考察回来后石桥纸厂于1978年开始试产国画纸。

据潘登海等口述，国画纸和传统白皮纸主要有以下不同之处：

原料：国画纸采用构皮麻

和梭草（龙须草）的混合原料。国画纸原料配比很有讲究，曾尝试过构皮麻与梭草按照80：20的比例混合，但吸水性差，墨点呈锯齿状；70：30的比例也不行；60：40的比例造出来的国画纸质量最好，墨滴下去扩散均匀，充分利用了构皮有拉力、梭草吸水性强的特点。

据潘登海等口述，石桥纸厂认为安徽宣纸第一，石桥国画纸第二。

21世纪20年代，石桥造纸引来众多国内外游客和文化研究者观赏和研究，从而打开了较为广阔的国内外市场，石桥造纸也逐渐出现很多新拓展和适应消费的新尝试，如近年来也采用了浇纸法制作草纸，同时还生产了各种色纸、麻丝纸、皱纹纸等新品种纸，有品种较为丰富的纸工艺品。

⊙ 22

⊙ 23

⊙ 22
村民自豪加自负的造纸标语
Couplets showing the local papermakers' pride and arrogance of their papermaking techniques

⊙ 23
浇纸法新品种
New paper type employing the fixed screen

(三)

丹寨苗族皮纸的性能分析

1. 丹寨白皮纸的性能分析

所测丹寨石桥村白皮纸的相关性能参数见表10.5。

表10.5 石桥村白皮纸的相关性能参数
Table 10.5 Performance parameters of white bast paper in Shiqiao Village

指标		单位	最大值	最小值	平均值
厚度		mm	0.130	0.080	0.098
定量		g/m²	—	—	27.8
紧度		g/cm³	—	—	0.284
抗张力	纵向	N	34.3	28.9	31.2
	横向	N	17.9	15.4	16.5
抗张强度		kN/m	—	—	1.590
白度		%	56.2	55.6	55.9
纤维长度		mm	9.16	0.82	3.22
纤维宽度		μm	34.0	3.0	13.0

由表10.5可知，所测石桥村白皮纸最厚约是最薄的1.63倍，相对标准偏差为1.70%，纸张厚薄差异较小。白皮纸的平均定量为27.8 g/m^2。所测白皮纸紧度为0.284 g/cm^3。

经计算，其抗张强度为1.590 kN/m，抗张强度值较大。

所测石桥村白皮纸白度平均值为55.9%，白度较高，白度最大值约是最小值的1.01倍，相对标准偏差为0.18%，差别较小。

所测石桥村白皮纸纤维长度：最长9.16 mm，最短0.82 mm，平均3.22 mm；纤维宽度：最宽34.0 μm，最窄3.0 μm，平均13.0 μm。丹寨白皮纸在10倍、20倍物镜下观测的纤维形态分别见图★1、图★2。

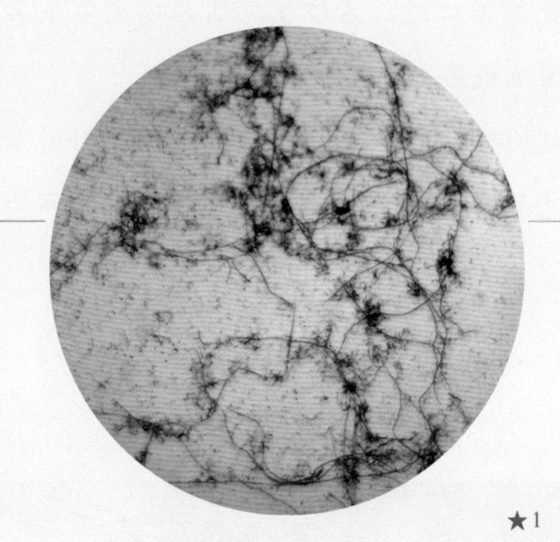

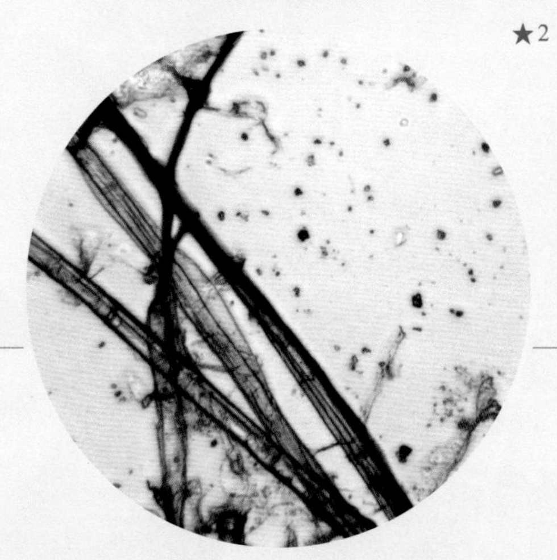

★1 石桥村白皮纸纤维形态图(10×) Fibers of white bast paper in Shiqiao Village (10× objective)

★2 石桥村白皮纸纤维形态图(20×) Fibers of white bast paper in Shiqiao Village (20× objective)

2. 丹寨花纸的性能分析

所测丹寨石桥村花纸的相关性能参数见表10.6。

表10.6 石桥村花纸的相关性能参数
Table 10.6 Performance parameters of flower paper in Shiqiao Village

指标		单位	最大值	最小值	平均值
厚度		mm	0.080	0.060	0.075
定量		g/m^2	—	—	14.6
紧度		g/cm^3	—	—	0.195
抗张力	纵向	N	15.1	10.2	12.6
	横向	N	13.4	10.0	11.8
抗张强度		kN/m	—	—	0.813
白度		%	70.0	67.7	69.0
纤维长度		mm	10.23	1.48	4.98
纤维宽度		μm	33.0	5.0	16.0

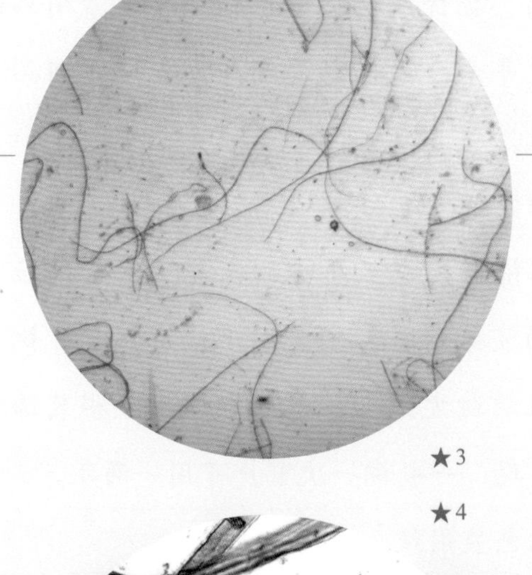

★3

★4

由表10.6可知，所测石桥村花纸最厚约是最薄的1.33倍，相对标准偏差为0.70%，纸张厚薄较为均匀。花纸的平均定量为14.6 g/m^2。所测花纸紧度为0.195 g/cm^3。

经计算，其抗张强度为0.813 kN/m，抗张强度值较小。

所测石桥村花纸白度平均值为69.0%，白度较高，白度最大值约是最小值的1.03倍，相对标准偏差为0.68%，差别较小。

所测石桥村花纸纤维长度：最长10.23 mm，最短1.48 mm，平均4.98 mm；纤维宽度：最宽33.0 μm，最窄5.0 μm，平均16.0 μm。所测花纸在10倍、20倍物镜下观测的纤维形态分别见图★3、图★4。

★3 石桥村花纸纤维形态图(10×)
Fibers of flower paper in Shiqiao Village (10× objective)

★4 石桥村花纸纤维形态图(20×)
Fibers of flower paper in Shiqiao Village (20× objective)

五
丹寨苗族皮纸的用途与销售情况

5
Uses and Sales of Bast Paper by the Miao Ethnic Group in Danzhai County

⊙1

(一) 丹寨苗族皮纸的基本用途

1. 制作书画用纸

白皮纸，尤其是传统白皮纸，较适宜书画创作，而曾经生产的由构皮、稻草混合制成的国画纸是较为高端的书画用纸。国画纸、传统白皮纸纸质纯白、细腻、柔韧，吸水性强，具有一定的层次感。据说以前书童蒙帖练字也常用石桥白皮纸。近年来，石桥也逐渐开发出一些新品种的书画用纸。

2. 制作装裱用纸

石桥白皮纸纸质柔韧，具有一定的拉力，是较好的装裱材料，主要用于糊窗户等。糊窗户一般用双层白皮纸，可于其中加上一些装饰物，使之更美观，一般糊一次窗户可用一两年。

3. 制作包装用纸

传统白皮纸主要用来包装茶、糖等日用品，而随着其他包装材料的逐渐普及，这方面用途逐渐萎缩。近年来，随着茶叶，尤其是普洱茶等的兴起，用石桥白皮纸包装茶叶的现象开始复苏。

4. 制作印刷用纸

以前曾用石桥白皮纸印书、发票等，调查时这一用途还未见复活。

5. 制作擦拭机器用纸

石桥白皮纸的纸性较适宜擦拭机器，一则可以抹净机器上的油污，二则不会残留细屑。

6. 制作捆绑用纸

石桥白皮纸拉力大，将双层或多层白皮纸裁小后，可用于捆钞及捆绑其他物品。

7. 制作灯草、爆竹的引线用纸

以前当地煤油灯所用灯草、爆竹的引线往往是将白皮纸裁小，拧细而成。

8. 制作刺绣用纸

旧日苗族女孩子学刺绣，先用白皮纸剪成或者画成特定图案，粘在布面上，然后照图施绣。

⊙1 白皮纸糊窗
Pasting a sheet of white bast paper over a lattice window

⊙2
制作花草纸
Making flower-grass paper

⊙1

此外，缝制衣服、帽子、鞋等，有时也先用白皮纸做成模，然后照模去做。

9. 制作祭祀用纸

这是目前最大的用途，一般逢年过节都需要用到白皮纸，有钱人家用400~500张，一般人家用100~200张。

(二) 丹寨苗族皮纸的销售与价格情况

虽然石桥皮纸从2009年以后开始具有较多用途，也不断开拓出中高端纸品销售市场，但普通造纸户所造的纸主要用于祭祀。如按2008年12月调查组前去调查时的祭祀纸计算，构皮价格为3.6~4元/kg，100 kg构皮可造10捆纸，即5 000张，按3.6元/kg算，买构皮需360元；造10捆纸，需要打2槽浆，共8元；而造1捆纸，需50 kg煤（包括煮、晒），共30元，则造10捆纸需300元；此外，100 kg构皮约需7 kg烧碱，烧碱的价格为4元/kg，共28元；约需0.5 kg纸药聚丙烯酰胺，聚丙烯酰胺的价格为24元/kg，共12元；约需5 kg漂白粉，漂白粉的价格为4元/kg，共20元。因此，100 kg构皮的造纸成本最少为728元。

2008年12月，白皮纸价格为0.25元/张，一般按捆卖，125元/捆，10捆能卖1 250元，净收入为522元。若一个人一天造1捆纸，则平均一天收入为53元，一个月大概造纸25天，有1 325元左右净收入。如人力不够，还需请工人，2008年底调查时，抄纸、晒纸各0.04元/张，如请人，则净收入更少，而一天抄纸5捆的工钱为100元。

买煤的钱约占到成本的30%，因此，调查时有些造纸户晒纸时，不用煤而改用自己砍的柴；另外，若中午阳光强烈，可利用余温和太阳的热量晒纸，这些都是减少支出的手段。

2008~2009年调查时，石桥皮纸仍大量用于祭祀，一般由造纸户在本村旅游观光店销售，或拉到南皋乡、舟溪镇、乌皋乡的集市，以及丹寨县城和凯里、雷山、黄和等地的销售店去卖。时常会有老板上门收购，但由于价格比自销低不少，因此造纸户一般不太愿意直接卖给转手获利的老板。

⊙1 花草纸工艺品 Crafts of flower-grass paper

六 丹寨苗族皮纸的相关民俗与文化事象

6 Folk Customs and Culture of Bast Paper by the Miao Ethnic Group in Danzhai County

(一) 造纸歌

虽然石桥白皮纸最先由汉族人抄造，但白皮纸技术传到石桥后，逐渐被石桥苗族的村民所掌握和发扬，造纸也逐渐成为当地苗族文化的一部分。如苗族民间流传的《造纸歌》《找书找纸歌》，歌唱历史上苗族先民制造出“竹纸”、“绵纸”（白皮纸）及造纸工具、工艺等。由此可见，传统造纸已经融入当地的民族文化中，成为日常生活中的一部分。

《造纸歌》[6]：

一架水车转辘辘，水车带动碓杆响。

碓嘴冲在碓窝上，冲烂皮麻作纸浆。

汉人传来苗家寨，张张皮纸做文章。

还有如下《造纸歌》[7]：

皮麻打成浆，杉根药水渗，和在水缸中，皮粉浓又黏。

这些苗族造纸民歌生动地反映了石桥皮纸的生产工艺和优良品质。

(二) 国际展演

⊙2

⊙2 杨大文于加拿大展演旧照 Photo of Yang Dawen's papermaking show in Canada

1982年9月，中国古代传统技术展演团到加拿大做展演，国内多个省市掌握传统技术的师傅随行展演。纸作为中国四大发明之一，传统造纸技术的展演也在其中，展演者是石桥纸厂的苗族造纸师傅杨大文。据杨大文口述，他于1982年3月14日到加拿大多伦多的科学中心做手工纸技术的展演，展演所需的工具是中国科学技术协会经天津海港通过海运运到加拿大的。

[6] 文朝汉．石桥白皮纸的民族工艺文化 [M]// 潘光华，龙从汉．贵州民间工艺研究．北京：中国民族摄影艺术出版社，1991:343.

[7] 文朝汉．石桥白皮纸的民族工艺文化 [M]// 潘光华，龙从汉．贵州民间工艺研究．北京：中国民族摄影艺术出版社，1991:347.

(三)
祭蔡伦

调查时，石桥村有祭蔡伦的习俗，其地点在始造皮纸的大岩脚石壁纸坊群，祭祀蔡伦的木牌放在凿出的天然石龛中，但旧俗详情未能获悉。

⊙1

⊙2

⊙3

⊙1
祭蔡伦的手写牌位
Handwritten tablet in memory of Cai Lun

⊙2/3
各种新的纸制品
Various new paper products

七
丹寨苗族皮纸的保护现状与发展思考

7
Preservation and Development of Bast Paper by the Miao Ethnic Group in Danzhai County

(一)
深入挖掘丹寨石桥苗族皮纸的历史与文化

目前关于丹寨石桥皮纸的各种宣传介绍，无一例外都称其为千年传统工艺或唐朝工艺，前已述及，石桥皮纸的信史或许没有那么长。当然，历史是否悠久并不是最关键的，也不需要为了配合旅游进行包装渲染。石桥皮纸虽然历史较短，但确有相当丰富的历史和文化内涵值得深入挖掘、整理。

丹寨石桥皮纸的起源有待更深入的挖掘和整理，同时也可对地方志书里提到的丹寨（八寨）此前不造纸的原因进行更深入的分析和研究。

丹寨石桥皮纸的发展历程也值得深入挖掘和整理，尤其是像杨大文随中国古代传统技术展演团到加拿大做展演的经历，应该值得在中国现代

⊙4

手工纸历史上留下印迹。

苗族工人学会造纸技术后，逐渐将其融入民族文化中，成为能歌善舞的苗族文化的一部分，这也非常值得深入挖掘、整理和研究。像苗族民间流传的《造纸歌》《找书找纸歌》等，在中国手工造纸文化中颇具特色，目前研究界了解到的还只是一小部分。

⊙5

(二)
开发更多具有民族特色和地方特色的纸工艺品

石桥因为造纸文化而吸引国内外众多游客前来参观旅游，从而创立了很多品牌。与之相应的是，石桥本村造纸户和外地入村的纸艺设计师自行设计或引进了一些纸工艺品，他们新推出的纸工艺品已有部分体现出民族特色和地方特色。如能在民族特色和地方特色上做成系列文章，则将会带来更大的市场和利润空间。

⊙6

(三)
恢复传统工艺，生产传统国画纸

虽然目前除了销量最广的祭祀用的普通白皮纸外，石桥一些造纸户也开发了多种书画纸、装饰用纸等，然而目前绝大多数品种的纸采用的是现代工艺，即采用了化学药品烧碱、漂白粉，以及机器打浆等工艺，对构皮纤维损伤较大。这些纸品种的细腻性、柔韧性等均不如传统白皮纸，更不如国画纸。

⊙4 国外游客参观抄纸过程
Foreign tourists watching papermaking show

⊙5 国内外游客观纸留言
Notes left by tourists from China and abroad

⊙6 中国石桥古纸上海书画名家笔会
Exhibition of paintings and calligraphy on bast paper in Shiqiao Village was held in Shanghai City

石桥部分造纸户已经逐渐开始恢复传统工艺，生产传统国画纸。应该说，这将极大地拓宽石桥皮纸的销售市场，同时还会产生较高的附加值，而且传统国画纸的市场需求非常大，这将会使整个石桥皮纸产生质的变化。

纸毯

Paper Blanket
by the Miao Ethnic Group
in Danzhai County

石桥村纸毯透光摄影图
A photo of paper blanket in Shiqiao Village seen through the light

丹寨苗族花纸

Flower Paper by the Miao Ethnic Group in Danzhai County

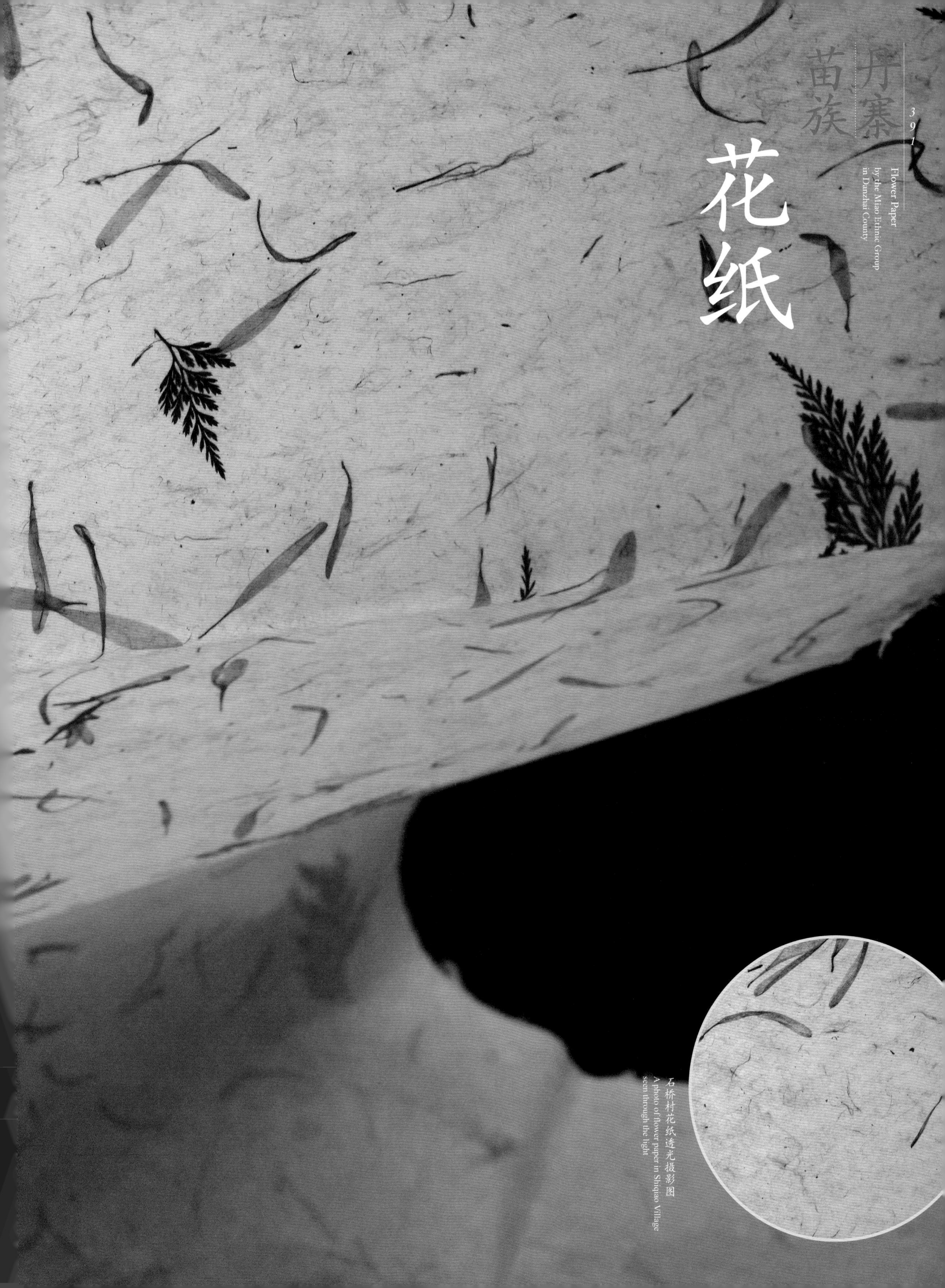

石桥村花纸透光摄影图

A photo of flower paper in Shiqiao Village seen through the light

丹寨苗族染色纸

Dyed Paper by the Miao Ethnic Group in Danzhai County

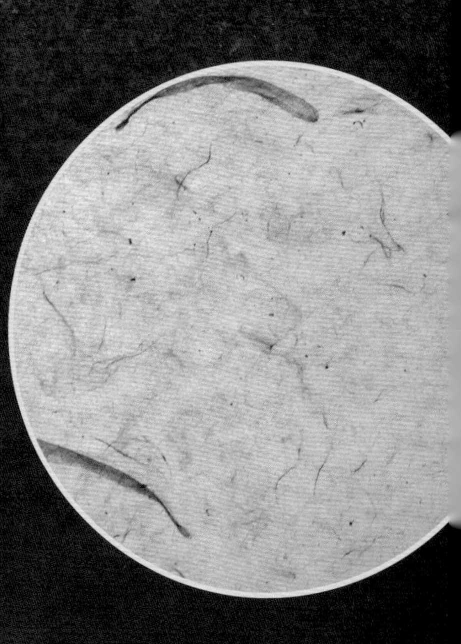

石桥村染色纸透光摄影图
A photo of dyed paper in Shiqiao Village seen through the light

丹寨苗族白皮纸

White Bast Paper by the Miao Ethnic Group in Danzhai County

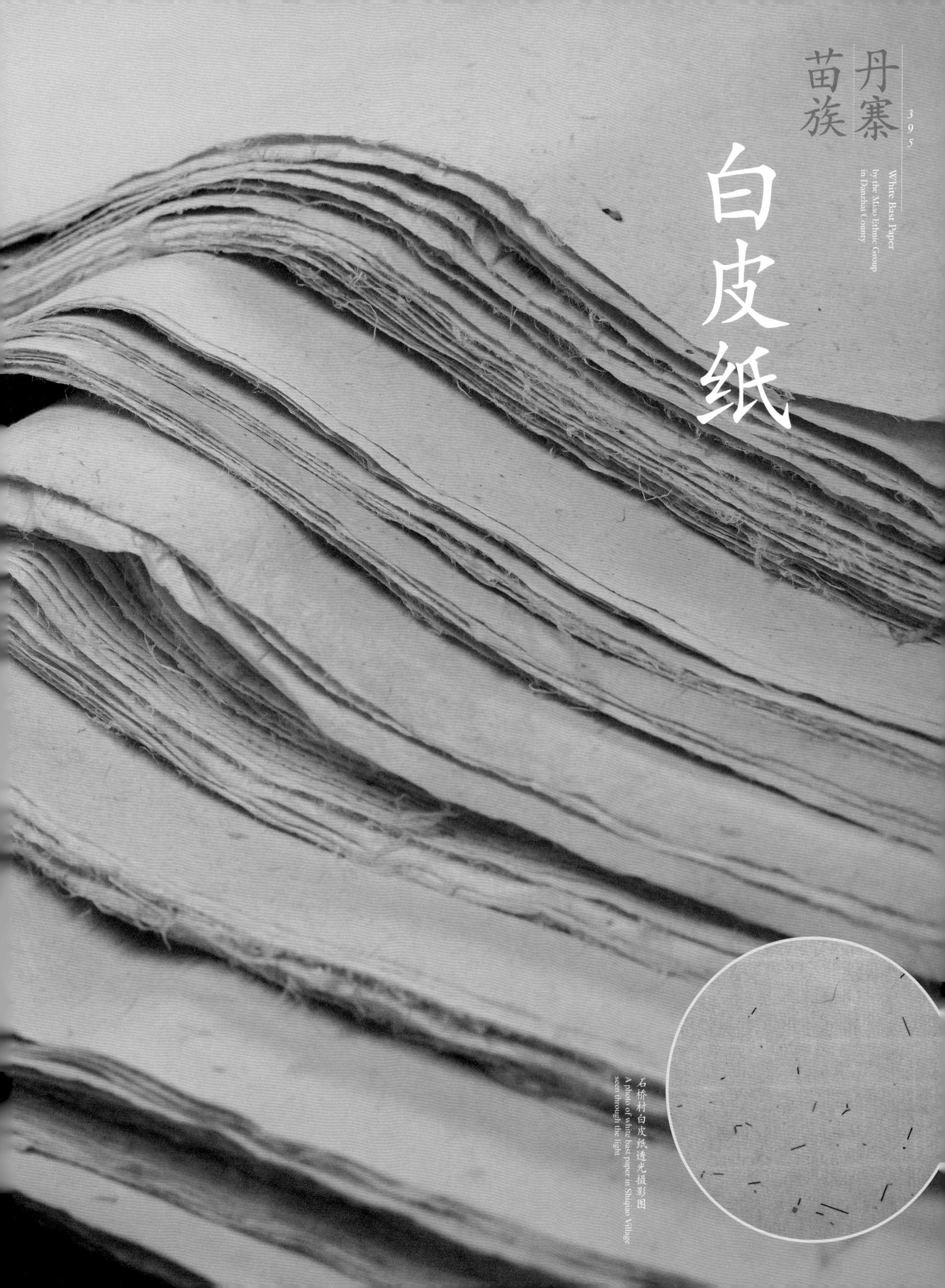

石桥村白皮纸透光摄影图

A photo of white bast paper in Shiqiao Village seen through the light

丹寨苗族书画纸

Calligraphy and Painting Paper by the Miao Ethnic Group in Danzhai County

石桥村书画纸透光摄影图

A photo of calligraphy and painting paper in Shiqiao Village seen through the light

第六节

榕江侗族皮纸

贵州省
Guizhou Province

黔东南苗族侗族自治州
Qiandongnan Miao and Dong Autonomous Prefecture

榕江县
Rongjiang County

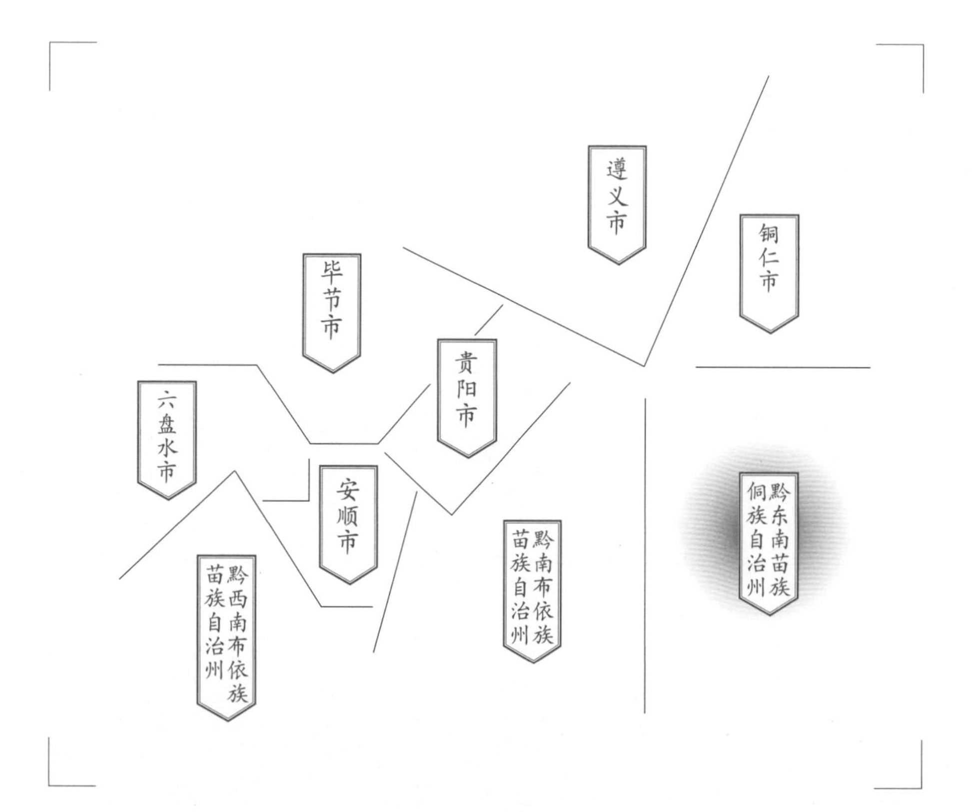

调查对象
乐里镇本里行政村
计划乡九秋行政村
侗族皮纸

Section 6
Bast Paper by the Dong Ethnic Group in Rongjiang County

Subject

Bast Paper by the Dong Ethnic Group in Jiuqiu Administrative Village of Jihua Town, Benli Administrative Village of Leli Town

一
榕江侗族皮纸的基础信息及分布

1
Basic Information and Distribution of Bast Paper by the Dong Ethnic Group in Rongjiang County

2011年7月16~18日调查组进入榕江县的山寨里田野访谈时，获得的信息是县域内还有着手工造纸工艺良好的活态生产和百姓生活消费需求。在距榕江县城约83 km的乐里镇本里村及其临近的计划乡九秋村所产皮纸就是现代榕江侗族手工造纸的典型传习样本。这种以构树皮为原料、采用传统浇纸法、纯手工制作的造纸工艺，在清末民国初年时传入本里和九秋侗寨，后一直流传至今。

乐里镇本里村及其相邻的计划乡九秋村制作的构皮纸洁白美观，坚韧耐用，放上百年也不易被虫蛀，在苗村侗寨的延伸用途很广，实用价值也很高：制作斗笠与纸伞（二者都需上桐油）以及刺绣用的剪纸、针线包，书写，包装衣服或贵重物品。在侗族山寨至今仍有着广阔的市场需求。

⊙1

⊙2

⊙1
皮纸制作的『时髦』针线包
"Modish" sewing kit wrappers made of bast paper

⊙2
洁白漂亮的本里皮纸
White and smooth bast paper in Benli Village

榕江侗族皮纸生产地分布示意图

Distribution map of the papermaking sites of bast paper by the Dong Ethnic Group in Rongjiang County

考察时间
2011年7月

Investigation Date
July 2011

路线图
榕江县城
↓
九秋村、本里村
Road map from Rongjiang County centre to the papermaking sites (Jiuqiu Village and Benli Village)

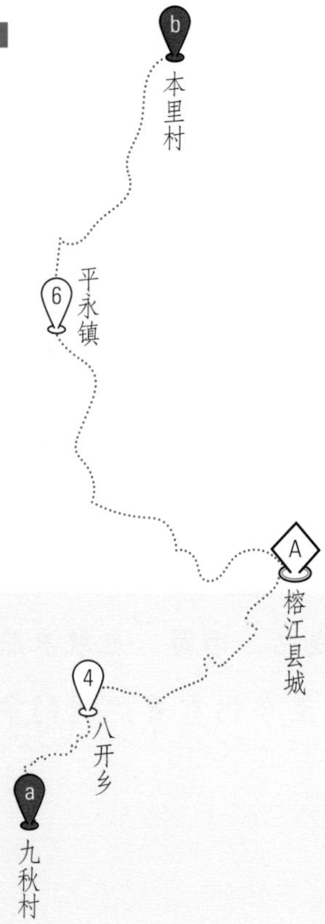

	A	①	②	③	④	⑤	⑥
地域名称	榕江县 古川镇	计划乡	乐里镇	寨蒿镇	八开乡	朗洞镇	平永镇
造纸点名称		a 九秋村 造纸点	b 本里村 造纸点				

位置分布

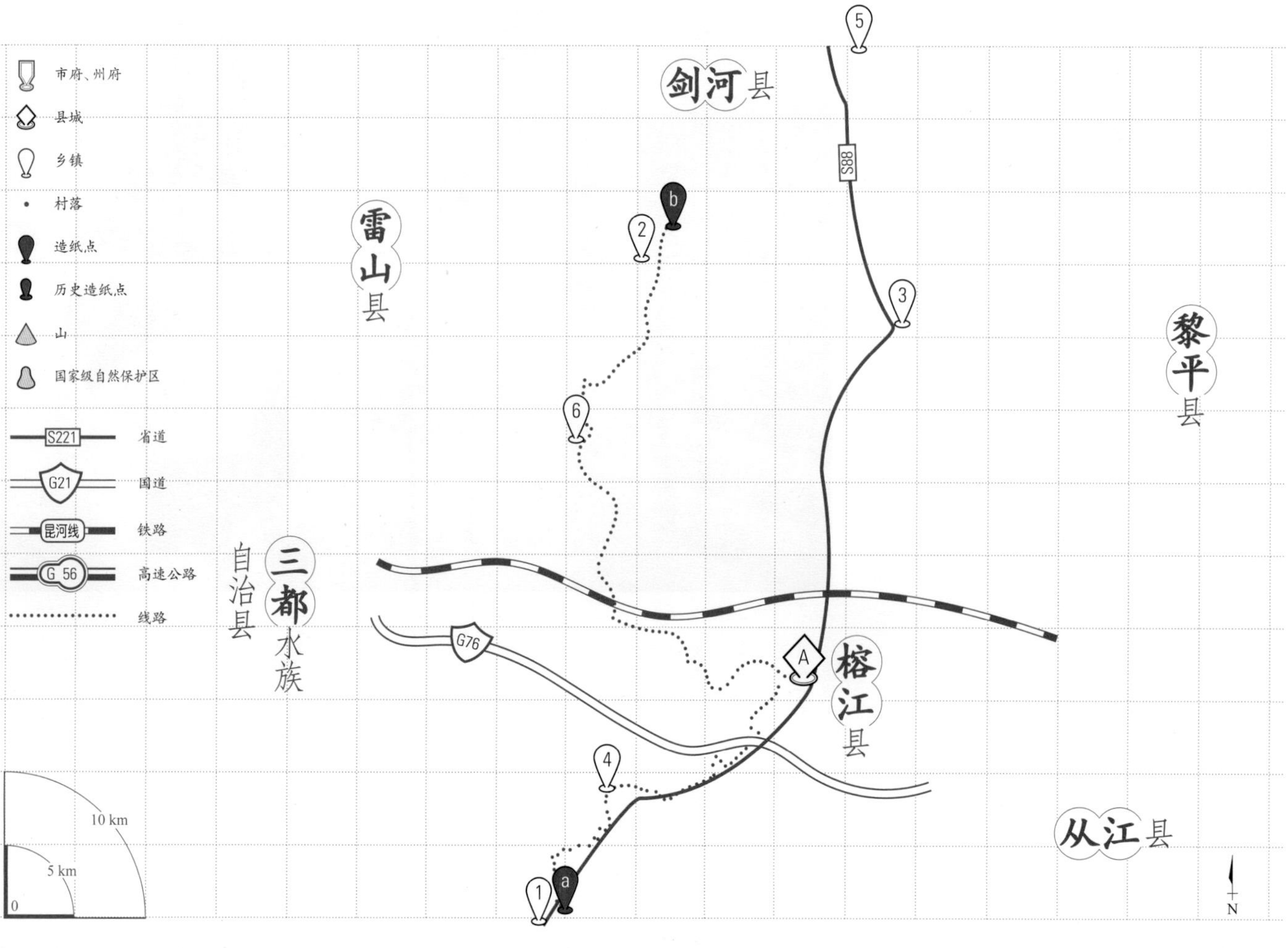

二 榕江侗族皮纸生产的人文地理环境

2 The Cultural and Geographic Environment of Bast Paper by the Dong Ethnic Group in Rongjiang County

榕江县位于贵州省东南部，黔东南苗族侗族自治州南部，总面积3 315 km²。据2010年第六次全国人口普查的资料，榕江县辖6个镇、13个乡（其中5个民族乡），总人口35万，其中少数民族人口占总人口的84.4%。

榕江旧称古州，春秋战国时期隶属楚国，后长期由土官实际管理。元至元二十年（1283年）置古州八万洞总管府。明洪武三年（1370年）设古州蛮夷长官司。清雍正八年（1730年）置古州厅，隶属黎平府，设流官，始为中央王朝直接治理。民国二年（1913年）改厅为县，因县城内古榕众多，且有都柳江、寨蒿河、平永河三江汇聚，故名榕江县，隶黔东道。由于水上交通优势得天独厚，榕江商贾纷至、商贸繁华，昔有“小南京”的美誉。

中华人民共和国成立后先后隶属独山专区（1950年）、都匀专区（1952年）、黔东南州（1956年），1959年从江县并入，1961年榕江县重新将从江县析出单设至今。

榕江历史文化悠久，民族文化底蕴深厚。境内的侗乡七十二寨保持着原始自然生态，村寨里传统的生产、生活习俗依然传承了远古的形式。“甜粑节”“乌米饭节”“吃新节”“侗年”等特有的民族节日，充分展示了侗寨奇异多姿的地方风情和丰富多样的民族食品。

榕江县侗家村寨大多坐落在古树参天、风光秀丽、竹林掩映、依山傍水的地方。房屋多为两层或三层不等的木质结构吊脚楼，吊脚楼一般依山势开辟成两层地基，下层屋基柱脚用筒柱支撑，有的因偏厦建在地基之外的村巷或鱼塘上而无支柱造成木楼空悬，吊脚楼因而得名。

⊙1

⊙2

⊙3

⊙1
九秋村侗寨的民宅
Local Dong ethnic residences in Jiuqiu Village

⊙2/3
本里村的侗家村寨吊脚楼
Dong ethnic Diaojiao (suspended) buildings in Benli Village

三
榕江侗族皮纸的历史与传承

3
History and Inheritance of Bast Paper by the Dong Ethnic Group in Rongjiang County

据1999年版《榕江县志》记载，清末民国初年，湖南、广东等省的造纸工匠相继进入榕江，落户开办家庭手工作坊，这是榕江造纸的开端。通过查阅《光绪古州府志》《贵州轻纺工业志》《榕江县志》等资料，调查组发现有明确记载的最早的造纸户是民国十二年（1923年）县城的一户黄姓人家，他们“迁居八开办造纸作坊，全家4人参加生产，另雇2人。以稻草和嫩竹作原料，土法生产包装纸，供制鞭炮、冥钱，年产7 500千克。主销县内和从江县。连续生产十余年”[8]。民国时期，黔东南纸张生产地中，“榕江县有八开、乐里、本里”[9]三个造纸点。

20世纪50年代，县城工商业者和八开供销社分别于1951年和1955年开办造纸厂，生产毛边纸、江东纸、黄皮纸和供雨伞社用作原料的白皮纸，也生产卫生纸和新闻纸，产品销往县内及邻县，20世纪60年代初先后停办。

1971年，榕江县竹器社在城郊仁育堡重建了集体造纸厂，以废纸作原料生产鞭炮用纸，1978年改革开放后，县财政拨出8.8万元无息贷款进行扩建和技改，由土法制造转为半机械化生产，日产500 kg，同时加产瓦楞纸，销往都匀、黄平和湖南等地。1981年，造纸厂从竹器社析出，成为独立核算、自主经营的集体企业。到1987年再次技改时，除切纸仍为手工操作外，其余工序均实现了机械化。

民国年间，本里村已有少数农民以构皮、稻草作原料，土法生产可供书写、裱糊的夹皮纸、白皮纸。据2011年7月16日调查组入村调查时本里村村长介绍，本里村自广西迁入侗寨即有手工造

⊙4

⊙4 旧日用皮纸抄写的风水书 Old Fengshui book transcribed on bast paper

[8] 贵州省榕江县地方志编纂委员会.榕江县志[M].贵阳:贵州人民出版社,1999:562.

[9] 贵州省地方志编纂委员会.贵州省志·轻纺工业志[M].贵阳:贵州人民出版社,1993:81.

⊙1

纸工艺，已在侗族传承约1 000年。20世纪六七十年代，造纸业盛行，全村几乎每户都从事造纸，成品主要销往乐里镇。而2011年村里共有826户人家，只有20户左右还在造纸。可见随着社会经济业态的变迁波动，21世纪初以来本里村构皮纸的生产广泛度已有明显收缩。

⊙2

侗寨造纸的整个流程全部由女性完成，而且技艺也有“传女不传男”的传统，造纸完全是女人的天下，男人只能做一些简单的协助工作。调查时据村民林世康描述，他的奶奶林老谋将造纸技艺传给他的母亲潘老乜，母亲又传给他的妻子林老梦，而现在他的儿媳林泽婵也在学习造纸。

⊙1
本里村造纸的侗族女村民
Female Dong papermakers in Benli Village

⊙2
本里村中『话构麻』
Enquiring papermaking techniques in Benli Village

四 榕江侗族皮纸的生产工艺与技术分析

4 Papermaking Technique and Technical Analysis of Bast Paper by the Dong Ethnic Group in Rongjiang County

(一) 榕江侗族皮纸的生产原料与辅料

1. 构树皮

据本里村造纸村民介绍，山上野生的构树有藤本和树本两种。相对来说树本构树更好，可以造出更多的原料。所采集的构树生长年限要在2年以上，否则枝干太细，得不到想要的原料，其纤维组织也达不到造纸的要求。

2. 纸药

制作纸药的原材料为猕猴桃树藤。将猕猴桃树藤捶烂后放于清水中，溶解出浆。造纸艺人凭经验调配浓度，至用手去搅拌就可以捞起长长的黏液即可。纸药是皮纸生产中的关键辅料，对于“撕纸”工序非常重要。

3. 草木灰

既可以防蛀，又可以使树皮更容易煮烂。每1.5 kg干的构树皮约加2.5 kg草木灰。

⊙3

⊙4

⊙5

⊙3 构树采样 Sample of paper mulberry tree

⊙4 『藤本构树』采样 Sample of *Broussonetia kaempferi Sieb.* var. *australis* Suzuki

⊙5 野生猕猴桃树 Wild Chinese gooseberry tree

(二)

榕江侗族皮纸的生产工艺流程

据2011年7月实地调查，总结榕江侗族皮纸生产工艺流程如下：

壹	贰	叁	肆	伍	陆	柒	捌
砍树	剥皮	晒皮	泡料	煮料	捞料	洗料	捶料

拾肆	拾叁	拾贰	拾壹	拾	玖
理纸	撕纸	晾纸	浇纸	搅料	过滤

壹 砍树

1

构树都是附近山上野生的，春、夏、秋都可以砍，春、夏为最佳，因为这个季节的构树已“上水”，更容易剥皮。

贰 剥皮

2 ⊙1

在山上剥皮，然后用筐装回家，再将外皮剥掉。剥外皮时要先用镰刀划一个口子，这样更容易剥掉外皮。

⊙1

叁 晒皮

3 ⊙2

将剥好的皮置于太阳下晒干，阳光强烈时，两天即可晒干。若天气不好，就得不到白的料，造出的纸质量也会不好。

⊙2

肆 泡料

4

通常是将晒干的构树皮在清水中浸泡2~3小时，有时也可再久些。新鲜树皮可直接煮，不用泡。

伍 煮料

5 ⊙3

将干树皮（1.6 kg）放入煮料锅里，并加入2.5~3 kg草木灰，加入的水量要能淹过皮料。边煮边用木棒翻，一直到树皮煮烂为止，煮料时间为4~5小时。

⊙3

⊙1 剥皮 Stripping the bark

⊙2 晒干的构树皮 Dried paper mulberry bark

⊙3 煮料 Boiling the papermaking materials

陆 捞料

6 ⊙4

用火钳捞出煮好后的料，因为温度很高，无法直接用手捞，如果不急着用也可等凉了以后再捞。

⊙4

⊙5

柒 洗料

7 ⊙5

用清水反复清洗煮好的料，尽量清洗干净，同时要把竹料里掺杂的黑色杂质清除掉，这是为了使制作出来的纸尽量白净。

捌 捶料

8 ⊙6

将洗好的料放在石板上，用棒槌反复捶打，捶得越碎越好。猕猴桃树藤也需用棒槌捶烂，将捶烂的猕猴桃树藤捆绑后再放于清水中浸泡。

⊙6

玖 过滤

9 ⊙7

将经过捶打、浸泡的猕猴桃树藤汁倒入一个布袋过滤，以免残渣夹杂在纸料中。

⊙7

⊙8

拾 搅料

10 ⊙8

把捶好的料放入装有清水的桶中，并用手搅拌均匀，再加入过滤后的纸药，一边加一边用手搅拌，直到桶中的浆液均匀为止。

⊙4 捞料 Picking out the papermaking materials

⊙5 洗料 Cleaning the papermaking materials

⊙6 捶猕猴桃树藤 Hammering Chinese gooseberry vine

⊙7 过滤猕猴桃树藤汁 Filtering Chinese gooseberry sap

⊙8 搅料 Stirring the papermaking materials

拾壹 浇纸

11 ⊙9⊙10

将搅拌均匀的料用小盆舀出并倒于立框的立布上一角，然后慢慢摇动。舀料的多少根据立布厚薄程度来决定，摇动的方向没有规定，只要能使料均匀分布在立布上即可。随后反复进行相同的动作。

⊙9

⊙10

⊙11

拾贰 晾纸

12 ⊙11

浇好的纸可以晒干或阴干，但不能烘烤。有阳光时最好晒干，将立框斜放于阳光下，立框后面用木棒支撑。如果是夏天，需晒4~5小时，晾晒的时间也因纸的厚薄而异。

拾叁 撕纸

13 ⊙12

待纸晒干后，先将纸从立框的左上角或右上角开始撕开一个缝，然后将手伸进纸与立框中间，从上往下慢慢将纸与立框分开，动作不能太快，以免将纸撕破。如果纸质很好且晒得很干，则撕纸时可以听见清脆的响声。

拾肆 理纸

14 ⊙13

将撕下的纸对折两次，且要折叠整齐，然后整理放好。

⊙9/10 浇纸 Pouring paper pulp onto the papermaking screen

⊙11 晾纸 Drying the paper

⊙12 撕纸 Peeling the paper down

⊙13 理纸 Sorting the paper

⊙12

⊙13

（三）

榕江侗族皮纸生产使用的主要工具设备

壹 垫石（青石板） 1

石质（对硬度要求很高），石板越大越好，便于捶料。

贰 棒槌 2

木质，圆柱形，实测样品长38 cm，棒头长20 cm，直径5 cm，手柄直径3.4 cm。

⊙1

叁 立框 3

74 cm ×74 cm的正方形木框，上边缝上立布。

⊙2

肆 盛放器具及仪器 4

草木灰盆（内径62 cm，外径65 cm，高17 cm）、煮料锅（直径91 cm）、火钳（煮料后捞料）、镰刀（用于剥树皮）、桶和盆（用于盛水和搅料）。

⊙3

⊙1 垫石和棒槌 Stone beating pad and wooden mallet

⊙2 立框 Frame for supporting the papermaking screen

⊙3 草木灰盆、火钳和煮料锅 Plant ash container, tongs and boiling wok

(四)

榕江侗族皮纸的性能分析

1. 榕江乐里镇本里村皮纸的性能分析

所测榕江本里村皮纸的相关性能参数见表10.7。

表10.7　本里村皮纸的相关性能参数
Table 10.7　Performance parameters of bast paper in Benli Village

指标		单位	最大值	最小值	平均值
厚度		mm	0.320	0.150	0.240
定量		g/m^2	—	—	51.0
紧度		g/cm^3	—	—	0.213
抗张力	纵向	N	31.1	19.7	25.5
	横向	N	29.4	18.5	24.2
抗张强度		kN/m	—	—	1.657
白度		%	31.4	30.2	30.9
纤维长度		mm	9.89	1.06	4.71
纤维宽度		μm	27.0	3.0	13.0

由表10.7可知，所测本里村皮纸最厚约是最薄的2.13倍，相对标准偏差为4.90%，纸张厚薄不均。皮纸的平均定量为51.0 g/m^2，所测皮纸紧度为0.213 g/cm^3。

经计算，其抗张强度为1.657 kN/m，抗张强度值较一般竹纸大。

所测本里村皮纸白度平均值为30.9%，白度较低，白度最大值约是最小值的1.04倍，相对标准偏差为0.53%，白度差异相对较小。这可能是因为构皮纸未经过较强漂白，加上表面比较粗糙，从而造成纸张白度较低。

所测本里村皮纸纤维长度：最长9.89 mm，最短1.06 mm，平均4.71 mm；纤维宽度：最宽27.0 μm，最窄3.0 μm，平均13.0 μm。所测皮纸在10倍、20倍物镜下观测的纤维形态分别见图★1、图★2。

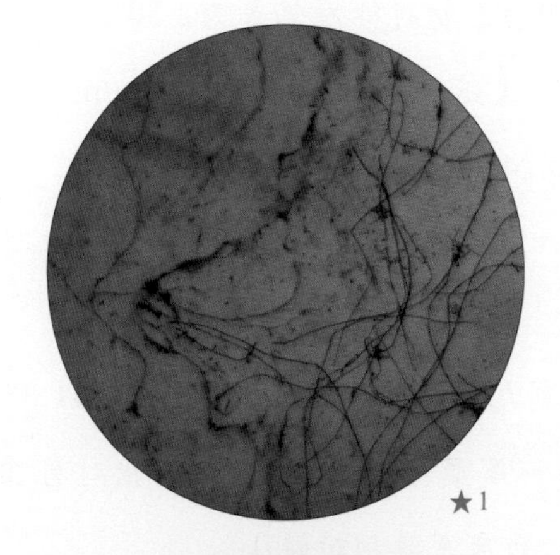

★1

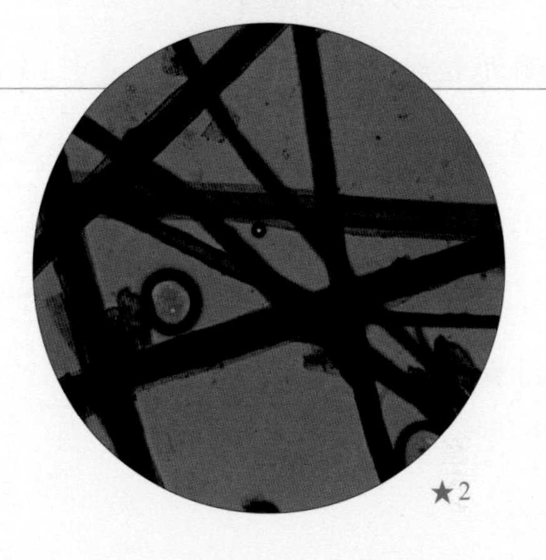

★2

★1 本里村皮纸纤维形态图(10×) Fibers of bast paper in Benli Village (10× objective)

★2 本里村皮纸纤维形态图(20×) Fibers of bast paper in Benli Village (20× objective)

2. 榕江计划乡九秋村皮纸的性能分析

所测榕江九秋村皮纸的相关性能参数见表10.8。

表10.8　九秋村皮纸的相关性能参数
Table 10.8　Performance parameters of bast paper in Jiuqiu Village

指标		单位	最大值	最小值	平均值
厚度		mm	0.260	0.180	0.220
定量		g/m^2	—	—	54.4
紧度		g/cm^3	—	—	0.247
抗张力	纵向	N	34.9	27.2	32.4
	横向	N	34.4	27.0	30.9
抗张强度		kN/m	—	—	2.110
白度		%	42.0	40.6	41.3
纤维长度		mm	14.88	1.23	4.66
纤维宽度		μm	34.0	3.0	15.0

由表10.8可知，所测九秋村皮纸最厚约是最薄的1.44倍，相对标准偏差为2.50%，纸张厚度较为均匀。皮纸的平均定量为54.4 g/m^2。所测皮纸紧度为0.247 g/cm^3。

经计算，其抗张强度为2.110 kN/m，抗张强度值较大。

所测九秋村皮纸白度平均值为41.3%，白度最大值约是最小值的1.03倍，相对标准偏差为0.51%，白度差异相对较小。

所测九秋村皮纸纤维长度：最长14.88 mm，最短1.23 mm，平均4.66 mm；纤维宽度：最宽34.0 μm，最窄3.0 μm，平均15.0 μm。所测皮纸在10倍、20倍物镜下观测的纤维形态分别见图★1、图★2。

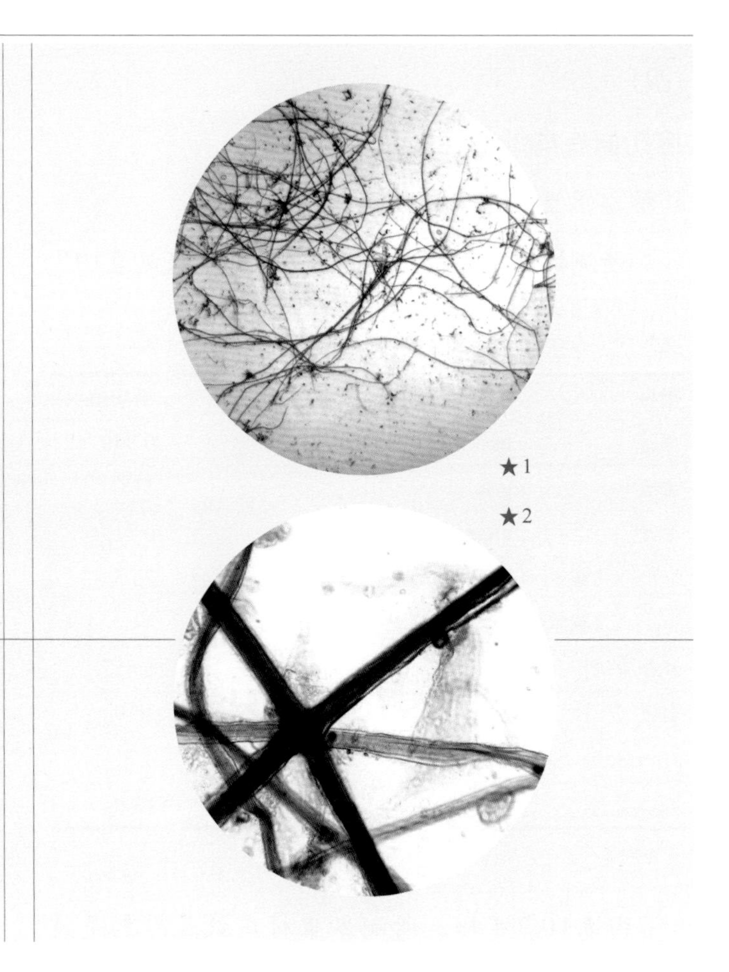

★1
九秋村皮纸纤维形态图(10×)
Fibers of bast paper in Jiuqiu Village
(10× objective)

★2
九秋村皮纸纤维形态图(20×)
Fibers of bast paper in Jiuqiu Village
(20× objective)

五 榕江侗族皮纸的用途与销售情况

5 Uses and Sales of Bast Paper by the Dong Ethnic Group in Rongjiang County

(一) 榕江侗族皮纸的用途

本里村制造的皮纸，涂上桐油后可用来制作斗笠和雨伞。据1999年版《榕江县志》记载，民国二十年（1931年），一位湖南艺人到榕江开办制伞作坊，到1950年，全县从事制伞业者37人，旺季月产上千把，产品有白花伞、大红伞、清凉伞等七种款式。中华人民共和国成立后，榕江雨伞曾三次获得省、州评比第一名，全国评比第四名。至20世纪70年代，因售价无法根据市场行情进行调整，加之机制轻便雨伞的冲击，纸伞逐渐被淘汰。此外，本里村的竹壳皮纸斗笠在当地也颇有名气。

大约100年前，涂了桐油的纸可当作雨衣。同时还可以用来包装衣服和贵重物品，因为它可以防霉、防蛀。传统的用途还包括写文书、契约、乡土演出的曲谱等。在侗、苗等少数民族中还有与服饰相关的刺绣、剪纸等用途。

邻近的计划乡九秋村生产的手工纸除了以上用途外，还可以用来做风筝。

⊙1

(二)
榕江侗族皮纸的销售情况

⊙2

20世纪六七十年代，本里村生产的成品构皮纸主要用来制作成斗笠卖给周边乡镇的人，当时周围地区有很多人到本里村购买斗笠，所以几乎家家户户都造纸。当时一张纸的价格只有几毛钱。现在则一般直接把纸卖给做斗笠和剪纸的人。调查组入村调查时，皮纸一般销往附近的村落和集市，也有少量卖给游客的，但囿于地理环境和交通，到当地旅游的人还比较少。

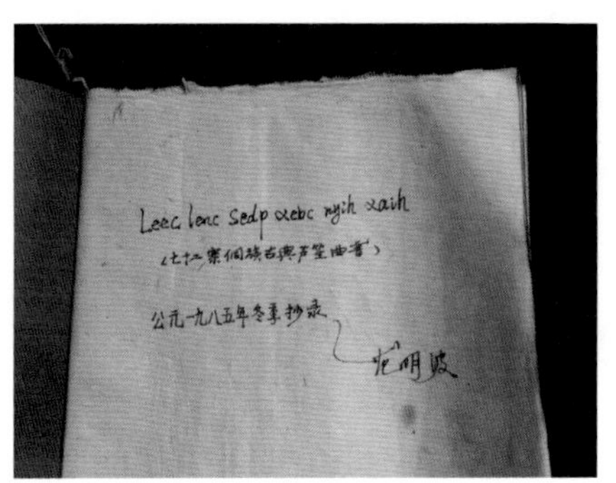

⊙3

现在，本里村的造纸户有20户左右，一张纸的价格是6元钱。但由于总体销量都不大，所以一个人一年只做500多张纸。从成本来分析，所用的料和纸药都不需要花钱购买，最多购买一些使用工具（50元左右），但这些工具都可以长期使用，消耗不是太大，所以最大的成本应该是人的工时。

⊙4

如果不包括前期备料的时间，按一个人一天可以造20张纸来算，若一年只做500张纸，那么造纸时间为25天，若每天的人工成本为20元，则总的费用为550元。因此，造纸的纯利润为2 450元，这应该算是不错的兼业收入。

⊙1 油纸斗笠 Hat made of oil paper

⊙2 剪纸 Paper cutting

⊙3 七十二寨侗族古典芦笙曲谱 Music score of traditional Dong ethnic musical instrument, Lusheng

⊙4 九秋村河边洗料的造纸人 A papermaker cleaning the papermaking materials in a river in Jiuqiu Village

六 榕江侗族皮纸的相关民俗与文化事象

6 Folk Customs and Culture of Bast Paper by the Dong Ethnic Group in Rongjiang County

调查组入本里村调查时，未发现当地侗族居民的家谱，也未发现任何和造纸家族历史相关的文字记载。据访谈中造纸的侗族村民介绍，他们也不是很清楚造纸术的由来，只知道他们的祖先自1 000多年前从广西迁居过来的时候，造纸术就已经在传承了。而在他们迁来之前这里居住的是苗族人。所以在造纸和许多民族习俗上，这两个民族有很多共同点。

在邻近的计划乡九秋村调研时，当地关于造纸术的由来有着这样一个传说：现在居住在这里的苗族居民，最早也是从广西迁移过来的（具体时间无法考证）。在广西的时候，由于当时的田地少，种出来的粮食不够吃，于是为了生存，有兄弟二人把一粒很大的米一分为二，一人一半，哥哥继续留在广西，弟弟则带着那半粒米来到了现在的九秋村。然而九秋村的气温偏低，村民们种不出纺布用的棉花，所以他们学习制作手工纸，用自己制作出来的纸到外边去换取纺布用的棉花。

⊙1

⊙1 进入九秋村的山道 Country road leading to Jiuqiu Village

七
榕江侗族皮纸的保护现状与发展思考

7
Preservation and Development of Bast Paper by the Dong Ethnic Group in Rongjiang County

（一）榕江侗族皮纸的传承与保护机制

调查中了解到，在本里村和九秋村，侗族皮纸的造纸工艺都是只传女不传男，村民们认为造纸是女人的工艺，男人只能做一些简单的协助工作。传承人既可以是本族、本村的人，也可以是亲戚，其余外人一概不能传。这一传习规则也给手工造纸工艺的当代传承带来了某种局限性。

在榕江手工造纸工艺的实地考察中，我们切身感受到：虽然本里村和九秋村的皮纸生产仍有一些聚集，但与十余年前相比，已经萎缩很多。古法造纸的现有传承人所掌握的传统造纸工艺活力不应在“固化”的保护中实现，只有“活化”的保护和发展才是激发传统手工造纸工艺活力的

⊙2

⊙2
山环水绕的榕江
Rongjiang County surrounded by mountains and rivers

⊙1

根本。然而，这种活化的保护仅仅靠传承人借力民族文化基因来维持是不够的。

国家重视非物质文化遗产保护的政策唤醒了地方政府和文化管理部门保护手工造纸工艺的意识。调查中了解到，榕江当地政府正努力推进当地的手工造纸工艺纳入非物质文化遗产保护的范畴。并且当地政府也有意将发展地方经济与保护民族文化结合起来，在招商引资和引入民间力量积极保护手工纸技艺、民族歌舞、民族节庆、民族服饰等文化遗产的基础上，培养非物质文化遗产的研究和传承人才，推进民族文化旅游，以带动民族地区的经济和文化持续与协调发展。

旅游也是带动手工造纸工艺的一种重要路径。本里村和九秋村有着独特的侗寨、苍天的大树、淳朴的村民、清新的空气、美丽的风景……发展旅游业，也能让更多的人目睹手工纸的制作流程，了解侗族手工纸文化。

（二）榕江侗族皮纸的发展思考

本里村与九秋村的手工造纸还停留在最传统的造纸方式上，没有采用任何的机械设备，整个流程是全手工制作，其传统的工艺基因保存得相当完好。但是，处于农业社会的全流程手工操作技艺对手工纸的当代生存空间也有抑制的一面。

⊙1
正在浇纸的侗族女人
A Dong woman making paper

⊙3

⊙2

由于传统皮纸用途的萎缩，必须开发新的产品或者拓展原有产品的用途，才能使造纸工艺得到传承和保护。建议在充分保存工艺原生性的前提下，适当加入一些半手工半机械加工原料的元素，提升效率，从而实现浇纸法技艺上的小批量生产。

⊙2 珍贵的国家一级保护树种
Trees under First-Grade State Protection

⊙3 辛苦的侗族造纸老人
An old Dong papermaker

榕江乐里侗族

419

皮纸

Bast Paper
by the Dong Ethnic Group
in Leli Town of Rongjiang County

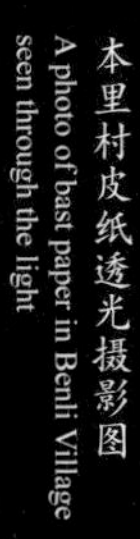

本里村皮纸透光摄影图
A photo of bast paper in Benli Village seen through the light

侗族皮纸

Bast Paper
by the Dong Ethnic Group
in Jihua Town of Rongjiang County

九秋村皮纸透光摄影图
A photo of bast paper in Jiuqiu Village
seen through the light

第七节

从江秀塘瑶族竹纸

贵州省
Guizhou Province

黔东南苗族侗族自治州
Qiandongnan Miao and Dong Autonomous Prefecture

从江县
Congjiang County

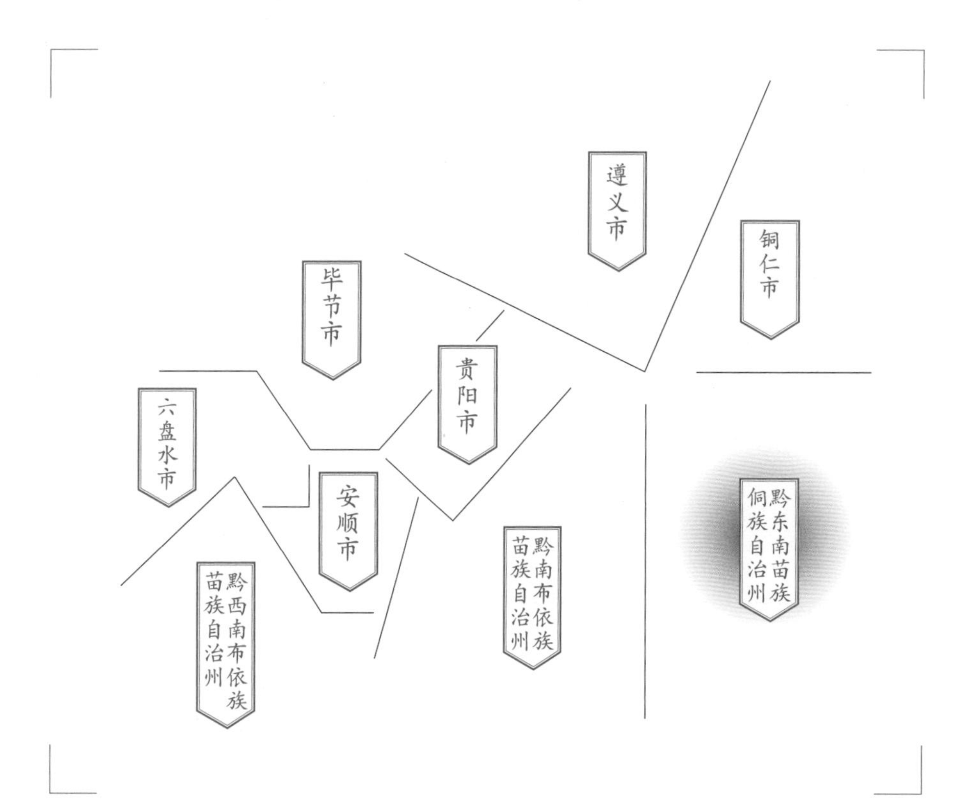

调查对象

秀塘壮族乡
打格行政村
瑶族竹纸

Section 7
Bamboo Paper
by the Yao Ethnic Group
in Xiutang Town of Congjiang County

Subject

Bamboo Paper by the Yao Ethnic Group
in Dage Administrative Village
of Xiutang Zhuang Town

一 从江秀塘瑶族竹纸的基础信息及分布

1 Basic Information and Distribution of Bamboo Paper by the Yao Ethnic Group in Xiutang Town of Congjiang County

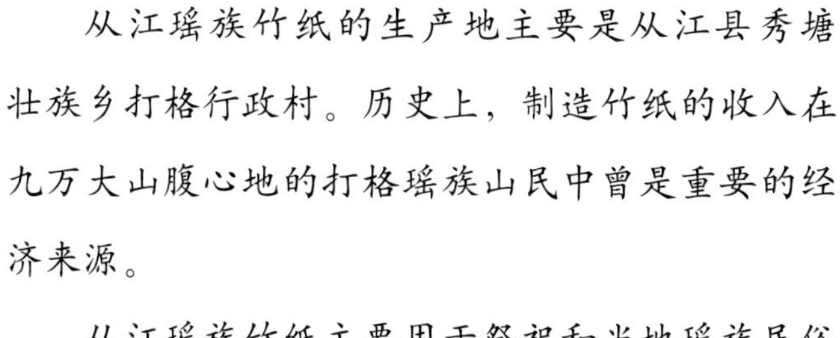

从江瑶族竹纸的生产地主要是从江县秀塘壮族乡打格行政村。历史上，制造竹纸的收入在九万大山腹心地的打格瑶族山民中曾是重要的经济来源。

从江瑶族竹纸主要用于祭祀和当地瑶族民俗度戒，在打格瑶族，需用竹纸、鸡蛋、谷穗等祭祖。度戒和盘王节是打格瑶族非常神圣的民俗文化，其间所使用的纸必须是手工竹纸，以表示对菩萨的尊重，足见打格瑶族竹纸与当地瑶族文化的密切关系。

2014年10月上旬，调查组来到秀塘乡，在乡政府相关人员的协助下，经过约3小时的路程抵达打格村第四村民组。通过与村民的访谈得知，打格村竹纸生产已完全停产近20年，打格瑶族竹纸的原料主要是当地瑶族所称的南竹，其植物纤维品质高，柔韧性与挺括性俱佳，成品纸呈米黄色，美观且可久藏。

⊙1

⊙1 前往打格村的山路 Country road leading to Dage Village

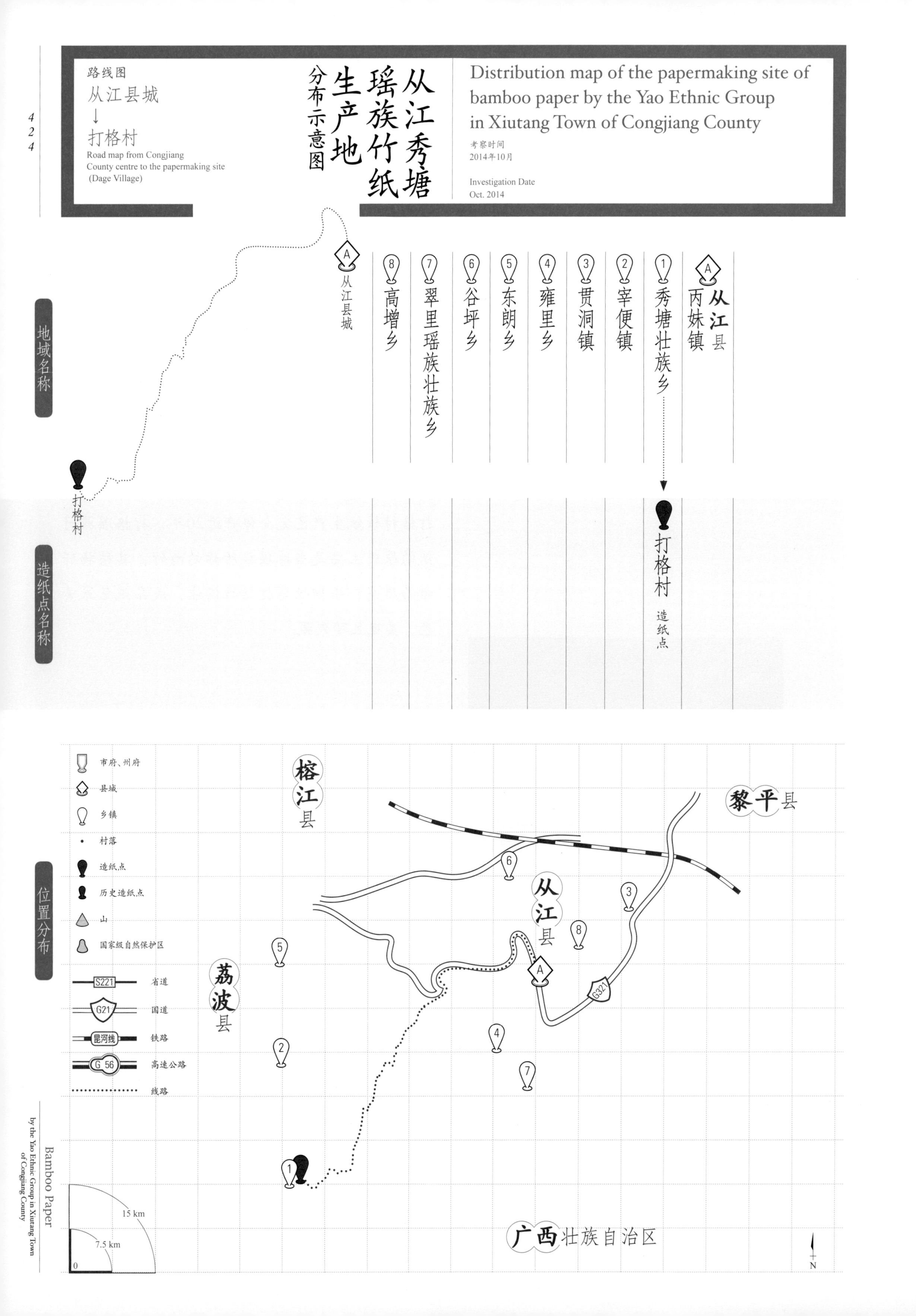

路线图
从江县城
↓
打格村
Road map from Congjiang County centre to the papermaking site (Dage Village)
从江秀塘瑶族竹纸生产地分布示意图
Distribution map of the papermaking site of bamboo paper by the Yao Ethnic Group in Xiutang Town of Congjiang County
考察时间
2014年10月
Investigation Date
Oct. 2014
地域名称
造纸点名称
位置分布
从江县
A 丙妹镇
1 秀塘壮族乡
2 宰便镇
3 贯洞镇
4 雍里乡
5 东朗乡
6 谷坪乡
7 翠里瑶族壮族乡
8 高增乡
A 从江县城
打格村
打格村 造纸点
市府、州府
县城
乡镇
村落
造纸点
历史造纸点
山
国家级自然保护区
S221 省道
G21 国道
昆河线 铁路
G56 高速公路
线路
榕江县
黎平县
从江县
荔波县
广西壮族自治区
G321
15 km
7.5 km
0
N

二
从江秀塘瑶族竹纸生产的人文地理环境

2
The Cultural and Geographic Environment of Bamboo Paper by the Yao Ethnic Group in Xiutang Town of Congjiang County

秀塘壮族乡位于从江县西南部，属于从江县的3个民族乡之一，政府驻地距县城92 km，地处月亮山和黔桂交界处九万大山腹地，总面积178.22 km^2，其中耕地面积约3 km^2，农业以水稻为主，林业以杉木为主，居住有壮、苗、汉、侗、瑶、水等民族，其中壮族人口占总人口的60%以上，故称。全乡辖13个行政村、68个自然村寨、79个村民组。截至2013年底，全乡有1 883户，8 165人。

乡内海拔在1 000 m以上的山有19座之多，其中，海拔最高为1 675 m，即堪称“从江最高峰”的打格尖山；海拔最低为650 m。年平均气温16.1 ℃，年平均降水量1 450 mm，无霜期240天左右。境内山河（溪）纵横交错，小峰林立，道路崎岖，交通闭塞，通信落后，是从江偏远的贫困乡（镇）之一。秀塘壮族乡的气候温暖、湿润，水质纯净，空气新鲜，森林覆盖率约60%，自然资源十分丰富，但是交通相当不便。

⊙1

⊙2

秀塘壮族竹纸的生产地在打格，打格乃壮语地名，是打格村十二盘、上甲路、下甲路3个自然村寨的统称。打格瑶族人多居住在山上的小溪边，海拔1 000 m左右的地方。汉语中今为赵姓的当地人自述其是从广西风昌县来塘洞一带并迁居打格的，距今约250年。

2014年调查时全村有173户，739人，已通电，有自来水，还有小学。历史上，瑶族人居住的地区多为亚热带，海拔一般在1 000~2 000 m，村寨周围竹木叠翠，风景秀丽。

横跨桂黔两省的九万大山，在贵州境内面积约858 km^2，地处从江县秀塘乡打格村的山峰，山高，坡陡，山岭连绵、起伏大，云雾缭绕。有大面积元古代前震旦系地层出露，是贵州省内出露的古老的地层之一。由于不曾受第四纪冰川的严重破坏，加上人为破坏较小，山里仍保存着许多古老和孑遗植物，以及上百种珍稀保护动物，这

⊙1 进入秀塘乡的壮族风格门楼 Unique Zhuang ethnic gate entrance to Xiutang Town

⊙2 南竹林环抱的打格村民居 Local residences in Dage Village, surrounded by *Phyllostachys pubescens* Mazel ex H. de Leh.

⊙1

⊙2

在很大程度上反映了贵州南部森林的原始自然本体。调查组于2014年10月6日前往九万大山深处的打格村时，还偶遇上山狩猎的当地村民。因具有独特的气候和地理环境，打格村寨附近盛产制作竹纸的原料——南竹。

三 从江秀塘瑶族竹纸的历史与传承

3 History and Inheritance of Bamboo Paper by the Yao Ethnic Group in Xiutang Town of Congjiang County

关于瑶族的来源，说法不一，或认为源于“山越”，或认为源于“五溪蛮”，或认为瑶族来源是多元的。但大多数人认为瑶族与古代的“荆蛮”“长沙武陵蛮”等在族源上有着渊源关系。

在瑶族的历史上，有一次重要的迁徙典故：元朝时，某官员派人前来征税，当地瑶民热情款待，以致官员忘了回衙，衙门误以为瑶民杀了前来征税的官员，故派兵前来剿杀，瑶民被迫迁徙分转各地，临前将牛角分成十二节由十二姓掌管，相约千年之后再回故地。

元、明、清时期，瑶族主要分布在广西、广东以及湖南西南部和云南、贵州的部分山区，社会经济发展极不平衡，发展较快的地区已接近汉族地区的水平，而偏远地区的瑶族“随溪谷群处”，甚至不从事农耕，以猎山兽为生，尚处于

⊙1 正在上课的打格小学学生 Students of Dage Elementary School having classes

⊙2 高山上的瑶族村寨 Yao ethnic residences on high mountains

原始社会发展阶段。明末清初，部分瑶族人逃至越南、老挝、泰国等国边境。至20世纪70年代，在越南、老挝、泰国等国的部分瑶族人分别迁徙到美国、法国、加拿大等国居住。

关于“瑶族”二字中“瑶”字的由来，有多种说法。其中以历史学家何光岳先生的著作《南蛮源流史》中所理线索较为可信：瑶族先民在新石器时代擅长制作瓦器、陶罐，故最早的瑶族先民被称为“窑民”；后来陶罐坯料制作由手工发展为旋转摇动制坯，窑民改称“摇民”；瑶族先民中四大姓之一雷氏，来源于发明养蚕缫丝的黄帝之妃方雷氏（嫘祖），故瑶民擅养蚕，又衍生出“繇民”一称。后蚩尤率领“三苗”和“摇民”与炎黄大战失败，“摇人”历经夏、商、周征伐，一部分被当成劳役奴隶，称之为“傜役”，即“傜人”（周去非《岭外代答》一书曰：傜人者，言其执傜役于中国也）；此后历代“傜人”不断反抗中原王朝压迫，啸聚山林、不缴赋税，至宋代又有“莫傜”之称，意即不缴赋税、不赋劳役之人；到了元代，统治者认为这种不赋傜役、刀耕火种的人群是野人，故将“傜人”的“傜”改为具有侮辱性的犬字旁的“猺”，谓之“猺民”；1949年中华人民共和国成立之后，征得本民族同意后，将瑶族作为该名族统一名称。

⊙3

⊙4

⊙5

⊙6

⊙3 打格村瑶民烹饪食物
A Yao people cooking food in Dage Village

⊙4 落满灰土的20年前的纸帘
Papermaking screen covered by dirt which hadn't been used for 20 years

⊙5 穿瑶族盛装服饰少女旧照
Old photo of maids wearing Yao ethnic clothing

⊙6 访谈赵金学夫妇
Interviewing local papermakers Zhao Jinxue and his wife

秀塘壮族乡打格村瑶族竹纸的历史传承记忆已经模糊，调查组于2014年10月5~6日进行田野调查时，已经无法获得更早的造纸传承记忆，重点调查的造纸人赵金学、赵金成均已回溯不出打格村造竹纸的起源与传承谱系，大约在1996年时村里就已经不再造竹纸，因此也没有造纸现场。调查时，赵金学55岁（1959年出生），赵金成64岁（1950年出生）。

调查时，村里会造纸的人几乎都是20世纪五六十年代出生的人，年轻人几乎都没有造过竹纸。在村长和若干热心村民的帮助下，经过反复动员，造纸人赵金学才拿出家里还存有的一些大约20年前造的竹纸，据说这是村里唯一的存纸，当年主要作度戒、盘王节时祭烧之用。

据赵金学自述，他当年学习造纸技艺并没有拜过师傅，自己的造纸手艺是通过观摩村里的人造纸而逐步学会的。打格村里都是男人造竹纸，但赵金学的大儿子赵有海和孙子赵正杰都没有学过造纸。赵金成的儿子赵有管也不会造纸。

据赵金学与赵金成回忆，大约20年前，全村有近70户人家，20%~30%的人都会造纸。手工造纸开始没落是在“文化大革命”时，“破四旧，立四新，不信鬼，不信神”，手工竹纸因主要用于瑶族民俗度戒、盘王节，所以被认为是迷信，村里的手工造纸及成品纸使用受到限制。因此，不能公开、规模化造纸，只能私下做，仅供自己家里祭祖使用。

四 从江秀塘瑶族竹纸的生产工艺与技术分析

4 Papermaking Technique and Technical Analysis of Bamboo Paper by the Yao Ethnic Group in Xiutang Town of Congjiang County

（一）从江秀塘瑶族竹纸的生产原料与辅料

秀塘壮族乡打格瑶族竹纸的生产原料为南竹。《现代汉语词典》第6版将“南竹”列为词条，并解释为“毛竹”。南竹被统称为毛竹，又别于毛竹，它实际上是毛竹中最名贵、最有使用价值和经济价值的一种实用竹。在中国300多种木本竹类植物引属中，南竹是生长最快、材质最好、用途最多、经济价值最大、种植面积最广的竹种。南竹生长快，适应性强，一棵18 m高的树需要生长60年，而一根18 m的南竹只需生长59天；大面积种植南竹能防止水土流失，调节局部气候，净化空气，美化环境，并且成林时间较木材而言大为缩短。

打格村造纸的水源为溪水，瑶族历来喜欢逐溪水而居。经现场检测，当地溪水的pH约为6.2。

⊙1

⊙2

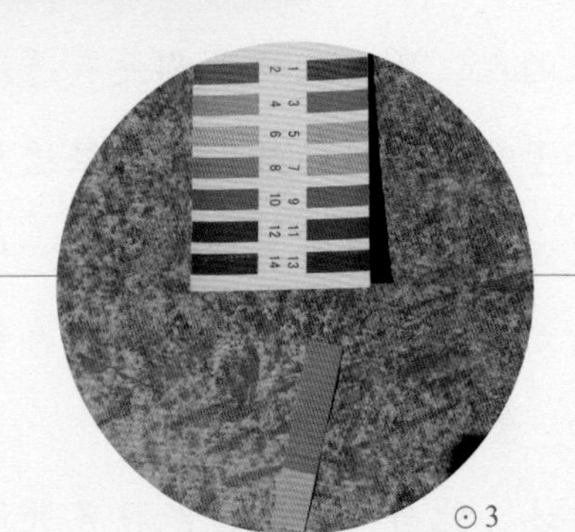

⊙3

（二）从江秀塘瑶族竹纸的生产工艺流程

调查组通过对打格村的实地考察，以及与赵金学、赵金成等人的交流，记录打格村瑶族竹纸的生产工艺流程如下：

壹	贰	叁	肆	伍	陆	柒
砍竹	浸泡	舂料	放料	抄纸	揭纸	晾纸

⊙1 打格瑶族竹纸的原料——南竹
Local papermaking raw material, *Phyllostachys Pubescens* Mazel ex H. de Leh.

⊙2 打格村旧日沿河造纸的山溪
Stream formerly used for cleaning the papermaking materials in Dage Village

⊙3 溪水的pH约为6.2
pH value of the stream was about 6.2

壹 砍竹

1 ⊙4

每年3月份，嫩竹向上长出5根枝条后，砍下整棵竹子。将砍下的竹子切成段，并剖成两指宽的竹条。

⊙4

贰 浸泡

2

将剖好的竹条一层层整齐码放于泡料池中，并在每层竹条上撒一层石灰。竹子与石灰的比例为10：1。浸泡的时间从3月开始，直到10月才开始出料造纸。

叁 舂料

3

10月开始造纸时，直接将泡好的竹料置于石头臼窝里舂碎。

肆 放料

4

将舂好的竹料放入舀纸塘。

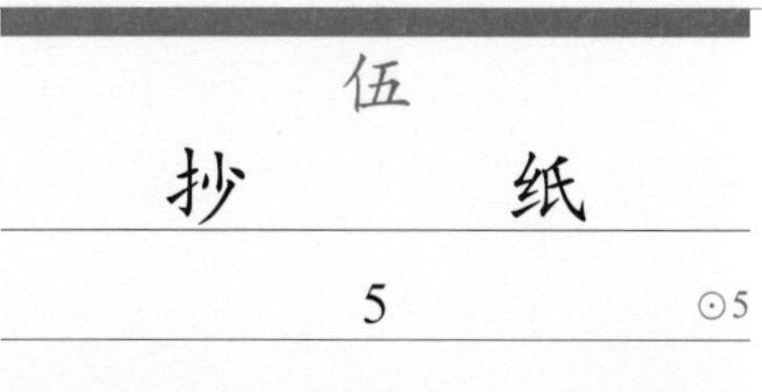

⊙5

伍 抄纸

5 ⊙5

将抄纸帘置于抄纸架上，右手握住抄纸架的手柄，左手拿住抄纸架的左侧，先往里朝自己身体方向舀纸，舀起后往外一晃，让纸料均匀地铺在纸帘上，再向右侧一荡，荡掉多余的水，然后放下整个抄纸架，双手捏住抄纸帘靠近自己身体的一边，提起纸帘，再用左手从下方捏住纸帘，双手配合将抄好的纸扣在待压榨的纸板上。一次最多能叠扣1 800张纸。压榨工序约需2小时。一天可以压2次。据赵金学回忆，当年村里最能干的纸工一天最多可以做5驮纸，一驮为500张。料如果不好的话，就只能做2~3驮。

陆 揭纸

6 ⊙6

将初步压好后的湿纸从右边缘一张张地向左翻开，并列排5张，5张纸一揭，称为一贴。揭起5张后，整贴对折。

⊙6

⊙4 赵金成讲述当年砍竹
Zhao Jincheng relating the bamboo-lopping experiences

⊙5 赵金学示意抄纸动作
Zhao Jinxue showing how to scoop and lift the papermaking screen out of water

⊙6 揭纸动作示意
Showing how to peel the paper down

柒

晾纸

7

⊙7

将整贴的纸置于用竹子做成的楼板上自然晾干，或者是一贴贴地放在家里楼板的竹竿上晾干。不需要晒，2~3天即可自然晾干。然后将晾干的纸以100贴为一驮打包存放待售。

⊙7

(三) 从村秀塘瑶族竹纸生产使用的主要工具设备

壹 抄纸帘 1

打格瑶族的抄纸帘来自广西融水同练瑶族乡，他们自己不会做纸帘。实测赵金学家旧纸帘外长约89.5 cm，内长约87 cm，宽约29.5 cm；抄纸架长约92 cm，宽约34 cm。

⊙8

⊙9

贰 石臼窝 2

石质，几乎保持着最原始的就地取材的形态。

⊙7 晾纸示意 Showing how to dry the paper

⊙8 旧抄纸帘与抄纸架 Old papermaking screen and its supporting frame

⊙9 溪边遗存的石臼窝 Stone mortar alongside the stream

(四)
从江秀塘瑶族竹纸的性能分析

所测从江打格村竹纸的相关性能参数见表10.9。

表10.9　打格村竹纸的相关性能参数
Table 10.9　Performance parameters of bamboo paper in Dage Village

指标		单位	最大值	最小值	平均值
厚度		mm	0.271	0.141	0.191
定量		g/m^2	—	—	38.7
紧度		g/cm^3	—	—	0.203
抗张力	纵向	N	13.9	10.2	11.9
	横向	N	6.9	3.9	5.2
抗张强度		kN/m	—	—	0.570
白度		%	18.4	17.6	18.1
纤维长度		mm	9.81	0.67	2.20
纤维宽度		μm	29.0	1.0	13.0

由表10.9可知，所测打格村竹纸最厚约是最薄的1.92倍，相对标准偏差为4.0%，纸张厚薄不均。竹纸的平均定量为38.7 g/m^2。所测竹纸紧度为0.203 g/cm^3。

经计算，其抗张强度为0.570 kN/m，抗张强度值较小。

所测打格村竹纸白度平均值为18.1%，白度较低，白度最大值约是最小值的1.05倍，相对标准偏差为0.25%，白度差异较小。

所测打格村竹纸纤维长度：最长9.81 mm，最短0.67 mm，平均2.20 mm；纤维宽度：最宽29.0 μm，最窄1.0 μm，平均13.0 μm。所测竹纸在10倍、20倍物镜下观测的纤维形态分别见图★1、图★2。

★1

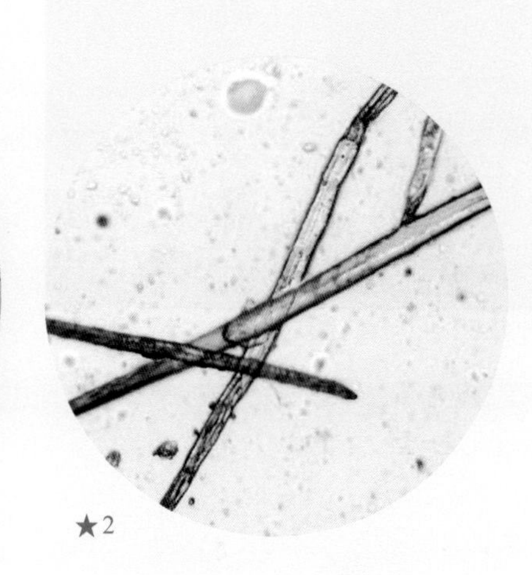

★2

★1 打格村竹纸纤维形态图(10×) Fibers of bamboo paper in Dage Village (10× objective)

★2 打格村竹纸纤维形态图(20×) Fibers of bamboo paper in Dage Village (20× objective)

五
从江秀塘瑶族竹纸的用途与销售情况

5
Uses and Sales of Bamboo Paper by the Yao Ethnic Group in Xiutang Town of Congjiang County

(一)
从江秀塘瑶族竹纸的基本用途

1. 祭祀

在打格瑶族，竹纸主要用于祭祀，祭祀用纸需要打孔，打15排，每排5个孔，寓意可在东南西北中五方使用。这是当地习俗，只有烧这样的纸祖宗才认得。打格村村民认为一张烧纸在阴间值5毛钱。

2. 度戒

度戒时，需要用竹纸制作成祖先像，包括盘王像在内一共十八像。供奉祖先要用竹纸，以示尊重。

3. 祭祖

在打格瑶族，一年四季都要祭祖。将装有祖先像的竹篓挂于神龛的右侧墙上，竹篓内装十八像（含盘王像），竹篓外挂白纸。神龛上放

⊙1

⊙2

⊙3

鸡蛋、谷穗和纸。祭祖时，师公用嘴将鸡的脖子咬破，使鸡血流出，再用纸沾上鸡血，贴于神龛上。纸表示祈求四季平安，血表示子孙家业千秋万代。被咬死的鸡只能由男人吃掉。

(二)
从江秀塘瑶族竹纸的销售状况

据造纸人赵金成、赵金学口述，大概在1965年时，纸的售价是8元/驮，1982年是12元/驮。到1996年不造纸时，售价达到30元/驮。造出的纸，除了供自家及村人使用外，多出的竹纸主要被卖到相邻的环江县和周家（地名）。

六
从江秀塘瑶族竹纸的相关民俗与文化事象

6
Folk Customs and Culture of Bamboo Paper by the Yao Ethnic Group in Xiutang Town of Congjiang County

1. 度戒

度戒是瑶族男人的成人仪式，是瑶族特有的一种习俗，是瑶族男人成长过程中不可或缺的神圣一课，比娶新嫁女还要隆重。瑶族不认为18岁是成人的年龄，在他们看来年龄无论大小，只要度戒过关，就是男子汉，就得到了神灵的保护，可以担任全寨的公职，获得男性人生的社会价值。没有度戒或度戒没有过关的男人就不能算作真正的有价值的男人，没有社会地位，得不到姑娘的爱慕，甚至找不到老婆。年长者如未举办度戒，必须找机会补办，否则就不被认为是瑶族人。至今，度戒在瑶族社会生活中仍然占据着重要的地位。

瑶族的度戒比较复杂，准备的时间也比较长。男孩长到10岁时父母就请识字先生推算吉

⊙1 打纸钱 Making joss paper

⊙2 打格村的纸钱 Joss paper in Dage Village

⊙3 神龛旁的祭物 Sacrificial offerings (bamboo basket and joss paper) by the niche

利年份来给他度戒，决定度戒年份后，父母提前一两年为度戒做准备，并在度戒当年确定度戒的具体时间。时间确定后又要请师父，师父越多越好。度戒前男孩要蒙被入睡5天，等到度戒仪式时才能出门。之前师父给男孩读经教授，并设立高桌反复练习度戒仪式上的动作，同时对男孩进行本民族传统道规的教育，使其在师父的训导下自觉修身。

度戒的整个仪式繁杂冗长，太阳尚未出来仪式便已开始。主师父先替男孩诵经做法事，然后身着红袍的引教师父指导男孩穿上同样的神袍，系红腰带，头戴瑶族的“上元”圣像。引教师父把一块红布条的一端系在自己的腰上，另一端系在男孩的腰上，象征着徒弟像婴儿一样尚未分娩离开母体。引教师父拿着神剑、念着神咒引领男孩走向度戒场院，路上引教师父还要拿出提前准备好的小钱给挡路人，以求买路放行。到了天台后，再领着男孩在天台下从左到右转3圈，然后手持树叶、念着神咒首先从梯的正面爬上天台，口中念念有词，下来后把男孩领到天台的戒桌旁，解开系着的红布条，表示婴儿从母体分娩，并给男孩讲誓言，拜叩完后将木梯撤掉。男孩对天发誓：不杀人放火，不偷盗抢劫，不奸女拐妇，不虐待父母，不陷害好人……发誓完毕后，师父会在男孩手上印一个红色的三角形印记，至此度戒仪式结束。

整个度戒仪式中有多个烧纸祭神祭祖环节，其用意为祈求神祖保佑度戒顺利。

2. 盘王节

盘王节是瑶族祭祀先祖盘古最盛大的节日，节日日期为农历十月十六。

盘王节具体多长时间举行一次根据各地瑶族传统习俗和谷物收成、人畜康泰的情况而定，有每3~5年举行一次的，也有每12年举行一次的。每逢盘王节，村寨房舍要打扫干净，男女老幼梳妆打扮，换上节日盛装，载歌载舞，尽欢而散。

盘王节可以一家一户进行，也可以联户或者同宗同族人集聚进行。但不管以哪种形式举办，都要杀牲烧钱祭祀，设宴款待亲友。时间一般为三天两夜，也有长达七天七夜的。

仪式主要分两大部分进行。第一部分是“请圣、排位、上光、招禾、还愿、谢圣”，整个仪式中唢呐乐队全程伴奏，师公跳《盘王舞》（《铜铃舞》《出兵收兵舞》《约标舞》《祭兵舞》《捉龟舞》等）；第二部分是请瑶族的祖

⊙4

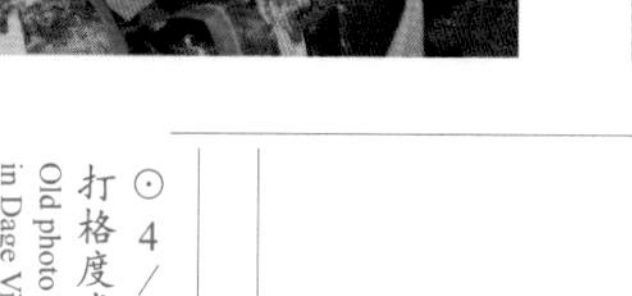

⊙5

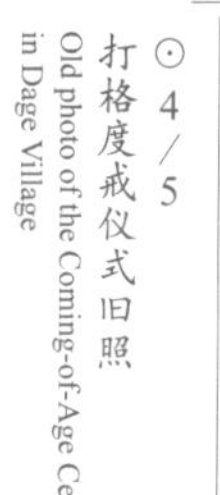

⊙4／5 打格度戒仪式旧照
Old photo of the Coming-of-Age Ceremony in Dage Village

先神和全族人前来“流乐”。流乐，瑶语意为玩乐。这是盘王节的主要部分，恭请瑶族各路祖先神参加盘王节的各种文艺娱乐活动，吟唱表现瑶族神话、历史、政治、经济、文化艺术、社会生活等内容的历史长诗《盘王大歌》。流乐仪式一般要持续一天一夜。

在“盘王节”仪式中，需烧竹纸钱祭祀盘王先祖。

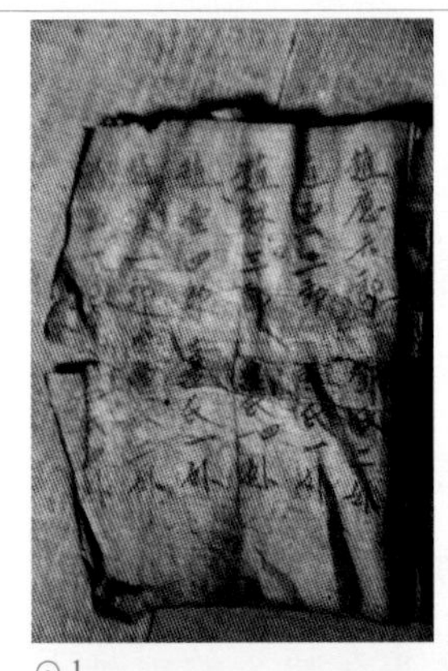

⊙1

⊙2

七 从江秀塘瑶族竹纸的保护现状与发展思考

7 Preservation and Development of Bamboo Paper by the Yao Ethnic Group in Xiutang Town of Congjiang County

(一) 从江秀塘瑶族竹纸的保护传承现状

秀塘壮族乡的主体民族和文化属壮族系统，瑶族文化只聚集于打格村的一个小区域，瑶族造纸技艺又传承于打格这一个行政村，且打格村处于秀塘壮族乡最偏远的东南角，深藏于九万大山深处，调查时交通仍极其不便。

打格村竹纸以南竹为原料，这是传统竹纸的标准原料，而且生产流程、工具、辅料也与传统竹纸基本一致，其技艺原生态的特性维系

⊙1 破旧的『盘王节』手抄祭本
Transcribed sacrificial book used for King Pan Festival

⊙2 打格村长讲述『盘王节』盛况
Head of Dage Village relating the grand occasion of King Pan Festival

得相当完整。

打格瑶族近20年从未造纸，造纸时使用的很多工具设备零星可见，活态实际上已终止。村里的造纸技艺只有中老年男人会，年轻人都不会，已完全呈现后继无人的状态。

⊙3

(二)
从江秀塘瑶族竹纸发展的若干思考

打格瑶族竹纸有其特别的民俗和文化价值，是贵州多民族手工纸技艺领域里的重要样式。但从调查组考察的情况看，目前，打格瑶族的竹纸已经停产近20年，交流中村民也没有再造纸的打算，其现状与发展空间均处于未知状态。

打格村得天独厚的造纸原材料、水源，以及较小的人口密度，为手工造纸业提供了良好的条件，可以保障在未来一段时间内不会有基础条件上的障碍。从抢救性保护和积极拓展的视角来看，如果要推动打格瑶族的手工造纸技艺重新复苏，需要设计可直接促进的事项。

⊙4

⊙5

⊙3
告别造纸村民时的合影
Photo of researchers and local papermakers before the departure time

⊙4
讲述造纸往事的赵金学
Zhao Jinxue telling the papermaking stories

⊙5
祭台
Worshipping niche

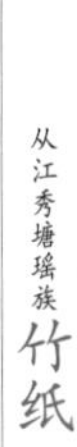

从江秀塘瑶族竹纸

Bamboo Paper by the Yao Ethnic Group in Xiutang Town of Congjiang County

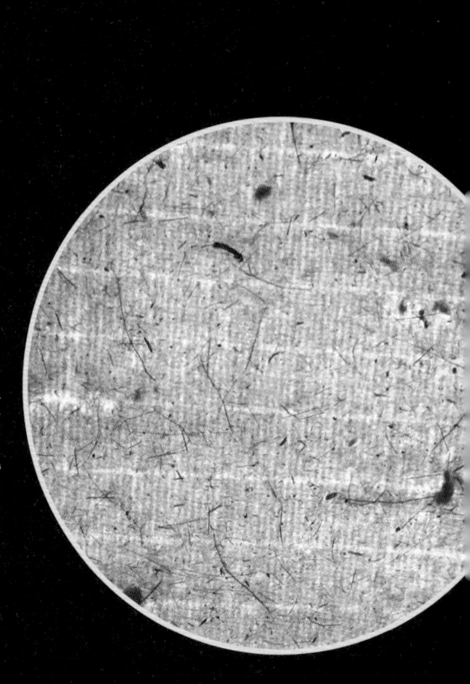

打格村竹纸透光摄影图

A photo of bamboo paper in Dage Village seen through the light

第八节

从江翠里瑶族手工纸

贵州省
Guizhou Province

黔东南苗族侗族自治州
Qiandongnan Miao and Dong Autonomous Prefecture

从江县
Congjiang County

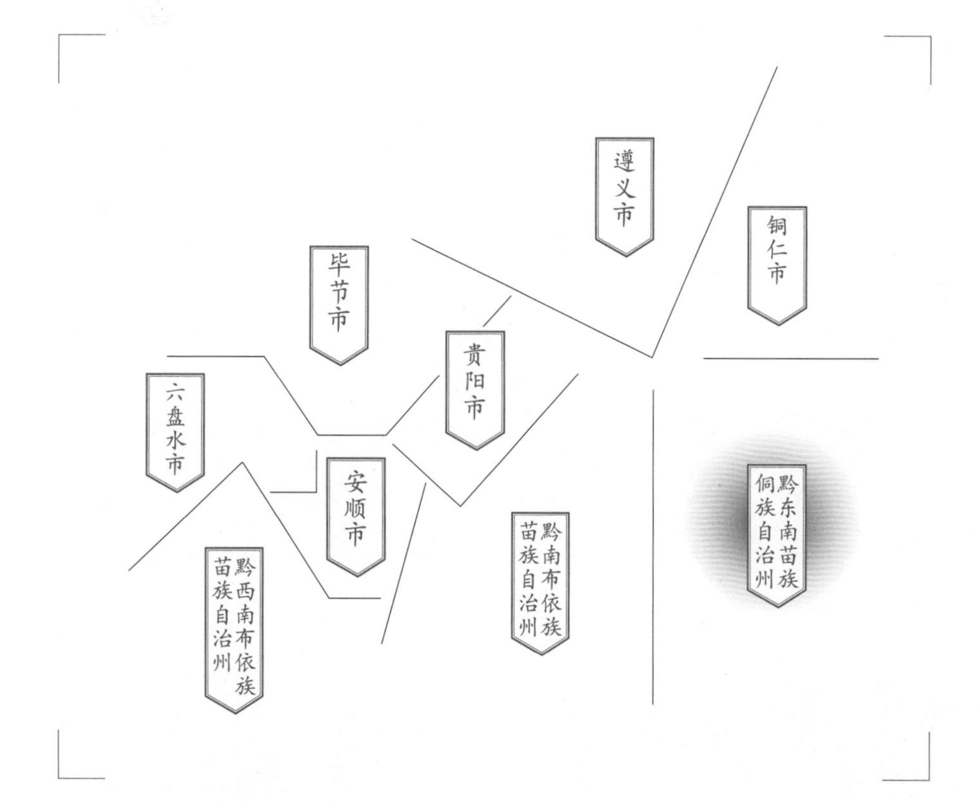

调查对象

翠里瑶族壮族乡
高华行政村
瑶族手工纸

Section 8
Handmade Paper by the Yao Ethnic Group in Cuili Town of Congjiang County

Subject

Handmade Paper by the Yao Ethnic Group in Gaohua Administrative Village of Cuili Yao and Zhuang Town

一 从江翠里瑶族手工纸的基础信息及分布

1 Basic Information and Distribution of Handmade Paper by the Yao Ethnic Group in Cuili Town of Congjiang County

从江翠里瑶族壮族乡现存造纸点集中在高华行政村，高华村属于中国黔东南边陲相当偏远的少数民族村寨，距从江县城约85 km，2011年7月和2014年10月调查组两次入村调查，山路均崎岖无比，难以前行。

高华村从事造纸的人均为瑶族山民，他们全部采用浇纸法纯手工造纸。皮纸与草纸的尺寸都是75 cm × 147 cm（因各个造纸户家中的“里白”大小有细微差别），而且每张纸都由两桶纸料浇成，属于贵州手工造纸技艺中不多见的浇纸法系的技术，但在调查时则未能获知其输入来源的任何信息。

高华村的手工皮纸以本地产的构树皮为主要原料，手工草纸则以本地产的糯稻秆为主要原料，成品的销售方式以附近的各民族百姓上门购买为主。皮纸主要用于书写、垫棺底、制作苗族“剪花”和雨衣等；草纸主要用于春节、清明节、七月半（中元节）祭祀，举行特别的法事活动时祭祀祖先，以及丧祭时作为纸钱焚烧祭奠逝者。

⊙1

⊙1 高华村田里收割完的糯稻秆 Glutinous rice stalk in Gaohua Village

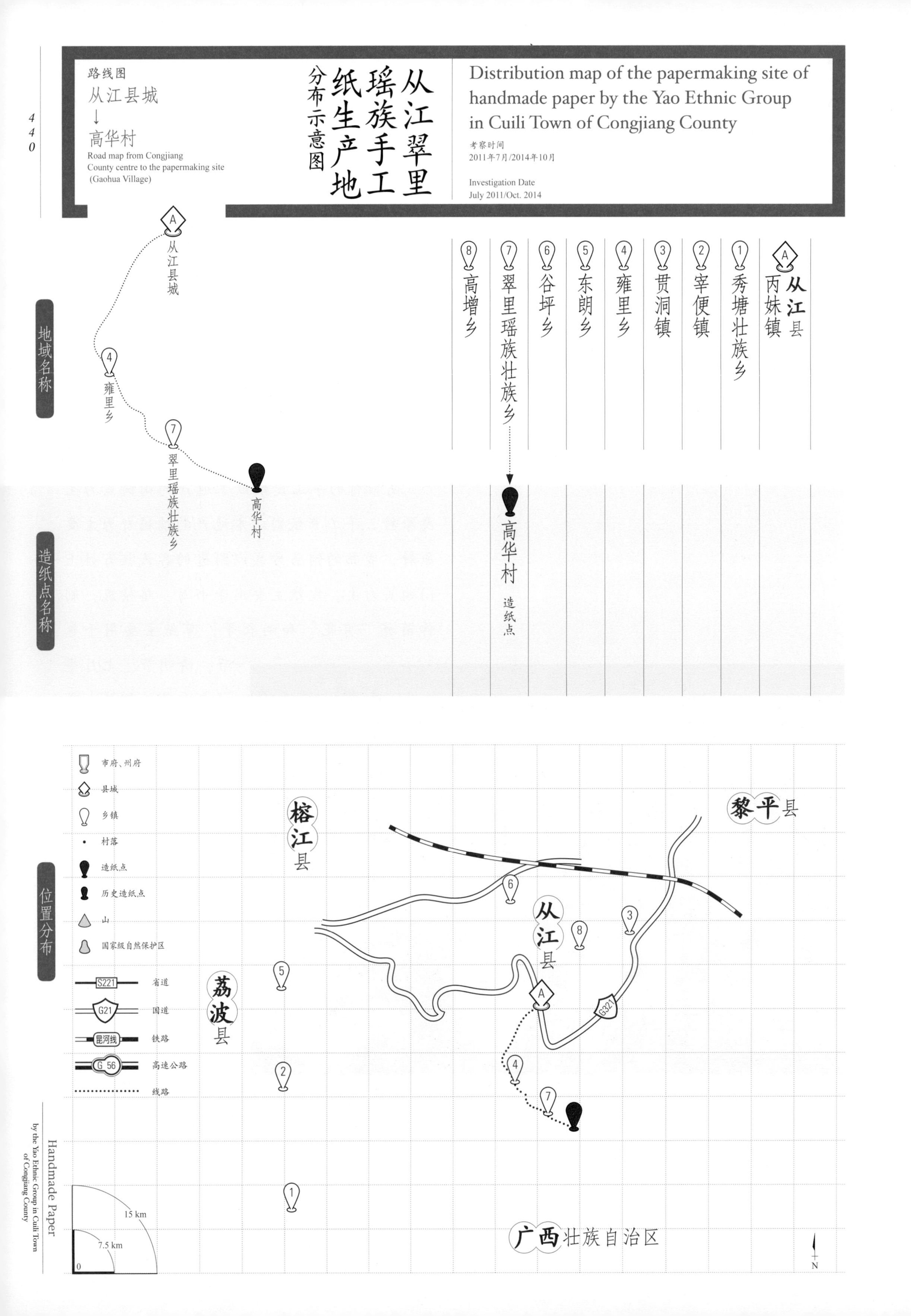
440
路线图
从江县城
↓
高华村
Road map from Congjiang County centre to the papermaking site (Gaohua Village)
从江翠里瑶族手工纸生产地分布示意图
Distribution map of the papermaking site of handmade paper by the Yao Ethnic Group in Cuili Town of Congjiang County
考察时间
2011年7月/2014年10月
Investigation Date
July 2011/Oct. 2014
地域名称
造纸点名称
位置分布
从江县城
雍里乡
翠里瑶族壮族乡
高华村
从江县
丙妹镇
秀塘壮族乡
宰便镇
贯洞镇
雍里乡
东朗乡
谷坪乡
翠里瑶族壮族乡
高增乡
高华村
造纸点
市府、州府
县城
乡镇
村落
造纸点
历史造纸点
山
国家级自然保护区
S221
省道
G21
国道
昆河线
铁路
G 56
高速公路
线路
榕江县
黎平县
从江县
荔波县
G321
广西壮族自治区
15 km
7.5 km
0
N
Handmade Paper by the Yao Ethnic Group in Cuili Town of Congjiang County

二 从江翠里瑶族手工纸生产的人文地理环境

2 The Cultural and Geographic Environment of Handmade Paper by the Yao Ethnic Group in Cuili Town of Congjiang County

从江县隶属黔东南苗族侗族自治州，位于贵州邻近广西的东南边界，东邻广西三江侗族自治县，南接广西融水苗族自治县和环江毛南族自治县，西连本州的榕江与黔南布依族苗族自治州荔波两县，北连本州黎平县，居都柳江中游，距省会贵阳450 km，距州府凯里252 km，距广西北海港600 km。县域面积3 244 km^2，总人口32万，少数民族人口占总人口的94%，以苗、侗、壮、瑶、水等少数民族为主要人口构成，是一个非常典型的多民族聚居县，县政府驻丙妹镇。

从江县地处东经108° 05′~109° 12′、北纬25° 416′~26° 05′，属中亚热带温暖型气候区，年平均气温18.4 ℃，冬温夏凉，年平均降水量1 195 mm。从江县为深山县区，全县海拔在1 200 m以上的山峰有270座，最高峰九万大山无头界峰海拔1 167 m，最低点位于都柳江出境处，海拔145 m。山地面积占县域总面积的91.34%，坝子占2%，河流滩涂占4%，真正称得上“九山半水半分田”（实际上远没有半分田），是一个传统农业生产资源十分贫乏之地。

⊙1

⊙1 流经从江县城的都柳江 Duliu River flowing through Congjiang County

从江县的主要河流为都柳江，由北往南纵贯全境，属珠江水系的支流。沿都柳江分布着丰富的山区河流，其中长度20 km以上的河流有7条，水能资源理论丰富，蕴藏量为20.88万千瓦。

从江地域早期历史湮没无闻，这无疑与其深山僻地的自然与交通环境有关。元朝时，该地域属思州军民安抚司统辖。明洪武三年（1370年），在其地设福禄永从蛮夷长官司，永乐五年（1407年），又设西山阳洞蛮夷长官司，归属思州宣慰司，正统六年（1441年），建永从县，归黎平府管辖。清乾隆三十六年（1771年），在其地域又设下江厅。民国二年（1913年），下江厅更名下江县，与永从县并属黔东道，民国三十年（1941年），合并永从、下江两县并各取后一字为从江县，其名沿用至今。

从江县2012年辖8镇13乡，其中翠里瑶族壮族乡、刚边壮族乡、秀塘壮族乡为民族乡。

从江县历史上以交通十分不便著称于黔地，甚至20世纪60年代前期，全县无一条公路，1964年由黎平县至从江县的第一条公路才修通，是贵州省最晚通公路的县。1992年，全县建成16条山区乡村公路，但无一条达到标准，而且全县无一座汽车站和水运码头，其进出不便闻名全贵州。2011年，随着从江高品质旅游资源的开发，已形成“321”国道南北贯通全境、都柳江水运通达两广（已建从江码头、八洛码头）的新局面。

从江县以文化旅游资源丰富多样而著称，其中具代表性的有：

被中华人民共和国文化部1996年命名为“中国民间艺术之乡”的高增乡小黄行政村，由小黄、高黄等7个自然村寨共20个村民组构成，村民全部为侗族人。小黄村是著名的国家非物质文化遗产“侗族大歌”的发源地与传承代表，全村人人能歌善舞，有很多支别具风采的民间歌队。该村的侗族大歌以多声部、复杂精妙组合的复调音乐享誉世界，是中国少数民族音乐的杰出代表。

被誉为“世外桃源”的高增乡占里村地处深山密林中，山清水秀，生态环境保护得非常好。全村包括8个村民组，均为侗族，据传其祖先是1 000多年前由广西梧州一带迁移而来的。占里村因独特的调节生育的绝技和控制人口增长的文化而知名于世。当地有非常有效的中草药，即“堵药”（避孕药）、“祛药”（终止怀孕药）、“换花药”（改变胎儿性别的药），以家庭为单位在寨里秘传，但只传女不传男。占里村控制人口的寨规是清代初年的寨老提议建立的，即一对夫妇只生两个孩子，时至今日，每年农历二月初一和八月初一，全寨人都要在鼓楼里听寨老训诫，并用侗歌唱寨规。1949年以来，占里村的人口自然增长率几乎为零，享有“中国人口文化第一村”之誉。

往洞乡的增冲村是以侗族为主形成的20多个小寨子，其以侗族鼓楼为中心的侗族乡村聚落文化闻名遐迩。增冲村的鼓楼始建于清康熙十一年（1672年），形如宝塔，五层十三檐为村寨中心，四周一层层被民居环绕，1988年被中华人民共和国国务院公布为全国重点文物保护单位。2008年增冲村被评为“中国历史文化名村”。

高华村位于翠里瑶族壮族乡东部，以瑶族人为主，少数苗族、侗族人因婚嫁来到高华村。调查组2014年10月入村时，村中共有89户人家，以赵姓为主，只有3户不姓赵。高华村居深山之中，古树满山，翠竹盖岭，森林覆盖率高达86%，空气十分清新，有“天然氧吧”之誉。茂密的大森林成为高华村村民的生命之源。高华村的山坳下有一条小河，名为乌峨河，造纸过程中使用的水都来自与乌峨河同源的山泉水。

高华村民居多以吊脚楼形式修建。一层为家畜家禽圈；二层为起居层，内有大厅、厨房、卧室、药池，每户家中大堂内均有一木质神龛，供祭祀先祖之用；三层为储藏层，但有不少人家把

三楼辟出一部分，装修成住房。当调查组2014年10月再次进村调查时，被调查的人家中还在使用树枝木柴作为主要燃料进行烹饪、烧水、取暖。

高华村虽为僻处深山的小村寨，但民族文化沉淀十分浓郁，瑶族药浴、瑶族长鼓舞、瑶族琵琶歌、瑶族藤编、瑶族手工造纸、瑶族个贷契约文书等均极具特色。

“瑶族药浴”是贵州黔东南瑶族人民世代相传的洗浴方式。2008年瑶族药浴已被列入国家级非物质文化遗产名录，有与土耳其浴、荷兰桑拿浴并称世界三大洗浴文化的民间说法。

“瑶族药浴”是瑶族民间用以抵御风寒、消除疲劳、防治疾病的传统知识文化传承。它是一门瑶族祖先独创、族内独有、传内不传外的保健技艺。“瑶族药浴”以多种植物药配方，经过烧煮成药水，再放入杉木桶中，人坐在桶内熏浴浸泡，让药液渗透五脏六腑、全身经络，以达到祛风除湿、活血化瘀、排汗排毒的功效。

⊙1
⊙2

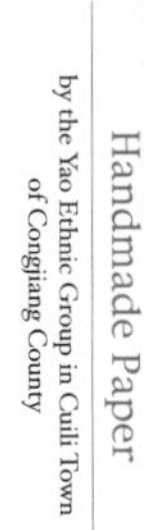

⊙3

⊙4

少吃少穿，而且长期居无定所，世人习称为“过山瑶”。瑶民长期游耕于高寒山区的深山密林之中，长年与瘴气、寒气打交道，加上气候多变，毒蛇、毒虫侵袭，环境十分恶劣，为适应这种恶劣环境，瑶民族祖先发明了“药浴”，它记录着该民族艰难生存的历史足迹。

高华村的瑶族长鼓舞雄壮激越，于2009年被列入贵州省级非物质文化遗产名录。

三 从江翠里瑶族手工纸的历史与传承

3 History and Inheritance of Handmade Paper by the Yao Ethnic Group in Cuili Town of Congjiang County

调查中并未了解到关于从江翠里瑶族手工纸的历史及起源。村民们都认为高华村瑶族皮纸是世代相传的，但起源的历史年代在当地未发现有任何记述，瑶族造纸户们均已记不清三代以上的情况，而当地的乡土史志文献也未见到与造纸有关的任何记载。此外，高华瑶族的造纸技术传女不传男，调查时曾就这一习俗问询村民赵成富等人，赵成富说，因为男人都“懒”，所以无法相信他们能将技术传好。

据2014年仍在从事手工纸生产的赵金妹老人描述，翠里乡高华村中目前只有6人长年造纸，分别为赵掌珠、赵金妹、赵瑞发、赵糖妹、赵客妹以及赵先花。2014年10月调查组二次进村时，重点选择了造纸人赵掌珠进行深入访谈，探寻了其家族造纸传承系谱的具体状况。

⊙1 茂林环抱的高华村 Gaohua Village surrounded by dense forest

⊙2 高华村的吊脚楼 Diaojiao (suspended) buildings in Gaohua Village

⊙3 柴灶铁锅熬煮的药浴药水 Boiling the herbs in an iron wok for bathing

⊙4 盛满药水的洗浴木桶 Bath tub for holding liquid herbs

⊙1

⊙2

⊙3

赵掌珠，1938年生，已在村里造纸60余年，有两个女儿一个儿子，儿子按习俗未得传授不会造纸。据赵掌珠回忆，她本人学会的造纸技艺是从广西嫁过来的母亲彭二妹（或冯二妹，调查中未能准确核实）处传下来的，母亲彭二妹（冯二妹）是从奶奶赵美凤处学会造纸的，母亲、奶奶都已去世，现在家中除了自己外，两个嫁到外地的女儿和孙女赵思琪也会造纸，如表10.10所示。从传承现状来看，五代女性全会造纸，而且代际传习一直正常延续。

表10.10　赵掌珠家五代造纸人传承表
Table 10.10　Papermaker Zhao Zhangzhu's genealogy of papermaking inheritors

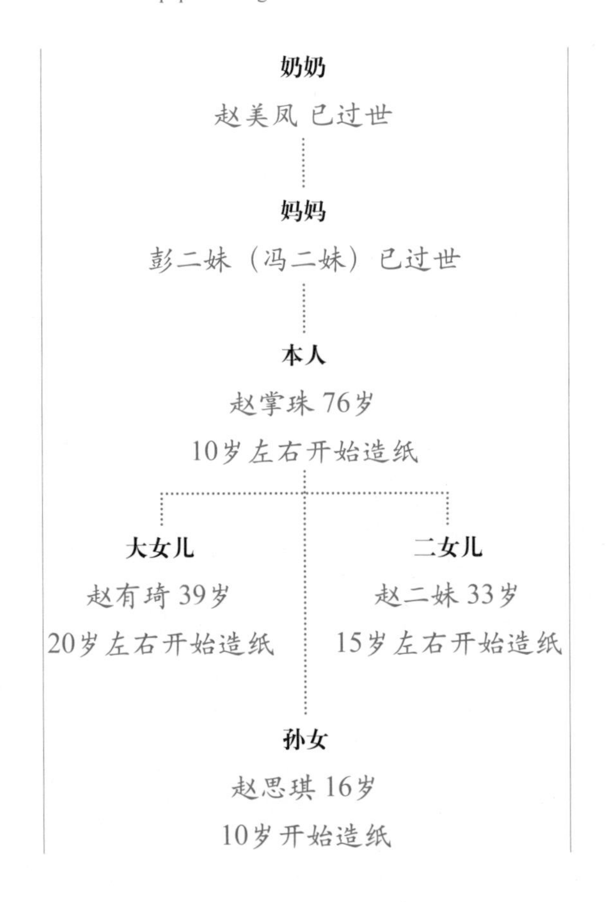

⊙1
访谈正在吃饭的赵成富
Interviewing papermaker, Zhao Chengfu, who was having dinner

⊙2
访谈赵金妹等造纸人
Interviewing Zhao Jinmei and other papermakers

⊙3
已剥好的构树皮
Processed paper mulberry bark

四
从江翠里瑶族皮纸的生产工艺与技术分析

4
Papermaking Technique and Technical Analysis of Bast Paper by the Yao Ethnic Group in Cuili Town of Congjiang County

(一)
从江翠里瑶族皮纸的生产原料与辅料

1. 构藤树皮

高华村瑶族之所以将构树称为构藤树，是因为当地该植物长得既像树，又像藤。砍构藤树的时间一般在农历二三月间，所砍构藤树以直径在3 cm以内为最佳，直径在3 cm以上的则易生虫。以1~2年生的构皮或者砍掉老树后第二年生长出来的新构皮为最好。

2. 纸药

当地造纸所用的纸药有3种，分别为“求之流”（即芙蓉树根的瑶语音译）、芙蓉树叶以及野棉花。当地瑶语音译所称的“求之流”实即当地野生猕猴桃树根部，砍下野生猕猴桃树根后，在水中泡1~2天就可浸出胶状黏液，该胶状液体就是纸药。调查组2014年10月再次入村调查时大多数造纸户已改用芙蓉树叶泡水后的液体作为纸药，只有少数造纸户仍然使用芙蓉树根泡水后的液体作为纸药。

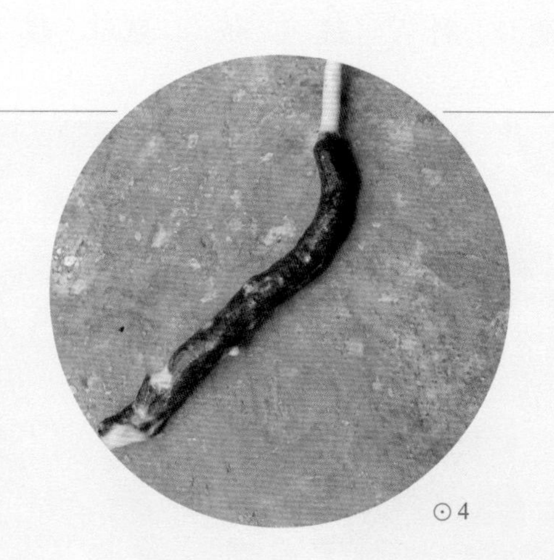

⊙4
芙蓉树根
Root of *Actinidia chinensis* (Planch.)

3. 火灰

火灰即草木灰，取自造纸户自家灶台，主要用于将构皮漂白。通常一锅可放2.5~3 kg干构皮，要用1.5 kg左右的火灰。

4. 水

高华村造纸过程中所使用的水为当地的山泉水，村边山坳中有一河名为乌峨河，乌峨河之水同源自山中山泉。

（二）

从江翠里瑶族皮纸的生产工艺流程

通过2011年7月8~9日和2014年10月3~4日入村考察及访谈，总结高华村皮纸的生产工艺流程如下：

壹	贰	叁	肆	伍	陆	柒	捌	玖	拾	拾壹
砍树	剥皮	晒皮	煮皮	洗皮	捶纸	摇纸料	浇纸	晒纸	扯纸	叠纸

壹 砍树

1

农历二三月间砍适宜用作造纸原料、直径为0.5~3 cm、1~2年生的新生构藤树。当地造纸户认为，太粗的构藤树的树皮造出的纸质量不好，而且造纸户们特别强调，在山上砍掉老构树后，第二年生长出来的嫩构树皮料最好。

⊙1

贰 剥皮

2

现场将砍下后的构藤树最外层黑皮剥掉，剩下的拿回家。回家后再将白皮剥出，最里层的树皮也不要，只要中间一层。如果黑皮不剥干净，造出的纸就会发黑。通常一个人一天可剥15~20 kg湿构皮。

叁 晒皮

3

将已剥出的湿构皮挂在屋檐下，太阳晒干或风干。晴天晒3~4天即干，阴天晾10天即干。如果天气不佳，就改用火烤（将构树皮放在屋内的火炕上用慢火烤干后，放1~2年都没问题）。

⊙2

⊙1 吊脚楼旁的构树 Paper mulberry tree by the Diaojiao (suspended) building

⊙2 晒好的干构树皮和皮纸 Dried paper mulberry bark and bast paper

肆
煮　皮
4 ⊙3

将干构皮放入锅内并加火灰和清水后，盖上盖子蒸煮。一锅可放2.5~3 kg干构皮，需1.5 kg左右的火灰。水烧开后，再煮约1.5小时即可，构皮煮好的判断标准是将其拿在手里一撕即烂。

⊙3

伍
洗　皮
5

将煮好的构皮用箩筐挑至小溪边，用手将其上的火灰洗干净。一个人一小时可洗完一锅。

陆
捶　纸
6 ⊙4

将洗好的构皮放在石板上，用木槌将其捶成浆状，折叠后再捶，反复多次，最后将捶好的料揉成团状。

⊙4

柒
摇　纸　料
7 ⊙5

将已浸泡好的纸药液倒入装有构皮的桶内，再用两根树枝搅拌几分钟使之均匀混融。

⊙5

捌
浇　纸
8 ⊙6

用瓢舀出纸浆，均匀地浇在里白上并放好即可。

⊙6

⊙3 煮构皮 Boiling paper mulberry bark

⊙4 捶纸 Hammering the paper

⊙5 摇纸料 Stirring the papermaking mucilage

⊙6 浇纸 Pouring paper pulp onto the papermaking screen

玖
晒纸
9 ⊙7 ⊙8

将里白放在太阳底下晾晒或通风处风干，不被雨淋即可。晴天2天左右能干，阴天需要4天左右才可阴干，如果不干或者希望干得快点，可用火烤，当地通常用小堆明火烤而不使用火墙烤。

⊙7

⊙8

拾
扯纸
10 ⊙9

用手将晒纸板上的纸扯下来。通常的做法是用牛骨轻轻刮开纸的右边并至1/2处，由右上往左上将纸刮开，然后再用手抓住纸上方左右两边角顺着往下撕。

⊙9

拾壹
叠纸
11 ⊙10~⊙12

扯下纸后，在纸架或晒板上直接将纸正面对叠，再将合口处往上叠2/3，最后将纸开口处往下叠1/3，这样就完成了成品纸的叠放步骤。

⊙10

⊙11

⊙12

⊙7 烤纸 Drying the paper by fire
⊙8 晒纸 Drying the paper
⊙9 扯纸 Peeling the paper down with an ox bone
⊙10/12 叠纸 Folding the paper

（三）从江翠里瑶族皮纸生产使用的主要工具设备

高华村的瑶族造纸户由于没有专用的手工纸加工作坊，因此造纸的场地显得很杂乱，使用的工具也相对简单。调查中实地观察所使用的工具通常有锅、木槌、石板、桶、牛骨、里白、纸帘架等。

壹 木槌 1

用来捶纸浆，可以将纸浆纤维捶得非常柔细，但需要不停地将纸浆饼进行折叠，然后再翻捶。实测赵掌珠家的木槌长约22 cm，圆槌直径约8 cm，高约7.8 cm。

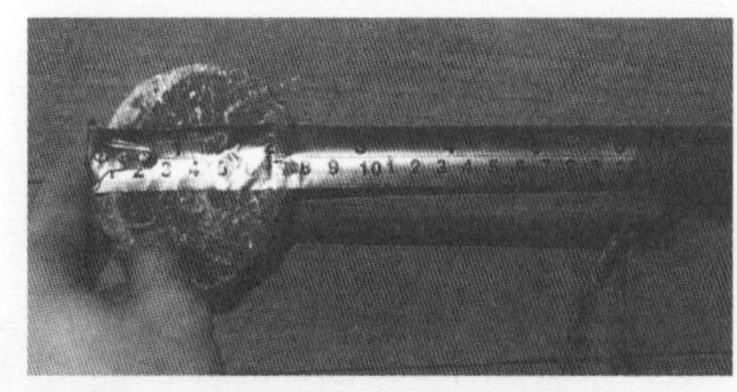

⊙13

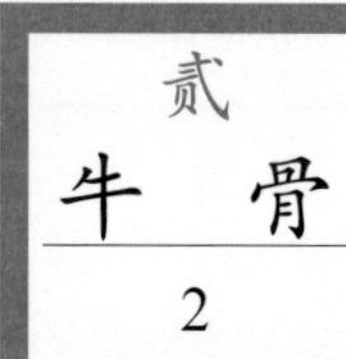

贰 牛骨 2

由牛肋骨制成，用于将晾晒在板上的纸揭开。实测牛骨长约15 cm。

⊙14

叁 里白 3

当地瑶语中普通白布的意思，浇纸时作为承受纸浆料的滤网使用，通常都紧绷在纸帘架上。实测里白长约157 cm，宽约77 cm。

⊙15

肆 纸帘架 4

由杉树制成，因为杉树碰水也不会烂，一个架子横向分为四格。实测纸帘架长约173 cm，宽约93 cm。

⊙16

⊙13 木槌 Wooden mallet

⊙14 牛骨 Ox bone for peeling the paper down

⊙15 里白 White cloth as a sieve on the papermaking screen

⊙16 纸帘架 Frame for supporting the papermaking screen

(四)

从江翠里瑶族皮纸的性能分析

所测从江高华村皮纸的相关性能参数见表10.11。

表10.11 高华村皮纸的相关性能参数
Table 10.11 Performance parameters of bast paper in Gaohua Village

指标		单位	最大值	最小值	平均值
厚度		mm	0.320	0.220	0.270
定量		g/m²	—	—	63.8
紧度		g/cm³	—	—	0.236
抗张力	纵向	N	33.9	24.7	31.4
	横向	N	26.7	18.6	22.8
抗张强度		kN/m	—	—	1.807
白度		%	23.4	21.3	22.3
纤维长度		mm	9.75	0.92	3.58
纤维宽度		μm	37.0	2.0	16.0

由表10.11可知，所测高华村皮纸最厚约是最薄的1.45倍，相对标准偏差为3.30%，厚薄差异较小。皮纸的平均定量为63.8 g/m^2。所测皮纸紧度为0.236 g/cm^3。

经计算，其抗张强度为1.807 kN/m，抗张强度值较大。

所测高华村皮纸白度平均值为22.3%，白度较低，白度最大值约是最小值的1.1倍，相对标准偏差为0.55%，白度差异较小。

所测高华村皮纸纤维长度：最长9.75 mm，最短0.92 mm，平均3.58 mm；纤维宽度：最宽37.0 μm，最窄2.0 μm，平均16.0 μm。所测皮纸在10倍、20倍物镜下观测的纤维形态分别见图★1、图★2。

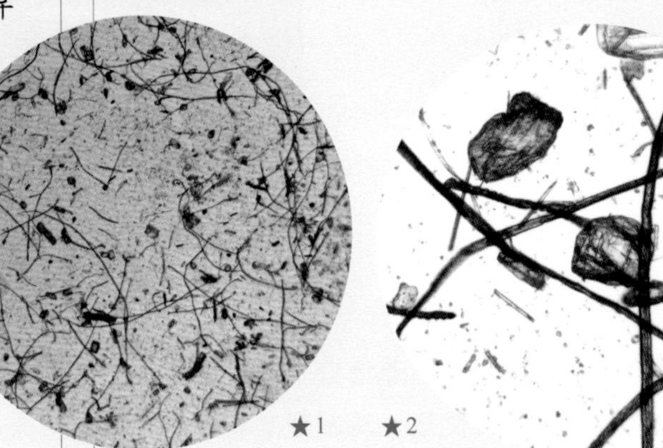

★1 高华村皮纸纤维形态图(10×) Fibers of bast paper in Gaohua Village (10× objective)

★2 高华村皮纸纤维形态图(20×) Fibers of bast paper in Gaohua Village (20× objective)

五 从江翠里瑶族草纸的生产工艺与技术分析

5
Papermaking Technique and Technical Analysis of Straw Paper by the Yao Ethnic Group in Cuili Town of Congjiang County

(一)

从江翠里瑶族草纸的生产原料与辅料

1. 糯稻秆

高华村造草纸所用原料来自本地，为当地村民自己种植的糯稻秆。每年糯稻成熟时，将糯稻收割后打下糯米，留稻秆作为造纸原料。造纸时去除叶子和穗，只用稻秆。

2. 纸药

当地造草纸与皮纸所用的纸药相同，均有三种，即“求之流”、芙蓉树叶以及野棉花。调查

⊙1

⊙2

组2014年10月入村调查时发现，因为芙蓉树叶泡水后可立即使用，简便实用，所以大多数造纸户已改用芙蓉树叶泡水后的液体作为纸药，只有少数造纸户将芙蓉树根泡水后的液体作为纸药。

3. 火灰

火灰即村民家炉灶烧过柴火后的草木灰，作用是软化禾草，有利于将禾草煮烂。煮料时，通常在煮料锅里放一层禾草撒一层火灰。据被调查人赵掌珠介绍，火灰一定要细，需要用簸箕进行筛选。

4. 水

高华村造纸户所使用的水是来自山上的泉水和乌峨河中的水，而乌峨河中的水也来自山上的泉水。

（二）从江翠里瑶族草纸的生产工艺流程

通过2011年7月8~9日和2014年10月3~4日入村考察及访谈，总结高华村草纸的生产工艺流程如下：

壹	贰	叁	肆	伍	陆	柒	捌	玖	拾
收禾	去穗去叶	煮料	挑料	舂料	下纸药	浇纸	晾干	剥纸	叠纸

⊙1
5把干糯稻秆
Five bundles of dried glutinous rice stalk

⊙2
纸药芙蓉树叶
Papermaking mucilage, leaves of *Actinidia chinensis* (Planch.)

壹 收禾 1

⊙1⊙2

每年9月中下旬，待糯稻成熟后即可收割。

⊙1

贰 去穗去叶 2

将收割回来的糯稻的稻穗打下，再去掉叶子和穗壳。

⊙2

⊙1/2 未收割的糯稻田与收割后的糯稻田
Glutinous rice field to be reaped and after reaping

叁 煮料

3

将煮料锅置于柴灶上，并将去掉叶子和穗壳的糯稻秆平铺于锅内，再把火灰铺在稻秆之上，一层稻秆一层火灰，然后加满水煮。一锅可放40把稻秆，大约需要3匀火灰。煮料时间需要一个白天。

肆 挑料

4

将煮好的草料用粪挑挑至乌峨河中冲泡，洗去料中的火灰及杂质，然后将煮好的草料放在河水中冲洗2~3天。

伍 舂料

5

将冲洗干净的草料挑回家中并用木舂舂碎，一次舂4把，需30分钟左右时间。

陆 下纸药

6 ⊙3

将舂好的纸料放在桶内加水（以前用木桶，现在改用塑料桶，桶大小不一，大桶一次可装4把稻秆的料，小桶可装2把稻秆的料，一把稻秆能浇一张草纸），然后加入纸药，用木棍将纸药和纸浆搅拌均匀。

⊙3

柒 浇纸

7 ⊙4⊙5

先用竹棍将纸浆搅匀，然后用瓢将纸浆缓缓地浇在纸帘上，使纸浆均匀地覆盖里白的每一部分。

⊙4

⊙5

捌 晾干

8 ⊙6

将浇好的纸连同纸帘架一起放在太阳底下或通风处，不被雨淋即可，使之慢慢晾干。晴天1天可晒干，阴天2天左右可干，雨天4天左右可干，如果不干或者希望干得快点，可用火烤。

⊙6

玖 剥纸

9 ⊙7

同从江翠里瑶族皮纸“扯纸”工序。

拾 叠纸

10

同从江翠里瑶族皮纸“叠纸”工序。

⊙7

⊙3 赵掌珠为调查组成员示范如何搅拌
Zhao Zhangzhu showing how to stir the papermaking materials

⊙4/5 赵掌珠正在示范浇纸
Zhao Zhangzhu showing how to pour paper pulp onto the papermaking screen

⊙6 屋外晾纸
Drying the paper outside the room

⊙7 赵掌珠正在示范剥纸
Zhao Zhangzhu showing how to peel the paper down

(三)

从江翠里瑶族草纸生产使用的主要工具设备

壹 煮料锅 1

用于煮糯稻秆。一锅能煮40把稻秆，实测煮料锅直径约90 cm。

⊙1

贰 粪挑 2

即箩筐和扁担，用于将煮好的料挑至乌峨河中冲洗。

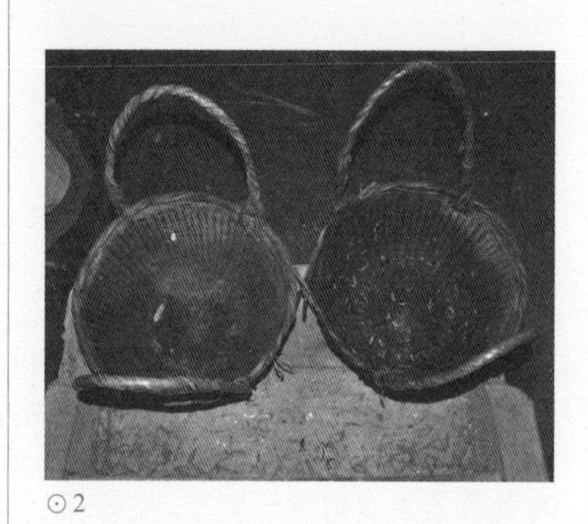

⊙2

叁 木舂 3

用于将冲洗干净的纸料舂碎、舂细成纸浆。实测木舂长约230 cm。

⊙3

肆 桶 4

用于盛放纸浆和纸药。

⊙4

伍 瓢 5

赵掌珠使用的瓢为葫芦所制，用于舀出舂好的纸浆并浇在里白上。

⊙5

陆 簸箕 6

用于筛选火灰。

⊙6

⊙1 煮料所用的大锅 Iron wok for boiling the papermaking materials

⊙2 用于挑料的「粪挑」 Bamboo basket and shoulder pole for carrying the papermaking materials

⊙3 木舂 Wooden pestle

⊙4 用于盛放纸浆的桶 Vat for holding paper pulp

⊙5 葫芦瓢 Gourd ladle

⊙6 用于筛选火灰的簸箕 Sieve for winnowing ashes

柒

纸帘架

7

由杉树制成，因为杉树碰水不会烂，一个架子横向分为四格，用于支撑里白。实测纸帘架长约173 cm，宽约93 cm。

⊙7

捌

牛　骨

8

由牛肋骨制成。用于将晾晒在板上的纸揭开。实测牛骨长约15 cm。

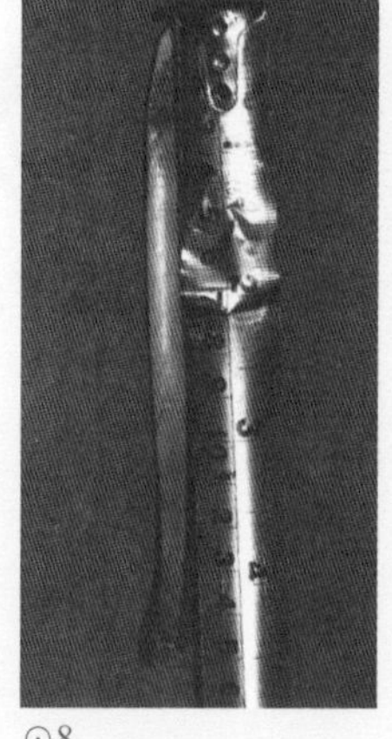

⊙8

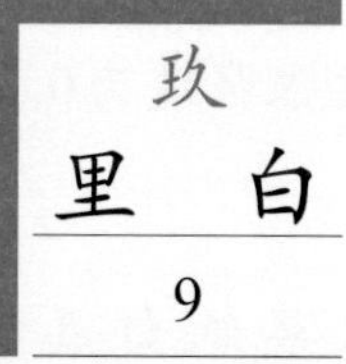

玖

里　白

9

当地瑶语对普通白布的叫法，用于浇纸时作为承受纸浆料的滤网使用，通常都紧绷在纸帘架上。实测里白长约157 cm，宽约77 cm。

⊙9

(四)

从江翠里瑶族草纸的性能分析

所测从江高华村草纸的相关性能参数见表10.12。

表10.12　高华村草纸的相关性能参数

Table 10.12　Performance parameters of straw paper in Gaohua Village

指标		单位	最大值	最小值	平均值
厚度		mm	0.120	0.100	0.110
定量		g/m^2	—	—	25.4
紧度		g/cm^3	—	—	0.231
抗张力	纵向	N	10.6	6.1	8.1
	横向	N	8.4	6.0	7.3
抗张强度		kN/m	—	—	0.513
白度		%	34.1	32.1	32.7
纤维长度		mm	3.02	0.45	1.25
纤维宽度		μm	17.0	2.0	7.0

⊙7 纸帘架 Frame for supporting the papermaking screen

⊙8 牛骨 Ox bone for peeling the paper down

⊙9 里白 White cloth as a sieve on the papermaking screen

由表10.12可知，所测高华村草纸最厚是最薄的1.2倍，相对标准偏差为3.08%，纸张厚薄差异较小。草纸的平均定量为25.4 g/m^2。所测草纸紧度为0.231 g/cm^3。

经计算，其抗张强度为0.513 kN/m，抗张强度值较小。

所测高华村草纸白度平均值为32.7%，白度较低，白度最大值约是最小值的1.06倍，相对标准偏差为0.65%，白度差异较小。

所测高华村草纸纤维长度：最长3.02 mm，最短0.45 mm，平均1.25 mm；纤维宽度：最宽17.0 μm，最窄2.0 μm，平均7.0 μm。所测草纸在10倍、20倍物镜下观测的纤维形态分别见图★1、图★2。

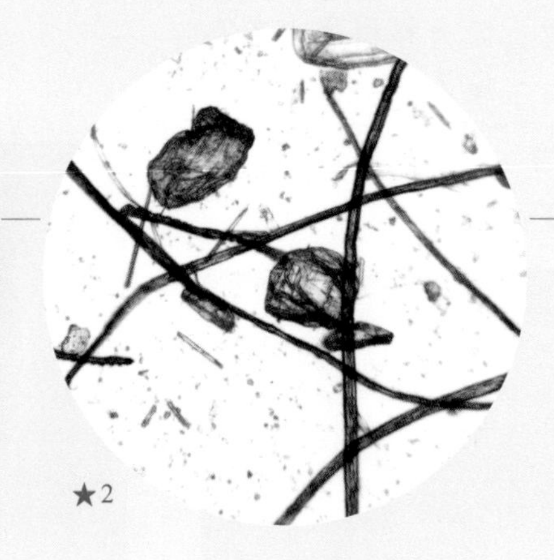

★1 ★2

六 从江翠里瑶族手工纸的用途与销售情况

6 Uses and Sales of Handmade Paper by the Yao Ethnic Group in Cuili Town of Congjiang County

(一) 从江翠里瑶族手工纸的用途

据调查组2014年10月深入访谈赵成富、赵掌珠等人所了解到的信息，高华村瑶族皮纸目前主要用于书写（例如，赵成富家中的书均用高华皮纸书写）、剪花绣花时作为内衬（主要是当地苗族和侗族使用）、做鞋垫以及老人过世后用来垫棺底（苗族、壮族也用）。据被调查人赵掌珠回忆，以前也用皮纸做雨衣（在两张皮纸表面抹上桐油，中间加一层渔网，自然晾干或者晒干），但是这一特别的用途现在已经消失，调查组曾专门寻找这一旧物，但最终未能找到。

草纸主要用于春节、清明节、七月半（中元节）、八月十五（中秋节）、法事活动时祭祀祖先以及临时丧祭时祭奠逝者。除扫墓外，其他时

★1 高华村草纸纤维形态图(10×)
Fibers of straw paper in Gaohua Village (10× objective)

★2 高华村草纸纤维形态图(20×)
Fibers of straw paper in Gaohua Village (20× objective)

⊙1

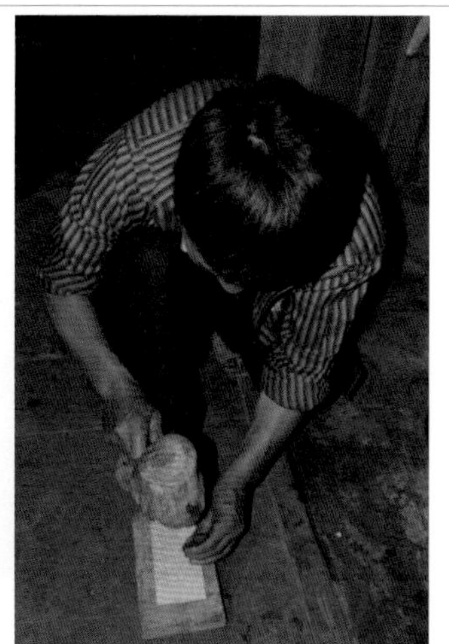
⊙2

⊙3

候烧纸都是在家中神龛前进行的。高华村祭祀时通常不使用原纸，而是将草纸加工成纸钱再用。其方法是将原纸用刀裁切成41 cm × 15.5 cm大小，3张裁好的纸合为1张，然后横向平均分成5份折叠打孔，3排或者5排，每排7孔。

(二)
从江翠里瑶族手工纸的销售情况

据被调查人赵成富口述，高华村所产皮纸、草纸较少外卖，基本上是本村自用。出售的纸只占其中很少一部分，购买的人也是附近的其他少数民族乡民。究其原因，除了可能有造纸自用的文化习俗外，最关键的还是历史上高华村与外界的交通极其不便。

据赵成富等人回忆，20世纪60年代，皮纸的销售价格为0.5元/张，草纸为0.1元/张；20世纪70年代，皮纸的销售价格为1元/张，草纸为0.2元/张；调查组2014年10月入村调查时，皮纸的销售价格为15~20元/张，草纸为2元/张，如表10.13所示。主要由高华村附近山寨里的苗族、侗族、壮族以及其他少数民族村民购买。

高华村造纸户以“张”为手工纸的计数单位，大小为75 cm × 147 cm（各个造纸户家中的里白大小有细微差别）。据被调查人赵掌珠口述，每年最多能做200张构皮纸，按照每张20元计算，一年造皮纸的收入为4 000元。由于销售外卖的纸并不多，每年最多卖出100张，毛收入最高为2 000元，所以造纸尚不能成为当地村民谋生的支柱产业。

⊙1/2
赵成富演示打纸钱
Zhao Chengfu showing how to make joss paper

⊙3
赵成富在家中的神龛前祭祖
Zhao Chengfu performing sacrificial ritual in front of the family niche

⊙1

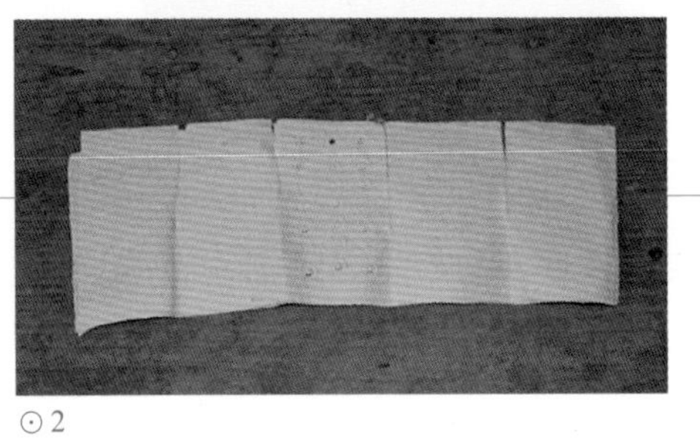
⊙2

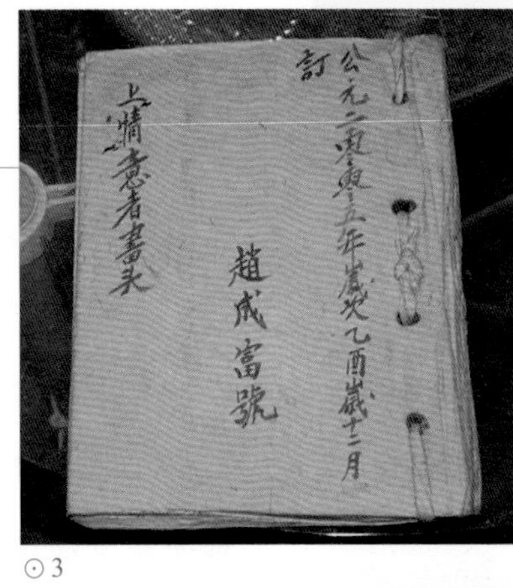

⊙3

⊙4

⊙5

⊙6

⊙7

表10.13　高华村皮纸、草纸销售价格表
Table 10.13　Price list of bast paper and straw paper in Gaohua Village

年代	皮纸销售价格（元/张）	草纸销售价格（元/张）
20世纪60年代	0.5	0.1
20世纪70年代	1	0.2
调查组入村时（2014年10月）	15~20	2

⊙1
展开的成品皮纸
Unfolded bast paper

⊙2
展开的成品草纸
Unfolded straw paper

⊙3/6
赵成富家中用皮纸抄写的书
Books transcribed on bast paper in Zhao Chengfu's house

⊙7
调查组成员正在向赵金妹买测试纸样
Researchers buying paper samples from papermaker Zhao Jinmei

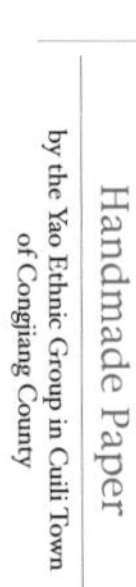

七
从江翠里瑶族手工纸的相关民俗与文化事象

7
Folk Customs and Culture of Handmade Paper by the Yao Ethnic Group in Cuili Town of Congjiang County

高华村在当地人的方言中称为“寨子”。高华村村民以赵姓为主，据村民赵成富介绍，当地村民的赵姓分为“大赵”“中赵”“小赵”“色赵”，虽然是同一个字，但姓的内涵却是不同的。大赵又是白赵，中赵又是杨赵，小赵又是蓝赵，色赵又是灰赵，色赵的颜色是不蓝不红不黑的。赵成富说，高华村这一支瑶族因为是从广东翻越一座座山过来的，所以称为过山瑶。

寨子里的造纸术传女不传男，每家的媳妇都会造纸，而男的都不会造纸，也不爱造纸。因为“造纸是女人该做的事情，而不是男人该做的事情，男人造纸会被别人笑话”。这是翠里乡瑶民造纸生产中的一大特色：女性包打天下。走进高华村的造纸作坊，忙活造纸的全是女人，而且全套造纸工序均由女性操作，偶尔有男性参与也只是打下手，承担打浆这类“粗活”。

⊙8

草纸祭俗。高华村在春节、清明节、七月半（中元节）、八月十五（中秋节）、扫墓时都会用到草纸。每年腊月二十八或二十九，高华村每户人家都会在自家门窗上贴草纸。过年期间，高华村村民每天早晚吃饭之前，都会在自家神龛中烧纸祭祖。而贴在门窗上的草纸直到正月十五那天才会撕下来，然后烧掉。当地山歌也唱到“十五烧了门前纸，烧完就要开工了”。清明节时，村民会在家中杀鸡、杀鸭、杀猪，备好贡品，请祖先回来，并烧4张大纸，而扫墓只烧3张。高华村瑶族也过七月半，七月半当天早上或者晚上会在自家神龛前烧5~6张大纸，八月十五当天也会烧1张大纸。高华村瑶族除扫墓以外，其余的祭祀活动都是在家中神龛前完成的。

⊙8 瑶族女人正在浇纸
A Yao woman pouring paper pulp onto the papermaking screen

祭祀“盘王”的祭俗。高华村瑶族人崇拜“盘王”，每户人家中都有供奉先祖和盘王的神龛，神龛内贴袱纸，放置香炉。据赵成富描述，每过20年会请“师傅”算卦，将盘王“请”回家，在神龛中供奉，然后杀2头猪还愿；如果家中人丁不旺，则每隔12~13年供奉一次。每次供奉盘王，需要烧100张大纸。赵成富为当地“师公”，负责主持村里祭祀活动。赵成富的这一身份是家族传承下来的，传至赵成富时已是第五代。在调查组入村调查的当晚，正巧赵成富要出门进行祭祀活动，调查组跟随赵成富参观了祭祀活动。在祭祀开始之前，赵成富的弟弟用篾条、草纸、谷草、树叶编织了一个外形像船的篮子，赵成富则用草纸、谷草和竹条做了幡。在祭祀过程中赵成富时而站立，时而下蹲，时而走动，时而吹牛角，时而拿斧子砍地，时而往酒杯里倒酒。整个祭祀活动共持续了3个多小时，赵成富一直用瑶语背诵祭词，其记忆力之强令人惊讶。在祭祀活动接近一半时，赵成富将事先准备好的草纸拿出来，放在地上堆放好，在屋门外搭建的台子上以及装有玉米的簸箕中放草纸和袱纸。在祭祀接近尾声时，赵成富点燃了火盆内、簸箕中以及屋门外台子上的草纸，待草纸燃烧殆尽后，整个祭祀活动就结束了。

⊙1

⊙2

⊙3

⊙4

⊙1/2
高华村村民家中的神龛
Niche in a papermaker's house in Gaohua Village

⊙3
用草纸做的篮子
Basket made of straw paper

⊙4
赵成富正在用草纸和竹子做幡
Zhao Chengfu making sacrificial flag by straw paper and bamboo

八
从江翠里瑶族手工纸的保护现状与发展思考

8
Preservation and Development of Handmade Paper by the Yao Ethnic Group in Cuili Town of Congjiang County

从调查组2011年和2014年两轮对高华村瑶族手工造纸的田野考察所了解的情况看，由于当地以瑶族为主的数个民族日常文化习俗中较好地保存着用纸的需求，因而高华村手工纸的生产业态依然发展良好，虽然较以往也有一定的收缩，但短期内未见技艺失传的危机。

伴随着交通的改善，高华村的瑶族药浴及民族乡村旅游开始发展，加上高华村一带美丽清新的自然生态旅游资源，高华旅游已初步呈现品牌传播的态势，由此对高华村瑶族手工皮纸与草纸的原生态带来的影响值得关注。

国家重视非物质文化遗产保护的政策唤醒了地方政府和文化管理部门保护瑶族手工皮纸的意识。调查组从从江县文化管理部门了解到，当地正计划将翠里乡的手工皮纸工艺纳入非物质文化遗产保护的范畴中，而翠里乡乡政府也有意将发展地方经济与保护民族文化结合起来。翠里乡和高华村的干部在访谈中均表示，该地区在大力发展与传承手工纸技艺、民族歌舞、民族节庆、民族服饰、药浴文化等文化遗产的基础上，还要积极推进民族文化旅游，从而带动民族地区的经济和文化发展持续进步。

今天，以高华村为代表的手工纸生产和消费在村域经济中还发挥着重要的支撑作用，这是一个基本良性的工艺消费与生产有机衔接的业态。但从非经济角度来说，保护乡土文化，恢复民族记忆，让人们体验自己的历史和文化，并进而传播自身的文化基因，也有着不可忽视的现代社会的历史文化价值。

⊙5

⊙6

⊙5 高华村的一处药浴传习基地
Herbal Bath Training Base in Gaohua Village

⊙6 调查组成员与赵掌珠老人
A researcher and local papermaker Zhao Zhangzhu

从江翠里瑶族皮纸

Bast Paper
by the Yao Ethnic Group
in Cuili Town of Congjiang County

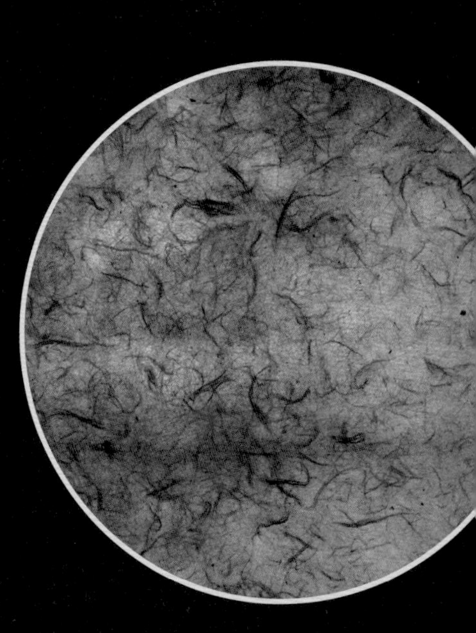

高华村皮纸透光摄影图
A photo of bast paper in Gaohua Village
seen through the light

草纸

Straw Paper
by the Yao Ethnic Group
in Cuili Town of Congjiang County

高华村草纸透光摄影图
A photo of straw paper in Gaohua Village
seen through the light

第九节

从江小黄侗族皮纸

贵州省
Guizhou Province

黔东南苗族侗族自治州
Qiandongnan Miao and Dong Autonomous Prefecture

从江县
Congjiang County

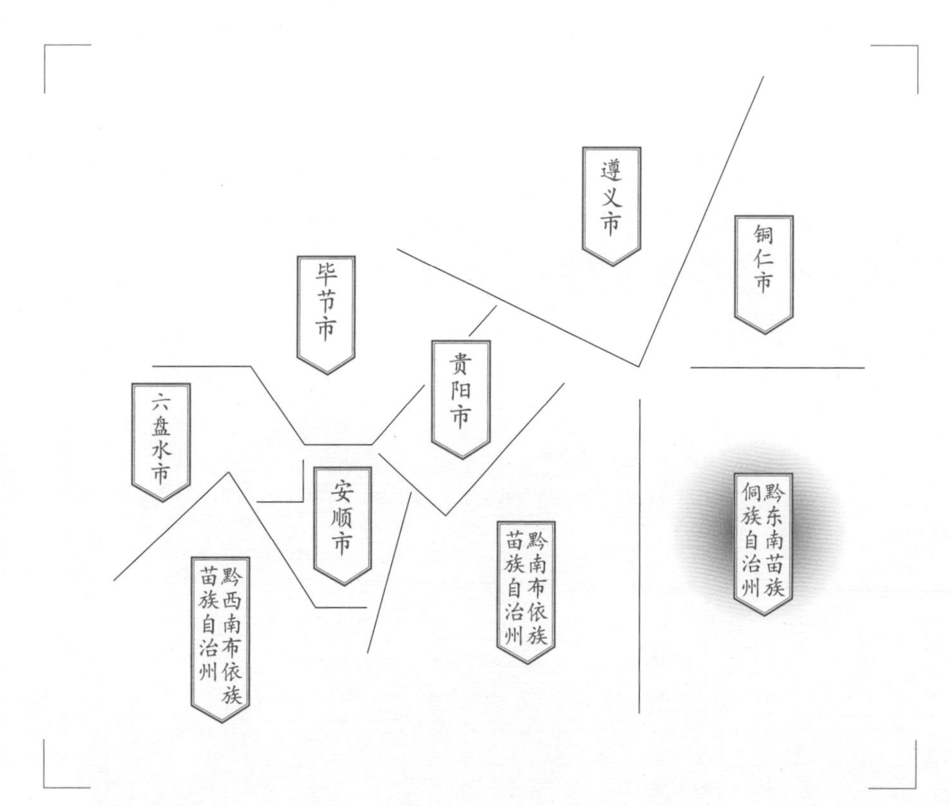

调查对象
高增乡
小黄行政村
侗族皮纸

Section 9
Bast Paper by the Dong Ethnic Group in Xiaohuang Village of Congjiang County

Subject
Bast Paper by the Dong Ethnic Group in Xiaohuang Administrative Village of Gaozeng Town

一 从江小黄侗族皮纸的基础信息及分布

1 Basic Information and Distribution of Bast Paper by the Dong Ethnic Group in Xiaohuang Village of Congjiang County

据调查组田野调查所获信息，从江县侗族造纸当代的主要活态生产点位于高增乡的小黄村、占里村，以野生的构树皮为原料，采用较为纯粹的浇纸法技艺生产。小黄侗族人称这种用类似"藤本构树"的野构树皮造出的纸为"洁纱"，侗语的意思就是用构皮造的纸。

2013年8月下旬和2014年10月上旬调查组两次进入小黄村考察，重点访谈了侗族造纸人吴奶保卫（奶的意思是妈，保卫是儿子的名字，当地侗族的习俗，女性有孩子后，要用第一个孩子的名字盖住自己的名字，从而显得孩子金贵，但吴保卫是吴奶保卫的二儿子）及潘奶桃，调研了小黄村构皮纸的完整生产过程及相关文化习俗。2014年入村调查时，全村5个村民组只有8户还在从事手工造纸，规模大幅萎缩。小黄皮纸的主要用途包括书写，制作斗笠、纸蓑衣，绣花、花盒，糊窗户、神龛等，与日常生活联系紧密。

⊙1

⊙1 类似『藤本构树』的构树种类 Local paper mulberry tree, similar to *Broussonetia kaempferi Sieb. var. australis Suzuki*

路线图

从江县城

↓

小黄村

Road map from Congjiang County centre to the papermaking site (Xiaohuang Village)

从江小黄侗族皮纸生产地分布示意图

Distribution map of the papermaking site of bast paper by the Dong Ethnic Group in Xiaohuang Village of Congjiang County

考察时间

2013年8月/2014年10月

Investigation Date

Aug. 2013/Oct. 2014

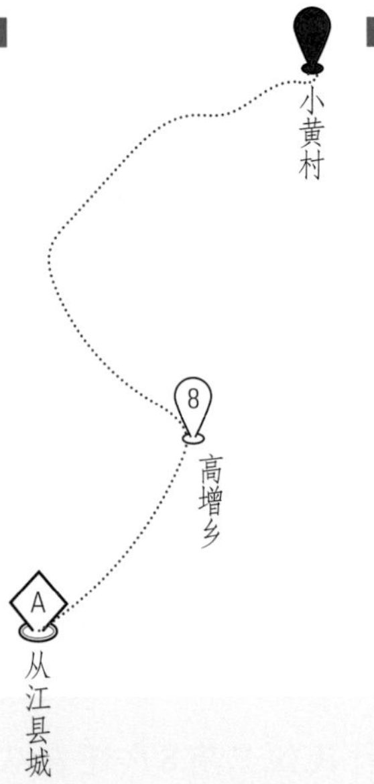

地域名称

A 从江县 丙妹镇

1 秀塘壮族乡

2 宰便镇

3 贯洞镇

4 雍里乡

5 东朗乡

6 谷坪乡

7 翠里瑶族壮族乡

8 高增乡

造纸点名称

小黄村 造纸点

位置分布

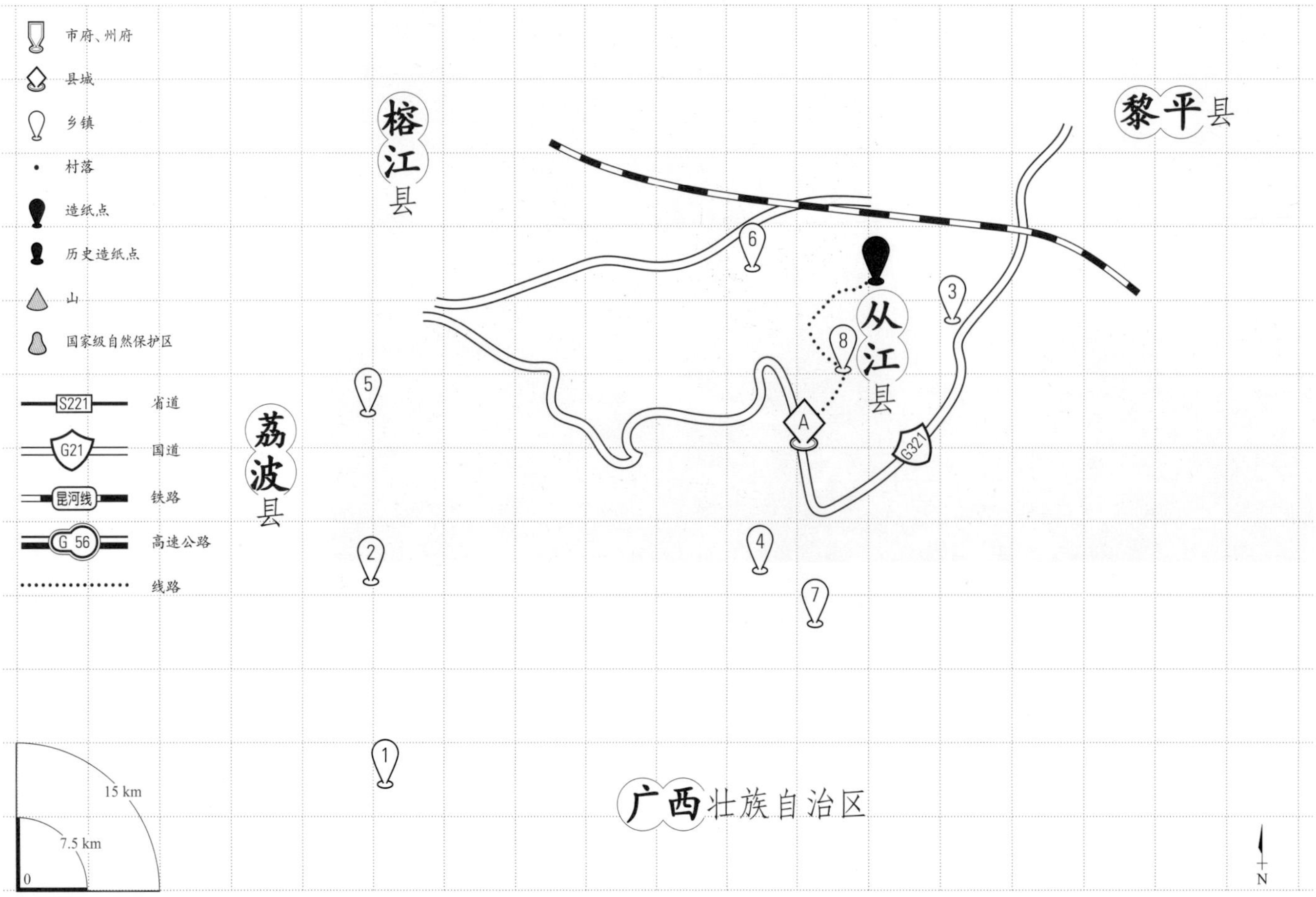

二 从江小黄侗族皮纸生产的人文地理环境

2 The Cultural and Geographic Environment of Bast Paper by the Dong Ethnic Group in Xiaohuang Village of Congjiang County

高增乡位于贵州省从江县东部，距县城8 km，东接贯洞镇，西邻丙妹镇和谷坪乡，南连广西梅林乡，北毗黎平双江乡。境内最高海拔1 107 m，最低海拔235 m。年平均气温18.4 ℃，年平均降水量1 190 mm，无霜期326天，属亚热带季风气候。全乡总面积148.78 km^2，辖12个行政村，35个自然村寨，89个村民组，侗族人口占总人口的98%，是一个以侗族为主的少数民族乡。

高增乡是贵州侗族民俗风情、文化传承的核心地带，是侗族大歌的源生地之一。侗族大歌起源于宋朝，是中国历史悠久的经典音乐样式。从20世纪50年代起，小黄村侗族大歌就多次参加全国汇演、调演并获奖；1986年，巴黎金秋艺术节

⊙1

⊙1 高增乡的侗寨 Dong ethnic residence in Gaozeng Town

⊙ 1

上，小黄村的侗族大歌在国际上首次亮相就被誉为“清泉般闪光的音乐，掠过古梦边缘的旋律”。

2006年，侗族大歌被列入第一批国家级非物质文化遗产代表作名录；2009年9月28日，侗族大歌被列入人类非物质文化遗产代表作名录，并被誉为“一个民族的声音，一种人类的文化”。

高增乡小黄村位于高增乡北部，距乡政府所在地17 km，距从江县城24 km。2013年8月入村调查时，全村有5个自然村寨718户，3 340人，均为侗族。

小黄村是极负盛名的“侗歌窝”，素有“歌的故乡”“歌的海洋”之称。小黄村人世代传下来的格言是“饭养身、歌养心”。这里无论男女老少，人人都有满肚子的歌，唱不完的歌。正如侗家乡谚所说：“汉家有读不完的书，侗家有唱不完的歌。”小黄侗歌被称为“嘎细王”（侗语“小黄歌”之意）。除了著名的大歌外，还有情歌、琵琶歌、蝉歌、采堂歌、拦路歌、敬酒歌、祭祖歌、劳动歌、叙事歌等。

2014年10月入村调查时，村里有少、青、中、老年歌队一共52组，队员2 000余人。在小黄

⊙ 2

⊙ 1 小黄鼓楼前的侗族大歌表演
Dong ethnic polyphonic music show persented in front of Xiaohuang drum tower

⊙ 2 男女老少齐合唱
Chorus of local residents including all ages and both genders

⊙1

⊙2

村，人人会唱，处处有歌，事事用歌，天天晚上唱歌。男女老少以唱歌为乐，以会歌为荣，不仅喜庆节日以歌相贺，青年男女相恋以歌为媒，就连贵客来临，也是以歌迎送。在一个偏远山村里有如此众多的村民自发性地参加唱歌活动，而且长期坚持，经久不衰，实属罕见。

"侗族大歌"是一种无乐器伴奏、无人指挥的多声部合唱，高、中、低音浑然一体，以其天然和声的完美协调、格调的柔和委婉、旋律的典雅优美著称于世。歌队由一人领唱，众人合唱，声音时而高亢宽广，时而低沉悠扬，把听众带到如诗如画、美妙和谐的大自然中。娃娃队童声单纯嘹亮，天真活泼；姑娘罗汉队歌声清纯如水；老年队歌声浑厚深沉，声音时而飘似仙乐，时而深厚如山，给人以神奇享受。

小黄村的民风纯朴，尤以"行歌坐夜"和"情人节"最为突出。每当夜幕降临，成双成对的情侣在鼓楼、花桥或家中火堂前，在寨边松、竹林里和溪边月光下，情意绵绵，牛腿琴、琵琶轻拉弹唱，倾诉衷肠，彻夜不散，直至凌晨方依依话别，夜夜如此。姑娘们的父母、兄嫂对此"行歌坐夜"的活动都主动回避，从不加以干预。"情人节"是在每年的农历六月初五、初六（又称"五六节"），40岁以上的已婚男女与自己青年时的旧相好在当天会刻意穿扮得年轻些，带上糯米饭、腌鱼，成双入对手拉手上山坡游玩、幽会，讲情话、忆当年、唱情歌、诉思恋，从早上直至次日月夜。因此，小黄村早有"小香港""不夜村"的誉称。

1956年，小黄村歌手潘凤高、潘井仁、石松花、石婢蝉等8人参加全省汇演，《小黄男声侗族大歌》荣获全省优秀节目奖；1957年，小黄村歌手吴大安、潘世才等人到云南参加全国文艺汇演，《珠郎娘美》荣获全国优秀节目奖；1964年8月，小黄村歌手吴大安等人到贵阳参加文艺调演，《嘎格罗》《唱丰收》获优秀节目奖。

1988年秋，小黄村根据当地民族特色，因地制宜，开始把民族文化引进课堂，增设侗歌、侗戏、刺绣等课程，给小黄村带来了新的活力。1994年9月，小黄小学潘美号等5名9~12岁的小姑娘代表侗族参加了"94北京国际少儿艺术节"；1995年，她们还应邀赴京参加中央电视台"六一"晚会，演唱多声部无伴奏侗族大歌；1996年，小黄小学吴培建等4名侗族女童组成侗族大歌队随中国民间艺术团赴法国演出，在巴黎演出13场，场场爆满，轰动"艺术之都"巴黎，受到法国外交部长夫人接见。2003年，桂林张艺谋艺术学校慕名到小黄村招生，22名9~12岁的小姑娘获得免费上学的机会。2003年，在小黄"侗

⊙1
小黄村民歌手汇演旧照
A photo of local singers in a show (Xiaohuang Village)

⊙2
侗歌表演中的诙谐小品
Funny show performed in Dong ethnic polyphonic music show

族大歌节”上成功举办了“侗族大歌千人大合唱”。2004年8月，小黄村的贾美兰等5名歌手参加全国民歌擂台赛，并荣获“歌王”金奖。

小黄村素有“歌的海洋，诗的家乡”之称，1993年被贵州省文化厅命名为“侗歌之乡”，1996年被文化部命名为首批“中国民间艺术之乡”。

⊙3

⊙3
小黄晒谷风俗
Unique custom of drying millets in Xiaohuang Village

三 从江小黄侗族皮纸的历史与传承

3 History and Inheritance of Bast Paper by the Dong Ethnic Group in Xiaohuang Village of Congjiang County

⊙1

⊙1 小黄村头迎客的风雨桥
Wind-Rain Bridge in Xiaohuang Village

关于从江小黄侗族的历史，2003年和2004年调查组从村民处获知墓碑、桥碑上有相关记载，但村民们说不出墓碑、桥碑的具体位置。小黄侗族由于没有旧传的家谱，其历史只能靠村民们一代代口传。当地吴姓的祖先来自江西吉安府泰和县，在南宋末（700多年前）大部分迁移到四川，小部分迁入贵州并有一支定居小黄；潘姓的祖先在宋末元初时由湖南迁入，沿都柳江逃难至此。

据2014年的调查数据，小黄行政村辖4个大寨，745户，3 400多人，全部为侗族，村民70%姓潘，20%姓吴，其余为陈姓、刘姓、杨姓（刘姓和杨姓以前属汉族，调查时均已改为侗族），从起源看，吴姓最初也应属汉族。至于村中什么时候开始手工造纸，手工造纸技艺是随移民输入还是后来学会的，田野调查和文献研究中均未获得明确的说法。

据2014年10月调查了解到的信息，小黄村历史上曾经家家造纸，但进入21世纪以来，由于手工造纸的效益比较低，造纸户已呈现逐年递减的衰落趋势，2013年8月访谈时了解到只有十几户造纸，而到2014年10月仅剩八九户造纸。按照侗族的习俗，造纸一直都是女性的技艺，是女人的活路，小黄村也如此，一般都是女儿跟妈妈学，没有发现男性村民传承手工造纸技艺的。

以重点调查与访谈的造纸人吴奶保卫为例，吴奶保卫，原名贾婢澳（根据发音写出并由其家人核准），1943年出生，调查时已72岁。据吴奶保卫老人讲述，她于22岁时跟着妈妈潘培计学造纸技术，而妈妈潘培计是跟外婆（由于吴奶保卫的外婆只有一个哥哥，她在大家族中也排行最小，所以大家都习惯叫她“小姑子”）学造皮纸的，外婆造了二三十年皮纸，于82岁时去世。

吴奶保卫育有两儿两女。大儿子叫吴保浩，2012年已去世，享年52岁，不会造纸。大女儿叫吴婢天，调查时52岁，会造纸，但调查时没在造纸。二女儿叫吴婢能，调查时50岁，也会造纸，调查时也没在造纸。二儿子叫吴保卫，调查时45岁，不会造纸，长年在小黄村里务农，种植蔬菜、红薯、稻谷等。如表10.14所示。门前摆一小摊，全家靠经营这个小摊得来的收入维持生计。吴保卫育有两个女儿，大女儿调查时22岁，没学过造纸，已结婚且育有一个10个月大的女婴，小女儿调查时7岁，在上小学。

吴奶保卫家四代同堂，她本人身体硬朗，自己上山砍野生构树，并背回家中剥皮造纸。据吴奶保卫说，她若从早上9点到下午5点造纸，天气正常一天可生产4张，阴天或雨天则只能生产1~2张。

吴奶保卫也是“侗族大歌”的积极参与者，且水平很高，2012年由县里的一位老师领队，吴奶保卫曾带着两位小黄小学的小朋友，代表黔东南州到北京唱敬酒歌。调查时，吴奶保卫的丈夫已85岁，交流时开朗健谈，曾于1964年到北京唱侗族大歌时见到过很多国家领导人，近年来也时常参加本村的“侗族大歌”表演活动。

另一位访谈对象是小黄村造纸人潘奶桃，她于1950年出生，2014年调查时65岁。据潘奶桃自述，她2009年开始造皮纸，技艺是妈妈所教。潘奶桃的妈妈从65岁开始造纸，到83岁停止造纸，2000年去世，享年88岁。潘奶桃的女儿名潘林德，没有学过造纸，调查时在家门口的路边上摆小摊，贩卖鸭子、蔬菜等。

潘奶桃一个人造纸，一年最多能生产300张皮纸，若当年农活太忙则最多只能造150张。2014年10月，潘奶桃所造皮纸每张能卖10元钱。

⊙2

表10.14　吴奶保卫家四代造纸人传承表
Table 10.14　Papermaker Wu Baowei's mother genealogy of papermaking inheritors

第一代
外婆：“小姑子”

第二代
妈妈：潘培计

第三代
吴奶保卫
（大名：贾婢澳）

第四代
大儿子：吴保浩
二儿子：吴保卫
大女儿：吴婢天
二女儿：吴婢能

⊙3

⊙4

⊙2 鼓楼底层玩石子游戏的男人们
Men playing stone games on the ground floor of the drum tower

⊙3 吴奶保卫与小孙女
Wu Baowei's mother (mothers are named after their son, based on local tradition), an old papermaker, and her granddaughter

⊙4 气宇轩昂的85岁老爷子
Wu Baowei's 85-year-old father, still in good spirits

四
从江小黄侗族皮纸的生产工艺与技术分析

4
Papermaking Technique and Technical Analysis of Bast Paper by the Dong Ethnic Group in Xiaohuang Village of Congjiang County

(一)
从江小黄侗族皮纸的生产原料与辅料

构皮纸在小黄侗族被称为“洁纱”，意为用构皮造的纸。小黄侗族构皮纸主要用野生构皮树作为造纸原料，构皮树在当地被称为“梅纱”，一般农历二三月的树皮质量最好。吴奶保卫告诉调查组，二月时的树枝最方便剥皮，而且造出来的皮纸质量也很好。到秋冬季节时树皮不太好剥，并且若天气较冷，强行剥下的构皮质量也不好。

小黄侗族造纸使用的纸药原料是杨桃藤枝，即野生的猕猴桃树，当地称纸药树的名称为“胶冻”。其加工方式是：用锤子将杨桃藤锤烂，然后放进桶里，加水搅拌，即可使用。若天晴，纸干得快，纸药用得也快。纸药一般可放2~3天，但只要不发酸就可以继续用。

小黄侗族构皮纸制作使用的碱主要是火灰（即灶灰），有的造纸户还将火灰和洗衣粉混在一起使用，并且一度认为用洗衣粉洗得更干净，但据吴奶保卫说，因为洗衣粉要花钱买，所以就不用或少用了。

小黄村造皮纸所用的水是流经村寨的河水，吴奶保卫把河水引到家旁的水池里，这样更方便老人造纸。实测吴奶保卫家造纸用水池中的水及河水pH均在6.5左右，相差很小。

⊙1

⊙1 访谈正在采构皮的吴奶保卫
Interviewing Wu Baowei's mother, who was collecting paper mulberry bark

(二)
从江小黄侗族皮纸的生产工艺流程

据调查组2013年8月底和2014年10月初的实地考察，以及对造纸老人吴奶保卫及潘奶桃的重点访谈，记录从江县小黄村侗族皮纸的生产工艺流程如下：

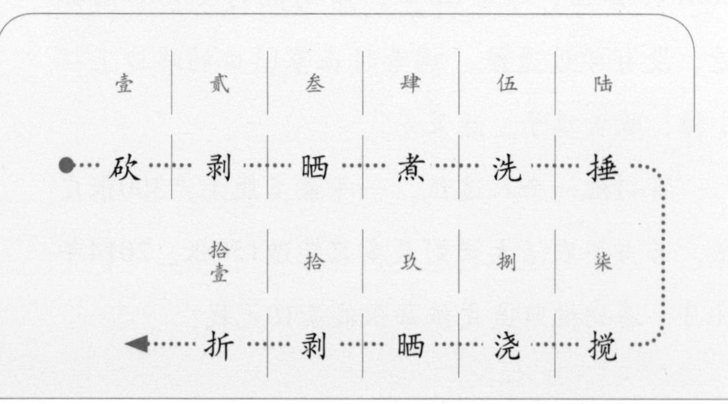

壹 砍

1

造纸户一般在农历二三月份用刀砍野生构树枝条，2~3年砍一次最好。调查时吴奶保卫已72岁，仍带着调查组到村外的山上去砍构树枝条。

贰 剥

2 ⊙2

用手直接将砍下的构树枝条的皮剥下来，如在山上剥不完，可带回家剥；通常是用砍柴刀将构树枝条割个口子，然后用手将白皮、黑皮分开。树皮都是自己剥，一个人一天可剥1~1.5 kg白皮。

⊙2

叁 晒

3 ⊙3

将树皮放在外面晒干，不能淋雨，否则造出的纸质量不好，因为淋雨后，构皮易被煮成线，不容易捣烂，造出的纸不白。树皮也可以不晒，直接煮。

⊙3

肆 煮

4

加上火灰、水，一次最多可煮7.5 kg，一般一次煮1.5~2.5 kg。1 kg干皮可造14张纸。火灰越多越好，一般1 kg干皮放2 kg火灰。若添加洗衣粉，则可减少火灰的用量。若多放洗衣粉，煮皮时皮可洗得更干净。放火灰和洗衣粉的目的是使构皮煮烂得快。

据吴奶保卫介绍，洗衣粉的作用是她偶然发现的。有一次她看火灰用完了，但构皮还不烂，随手放了一些洗衣粉，皮料很快就烂了。煮好后，如天气不好，可先不洗，这样原料就不会变坏，洗了反而易坏。一般是浇多少纸就洗多少料。

伍 洗

5 ⊙4⊙5

将煮过的皮料拿到河里，直接用手上下抖动构皮，冲洗干净料上的火灰，同时除掉黑壳等杂质（2014年10月调查时，吴奶保卫老人已在专门砌的大流水池里洗料）。洗干净后，将其捏成一个个料团。

⊙4

⊙5

⊙2 剥皮 Stripping the bark

⊙3 干构皮 Dried paper mulberry bark

⊙4 洗料水池 Pool for cleaning the papermaking materials

⊙5 捏料团 Balling up the papermaking materials

陆 捶 6

⊙6

将料团放在石板上捶，捶烂为止，通常几分钟即可。

⊙6

柒 搅 7

⊙7

将捶烂的料放进桶内，加水后即称为纸水，再加入杨桃藤汁，并用棍搅拌均匀。

⊙7

捌 浇 8

⊙8~⊙10

用小盆盛出纸水并浇到帘子上，然后摇动帘子，使纸水均匀分布在帘子上。

⊙8

⊙9

⊙10

玖 晒 9

将帘子斜靠在墙边晒，阳光充足时，一天即可晒干，且可晒两次得两张纸，阴天一天只能得一张纸。如下雨，则不造纸，将浇好的纸放在室内慢慢阴干。吴奶保卫说，太阳晒干的纸容易剥下来且质量好。

拾 剥 10

用手先剥开已晒干纸的上方一角，然后将手插进去，从上方揭开，再由上往下剥下整张纸。

拾壹 折 11

⊙11

先将剥下的整张纸对折，然后沿另一方向对折两次，即折成原来纸张的1/8大小，此即成纸包装规格。

⊙11

⊙6 捶料团 Hammering the papermaking material balls

⊙7 浸泡好的纸药液 Soaked papermaking mucilage

⊙8/10 浇纸 Pouring paper pulp onto the papermaking screen

⊙11 折纸 Folding the paper

(三)
从江小黄侗族皮纸生产使用的主要工具设备

壹 帘子 1

即浇纸的纸帘。小黄侗族造纸使用的纸帘由4根木条作框，将一块纱布绷平，一般长约80 cm、宽约80 cm。通常造纸户会有数个帘子，如吴奶保卫家有6个。

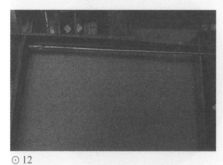
⊙12

贰 槌 2

木质，实测吴奶保卫使用的槌的尺寸为：手柄长约27 cm，圆柄长约21 cm，大的一头直径约7 cm，小的一头直径约5 cm。捶打时重心在下，不易翻转。

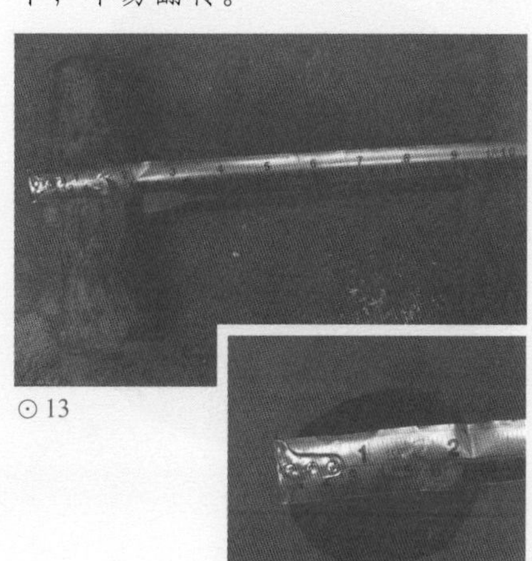
⊙13

⊙14

叁 煮料锅 3

实测吴奶保卫家煮料锅的尺寸为：直径40~41 cm，高约24 cm。

⊙15

(四)
从江小黄侗族皮纸的性能分析

所测从江小黄村皮纸的相关性能参数见表10.15。

表10.15 小黄村皮纸的相关性能参数
Table 10.15 Performance parameters of bast paper in Xiaohuang Village

指标		单位	最大值	最小值	平均值
厚度		mm	0.210	0.170	0.190
定量		g/m^2	—	—	39.2
紧度		g/cm^3	—	—	0.206
抗张力	纵向	N	32.0	21.4	27.8
	横向	N	29.3	21.1	26.2
抗张强度		kN/m	—	—	1.800

⊙12 帘子 Papermaking screen

⊙13/14 木槌 Wooden mallet

⊙15 煮料锅 Wok for boiling the papermaking materials

续表

指标	单位	最大值	最小值	平均值
白度	%	26.5	15.3	25.9
纤维长度	mm	11.68	1.21	4.64
纤维宽度	μm	37.0	4.0	14.0

由表10.15可知，所测小黄村皮纸最厚约是最薄的1.24倍，相对标准偏差为1.40%，纸张厚薄较为均匀。皮纸的平均定量为39.2 g/m^2。所测皮纸紧度为0.206 g/cm^3。

经计算，其抗张强度为1.800 kN/m，抗张强度值较小。

所测小黄村皮纸白度平均值为25.9%，白度较低，白度最大值约是最小值的1.73倍，相对标准偏差为3.37%，白度差异较大。

所测小黄村皮纸纤维长度：最长11.68 mm，最短1.21 mm，平均4.64 mm；纤维宽度：最宽37.0 μm，最窄4.0 μm，平均14.0 μm。所测皮纸在10倍、20倍物镜下观测的纤维形态分别见图★1、图★2。

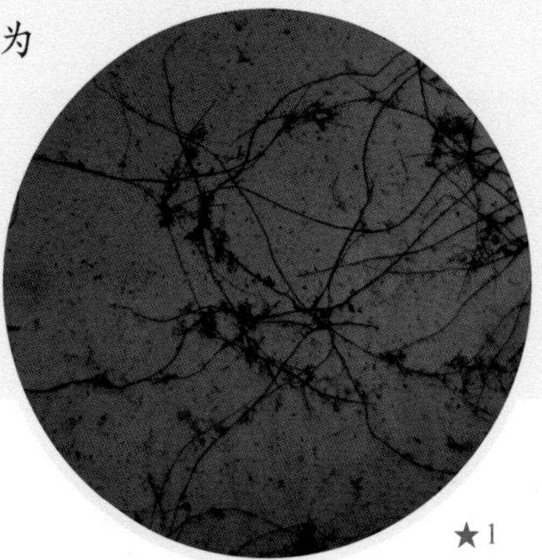

★1

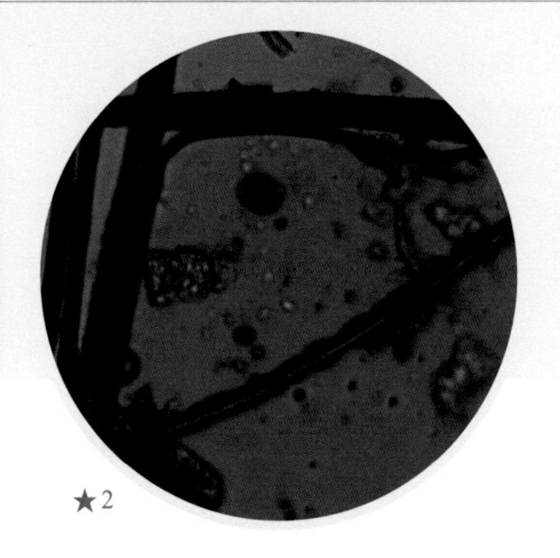

★2

★1 小黄村皮纸纤维形态图(10×) Fibers of bast paper in Xiaohuang Village (10× objective)

★2 小黄村皮纸纤维形态图(20×) Fibers of bast paper in Xiaohuang Village (20× objective)

五 从江小黄侗族皮纸的用途与销售情况

5 Uses and Sales of Bast Paper by the Dong Ethnic Group in Xiaohuang Village of Congjiang County

(一) 从江小黄侗族皮纸的用途

经实地采样考察，小黄侗族皮纸长、宽均为80 cm，主要有以下用途：

1. 书写

小黄侗族皮纸绵软受墨、经久耐用，历史上曾广泛用于书写，包括抄书、写歌本、红白喜事记礼金等。中华人民共和国成立前，私塾曾流行用墨汁配合毛笔在皮纸上书写，现在村里仍有少部分人用墨书写，包括写对联。

2. 绣花样纸与绣花鞋

当地侗族女性衣服上的精美图案，都是先用皮纸刻图，即用笔在纸上画好特定的图形，再用刀刻，并用米浆粘上，然后贴于布上，最后根据其式样剪出来的。传统时期当地侗族女孩一般14岁左右就开始学习刻花和绣花，调查时因现代衣装已在小黄村流行，故会刻花手艺的年轻女性已不多。

调查组有幸得到一双吴奶保卫在2012年到北京参加唱歌比赛时穿过一次的绣花鞋，虽有破损，但很有纪念意义。据吴奶保卫介绍，她自己就会做这种夹有皮纸的绣花鞋，整双鞋全部绣满图案需要一个月的时间，约500元/双。问及鞋底中间夹皮纸有什么特殊功用，吴奶保卫老太太却说并无讲究，只是方便穿针而已。雨天绣花鞋不能穿出去，以免鞋底被水浸泡，腐烂皮纸。

3. 制作绣花针线盒

绣花时，需要用到多种针线以及绣花图案，为便于保管，小黄村村民会用构皮纸做成精美、复杂的绣花盒来盛放工具，不同层放不同颜色的线。为了更好地保存，绣花盒外侧会被涂上几层桐油，可防水防潮。

4. 制作纸斗笠与纸蓑衣

小黄村一带在历史上属偏远崎岖之地，雨伞、雨衣不易获得，当地主要用纸斗笠、纸蓑衣来遮风挡雨。先在构皮纸上涂5次以上桐油，然后制成雨天用的器具。调查中未发现旧日使用的纸斗笠、纸蓑衣实物。

5. 糊窗户

以前窗户没有玻璃，结实透光的构皮纸是很好的糊窗户材料。为了延长使用时间，室外接触

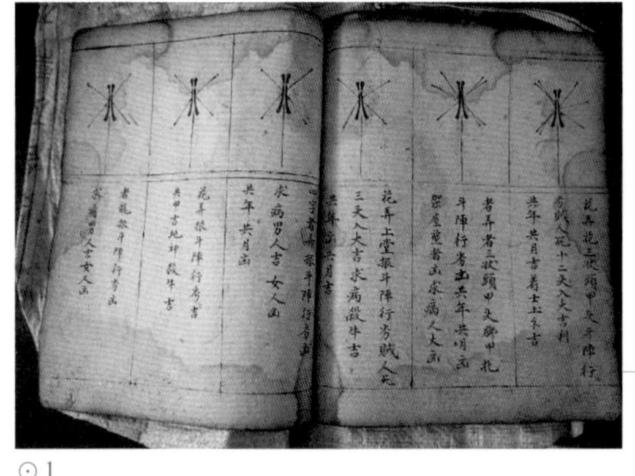

⊙1

⊙3

⊙4

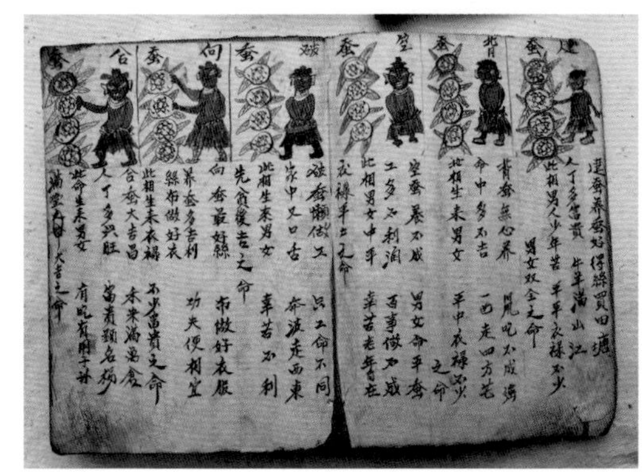

⊙2

⊙5

⊙6

⊙1/2 村里旧日传下的皮纸手抄风水书 Old Fengshui book transcribed on local bast paper

⊙3 有皮纸夹层的绣花鞋 Embroidered shoes with bast paper inside

⊙4 针线盒外包装 Sewing kit wrapped by bast paper

⊙5/6 展开的精美针线盒 Exquisite unfolded sewing kit wrapper

风雨的一面要涂桐油，室内的那一面则不用。

6. 制作神龛

这一传统用途在调研中未能了解到具体用法。

(二) 从江小黄侗族皮纸的销售情况

2013年8月和2014年10月两次调查了解到：小黄村所造的皮纸主要在本村销售，有村民上门购买的，有在门前村里主街上摆摊售卖的，也有送给乡邻使用的。2014年10月调查组采样时，80 cm × 80 cm的皮纸每张售价为10~12元。

六 从江小黄侗族皮纸的相关民俗与文化事象

6 Folk Customs and Culture of Bast Paper by the Dong Ethnic Group in Xiaohuang Village of Congjiang County

1. “女人的活计”

在小黄村，造纸被认为是女人的活计，全流程都由女人负责。调查中了解到，以前村里的女人都会造纸，造纸是女性的一项基础技能，一般都是母亲将技艺传给女儿。小黄村侗族没有男性造纸的传统，男人很少会插手帮忙。

2. 祭祀用纸有讲究

老人过世时，要在棺材的4个角落各放2张共8张构皮纸，据当地村民介绍，这是送死人的钱，让其去阴间有钱花。此外，还要用3张构皮纸盖在死者身上，且两只手里也塞上纸，盖棺时，要将左右手的纸互换。有的也会用皮纸剪出人脸的轮廓盖在死者的脸上。

拜谢土地神的时候也要用构皮纸，但需要将纸裁成小块，并剪成条。木工户祭祖师爷鲁班

⊙1

⊙2

时也要烧纸，反倒是造纸户不用烧纸祭神或祖师爷。

七 从江小黄侗族皮纸的保护现状与发展思考

7 Preservation and Development of Bast Paper by the Dong Ethnic Group in Xiaohuang Village of Congjiang County

从江小黄村在农历三月、八月造纸较多，秋收后天气变冷，造纸较少，冬天完全不造纸。小黄村传统上是为了满足本民族本村落的文化习俗需要并传承这些自家生产纸张的应用方式而造纸的，并不指望以此带来经济效益。近年来，随着小黄村“侗族大歌”名声越来越大，更多的游人和文化人到此旅游，也有人会购买一些构皮纸，但是新的经济和生活要素对传承习惯的影响并不明显。通过进一步分析，可以发现小黄村构皮纸生产文化所具有的原生价值。

(一) 从江小黄侗族皮纸的保护现状

总体而言，小黄侗族皮纸的技艺文化基因保护得较好，呈现出良好的原生态文化。

⊙1 『女人的活计』 Papermaking practice limited to female papermakers only

⊙2 祭祀焚烧的纸钱 Joss paper used in sacrificial ritual

1. 独特的名称

小黄村造纸中有一些独特的名称，如工序名称都用一个字表示，纸帘称为帘子，搅料时所加的水称为纸水等。

2. 独特的民俗

调查时发现，小黄村几乎是一个女人承担造纸全流程的活，而且多数是老太太；祭祀用纸的讲究也相当独特。

3. 独特的销售

小黄村生产的构皮纸几乎全部在村内买卖或赠送。据调查，小黄村没有到村外市场进行销售的行为。皮纸是为了满足本民族小群落的自有需要而生产的。

4. 独特的工艺

小黄村生产构皮纸最独特的工艺是采用浇纸法造纸，并使用纸药。通常采用浇纸法造纸都不用纸药，采用抄纸法才用，因为浇纸是单张，而抄纸是一叠多张，必须用纸药来解决分张难题。采用浇纸法却使用纸药的造纸技术在中国手工纸发展过程中并不多见。可以认为，小黄村手工造纸的原生态传承与发展具有重要的文化基因保护价值。

(二)
从江小黄侗族皮纸传承与发展的思考

1. 申报省级和国家级非物质文化遗产项目

小黄侗族构皮纸最具价值的是其独特的采用浇纸法却又使用纸药的造纸技术，迫切需要多机构合作进行更为深入的研究，进一步挖掘该技术的历史和传播路径以及功能内涵。在研究的基础上，提炼总结其特殊的造纸技术内涵，将小黄侗族皮纸制作技艺申报为省级和国家级非物质文化遗产项目，使其传承与保护纳入非物质文化遗产保护体系，同时使小黄村手工造纸业态获得更广泛的关注性传承。

2. 适当进行旅游开发

从江有相当丰富的文化旅游资源，对小黄村侗族造纸进行适当的旅游开发，不仅可以丰富从江旅游的技艺文化内涵，使小黄村侗族构皮纸得以传承，还能促进相关经济的发展，如可以考虑将当地

⊙1

⊙2

⊙1
调查组成员在买皮纸
A researcher buying bast paper from the papermakers

⊙2
在吴奶保卫家聚会的造纸老人们
Old papermakers gathering in Wu Baowei's mother's house

侗寨手工纸制品做成旅游纪念品进行开发。

3. 开发小黄村侗族构皮纸的新品种、新用途

在保留传统造纸技艺的基础上，考虑进行技术改进及产品开发。比如，延长捶料时间，以得到更为细腻、均匀的纸浆，造出更为平滑、柔韧的纸张，逐渐引入更高端的书写、绘画、印刷等用途。

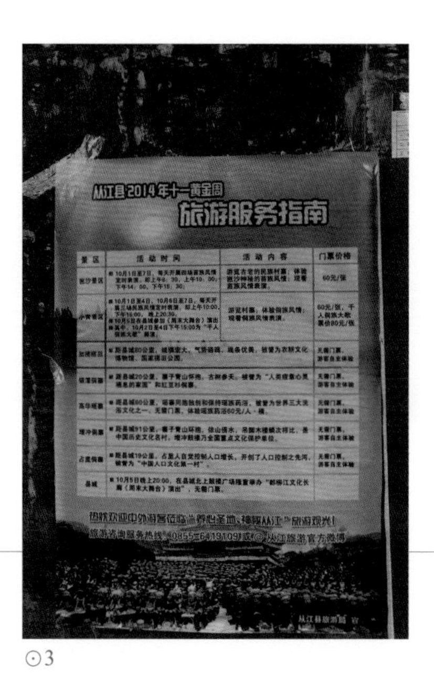

⊙3

⊙4

⊙5

⊙3 从江黄金周旅游海报
Tourism poster for the Golden Week in Congjiang County

⊙4 小黄小学的『非遗』传习基地
Intangible cultural heritage training base set in Xiaohuang Elementary School

⊙5 调查组成员向造纸老人敬酒
A researcher making a toast to the old papermakers

从江小黄侗族皮纸

Bast Paper
by the Dong Ethnic Group
in Xiaohuang Village of Congjiang County

小黄村皮纸透光摄影图
A photo of bast paper in Xiaohuang Village seen through the light

第十节

从江占里侗族皮纸

贵州省
Guizhou Province

黔东南苗族侗族自治州
Qiandongnan Miao and Dong Autonomous Prefecture

从江县
Congjiang County

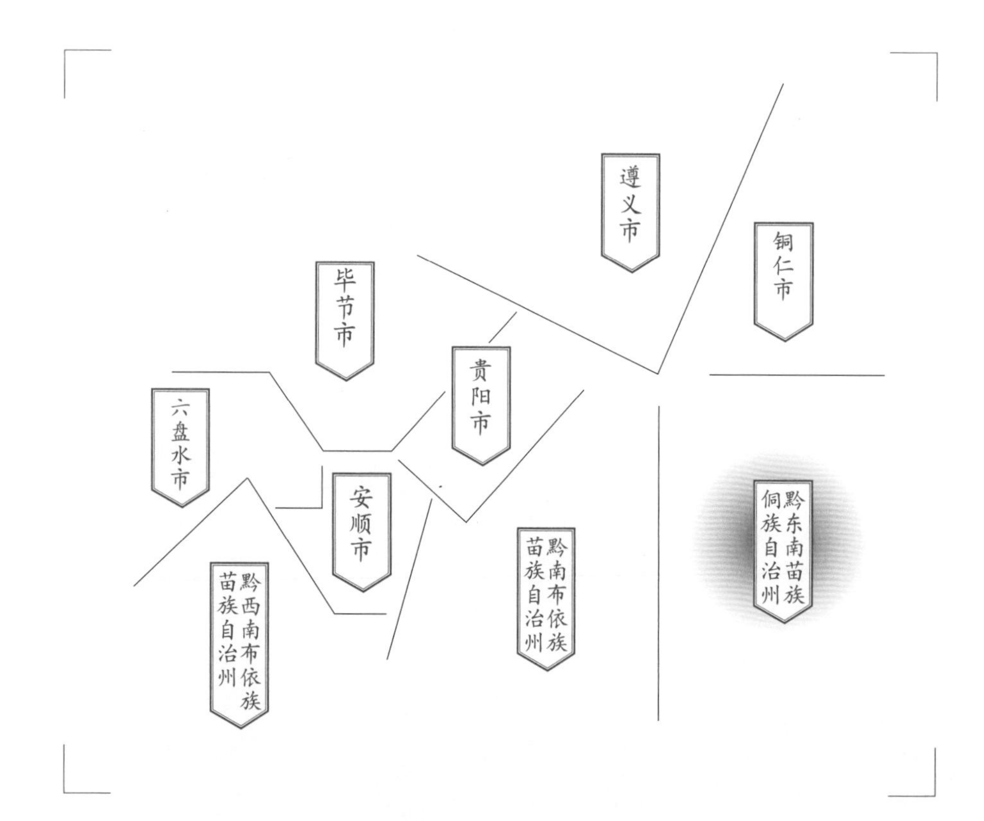

调查对象
高增乡
占里行政村
侗族皮纸

Section 10
Bast Paper by the Dong Ethnic Group in Zhanli Village of Congjiang County

Subject

Bast Paper by the Dong Ethnic Group
in Zhanli Administrative Village
of Gaozeng Town

一 从江占里侗族皮纸的基础信息及分布

1 Basic Information and Distribution of Bast Paper by the Dong Ethnic Group in Zhanli Village of Congjiang County

占里行政村位于从江县高增乡，距县城约25 km，全村共有8个村民组，2014年10月4~5日调查组入村考察时，全村有侗族居民180多户，不到1 000人。占里村的盛名源于其独特的生育调节秘方，自1949年以来，人口自然增长率几乎为零，被誉为“中国人口文化第一村”。该村清初制定的寨规就要求每家每户生育须一男一女且不能超生，由村里的寨老负责督察落实，调查中据村民们反映，300余年来占里村确实令人难以置信地坚持做到了这一点。

占里村的村民都为侗族，以吴姓为主，造纸点分布在每家每户的住宅内，以独立的家中操作模式为主，未见村边或村外设纸槽、纸坊的现象。占里村的手工造纸为皮纸，造纸原料为构树皮，使用的是贵州较为少见的浇纸法。占里村皮纸主要用于纳鞋底、老人过世后垫棺底、画画以及居家辟邪、驱鬼等法事仪式。

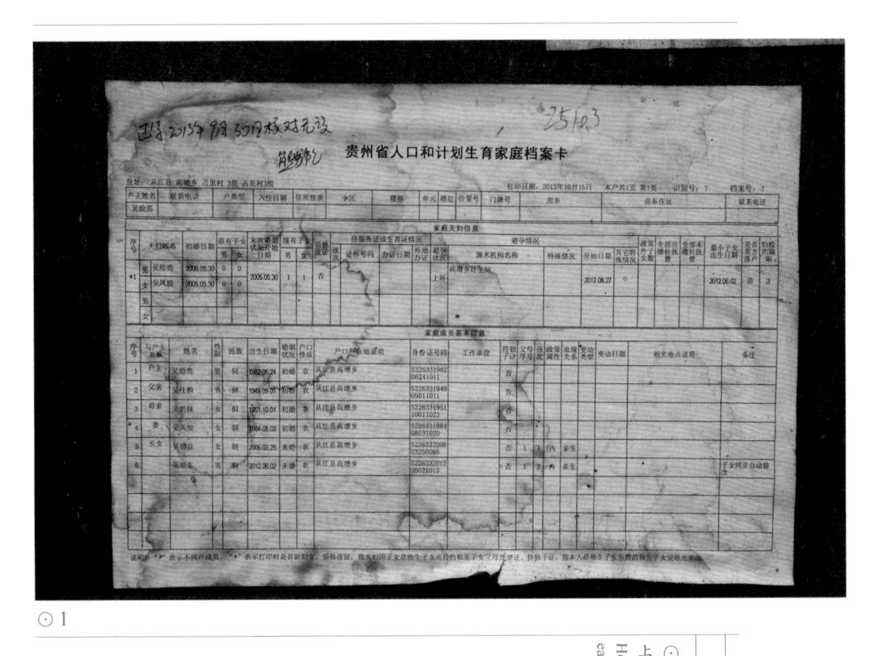

贵州省人口和计划生育家庭档案卡

住址：从江县 高增乡 占里村 3组 占里村3组

打印日期：2013年10月15日　本户共1页 第1页　识别号：7　档案号：7

家庭成员基本信息

序号	与户主关系	姓名	性别	民族	出生日期	婚姻状况	户口性质	户口所在地说明	身份证号码	工作单位	持独子证	父母序号	孩次	政策属性	血缘关系	变动类型	变动日期	相关地点说明	备注
1	户主	吴给亮	男	侗	1982.06.24	初婚	农	从江县高增乡	522633198206241011		否								
2	父亲	[illegible]	男	侗	1949.09.01	初婚	农	从江县高增乡	522633194909011011		否								
3	母亲	[illegible]	女	侗	1951.10.01	初婚	农	从江县高增乡	522633195110011023		否								
4	妻	[illegible]	女	侗	1984.08.03	初婚	农	从江县高增乡	522633198408031020		否								
5	长女	[illegible]	女	侗	2006.03.25	未婚	农	从江县高增乡	522633200603250085		否	1	1	内	亲生				
6		[illegible]	男	侗	2012.06.02	未婚	农	从江县高增乡	522633201206021013		否	1	2	内	亲生				子女同步自动修改

⊙1

⊙1 占里村每户门前贴的人口档案卡 Household population file card attached on each house gate in Zhanli Village

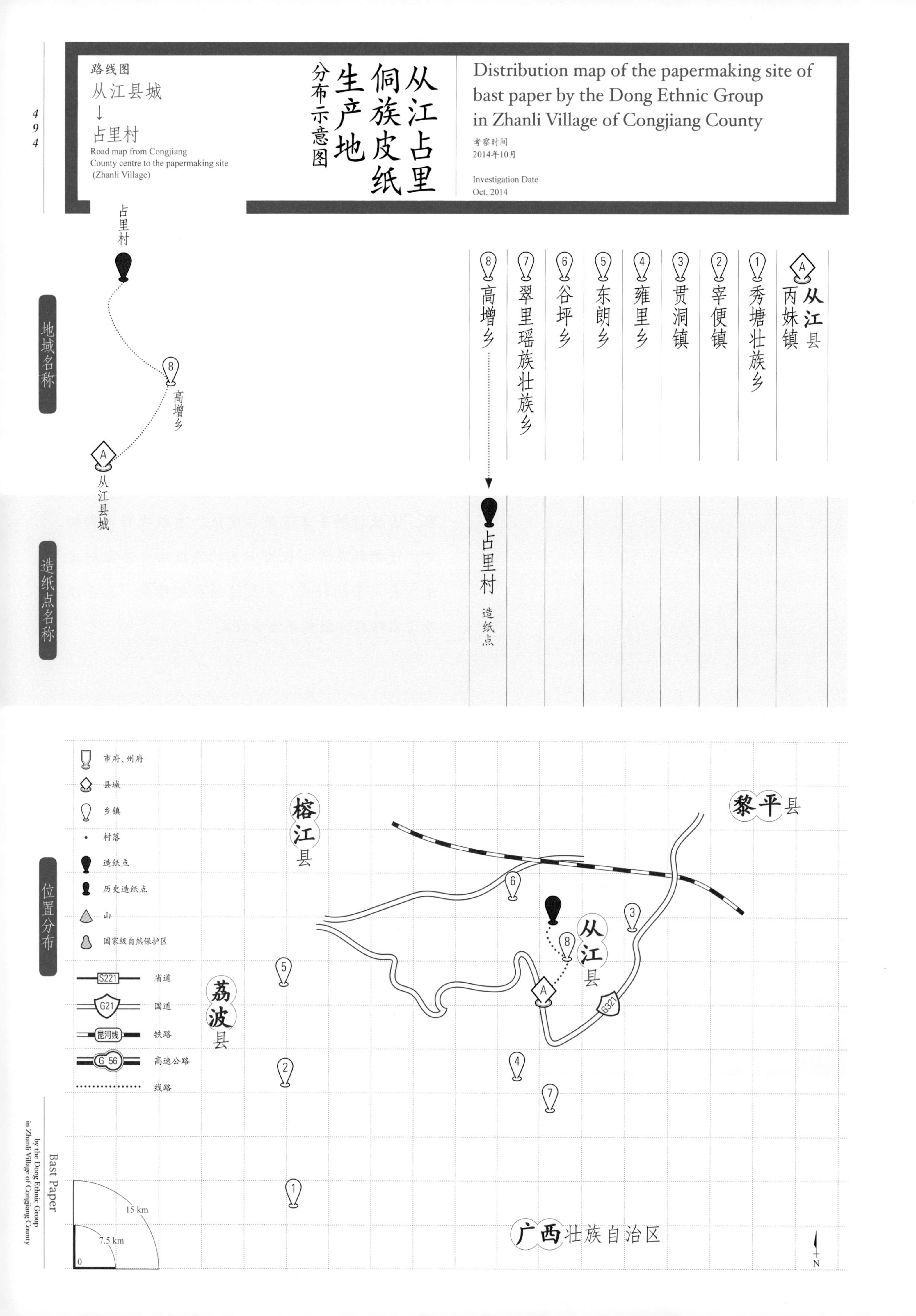

路线图
从江县城
↓
占里村
Road map from Congjiang County centre to the papermaking site (Zhanli Village)
从江占里侗族皮纸生产地分布示意图
Distribution map of the papermaking site of bast paper by the Dong Ethnic Group in Zhanli Village of Congjiang County
考察时间
2014年10月
Investigation Date
Oct. 2014
地域名称
造纸点名称
位置分布
占里村
高增乡
从江县城
从江县
丙妹镇
秀塘壮族乡
宰便镇
贯洞镇
雍里乡
东朗乡
谷坪乡
翠里瑶族壮族乡
高增乡
占里村
造纸点
市府、州府
县城
乡镇
村落
造纸点
历史造纸点
山
国家级自然保护区
S221 省道
G21 国道
昆河线 铁路
G 56 高速公路
线路
榕江县
黎平县
荔波县
从江县
G321
广西壮族自治区
15 km
7.5 km
0
N

二
从江占里侗族皮纸生产的人文地理环境

2
The Cultural and Geographic Environment of Bast Paper by the Dong Ethnic Group in Zhanli Village of Congjiang County

占里村坐落在深山密林中，是一个植被良好、民族风情浓郁的美丽侗寨，村外苍山叠翠，门前流水淙淙，一幢幢吊脚楼错落有致，鼓楼、风雨桥历历在目。据乡土文献记述，占里村村民的祖先为躲避战乱，在1 000多年前由广西梧州迁徙而来，他们溯都柳江而上，最后在此定居下来。

占里村虽地处边远山区，却具有纯朴、先进的人口控制意识和调节生育的秘方，因此，300余年来，当地人口一直保持着近乎不增不减的稳态，与当地自然资源的农业型利用保持平衡。占里侗族村民丰衣足食、富足安康，人均耕地面积约870 m^2，比全县人均稻田面积高出一倍多，人均占有粮食量远高于全县平均水平，家禽饲养量也居全县前列，这对于一个处在深山僻地的小村来说相当不易。

2000年8月，国家计生委科技司副司长肖绍博专程到占里考察其独特的生育文化。中央电视台为此也拍摄了专题片《大山深处》，中央人民广播电台、《中国青年报》、《中国社会报》等多家媒体及中国台湾的《汉声》杂志都对占里村的人口调控做了专题报道。

⊙1

⊙2

⊙1 从江深山里的小村寨
A small village in mountain area of Congjiang County

⊙2 村中心的文化广场——侗家鼓楼
Dong ethnic drum tower at the cultural square in the village centre

⊙1

⊙2

三 从江占里侗族皮纸的历史与传承

3 History and Inheritance of Bast Paper by the Dong Ethnic Group in Zhanli Village of Congjiang County

据2014年10月4~5日的访谈了解，村民的口传记忆是：占里人的祖先于明洪武元年（1368年）从江苏逃难，长途跋涉后，经广西梧州辗转最终定居此地，“占里”二字就来源于其最初定居的祖先“吴占”和“吴里”两人。从传说的来源看，似乎占里人的祖先是汉族人，但如何既保持了源自江南的吴姓，又全部变成了侗族，调查中未能获得解答。由于自然地理环境好，祖先们在这里定居以后，开垦了很多土地，生活得十分富裕，享受着与世无争的幸福生活。于是，这里的

⊙1 安宁的占里侗族小街
Peaceful Dong ethnic alley in Zhanli Village

⊙2 占里村雨中连排的晒谷架
Rows of drying rack in Zhanli Village in the rain

⊙3

⊙4

⊙5

人口逐步从最初的2户发展到了100余户。

随着人口不断增加，人多地少的矛盾也开始突显。到了清朝初期，这一矛盾已经使得曾经富裕的占里村开始衰败。村里口耳相传的“信史”：最初，一位叫吴公力的村民挺身而出，找到寨中的长老们不断商议，寻找解决办法。最终制定出一条独特的寨规：“有50担稻谷的夫妇可生育两个孩子，但必须是一男一女；有30担稻谷的夫妇只能生育一个孩子。如有违规，轻者会将其饲养的牲畜强行杀掉；重者则将其逐出寨门。”

据村人的说法，寨规在占里一直沿袭了300余年，至今还未有人违反过。每年的农历二月初一和八月初一，全寨人都要聚集到鼓楼里听寨老训诫，并用侗歌传唱寨规；青年男女行房时，也被要求先唱控制人口增长的寨规歌，从而确保这一习俗一代代地延续下去。

当然，仅靠寨规并不能保证新生育人口的精准调控。实际上，占里村在口碑传播中流传更广的是神奇的控制生男生女及生育数量的家传秘方。据文献调研及与村民们的交流得知，村里一直有一户人掌握着这一秘方，从吴奶卫和吴奶妹及其家人处了解到的信息是，这一秘方是一种特殊配制的侗药，若按照“神医”的要求择时服用，便可控制生男生女。在调查组的一再探问下，吴奶妹告诉调查人员最近的“神医”是位老太太，住在村尾的山上，但因年事已高，据说已将秘方传给了儿子。调查组曾提出能否请她带路前去拜访，但未获同意。

据调查中重点访谈的造纸人吴奶卫和吴奶妹口述，2014年村里的180余户人家中，有10余户还在造纸，造纸人多为60岁以上的老人。造纸没有特殊的传承方式，本村女性当了奶奶或者外婆后，有想要学造纸的，在一旁看本村的老人造就能学会。至于为什么要有了第三代并且年纪较大

⊙3 秀美富饶的占里侗寨
Rich and beautiful Dong ethnic residence in Zhanli Village

⊙4 占里村的夜色
Night scene of Zhanli Village

⊙5 调查组成员与吴奶卫和吴奶妹合影
Researchers with Wu Naiwei and Wu Naimei

时才开始造纸，却未能探寻出特别明确的说法。

占里村的造纸世代相传，但是造纸户均说不清其历史起源和传承谱系。表10.16是造纸人吴奶卫和吴奶妹的家谱。

表10.16 吴奶卫和吴奶妹的家谱
Table 10.16 Genealogy of Wu Naiwei and Wu Naimei

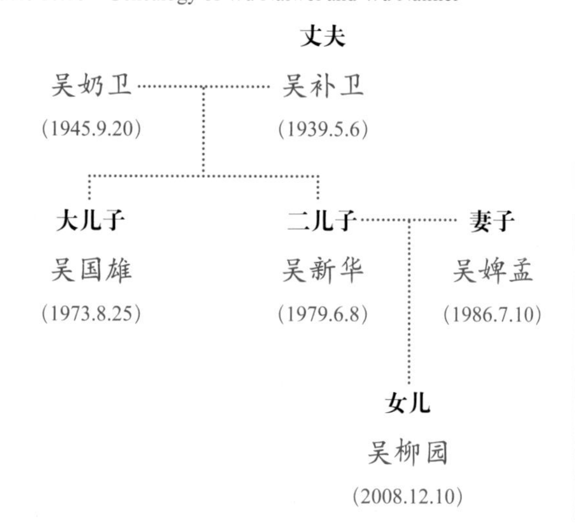

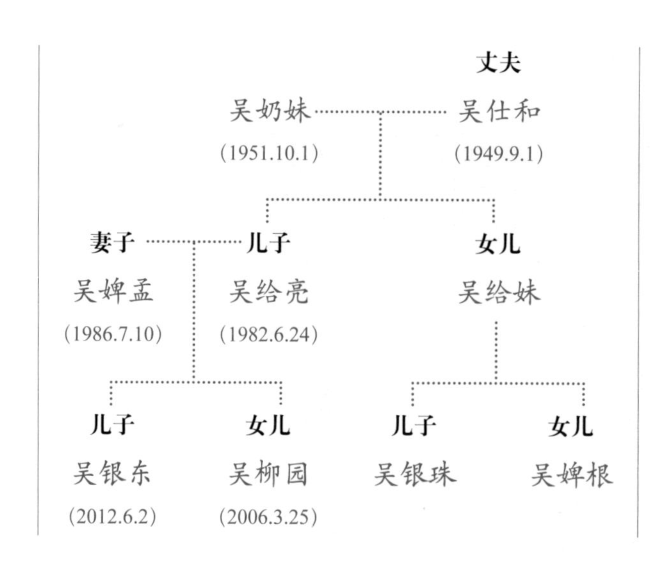

四 从江占里侗族皮纸的生产工艺与技术分析

4 Papermaking Technique and Technical Analysis of Bast Paper by the Dong Ethnic Group in Zhanli Village of Congjiang County

(一) 从江占里侗族皮纸的生产原料与辅料

1. 构树皮

占里村皮纸生产的原料为构树皮。构树是桑科构树，属落叶乔木，又名谷浆树，古名楮。构树皮纤维长而柔软，为优质皮纸原料。占里当地砍构树皮的时间一般在农历二三月间。

2. 洗衣粉和灶灰

2014年入村调查时，占里村皮纸生产已采用洗衣粉来漂白，1.5 kg干构树皮需配一袋（455 g）

洗衣粉和一碗灶灰来熬煮。

占里村造纸的水源为当地的山泉水。全村寨伴溪而居，就用流经家门口的溪水造纸。实测溪流的山泉水pH为6.5~6.6。

⊙1

⊙2

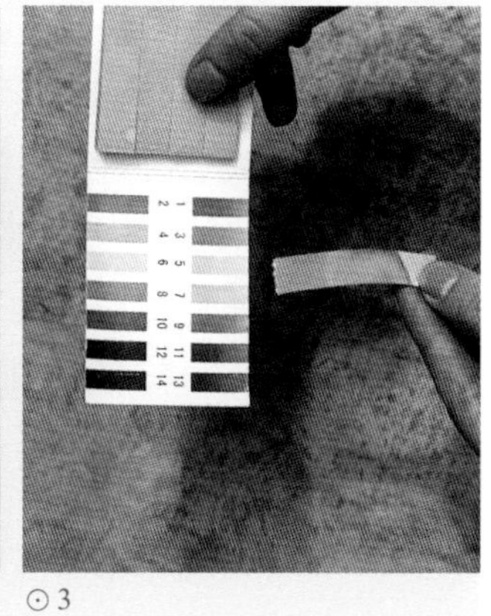

⊙3

(二)

从江占里侗族皮纸的生产工艺流程

调查组通过对占里村的实地考察，并与吴奶妹、吴奶卫等人交流，记录占里村皮纸的生产工艺流程如下：

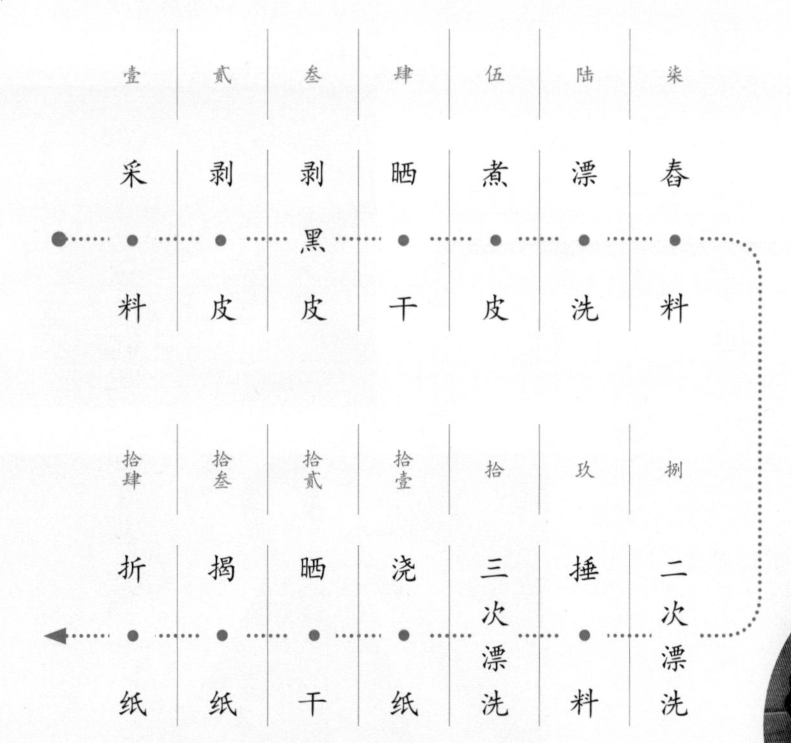

壹 采料

1

每年农历二三月间，造纸户都会上山砍构树。

贰 剥皮

2

砍下的枝条要及时剥皮，否则随着枝条水分挥发，树皮附着很紧，难以剥下，只能泡水后再剥，费工费时。

叁 剥黑皮

3

⊙4

将剥下的树皮再用小刀剥去黑褐色表皮，只留白色内皮。

⊙4

⊙1 造纸使用的干构皮 Dry paper mulberry bark for papermaking

⊙2 造纸使用的灶灰 Stove ash for papermaking

⊙3 实测山泉水pH pH value of the local spring water

⊙4 吴奶妹在剥黑皮 Wu Naimei stripping the bark

肆
晒干
4 ⊙5

将剥下的白皮置于太阳下晒干，天气晴好时，晒一天即可。将晒干的构树皮放置干燥处存放，等到每年农历十至十一月造纸时取出来使用，干树皮多的话，就会做到农历十二月。

⊙5

伍
煮皮
5 ⊙6

将晒干的1.5 kg干构树皮放进锅内，并加入清水用柴火煮。煮时凭经验加入一碗灶灰和一袋洗衣粉（455 g），盖上锅盖，通常煮一天。当皮料轻轻一拉即可拉断，且又稍稍有些韧性时即可。

⊙6

陆
漂洗
6 ⊙7

将煮好的构皮装进箩筐，并拿到溪流里漂洗，洗去里面的灶灰等杂质。

⊙7

柒
舂料
7 ⊙8

将洗干净的构皮再次装进箩筐，放到石舂里面舂碎。人站至舂料棰的另一端，一上一下做“跷跷板”运动，通过舂料棰的重力作用，将料舂碎。

⊙8

捌
二次漂洗
8

将舂好的料装入箩筐，再次漂洗。

玖
捶料
9 ⊙9

洗好碎料后，放在石板上，用木槌反复捶打，至料成浆。

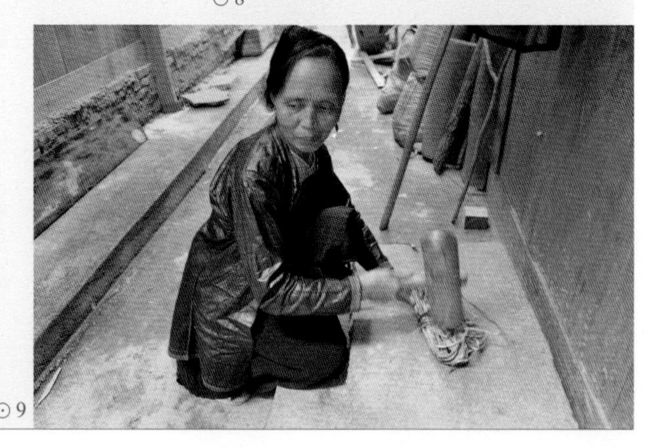

⊙9

⊙5 吴奶妹拿出存放的干构皮
Dry bark for papermaking stored by Wu Naimei

⊙6 吴奶妹示范煮皮
Wu Naimei showing how to boil the bark

⊙7 洗料的竹编箩筐
Bamboo basket for cleaning the papermaking materials

⊙8 吴奶妹在示范舂料
Wu Naimei showing how to beat the papermaking materials

⊙9 吴奶妹在示范捶料
Wu Naimei showing how to hammer the papermaking materials

拾 三次漂洗

10

将捶好的料再次入水漂洗。

拾壹 浇纸

11 ⊙10

将料放入桶内，并加入清水，再加入一瓢泡好的野生猕猴桃藤汁，并搅拌均匀，用瓢将纸浆舀起并浇在纸帘上。双手握住纸帘并逆时针晃动，使纸浆均匀地铺在纸帘上。如此反复，将一桶料用完。

⊙10

拾贰 晒干

12

将纸帘放在太阳底下晒干或通风处风干，不被雨淋即可。

拾叁 揭纸

13

将晒干的纸从边角处揭下。

⊙11

拾肆 折纸

14 ⊙11~⊙13

先横向对折，再竖向对折，最后均匀折叠两次，使整张纸分为12等份。

⊙12

⊙13

⊙10 浇纸装帘 Fixing the papermaking screen

⊙11/13 折纸 Folding the paper

(三)

从江占里侗族皮纸生产使用的主要工具设备

壹 纸帘 1

实测吴奶妹家的纸帘：内框尺寸为70.5 cm × 70.5 cm，外框尺寸为77 cm × 77 cm。

⊙1

贰 石板和槌 2

实测吴奶妹家的石板长约70 cm，宽约49 cm。槌柄长约37 cm，槌头长约22 cm，直径约8 cm。

⊙2

叁 石舂 3

用于舂碎构皮料的石木结构窝状工具。

⊙3

⊙1 装纸帘示意 Showing how to fix the papermaking screen

⊙2 吴奶妹家的石板和槌 Wu Naimei's stone board and mallet

⊙3 石舂 Stone pestle

(四)

从江占里侗族皮纸的性能分析

所测从江占里村皮纸的相关性能参数见表10.17。

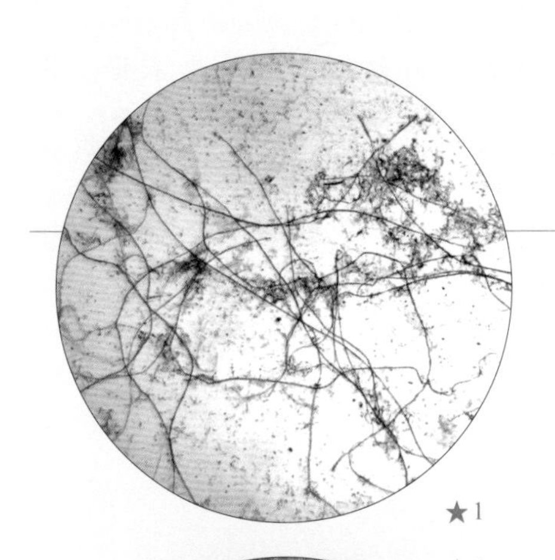

★1

★2

表10.17　占里村皮纸的相关性能参数
Table 10.17　Performance parameters of bast paper in Zhanli Village

指标		单位	最大值	最小值	平均值
厚度		mm	0.280	0.100	0.220
定量		g/m^2	—	—	38.3
紧度		g/cm^3	—	—	0.174
抗张力	纵向	N	13.2	9.4	11.5
	横向	N	13.1	8.9	10.8
抗张强度		kN/m	—	—	0.743
白度		%	43.6	41.5	42.7
纤维长度		mm	12.23	1.14	4.16
纤维宽度		μm	30.0	1.0	11.0

由表10.17可知，所测占里村皮纸最厚是最薄的2.8倍，相对标准偏差为5.50%，纸张厚薄不均。皮纸的平均定量为38.3 g/m^2。所测皮纸紧度为0.174 g/cm^3。

经计算，其抗张强度为0.743 kN/m，抗张强度值较小。

所测占里村皮纸白度平均值为42.7%，白度较高，白度最大值约是最小值的1.05倍，相对标准偏差为0.61%，白度差异相对较小，可能是因为占里村吴奶妹家皮纸加工时没有经过较强的漂白。

所测占里村皮纸纤维长度：最长12.23 mm，最短1.14 mm，平均4.16 mm；纤维宽度：最宽30.0 μm，最窄1.0 μm，平均11.0 μm。所测皮纸在10倍、20倍物镜下观测的纤维形态分别见图★1、图★2。

★1
占里村皮纸纤维形态图(10×)
Fibers of bast paper in Zhanli Village (10× objective)

★2
占里村皮纸纤维形态图(20×)
Fibers of bast paper in Zhanli Village (20× objective)

五 从江占里侗族皮纸的用途与销售情况

5 Uses and Sales of Bast Paper by the Dong Ethnic Group in Zhanli Village of Congjiang County

占里村皮纸主要用于纳鞋底及老人过世后垫棺底，但吴奶妹说外来的人也会买去画画用，至于是什么人买去画什么画则未知详情。占里村皮纸的另一种用途是将皮纸剪成一定形状挂在门头驱鬼。

调查时，占里村皮纸的尺寸基本上是70 cm × 70 cm，呈方形。调查组在采购吴奶卫和吴奶妹的测试纸样时，价格10~15元/张不等。占里村皮纸通常以村人自用为主，因而其外售价格并不是很标准。

⊙1

⊙2

⊙3

⊙4

⊙1/2 占里村流行的住家门头驱鬼用纸习俗
Paper hanging on door for expelling ghosts, a local custom in Zhanli Village

⊙3 吴奶妹卖纸前数纸
Wu Naimei counting the paper before it was sold

⊙4 『生意』达成
A deal was made

六
从江占里侗族皮纸的相关民俗与文化事象

6
Folk Customs and Culture of Bast Paper by the Dong Ethnic Group in Zhanli Village of Congjiang County

(一)
从江占里侗族皮纸技艺传承的独特习俗

据造纸人吴奶卫及吴奶妹口述，占里村造皮纸的传习方式非常特别，只有60岁以上当了奶奶或者外婆的女性才能造纸。至于为什么会有这样的规定，调查组未能了解到其历史或文化缘由。通常的学习过程是：当地女性当了奶奶或者外婆后，如果喜欢造纸，就去观摩其他造纸老人造纸，造纸老人也会乐意教，在一旁看着自然就学会了；如果不喜欢造纸，即使年纪到了也可以不学。占里村造纸没有家庭或家族传承的习惯，基本上是全村年长女性开放传习的传统，调查组在云南、四川、广西、贵州等西南地区的考察中还是第一次发现这种传统。2014年10月访谈时，全村会造纸的老太太有10余人。

⊙5

⊙5 吴奶卫展示浇纸帘 Wu Naiwei showing the papermaking screen

(二)
占里侗民与婚育文化相关的其他村规民俗

1. 早恋晚婚晚育

占里村的侗族民俗平和开放，青年男女的交往十分自由，一般十八九岁就开始谈恋爱。但占里村同时又遵循着晚婚晚育的习俗，因此姑娘们大多在23岁以后才结婚。在占里人看来，结婚晚就老得晚，早要孩子就意味着会早当老人，寿命也就短了。正是这种纯朴的生命观使占里的青年不愿意过早结婚，而生育期的缩短客观上降低了妇女的终身生育率。

除此之外，占里村还有一个独特的习俗使得村里的青年即使早婚也不会早育。在占里村，男女结婚后，新娘会仍住在娘家——“不落夫家”。农忙时，男女可以到对方家里帮忙干

⊙1

⊙2

活——“夫妻互助”，待姑娘年纪稍大一些——27~28岁时，才会住到夫家去。

2. 集体婚礼

有趣的是，几百年来，占里人一直都坚持举办“集体婚礼”——在占里村，只有在正月初五和初六的日子才能举行订婚仪式，而婚礼则只能在农历二月十六或十二月二十六举行，其他日子都不允许结婚。

除了对结婚严格限制外，占里人对离婚者还会进行严厉的惩罚：提出离婚者需上缴“稻谷150 kg、白酒25 kg、肉50 kg”，而这些东西都会用来充公，并且提出离婚者不得继续留在村里生活。

3. 男女平等

在占里村，还有一条规定很特别：女儿继承棉花地，儿子继承稻田。此外，父母还要给女儿一份“姑娘田”，谁家若不给女儿“姑娘田”，不仅会遭人取笑，还会被男方退婚。在老人的财产继承上，山林、菜园实行男女对半分成，房基、家畜归儿子，而金银首饰、布匹则让女儿带到夫家。因此，在占里村，中国传统的“重男轻女”观念并不存在，而是真正的男女平等。占里人也没有“多子多福”这种观念，他们说，孩子一多，每个人分的就少了，人家就不愿嫁给你或娶你了，因此，谁也不愿意多生孩子。

4. 换花草之谜

占里侗寨几乎98%家庭的孩子均为一男一女，很少有双男或双女的现象。这里面隐藏着一个近乎神话的秘密，那就是用一种叫作“换花草”的草药来平衡胎儿的性别。在整个占里侗寨，并非每个人都知道“换花草”的庐山真面目，能有资格掌握秘方的始终只有一人，这个人被寨里人称为“药师”，并且“药师”通常都是女性，除非要传秘方时女儿已离世。调查中据村民的说法，现在村里的“药师”是一位年事已高的老太太，传闻已将秘方传给她的儿子了。村人认为这种“药师”一脉单传的约定是自吴姓迁来此地时开始的。

据说，倘若女人生的第一个小孩是男孩，那么“换花草”就会让她的第二胎怀上一个女孩；倘若女人生的第一个小孩是女孩，则第二胎也就必定会怀个男孩。据说“换花草”是一种藤状的植物，但根部却不相同。

⊙1 提倡控制生育的占里古歌
Zhanli ballad on birth control

⊙2 占里鼓楼底层戏耍的男人
Men enjoying their time on the ground floor of Zhanli drum tower

七 从江占里侗族皮纸的保护现状与发展思考

7 Preservation and Development of Bast Paper by the Dong Ethnic Group in Zhanli Village of Congjiang County

占里村具有独特的手工纸制作技艺与传承习俗，家庭与家族传承谱系完全无法构成，基本上是开放自由的学习机制，而且按照习俗年轻女性也不可学习；同时，占里皮纸主要供本村村民生活与民俗消费使用，因交通不便和交流不畅等诸原因外销一直很少。综合上述背景，虽然目前村里只有10余位皮纸技艺承担人，但这并不算少，完全能够满足村民的皮纸消费需求，从造纸人只是闲暇或有需求时才造纸的作业习惯，以及调查组采样时吴奶卫和吴奶妹家中仍存有足量余纸来看，占里村的侗族皮纸传习状态较为良好，保持着原生文化态的生产与消费。

21世纪的第二个十年，国家四级非物质文化遗产保护体系正在逐步完善，多项促进措施开始发布及落地执行，特别是贵州省，非遗保护的推动与多彩贵州发展战略紧密结合，非遗文化资源的培育呈现生机勃勃的局面。但是，占里村的侗族皮纸由于内向自足的体系及传播交流不足，尚未纳入非遗技艺文化保护系统，传承人群也未能获得保护与发展资源的支持。申报从县到省的非遗保护项目，为造纸老太太们申报各级侗族皮纸代表性传承人，是占里侗族皮纸需要尽快推动的事。

占里村作为“中国人口文化第一村”，其生育习俗和延伸文化价值很高；同时，占里也被誉为“世外桃源”，其自然人文资源丰富。随着知名度的快速扩大，外来旅游人口的不断增多，交通条件的显著改善，将占里村手工皮纸与旅游产品和民族文化产品关联开发的空间也快速打开。

⊙3

⊙4

⊙3 占里村风雨桥头 One end of the Wind-Rain Bridge in Zhanli Village

⊙4 造纸老人吴奶卫 Old papermaker Wu Naiwei

皮纸

Bast Paper
by the Dong Ethnic Group
in Zhanli Village of Congjiang County

占里村皮纸透光摄影图
A photo of bast paper in Zhanli Village
seen through the light

第十一节

黎平侗族皮纸

贵州省
Guizhou Province

黔东南苗族侗族自治州
Qiandongnan Miao and Dong Autonomous Prefecture

黎平县
Liping County

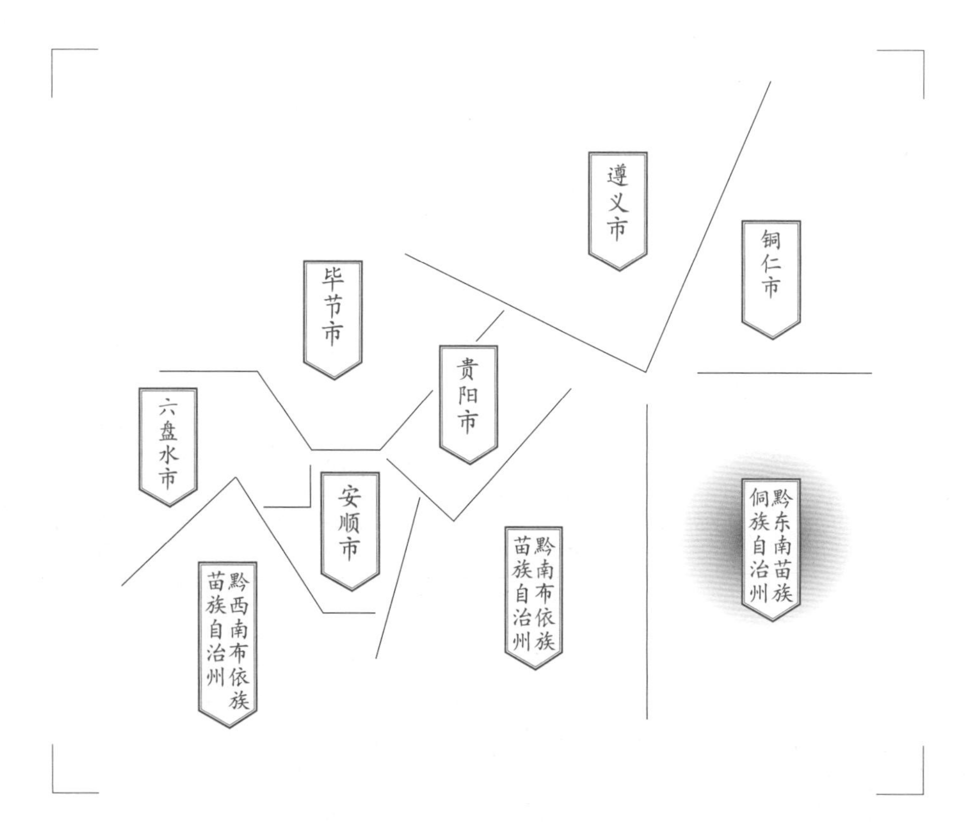

调查对象

茅贡乡
地扪行政村
侗族皮纸

Section 11

Bast Paper by the Dong Ethnic Group in Liping County

Subject

Bast Paper by the Dong Ethnic Group in Dimen Administrative Village of Maogong Town

一 黎平侗族皮纸的基础信息及分布

1 Basic Information and Distribution of Bast Paper by the Dong Ethnic Group in Liping County

据调查组2013年8月田野调查了解的情况，黎平县当代手工纸的活态主要是侗族构皮纸的制作，集中分布在茅贡乡的地扪行政村。地扪村的侗族构皮纸采用野生纯构树皮为原料，以浇纸法制作而成。

地扪村所生产的皮纸至调查时仍有较丰富的日常生活用途，如书写、刺绣、制作针线盒、包装等。传统销售渠道主要满足本村及茅贡乡周边的需求，近年来，随着黎平侗族乡国家级名胜风景区的建设，入境游客不断增多，地扪侗族皮纸观光性的流通也开始活跃起来。

⊙1

⊙1 浇纸的老人 An old woman pouring paper pulp onto the papermaking screen

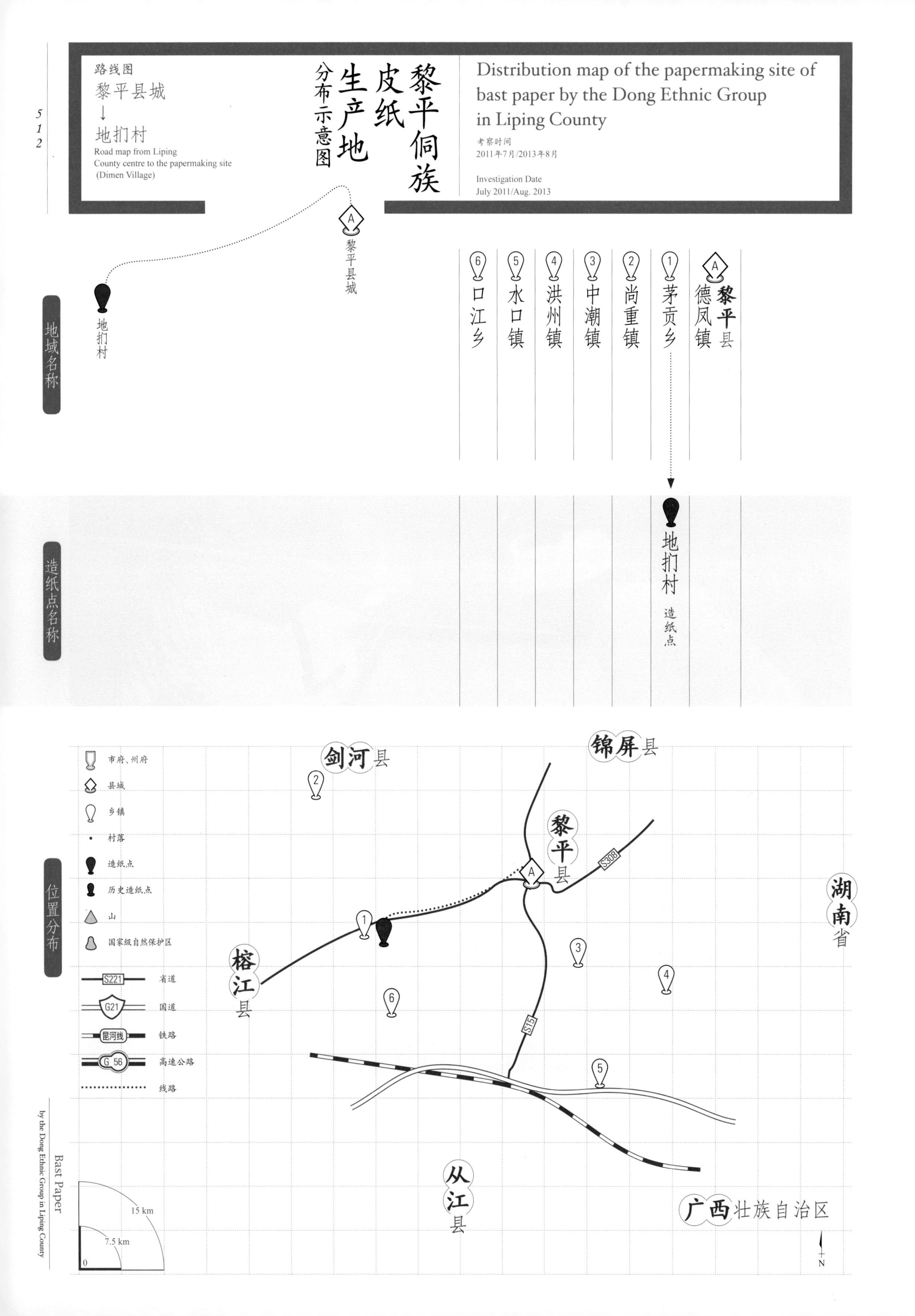

黎平侗族皮纸生产地分布示意图
Distribution map of the papermaking site of bast paper by the Dong Ethnic Group in Liping County
考察时间
2011年7月/2013年8月
Investigation Date
July 2011/Aug. 2013
路线图
黎平县城
↓
地扪村
Road map from Liping County centre to the papermaking site (Dimen Village)
黎平县城
地扪村
地域名称
黎平县
德凤镇
茅贡乡
尚重镇
中潮镇
洪州镇
水口镇
口江乡
造纸点名称
地扪村
造纸点
位置分布
市府、州府
县城
乡镇
村落
造纸点
历史造纸点
山
国家级自然保护区
S221
省道
G21
国道
昆河线
铁路
G 56
高速公路
线路
剑河县
锦屏县
黎平县
S308
S15
榕江县
湖南省
从江县
广西壮族自治区
15 km
7.5 km
0
N

⊙1

二
黎平侗族皮纸生产的人文地理环境

2
The Cultural and Geographic Environment of Bast Paper by the Dong Ethnic Group in Liping County

黎平县位于贵州省东南部，黔东南苗族侗族自治州南部，东经108° 31′~109° 31′、北纬25° 41′~26° 08′，地处黔、湘、桂三省（区）交界及云贵高原向江南丘陵过渡地区。黎平县属中亚热带季风湿润气候区，年平均气温16 ℃左右。

黎平县东南面与湖南省靖州县、通道县及广西壮族自治区三江县交界，西南面与贵州省榕江县、从江县毗邻，东北面与贵州省剑河县、锦屏县接壤，是贵州东进两湖、南下两广的桥头堡，境内有一个已经通航的民用机场。

⊙1
黎平侗乡小景
Landscape of Dong ethnic residence in Liping County

黎平古城已有约1 300年的历史，其地域唐代为龙标县治所，时名五脑寨；元至治二年（1322年），设上黎（里）平长官司；明洪武二十年（1387年）置中潮、黎平二千户所，属五开卫，永乐十一年（1413年）置黎平府；民国二年（1913年）改黎平府为黎平县。1950年属独山专区，1952年属都匀专区，1956年划归黔东南苗族侗族自治州至今。

调查时，全县辖15个镇、8个乡、2个民族乡，县辖面积4 439 km^2，总人口55万，是黔东南苗族侗族自治州面积最大、人口最多的县（市）。县内居住着侗、汉、苗、水等众多民族，其中侗族人口约占全县总人口的71%，是全国侗族人口最多的一个县，也是侗族文化的主要发祥地，因而有“侗乡之都”的誉称。著名的肇兴侗寨为全国最大的侗寨，2005年被《中国国家地理》评选为“中国最美的乡村古镇”，2007年被《美国国家地理》和中国《时尚旅游》杂志社联合评选为“全球最具诱惑力的33个旅游目的地”之一。

黎平是省级历史文化名城，是红军长征进入贵州的第一城，有“杉海粮仓油壶”“侗族大歌之乡”“鼓楼之乡”等称号。黎平县还是中国名茶之乡，截至2012年底，黎平县茶叶种植面积已经达到338 km^2。黎平有世界上最大的天生石拱桥——高屯天生桥，有世界上最大的侗寨鼓楼群——肇兴鼓楼群，同时也是“侗族大歌”申报人类非物质文化遗产代表作名录的主申报地。

茅贡乡位于黎平县西部，距县城42 km，距榕江县城56 km，东与坝寨乡接壤，南与岩洞镇交界，西与九潮镇为邻，北与孟彦镇毗连，“308”省道穿境而过，沿线村寨俗称“十里侗寨”“十里画廊”。全乡总面积172 km^2，辖15个行政村、1个居委会、78个村民组，共4 030户16 845人，其中侗族人口占96%。

⊙1

⊙1 肇兴侗寨 Dong ethnic residence in Zhaoxing Town

◎2 鼓楼内部仰望 Inside view of the drum tower

⊙1

⊙2

⊙1 茅贡乡间的桥
Bridge in Maogong Town

⊙2 侗寨花桥
Flower Bridge in Dong ethnic residence

茅贡乡是享誉海内外的侗戏发祥地，腊洞村人吴文彩是侗族人民公认的侗戏鼻祖。乡内侗族风情浓郁，斗牛、吹芦笙、踩歌堂等习俗别具风格。侗寨的鼓楼、花桥、戏台和侗族民居是侗族智慧的结晶，高近田间风雨桥、古戏台和腊洞吴文彩墓被列为贵州省级重点文物保护单位；地扪村的花桥群、鼓楼群、寨门，登岑村的中日友谊鼓楼等建筑是侗族建筑的精华；茅贡乡村村有侗戏队，人人爱看侗戏，个个会唱侗歌。茅贡乡早在1984年即被贵州省文化厅命名为“侗戏之乡”。

三
黎平侗族皮纸的历史与传承

3
History and Inheritance of Bast Paper by the Dong Ethnic Group in Liping County

调查中，据地扪侗族造纸世家的文化人——茅贡中学的吴永峰老师介绍，该地造纸历史传承信息没有文字记述，是一代代口传下来的。吴姓祖先在唐末从江西吉安府太和县的珠子巷迁到今贵州天柱县的远口村，带来9个孩子，因人多地少，其中2个孩子随父迁到今榕江县东江乡；后因气候问题，雾太大，难辨天晴、下雨，又无泉水可喝，宋朝时又搬到今黎平县九潮镇大榕村；后因村子太窄，又带孩子上五岭山，下十字冲，到“边心”，顺河而下到地扪，看到此地山水秀丽，扪心可留，于是留在此地安居乐业，这也是“地扪”村名来历的一种说法。调查时据吴永峰介绍，乾隆十五年（1750年）立的墓碑上有相关记述，但现在已看不见（调查组成员现场未弄清是墓碑已毁、看不清，还是地点偏远）。据说在明代时就有立碑叙述来历。

地扪侗族造构皮纸技术的源头，吴永峰认为是祖先从江西带来的，吴家代代都造纸，家族内认为有20代以上，所能记住的传承谱系如下：××－吴银桥（太公）妻子－吴有贵（曾祖父）妻子－××－吴××－吴化寅（奶奶）－吴红化（妈妈）－吴英亮（大姐，1954年出生）、吴妹叶（二姐，1955年出生）、吴美行（三姐，1961年出生）、吴桃根（吴永峰妻子，1965年出生），技艺传承全部在家族女性人群中进行。2006年寨上不慎发生大火，把吴永峰家的造纸工具、织布机烧了，之后就没有再造纸了。

2013年8月调查时，地扪村有650多户人家，400多户造纸，基本上有老人的家庭都造纸，而且造纸户全部是侗族村民。从业态传习的范围来看，仍有相当大的覆盖面和普及度，不过据吴永峰说，以前一家几个人做，媳妇帮婆婆做，现在一般一户只有一位老年妇女造纸，每户平均每年造200~300张皮纸，少的只造几十张。

⊙1

⊙2

⊙1
调查组成员与吴永峰
A researcher and local papermaker Wu Yongfeng

⊙2
地扪村水塘边的民居
Local residences by a pond in Dimen Village

四 黎平侗族皮纸的生产工艺与技术分析

4 Papermaking Technique and Technical Analysis of Bast Paper by the Dong Ethnic Group in Liping County

（一） 黎平侗族皮纸的生产原料与辅料

地扪村侗族皮纸的生产原料为野生构树皮，造纸人大多是60岁以上的老年妇女，在农历二月至清明节前，去菁山（当地指草木浓密的山）用柴刀砍构树，通常生长了一年左右的构树枝条都可以砍。过了清明节也能砍，但当地湿热，蛇及各种虫开始出动，不便上山砍。具体的方式是今年砍下来明年用，砍树时留下树桩，来年再生，次年又可以砍。野生构树长得快，大部分造纸人用一年生的枝条，但村民说，其实2~3年生的构树皮更好，皮厚且出料率高，不过若树太大则较难剥皮。

地扪村侗族造皮纸所用的纸药为杨桃藤，以当年从老藤中间分出来的枝且已生长6~7个月的为最佳，太嫩的没浆，太老的不好捶，最多只能用生长2年的藤。

⊙3 ⊙4 ⊙5

⊙3 杨桃藤原料 *Averrhoa carambola* vine, raw material of papermaking mucilage

⊙4 捶纸药料 Hammering the papermaking mucilage

⊙5 揉出纸药液 Squeezing out the papermaking mucilage sap

(二) 黎平侗族皮纸的生产工艺流程

据调查组成员于2011年7月、2013年8月两次到地扪村实地考察，总结地扪侗族构皮纸的生产工艺流程如下：

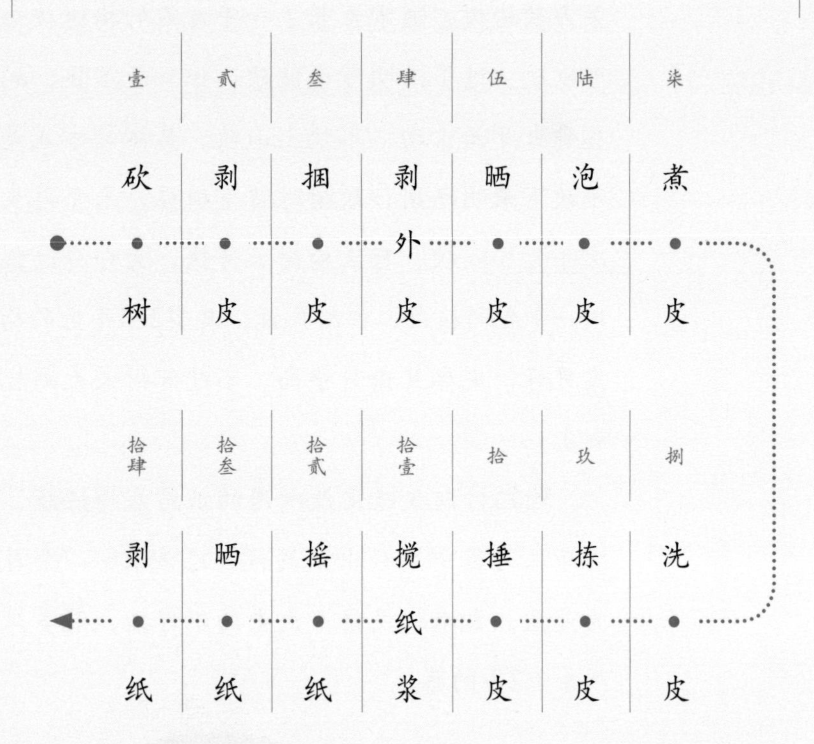

壹 砍树

1

造纸户在农历二月至清明节前到菁山用柴刀砍一年生的野生构树枝。

贰 剥皮

2

用手将鲜树皮剥下来。

叁 捆皮

3

将2~2.5 kg剥下的鲜皮捆成一把，一般一个人一天最多可砍树、剥毛皮13~14把，即一个人一天最多可获30~35 kg毛皮，然后挑回家。

肆 剥外皮

4 ⊙1⊙2

当天晚上在家里将毛皮的外皮剥掉，只留下白色内皮。剥外皮需及时，如当天晚上来不及剥，第二天一定要剥；若等到第三天，则必须浸水后方可剥开，再久就只能扔掉。调查中造纸户认为白皮和外皮中间有一种黏液，如构皮干了，黏液就没有了，也就剥不开了。

⊙1

⊙2

⊙1/2 剥构树外皮 Stripping paper mulberry bark

伍 晒皮

5 ⊙3

将白色内皮一根根摊开放在竹竿上，置于太阳下晒干。若阳光强烈，2~3天可晒干；若阳光不强烈，5~6天才可晒干。如没阳光则需在火塘边烤干。白皮如被水淋则会烂，很难出纸浆。晒好后，拿回家捆好，一捆约0.5 kg，一般放在火塘边或竹竿上晾。

⊙3

陆 泡皮

6 ⊙4

⊙4

造纸前，将干构皮置于小溪中浸泡1~2天，不能超过3天，否则就被泡烂了。泡皮时，构皮一般要用石头压紧，以免在水中浮起来。

柒 煮皮

7 ⊙5⊙6

将浸泡好的构皮从溪水中捞出，并放在家中的锅里煮，通常煮6小时左右，至皮料涨起来即可。如果晚上煮，则放置一夜后，第二天早上再稍煮一下，至有气泡冒出来即可，不能煮沸腾，否则皮会被煮烂。大锅可煮约2.5 kg干皮，小锅可煮1~1.5 kg，如煮2.5 kg干皮，约需加0.5 kg石灰。

捌 洗皮

8 ⊙7

煮好后，用捞棒将皮捞出来，先在锅里，然后在溪水或塘里将灰洗掉。

玖 拣皮

9 ⊙8⊙9

洗干净后，用手拣掉皮里的黑壳等杂质，然后挤掉皮里的水，并将其捏成团。

⊙7

⊙8

⊙6

⊙9

⊙5

⊙3 干皮 Dried bark

⊙4 浸泡好的纸药液 Soaked papermaking mucilage

⊙5/6 煮皮 Boiling the bark

⊙7 洗皮 Cleaning the bark

⊙8 拣皮 Picking out the impurities

⊙9 捏团 Balling up the papermaking materials

⊙10

拾 捶皮

10 ⊙10

将料团放在石板或水泥地上，用棒槌将其捶融，然后再揉成团。

拾壹 搅纸浆

11 ⊙11⊙12

将捶好的成团纸浆放入桶里，加水，用竹棍将其搅拌均匀，然后将杨桃藤汁经布、撮箕过滤后，加入桶里，再次将其搅匀。

⊙11

⊙12

拾贰 摇纸

12 ⊙13⊙14

亦称浪纸。先将浪纸架在水里泡湿，然后挂在一钩上，根据经验，加入适量搅匀的纸浆，手持浪纸架将其摇均匀。

⊙13

⊙14

拾叁 晒纸

13 ⊙15

静置片刻，确定基本上没有水再流出来后，将浪纸架斜靠在墙上或其他物体上，晒干。阳光强烈时，一个多小时即可晒干，阴天约需半天，雨天则置于室内自然晾干，约需两天。

⊙15

⊙10 捶皮 Hammering the bark

⊙11 搅纸浆 Stirring paper pulp

⊙12 加杨桃藤汁 Adding in the sap of *Averrhoa carambola* vine

⊙13/14 摇纸 Shaking the papermaking screen to even paper pulp

⊙15 晒纸 Drying the paper

拾肆
剥　纸

14　⊙16~⊙19

纸干后，先剥开上沿一角，然后将上沿整体剥开，再由上往下剥，从中间对折，沿另一侧对折再对折，即折成原纸的1/8大小。

⊙16

⊙17

⊙18

⊙19

⊙16/19
剥纸和折纸
Peeling the paper down and folding it

（三）

黎平侗族皮纸生产使用的主要工具设备

壹 木槌 1

用于捶杨桃藤。

⊙1

贰 葫芦瓢 2

原用杉木，1980年后改用塑料请木匠师傅制作而成。实测葫芦瓢直径约23 cm，总长约48 cm。

⊙2

伍 撮箕 5

平时用于捞鱼。实测尺寸为：口宽37.5 cm，长（内深）36 cm，弧形，竹质，一般为斑竹和水竹。中间起固定作用的木头易折而不断。

⊙3

叁 浪纸架 3

用于摇纸。含挂钩，木质，实测长约30 cm。亦有非木质的挂钩。

肆 竹棍 4

用于搅匀杨桃藤汁。实测长约24 cm。

⊙1 木槌 Wooden mallet

⊙2 葫芦瓢 Gourd ladle

⊙3 撮箕 Winnowing basket

（四）

黎平侗族皮纸的性能分析

所测黎平地扪村皮纸的相关性能参数见表10.18。

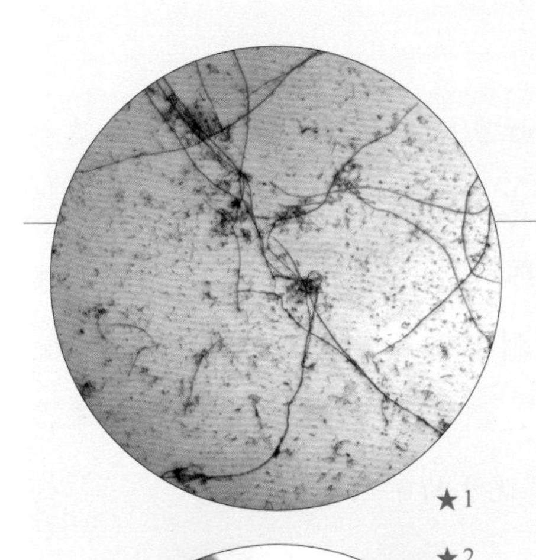

★1

★2

表10.18　地扪村皮纸的相关性能参数
Table 10.18　Performance parameters of bast paper in Dimen Village

指标		单位	最大值	最小值	平均值
厚度		mm	0.260	0.155	0.207
定量		g/m^2	—	—	46.7
紧度		g/cm^3	—	—	0.226
抗张力	纵向	N	26.0	20.7	20.8
	横向	N	22.8	14.4	14.7
抗张强度		kN/m	—	—	1.183
白度		%	40.3	39.2	39.7
纤维长度		mm	11.72	0.76	4.72
纤维宽度		μm	31.0	1.0	14.0

由表10.18可知，所测地扪村皮纸最厚约是最薄的1.68倍，相对标准偏差为3.80%，厚薄差异相对较小。皮纸的平均定量为46.7 g/m^2。所测皮纸紧度为0.226 g/cm^3。

经计算，其抗张强度为1.183 kN/m。

所测地扪村皮纸白度平均值为39.7%，白度最大值约是最小值的1.03倍，相对标准偏差为0.45%，白度差异相对较小。

所测地扪村皮纸纤维长度：最长11.72 mm，最短0.76 mm，平均4.72 mm；纤维宽度：最宽31.0 μm，最窄1.0 μm，平均14.0 μm。所测皮纸在10倍、20倍物镜下观测的纤维形态分别见图★1、图★2。

★1
地扪村皮纸纤维形态图（10×）
Fibers of bast paper in Dimen Village (10× objective)

★2
地扪村皮纸纤维形态图（20×）
Fibers of bast paper in Dimen Village (20× objective)

五 黎平侗族皮纸的用途与销售情况

5 Uses and Sales of Bast Paper by the Dong Ethnic Group in Liping County

（一）黎平侗族皮纸的用途

1. 书写

地扪侗族构皮纸原料为纯构皮，使用较纯粹的手工工艺，所造出的纸可保存很长时间。因此，当地传统上广泛将其用于毛笔书写，包括写契约（如房屋契约、卖山卖地契约、过继小孩契约）、抄写书本等。

2. 刺绣

当地侗族女性衣服上有较多精美图案，这些精美图案要先用皮纸刻花，即先用笔在皮纸上画好特定的图形，再用刀刻，一般刻2~5层后，用米浆粘上，之后贴到布上，再根据其示样剪出来，当地称之为刻花。调查中了解到，在黎平一带，以前女孩13~14岁就开始学刻花。

3. 制作针线盒

地扪村用本村生产的皮纸制成朴素而又有民族风情的针线盒，这些针线盒分成相当多的盛物空间，可用来摆放做刺绣或其他女红活时的绣线与绣针。

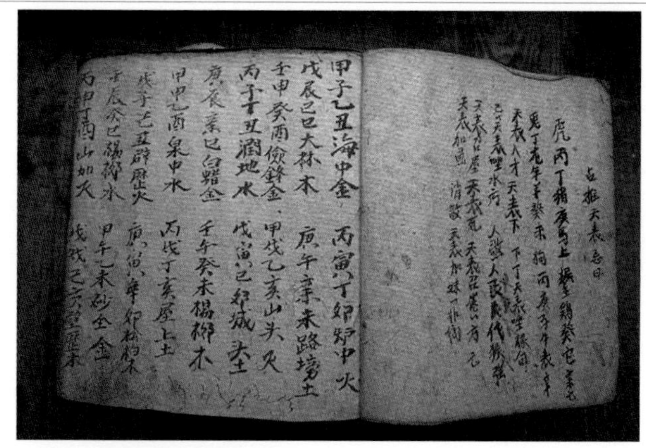
甲子乙丑海中金　丙寅丁卯炉中火

⊙1

⊙2

⊙3

⊙4

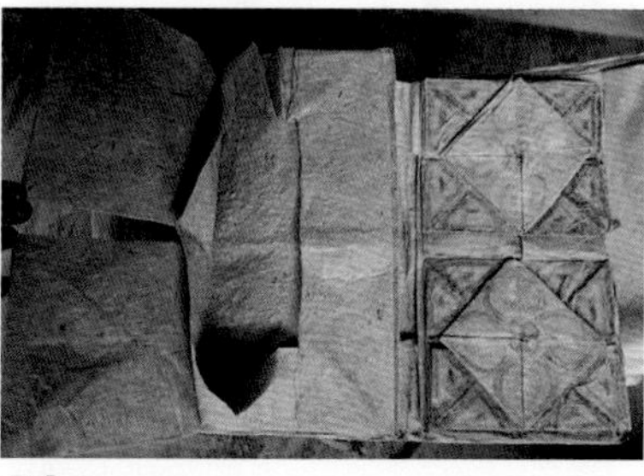

⊙5

⊙6

⊙1 地扪手抄皮纸书 Book transcribed on bast paper in Dimen Village

⊙2 粉红色绣领下的刺绣纸 Embroidery paper covered under the pink neckline

⊙3/6 皮纸针线盒 Sewing kit wrapper made of bast paper

4.包装

地扪侗族皮纸结实且干净，可用来包装多种物品。

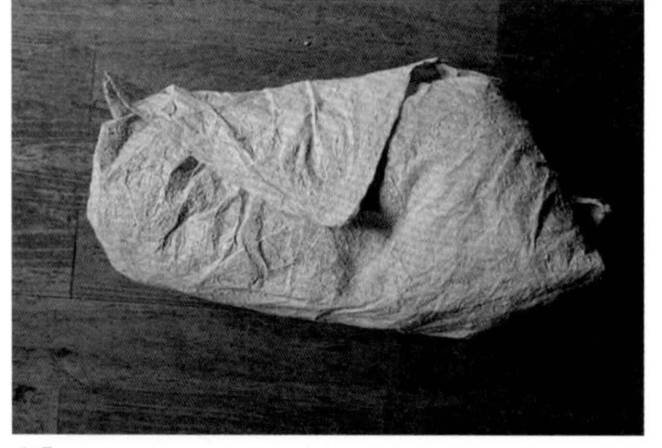

⊙7

⊙8

（二）黎平侗族皮纸的销售情况

1980年以前，地扪侗族构皮纸主要在地扪本村销售，只有少数老人会将其挑到不远的岩洞镇上，以换取酒曲、鸡、鸭、衣服等，一般不卖钱。20世纪70年代，乡镇供销社也来地扪村收过竹纸，当时价格为0.05元/张（如表10.19所示）。

由于地扪村的构皮纸基本不外卖，且每家一年也就用50~60张，因此历史上地扪的造纸人往往一年只做几十张自用，极少数不造纸的村人就到造纸户家里去买。2004年，国务院批准地扪为国家级旅游区，地扪吸引着越来越多的游客前来观光旅游，其皮纸成为游客爱买的侗族记忆文化产品之一。

2011年和2013年入村调查时，地扪侗族构皮纸的销售业态已发生了较大改变，首先是卖给络绎不绝的游客，其次是留给自己用，最后若有剩余才会拿到岩洞镇去卖。目前造纸多的老太太一年可造600~700张纸，按5元/张算，一年可有3 000~3 500元收入，这对当地老年妇女来说是一笔不错的收入。

表10.19　不同年份地扪侗族构皮纸价格
Table 10.19　Price list of bast paper by the Dong Ethnic Group in Dimen Village in different years

年份	1978年以前	1978年	1980年	1995年	2000年	2005年	2008年	2010年	2011年
单价（元/张）	0.05	0.2	0.5	1	1.5	2	3	4	5

⊙7 包装用纸 Packing paper

⊙8 地扪吸引了大批艺术家 Dimen Village appeals to the artists

六
黎平侗族皮纸的相关民俗与文化事象

6
Folk Customs and Culture of Bast Paper by the Dong Ethnic Group in Liping County

(一) 造纸习俗

在地扪村，造纸被认为是女人的活计，以前村里的女性十三四岁就开始学习造纸，调查时已改为40岁左右或更大年龄时才开始学习造纸，年轻且已辍学的女孩大部分选择外出打工。地扪村的男人不造纸，因为地扪人认为男人的手太硬，所造的纸不均匀。

不过造完纸后，除刻花外，对纸进行剪裁、加工，以及祭祀烧纸等活计，却都由男人来做。

(二) 民俗用纸

1. 过世时用纸

当地人过世，会在其身下垫三张纸（脚、腰、头各一张），两侧一般不放纸，讲究的人家会将一张纸分成两节，两侧各放一节，还有一些人家会用一张纸包住遗体的双脚。据当地传说，用这种纸来制作钱纸，并垫在去世的人身下，送给阎王爷，走得快些，抬遗体的人感觉轻些。敬鬼神的物品必须是单数，纸、香均如此，但纸的数量一般为三张，因为即使棺材够长，五张纸也会盖住棺材，看不到木头。

将棺材抬到墓穴边，并在墓穴里放三张纸（不用烧，是垫背钱），道长烧完构皮纸并将墓穴打扫干净后才能放棺材。安墓碑时，用纸包碑脚（即底部），穿下去（下有座子）。纸的数量由道长定。

2. 鬼娃娃

当地女性怀孕六个月后，用纸剪成鬼娃娃形状，并用竹钉将鬼娃娃钉在卧室的墙上，再将夫妻双方的祖宗都接到家里，以保佑小孩健康成长。

3. "门后"

当地女性生完第一个孩子后，将构皮纸剪成长条形，挂在门旁边，以保佑孩子皮肤好。每年正月买一块猪肉，烧纸钱、香，敬"门后"，七

⊙1 鬼娃娃剪纸 Ghost kid paper cutting for protecting kids in the family

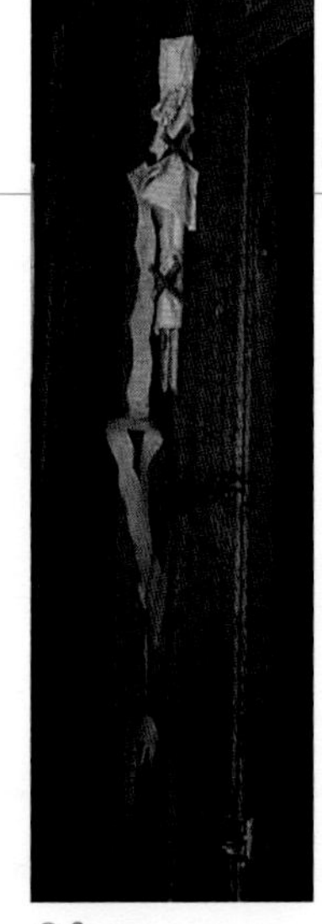

⊙2 皮纸长串串 A string of bast paper

月再买三条鱼，烧构皮纸、香，谢“门后”。

4. 小孩生病用纸

小孩生病，如感冒、拉肚子等，如果吃药不见好，就去找道长做法事。道长做完仪式后，说是某个祖宗找小孩要吃的（如奶奶找孙子、孙女要吃的），需要在祖宗的神位前供一头猪、一个猪脑壳等，还要烧纸，纸的数量不限。

5. 纸钱

清明节时，不仅要给自家已去世的老人烧纸钱，还要给亲戚家已去世的老人烧纸钱。当地纸钱三排七列，以前都由构皮纸制成，大约20世纪80年代后改用竹纸。

⊙3

⊙3 竹纸钱 Bamboo joss paper

七 黎平侗族皮纸的保护现状与发展思考

7 Preservation and Development of Bast Paper by the Dong Ethnic Group in Liping County

（一）黎平侗族皮纸的传承现状与技艺价值

虽然地扪侗族的造纸户已不像过去那么多，但访谈时村民们都认为：造纸会一直传下去。至今造纸技术没有发生变化，以后也不太可能会变。虽然年轻人都去打工了，但年纪大的妇女都会做，造纸技术的传承主要靠中老年妇女。

黎平地扪侗族皮纸的造纸现状就像民间社区的博物馆，为了本民族的文化习俗需要定时有人造纸，传承并使用这些自家生产的纸张，基本上不指望通过造纸带来经济效益。地扪村造纸技艺的传承与中国其他地区或民族很多造纸点的现状不同，经济和生活的要素并不影响传承，主要的支撑来自乡土习俗与文化目标。通过进一步分析，可以发现地扪侗族皮纸具有独特的价值。

1. 独特的民俗

女人造纸，男人用纸；纸钱有讲究；作为节日及生病用纸等，民俗内涵相当丰富且独特。

2. 独特的销售

地扪村生产的构皮纸，主要通过买卖或赠送的方式在村内进行。据调查，很少有人拿到市场上销售，也基本上没人去买。纸是为了满足本民族小群落的需要而生产的。

3. 独特的工艺

地扪村生产皮纸采用的是独特的浇纸法，并使用纸药，而通常的浇纸法，如傣族传统生产的白绵纸，因浇一张纸就用一个固定式纸帘，故不用纸药；纸药起到悬浮、阻聚、阻滤、增滑等作用，抄纸法往往需用纸药，一般认为没有纸药就抄不了纸。纸药是中原系统发明造纸的关键，蔡伦发明造纸时就已经使用纸药。

⊙2

⊙1

地扪村手工皮纸采用浇纸法并使用纸药的造纸技术，是中国手工纸发展过程中重要的活化石。此外，其独特的用纸民俗、自我消费等对深入研究黎平侗族的历史、文化等也具有重要的作用。

⊙3

⊙1 穿侗装的造纸老太与调查组成员
A researcher and an old Dong papermaker wearing ethnic clothing

⊙2 河边挂钩上摇纸
Shaking the papermaking screen with a pothook by the river

⊙3 黏稠的纸药液
Sticky papermaking mucilage

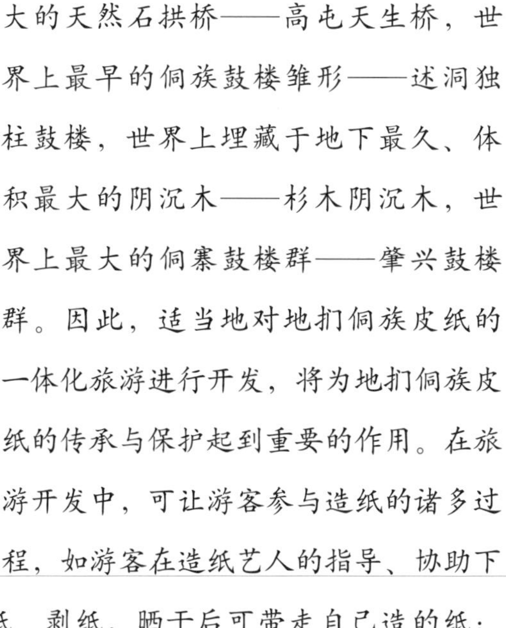

⊙4

（二）地扪村皮纸的传承与发展思考

1. 申报省级和国家级非物质文化遗产项目

地扪侗族皮纸最具价值的是其独特的采用浇纸法并使用纸药的造纸技术，且其浇纸方式与云南傣族、纳西族有相当明显的差异；此外，其独特的名称、独特的民俗、独特的销售等，需要多机构合作，对地扪村侗族皮纸进行更为深入、细致的研究，以进一步挖掘该技术的历史、传播路径及内涵。

在上述研究的基础上，当地政府部门努力将地扪侗族构皮纸申报为省级以及国家级非物质文化遗产，不但可以丰富非物质文化遗产名录，同时也可将地扪侗族构皮纸的传承与保护纳入非物质文化遗产保护体系，使地扪侗族构皮纸获得更广泛的文化关注性传承。

2. 适当进行旅游开发

黎平有较多的旅游资源，如八舟河国家级风景名胜区、肇兴侗寨、弄相山森林公园等，而地扪本就是黎平侗乡国家级名胜风景区的核心地。此外，黎平有多项世界吉尼斯之最：世界上最大的天然石拱桥——高屯天生桥，世界上最早的侗族鼓楼雏形——述洞独柱鼓楼，世界上埋藏于地下最久、体积最大的阴沉木——杉木阴沉木，世界上最大的侗寨鼓楼群——肇兴鼓楼群。因此，适当地对地扪侗族皮纸的一体化旅游进行开发，将为地扪侗族皮纸的传承与保护起到重要的作用。在旅游开发中，可让游客参与造纸的诸多过程，如游客在造纸艺人的指导、协助下摇纸、晒纸、剥纸，晒干后可带走自己造的纸；也可将用侗族皮纸制成的针线包、服饰等开发成各种独具侗族文化特色的旅游纪念品。

3. 开发地扪侗族皮纸的新品种、新用途

在保留传统造纸技艺的基础上，进行适当的技术改进及产品开发，如制作书画、装饰等用纸。结合地扪侗族皮纸独特的工艺与文化、特有的质感，同时借力高品质的旅游资源，有利于激发地扪侗族皮纸更优质的市场需求。由需求拉动生产，使生产性保护成为地扪侗族皮纸积极传承的重要方式。

⊙5

⊙4 地扪村的风雨桥文化 Wind-Rain Bridge culture in Dimen Village

⊙5 地扪村的旅游宣传牌 Tourism billboard in Dimen Village

皮纸

Bast Paper
by the Dong Ethnic Group
in Liping County

地扪村皮纸透光摄影图
A photo of bast paper in Dimen Village
seen through the light

Introduction to Handmade Paper in Guizhou Province

1 History of Handmade Paper in Guizhou Province

1.1 Developmental Features of the Regional Culture in Guizhou Province

Guizhou Province is a province within the People's Republic of China, located in the southwestern part of the country. It possesses a unique regional culture stemming from the geographical configuration, with about 90% local area covered by hills and mountains. The hills and mountains blocked regional development, and subsequently caused an underdevelopment of the region.This promoted a sort of striving towards self-sufficiency for a long time. Being cut off from the outside world, the 49 ethnic groups in the area (with 17 native groups) frequently communicated and interacted among themselves, hence forming a mixed regional culture, while each maintaining their own identities. Therefore, the culture profile of the area can be compared to a land composed of a thousand islets, each one enjoying a unique culture. This multiplex Guizhou culture is treasured as an important component and special heritage of Chinese culture.

The other typical feature with the development of Guizhou culture is its discontinuity. Archaeological studies show that Guizhou culture interacted with Central Chinese Culture from the Qin Dynasty and the Han Dynasty. Its developmental journey was blocked now and then, and continued on and off for over a thousand and six hundred years till the Ming Dynasty and the Qing Dynasty, which finally became relatively steady. On occasion Wang Yangming, grand master of the Yangming School, was demoted and exiled to the Longchang Post in Guizhou Province. While there he gained insight into the truth and gave lectures to his students, directly promoting the formation of "Yangming's Theory of Mind", marking its cultural culmination.

Portrait of Wang Yangming

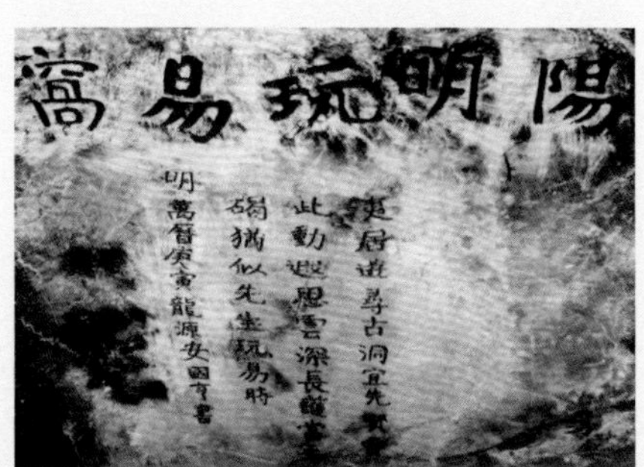

Yangming Wanyiwo stone cave monument in Longchang Town of Xiuwen County

In the early period of thousand years cultural development in Guizhou Province, the Central Chinese Culture failed to dominate, while after the Ming Dynasty, Central Chinese Culture became overwhelming. Meanwhile, the native subcultures remained flourishing, and made Guizhou Province an area accommodating various cultures.

History of handmade paper in Guizhou Province is greatly influenced by its cultural development. Before the Ming Dynasty, few handmade paper record could be traced, just as the area's obscure political, economic, or cultural resources. In the Ming Dynasty, the central government began to intensify its domination over Yi areas, namely southwest areas, by large-scale migration. From then on, governed by the central government for about six hundred years, Guizhou enjoys its own voice in terms of politics, economy and culture as an independent province. Handmade paper in Guizhou Province can be traced back to the Sui Dynasty and Tang Dynasty, according to a local scholar named Zu Ming, who believes that handmade paper is essential for the local governments to communicate with each other. However, this is only a hypothesis without sound documentation or proof. Before the Ming Dynasty, only few unverifiable records or folk tales could be found, and after the Ming Dynasty, handmade paper history became visible in historical literature.

1.2 Four Phases of Handmade Paper Development in Guizhou Province

Handmade paper development in Guizhou Province can be divided into four phases since the Ming Dynasty, based on the literature review and field investigation by the research group.

1.2.1 Emergence of Handmade Paper in Guizhou Province in the Ming Dynasty

Handmade paper in Guizhou Province could have enjoyed a long history. However, we can only verify that the emergence and flourishing of handmade paper happened in the early Ming Dynasty, when the first emperor of the Ming Dynasty, Zhu Yuanzhang, treated Guizhou Province as the strategic place to fortify his rule of Yunnan Province, and other southwest areas. He had military forces marched into the area, to fight against the local armed forces, strengthening the local government, and finally taking up the garrison duty. This military action is called "North-to-South Migration for the purpose of conquering". Later on, civilians, mostly from South of the Yangtze River and the Central Plains, were also transferred to the area under the government arrangement. This is illustrated in a book written by Chen Guosheng, named *Agricultural Geography in Yunnan, Guizhou and Sichuan in the Ming Dynasty*, which states that after the Ming Dynasty, "the number and distribution of Han people in Guizhou is unprecedented". The large population of Han people helped to cultivate the wild in the area.

With this population migration, the techniques were also brought to the area, together with the numerous craftsmen and soldiers (who were born craftsmen) from South of the Yangtze River and the Central Plains, where handmade paper had become a popular practice. These histories were recorded together with the policies like

"North-to-South Migration for the purpose of conquering" and "North-to-South Migration for the purpose of occupying", became the heritage of many local families, especially papermaking families in the Guizhou Province.

During the Ming Dynasty, some cities and counties already started the practice of papermaking. It is recorded in *Guizhou Chorography* that in 1553, local government built a papermaking mill in Guiyang, and papermakers from Jiangsu and Zhejiang Provinces were hired to help local people make paper. "Local people learned papermaking techniques and profited from the practice." Places like Duyun Fu, Pingyue Wei (now Fuquan City), Longli Wei (now Longli County), Anzhuang Wei (now Zhenning County) also have records of papermaking. For instance, Longli Wei has a dam named "Paper", and Huangjing of Anzhuang Wei used to be a place where people made paper.

Based on the literature review and field investigation, current papermaking sites in Guizhou Province can be traced back to the Ming Dynasty (or even earlier according to the oral history). The papermaking sites in the Ming Dynasty, the Qing Dynasty and later periods in Guizhou Province are listed in table1.1.

Table 1.1 History of current handmade papermaking sites in Guizhou Province

Serial number	Paper type	Source	Period
1	Handmade paper in Renhuai City	Oral history, local chronicles, genealogy	Around the Ming Dynasty according to the oral history; early Qing Dynasty according to the local chronicles, and Daoguang Reign of the Qing Dynasty according to the genealogy
2	Bamboo paper in Wudang District	Oral history, local chronicles	Early Ming Dynasty according to the oral history; Ming Dynasty based on the local chronicles
3	Bast paper by the Gelo Ethnic Group in Wuchuan County	Genealogy, local chronicles	Daoguang Reign of the Qing Dynasty according to the genealogy; late Qing Dynasty according to the local chronicles
4	Baboo paper by the Gelo Ethnic Group in Wuchuan County	Oral history	Late Qing Dynasty
5	Handmade paper in Zheng'an County	Oral history, local chronicles	During the Yuan and Ming Dynasties according to oral history; mid-Qing Dynasty according to the local chronicles
6	Bamboo paper in Yuqing County	Local chronicles	Ming and Qing Dynasties
7	Bast paper by the Gelo Ethnic Group in Shiqian County	Local chronicles	Before the early Qing Dynasty
8	Bamboo paper by the Gelo Ethnic Group in Shiqian County	Oral history	Qing Dynasty
9	Bast paper by the Tujia Ethnic Group in Yinjiang County	Oral history	Early Ming Dynasty or between the late Ming and early Qing Dynasties
10	Handmade paper by the Tujia Ethnic Group in Yinjiang County	Oral history	Late Qing Dynasty and early period of the Republican Era of China
11	Bamboo paper by the Tujia Ethnic Group in Jiangkou County	Oral history	Kangxi Reign of the Qing Dynasty
12	Bast paper by the Miao Ethnic Group in Danzhai County	Folklores, local chronicles	Tang Dynasty according the folklores; late 1920s according to the local chronicles
13	Bamboo paper by the Dong Ethnic Group in Sansui County	Genealogy	Hongwu Reign of the Ming Dynasty

(Continued)

Serial number	Paper type	Source	Period
14	Bamboo paper by the Miao Ethnic Group in Kaili City	Oral history	Late Qing Dynasty and early period of the Republican Era of China
15	Bamboo paper by the Miao Ethnic Group in Huangping County	Oral history	Qing Dynasty
16	Bast paper in Duyun City	Local chronicles	Tongzhi Reign of the Qing Dynasty or early period of Guangxu Reign of the Qing Dynasty
17	Wax paper in Duyun City	Local chronicles	Early 1950s
18	Bamboo paper in Longli County	Local chronicles	Jiajing Reign of the Ming Dynasty
19	Bast paper by the Dong Ethnic Group in Rongjiang County	Folklores	During the Tang and Song Dynasties
20	Handmade paper by the Yao Ethnic Group in Cuili Town of Congjiang County	Oral history	Late Ming Dynasty or early Qing Dynasties; or before the late Qing Dynasty
21	Bamboo paper by the Yao Ethnic Group in Xiutang Town of Congjiang County	Oral history	Uncertain
22	Bast paper by the Dong Ethnic Group in Zhanli Village of Congjiang County	Oral history	Before the Republican Era of China
23	Bast paper by the Dong Ethnic Group in Xiaohuang Village of Congjiang County	Oral history	Before the late Qing Dynasty
24	Bast paper by the Dong Ethnic Group in Liping County	Folklores	Song Dynasty
25	Bast paper in Zhenfeng County	Genealogy	Qianlong Reign of the Qing Dynasty
26	Bamboo paper by the Bouyei Ethnic Group in Zhenfeng County	Oral history	Hongwu Reign of the Ming Dynasty
27	Bamboo paper by the Bouyei Ethnic Group in Anlong County	Oral history	Before the late Qing Dynasty
28	Bamboo paper by the Bouyei Ethnic Group in Libo County	Oral history, publicity materials of tourism	Mid-late Qing Dynasty according to the oral history; history of a thousand years according to publicity materials of tourism
29	Bamboo paper by the Dong Ethnic Group in Cengong County	Oral history, local chronicles	Late Ming Dynasty

(Continued)

Serial number	Paper type	Source	Period
30	Bast paper in Panxian County	Oral history	Late Qing Dynasty
31	Bamboo paper in Panxian County	Oral history, genealogy	Early Ming Dynasty according to the oral history; Kangxi Reign of the Qing Dynasty according to the genealogy
32	Bast paper in Pu'an County	Oral history	Mid-Ming Dynasty or early Qing Dynasty
33	Bast paper by the Yi, Miao and Gelo Ethnic Groups in Liuzhi Special Area	Local chronicles	Early Qing Dynasty
34	Bast paper in Guanling County	Oral history	Early Ming Dynasty
35	Bamboo paper by the Bouyei and Miao Ethnic Groups in Guanling County	Oral history	Start from the Qing Dynasty
36	Bast paper in Changshun County	Oral history, local chronicles	Early Ming Dynasty according to the oral history; mid-Ming Dynasty according to the local chronicles
37	Bamboo paper in Huishui County	Oral history, local cultural and historical materials	Late Ming and early Qing Dynasties according to the oral history; Qianlong Reign of the Qing Dynasty according to the local cultural and historical materials
38	Bast paper in Zhenning County	Local chronicles, genealogy	Jiajing Reign of the Ming Dynasty according to the local chronicles; early Ming Dynasty according to the genealogy
39	Handmade paper in Ziyun County	Oral history, local chronicles	Late Qing Dynasty according to the oral history; Guangxu Reign of the Qing Dynasty according to the local chronicles
40	Bast paper in Nayong County	Oral history	Qing Dynasty
41	Bast paper in Jinsha County	Oral history	The Republican Era of China
42	Bamboo paper in Jinsha County	Oral history	Before the late Qing Dynasty

Literature review and field investigation revealed that events like "North-to-South Migration for the purpose of conquering" and "North-to-South Migration for the purpose of occupying" in the early Ming Dynasty shed huge effects upon the handmade paper industry in the Guizhou Province. Most papermakers in the area (e.g., Guanling County, Wudang District, Zhenning County, etc.) referred to their ancestors as immigrants during that period. One example is about the source of handmade paper in Wudang District of Guiyang City. Based on a local folklore, during the Hongwu Reign of the Ming Dynasty, only one Peng family was able to produce paper, as sacrificial offerings to the soldiers, who fought and died for the revered Lord Wang of the Yue State. The family genealogy verifies the legend. During our investigation, the family still worshipped the revered Lord Wang of the Yue State in the family niche. It is a hypothesis that the revered Lord Wang of the Yue State might also be Lord Wanghua of the Yue State, during the Sui and Tang Dynasties, when the Yue State was located in regions south of the Yangtze River. Therefore, it is quite possible that the Wudang papermaking families, together with their papermaking techniques, might have their origins from the regions south of the Yangtze River. The other proof comes from a 70-year-old man called Wang Changlun, a native of Longjiao Village, located in the Wudang District. He claimed that his ancestors came to Xiangzhigou with a troop, and then settled down to make paper following the traditional way.

Although 22 Fu (ancient name for city and prefecture) in the Ming Dynasty were

involved in the papermaking practice, the scale was quite small as a farm sideline, and the output was not high. Handmade papermaking techniques in the Ming Dynasty are eluding nowadays, yet we can still see the recognition of quality from *Guizhou Chorography* published during the Republican Era of China. For instance, white paper in Yinjiang County is depicted as "smooth and white as jade, perfect for calligraphy and printing". Due to its high quality and low price, the handmade paper of Guizhou Province was sold to areas like Yunnan Province and Sichuan Province. Though several places in Yunnan Province produced paper during the Ming Dynasty, paper from Guizhou Province was better in quality and higher in quantity. Thus, the capital city of Yunnan Province usually used paper from Guizhou Province, which was overall popular in the southwestern area of China.

1.2.2 Stable Development of Handmade Paper in Guizhou Province During the Qing Dynasty

Development of handmade paper in Guizhou Province shows a stable status after the great political changes in the transition from the Ming Dynasty to the Qing Dynasty. In the early Qing Dynasty, ten prefectures were involved in the papermaking practice; However, another source, *The History of Industrial Development in Guizhou Province*, claimed that 15 cities were involved in papermaking. They mainly located in the northern area of the province, e.g. Zunyi Fu, middle area like Guiyang Fu, Anshun Fu, or southeast area like Xingyi Fu, producing bast paper, bamboo paper and straw paper.

According to *The History of Industrial Development in Guizhou Province*, about 300 family-based papermaking mills, and more than 2 000 people were involved in papermaking in the early Qing Dynasty, while to the end of the Qing Dynasty, over 400 family-based papermaking mills with over 2 500 people were practicing papermaking. In the period, there are also some places that gathered papermaking mills. For instance, Wenggui Town of Guangshun Fu possessed about 82 families and 500 people in papermaking in Guangxu Reign of the Qing Dynasty, with an annual output value amounting to 60 thousand liang (an old Chinese unit of weight) silver. They produced high-quality paper called Gaigong paper, which was used exclusively for local imperial examination or government notice. A broken monument was still erected in Wenggui Village of Changshun County, with the words of "Permanent Stipulation". Based on the incomplete inscription, local government and papermakers used to have controversies over the papermaking tax when profits from the local papermaking industry increased. The local papermakers united to fight against the increasing tax imposed on them, and the government finally compromised with the papermakers, agreeing on a fixed tax rate, which was inscribed on the monument as a solid proof. This monument, as a historical proof, truly recorded the evolutionary development of papermaking history in Wenggui Village.

Based on our literature review, we made a list of all the relevant materials concerning handmade paper in the Qing Dynasty in various places of Guizhou Province.

Zunyi Fu: It had flourishing papermaking practices. Both the prefecture and various affiliated counties were involved. ... Paper made from paper mulberry bark was called bast paper, and bamboo paper for that made from bamboo, and straw paper for that made from straws and weeds, which was rough and usually employed for sacrificial purposes. Bast paper made in Shangxichang of Zunyi Fu, was even better, and the good-quality paper also came from Huangnijiang of Suiyang County, which was also better than that of Shangxichang in terms of whiteness and tenacity. The best paper was usually sold to central areas of Sichuan Province, which was called either Xiaodigou paper or Dacao paper according to the papermaking locations. Bamboo paper was exclusively produced in Suiyang County; A special kind of bamboo, *Phyllostachys heteroclada* Oliver, was used as raw material. It was very exquisite and white, even better than bast paper. Straw paper was of different quality levels, with that produced in Maoya Town of Suiyang County called Maoya paper, which was the best. According to local chronicles of Zheng'an Prefecture, bamboo paper made of *Phyllostachys parvifolia*, or *Phyllostachys heteroclada* Oliver, in Fuyanping could compete with red paper and raw-edged paper produced in Sichuan Province. *The Annals of the Southern Guizhou Province* recorded that Suiyang County produced indigo, bast paper, raw lacquer, tea, Chinese wood oil, white wax, and gallnut. In the fourth volume on *Foods and Goods* in *Suiyang County Annals*, papermaking was also included. In Tongzi County, people cut down paper mulberry trees every three years and used its bark for papermaking, while *Phyllostachys heteroclada* Oliver was used to make white paper. In Zhanmuqiao area of Zhenzhou, people made bast paper which was thick and somber black. When plant ash was added, it became white and smooth, which was even better than that made in Zunyi Fu. Yelouli made white paper called Huo paper from *Neosinocalamus affinis* and *Phyllostachys heteroclada* Oliver. Some papermakers used straw and bamboo to

Monument of "Permanent Stipulation" erected during Guangxu Reign of the Qing Dynasty, in Wenggui Village of Changshun County

produce rough paper called straw paper for sacrificial purposes, which was quite popular then. Paper made in Tongzi County was usually packed and sold to Sichuan Province.

Guiyang Fu: According to *The Annals of the Southern Guizhou Province*, both Dingfan Prefecture and Guangshun Prefecture made white paper. Papermaking in the early years of Guangxu Reign of the Qing Dynasty flourished in Wenggui Town of Guangshun Prefecture, with about 82 family-based papermaking mills, making a high quantity of white paper. Volume 47 of *The Annals of Guiyang Fu* stated that the place was paper-mulberry-tree-rich, and the bark of which could be used for papermaking. Paper made from paper mulberry trees in the Wenggui Town of Dingfan Prefecture. It was white and durable like cotton cloth. Therefore, it was named Mian paper (cotton paper), which could be used to produce mosquito-proof nets. The high-quality paper made in the area was also good enough to be used for calligraphy. Multi-layered paper called Jia paper was used to make umbrellas or for wrapping. Conversely, paper made in Baisuo Town or Gouchangying Village was not that high in quality.

Anshun Fu: Langdai Ting (grassroots administrative office) made straw paper, while Bandang Town of Guihua Ting made Mian paper. In Anping County, various places in Xibao made straw paper, with no less than a hundred papermaking families. Based on *The Annals of Pingba County* compiled during the Republican Era of China, the fifth volume on local industries, Maiweng area of the county made straw paper, while Qiushaohe area of the county made single-layered white paper. In Zhenning County, about ten papermaking mills, located three miles away from Jianglong could produce white paper from paper mulberry bark. According to *Anshun Volume* of *The Annals of Anshun Fu* (*Revised Version*), local places like Yunpan Town of Langdai County, Bandang Town of Ziyun County, Shaying Town and Longchang Town of Zhenfeng County all made white paper. Among all, paper made in Shaying Town and Bandang Town was the best. Each bundle of paper contained 25 dao (each dao contains 80 pieces) paper. Straw paper produced in Yanjiao Town of Langdai County, Liuzhi Town, Danong Town and Yingpan Town was transported to the county centre in unit of 60 jin (Chinese unit of weight), and used for various purposes.

Xingyi Fu: It is recorded that paper produced in the areas directly reported to the local government, and that produced in Liaojiqing of Annan County was even better. For instance, paper made in Shiqian County was thick and smooth, which could be used for calligraphy. While paper, white and durable, made in the previously mentioned areas was much better. In Xingyi Fu, paper mulberry tree bark was employed in papermaking, because the area is abundant with paper mulberry trees. Bamboo was also used for papermaking in the area, and most bamboo paper was sold to various places in Yunnan Province. One local street was named White Paper Street ever since then. Altogether, white paper made from paper mulberry bark and straw paper made from bamboo amounted to 3 800 bundles.

Other cities and counties in Guizhou Province, such as Duyun, Yinjiang, Xifeng, Qianxi, etc., also enjoyed a prosperous papermaking practice. For instance, white paper produced in Duyun was abundant in production and of high quality. In Guangxu Reign of the late Qing Dynasty, the area had more than 50 family-based papermaking mills, and in one street located at the northern area of the city, called Guanxiang Street, ten families were involved in papermaking practice. The paper they made was sold to various counties, such as Dushan, Pingzhou, Bazhai, Libo, Rongjiang. They could hardly keep up with the market demand because merchants from outside the province also came for their paper. White paper produced in Yinjiang was white and smooth like white jade, which was similar to white Mian paper produced during the Ming Dynasty, and was perfect for writing and printing. Together with the paper made in Dingfan Prefecture (now Huishui County), Yinjiang paper was used exclusively for imperial examination of the area. While Dingfan paper refers to paper made in Dingfan Prefecture and neighboring places like Wenggui, Yingpan, Mingzhong. Bast paper produced in Danzhai, and bamboo paper produced in Xifeng and Qianxi was also well known at that time.

1.2.3 Handmade Paper Boom in Guizhou Province During the Republican Era of China

The handmade paper boom during the Republican Era of China should be attributed to the expansion of market. In addition, handmade papermaking techniques were successfully spread to other areas during the period.

The evolution of the printing industry made increasing demands on paper, which led to the pinnacle of handmade paper in the period. According to the local statistical data, in the early period of the Republican Era of China, 28 counties were involved in papermaking, and they produced over 10 varieties of handmade paper. It is then that the handmade paper practice experienced some changes, such as semi-industrialization, that is, part of the papermaking procedures employed machines. Also during the period, some farmers who used to make paper as an income-earning sideline, now became professional papermakers. Some even hired employees and eventually evolved into the earliest small-sized paper factories. A Manufacture Exhibition Brochure in 1937 described that: "Various counties in the province, such as Langdai, Duyun, Yinjiang, were abundant with paper mulberry trees, the bark of which could be used to produce paper of thin and durable quality. Most local citizens were involved in the papermaking practice, and small-scaled factories were set up to produce paper of various types. The paper produced was sold to places all over the province. Textbooks sold in shops and used by schools were mostly printed on paper made from paper mulberry tree bark, which was popular and the output value could amount to 100 thousand yuan per year."

In 1947, the economic bureau of the government compiled *Economic Development of Guizhou Province in a Decade*, and it recorded the handmade paper output in various counties of the province. The details can been seen in Table 1.2.

During the Anti-Japanese War, with the coastal cities being conquered one after another, the local governments, schools and factories moved to southwest areas, and floating population in Guizhou Province increased dramatically. Consequently, demands for printing industry and paper greatly increased, while the political status wouldn't allow transportation of paper from other areas. This led to the prosperity of handmade paper in Guizhou Province. Another 17 nongovernmental handmade paper factories were also established during the period. By the year of 1947, over 50 handmade paper factories existed in Guizhou Province, and annual production amounted to 9 088 ton. However, this prosperity lasted less than 9 years, which ended after the war, because the outmigration of governments, factories and schools, causing a decrease in demand for handmade paper.

1.2.4 Handmade Paper in Guizhou Province After the Founding of the People's Republic of China

After 1949, handmade paper industry in Guizhou Province was featured by collectivization. Cooperative form promotes the local papermaking development, and by 1958, the annual output amounted to 8 012 ton. However, after 1960, handmade paper was hit by mechanized paper production, and was gradually replaced by machine-made paper, but white bast paper, and certain bamboo and straw paper varieties, like deckle-edged bamboo paper made in Laochang Town of Panxian County continued to be handmade.

Table 1.2 Paper output of various counties in Guizhou Province in the mid-late period of the Republican Era of China

County	Annual output (dao)	Paper type	County	Annual output (dao)	Paper type
Jinping	200 000	Bast paper	Zunyi	200 000	Bast paper
				3 790 000	Straw paper
Yinjiang	4 356 000	Bast paper	Hezhang	100 000	Bast paper
	100 000	Straw paper		90 000	Straw paper
Yuping	2 000 000	Straw paper	Tongzi	100 000	Bast paper
	342 000	Bast paper		90 000	Straw paper
Dushan	100 000	Straw paper	Renhuai	20 000	Bast paper
	257 000	Bast paper		100 000	Straw paper
Duyun	100 000	Straw paper	Suiyang	30 000	Bast paper
	3 486 000	Bast paper		100 000	Straw paper
Panxian	2 400 000	Straw paper	Pingyue	100 000	Bast paper
	250 000	Bast paper		45 000	Straw paper
Langdai	200 000	Straw paper	Cengong	200 000	Bast paper
	300 000	Bast paper		140 000	Straw paper
Zhenning	200 000	Straw paper	Zhenyuan	100 000	Bast paper
	1 000 000	Bast paper		120 000	Straw paper
Zhenfeng	150 000	Straw paper	Longli	120 000	Bast paper
	36 000	Bast paper		45 000	Straw paper
Bijie	105 000	Straw paper	Xifeng	130 000	Bast paper
	79 000	Bast paper		500 000	Straw paper
Zhijin	110 000	Straw paper	Weng'an	350 000	Bast paper
	15 000	Bast paper		10 000 000	Straw paper

Note: each dao contains 80 pieces. Output is not large because paper was produced only during the agricultural slack season.

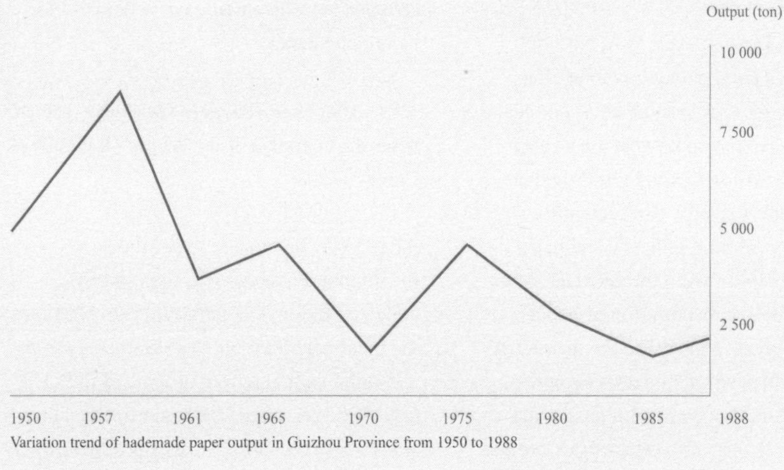

Variation trend of hademade paper output in Guizhou Province from 1950 to 1988

Handmade paper in Guizhou Province was featured by diverse varieties. Its function and application also extend with its development. In 1950s, in response to market demands, Duyun City produced wax paper. In 1978, the Shiqiao Papermaking Factory of Danzhai County began the trial production of traditional Chinese painting paper, after investigating Xuan paper in Jingxian County of Anhui Province. Our investigations reveal that handmade paper made in areas of Danzhai, Zhenfeng, Wudang, Cengong, Guanling, Yinjiang, Congjiang and Rongjiang was quite influential and well accepted.

1.3 Expansion and Dissemination of Handmade Papermaking Techniques in Guizhou Province

Handmade papermaking Techniques in Guizhou Province followed the following sequence, except for the huge influence exerted by the immigrants from Central China and South of the Yangtze River:

It is recorded that handmade papermaking in Guizhou Province started in the Ming Dynasty. In the early period of the Ming Dynasty, various places like Duyun, Longli, Changzhai, Pingyue started papermaking practice. In the Qing Dynasty, Dingfan and Pingzhou also witnessed the immigration of papermakers. According to *Industrial Development in Guizhou Province*, in Jiajing Reign of the Ming Dynasty, various areas in Guizhou Province produced handmade paper, e.g., Chengfan, Duyun, Longli, Xintian, Pingyue, Qingping, Xinglong, Weiqing, Puding, Anshun, Pingba, Anzhuang, Zhenning, Yongning, Annan, Pu'an, Bijie, Wusa, Chishui and Huangping. Toward the end of the Ming Dynasty, handmade papermaking practice also extended to Yinjiang.

Literature from the Qing Dynasty and the period of the Republican Era of China reveals the handmade papermaking technique development through the details.

Wenggui Village and Yingpan Village of Changshun County started papermaking in the Ming Dynasty. According to oral history and the local chronicles, residents from both of them consented to the fact that their ancestors came from the northern China. However, they wouldn't reach consensus on who started papermaking first and was more influential. One thing for sure is that Wenggui Village of Changshun County finally became one of the most important papermaking bases of Guizhou Province.

In the late Qing Dynasty, the Wenggui Village became the source of handmade paper techniques in the Guizhou Province. In 1862, Zhang Youyin, a local papermaker used paper mulberry bark to make white paper, but with poor quality his paper didn't sell well at all. He had to travel through the towns and villages to find a niche for his paper. In 1864, he finally found Duyun, where paper was in high demand but with humble output, and people there were less demanding concerning the paper quality. When Zhang Youyin found the paper mulberry trees in abundance and limpid stream located at the Guanxiang Street of Duyun, he decided to open a papermaking factory there. He employed papermaker Jian Chenggui and papermaking screen producer Tao Yi, and started his papermaking career by a river in Tudimiao Lane of Guanxiang Street in 1865. The paper he made was called White paper in Duyun City. In 1890, the three founders died and their descendants inherited the papermaking practice and further developed the business by raising money in the neighborhood. Local people also opened papermaking factories, and by 1930, there were more than 20 papermaking families in Duyun with over 50 papermaking troughs and over 300 papermakers. Paper produced there was sold all over the country, which made Duyun the source of the paper mulberry bark papermaking technique.

Based on the *The Annals of Bazhai County* compiled by Guo Fuxiang in 1931, literature review and our field investigations, it is concluded by the research group that in the late 1920s, papermakers migrated from Duyun to Danzhai, located at the southeast area. It marked the emergence of handmade paper in Danzhai County. The route for the transmission of handmade papermaking techniques development spreading in Guizhou Province, went from Changshun to Duyun, then to Danzhai.

In 1949, a tragedy happened in Wenggui, the source of handmade papermaking technique in Guizhou Province, and caused the sad ending of the area as the centre of handmade paper. In that year, armed conflict between the bandits from Dawei area of Huishui County, and the town government of Baiyun Town located at Wenggui enraged the bandits who lost a lot of members in the fighting. They set fire to the village buildings, and robbed the local people, except few who were related to the bandits. For a better and safer life, local papermakers moved to places like Pingtang, Luodian, Duyun, Guiyang and Anshun, which was a blessing to these

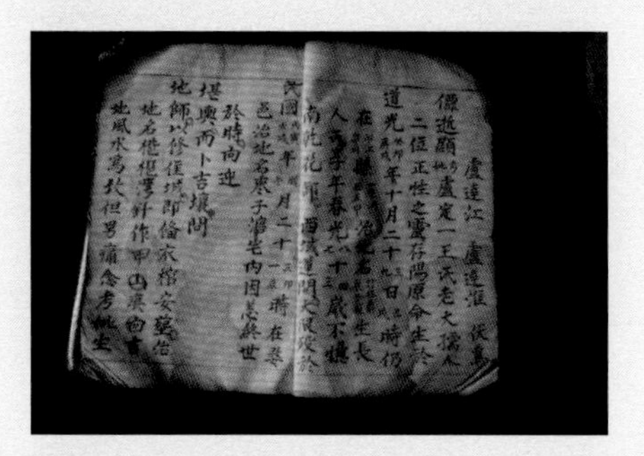
History of the Lus, handwritten version kept in Zaozhitang Village of Wuchuan County

places. From then on, these places started the papermaking practice, while to Wenggui Village, this accident tragically shattered the place as the papermaking centre.

Yinjiang County is another famous source of handmade paper in Guizhou Province. It started the practice in the late Ming Dynasty. In *County Annals of Shiqian* published in 1992, that in 1869, Xu Zhiwen and Yang Zhiquan from Yinjiang County came to Tiangou Village of Baisha Town in Shiqian County and opened a papermaking mill. The paper made was sold to the southeast part of Guizhou Province. In *Chorography of Guizhou Province* compiled by Tian Wen in 1690, paper made in Shiqian County was thick and smooth, which could be used for calligraphy. Though it's hard to tell where exactly the papermaking place is, it is almost certain that paper was produced at least during Kangxi Reign of the Qing Dynasty. Though local paper was made in Shiqian County from that time, papermakers from Yinjiang County also found a niche for themselves there. In 1927, it is recorded in local literature, *History of the Lus*, Zaozhitang (papermaking) Village of Wuchuan County, was once called Zaozitang. During Daoguang Reign of Qing Dynasty, Lu Dingyi moved to the place from Yinjiang County of Tongren City, and made a living by making paper. After that, more papermakers, e.g., Ran's family, Hu's family, Shi's family and Ruan's family moved in from Yinjiang County. So Yinjiang County was an epicentre for papermaking from which the papermaking techniques spread to other places, such as Shiqian County and Wuchuan County.

Comparatively, the history of papermaking in Guizhou Province isn't long, but ever since the Qing Dynasty, the place had produced regionally known and influential paper. For instance, in *Local Chronicles of Xingyi Fu During Xianfeng Reign*, it is said that paper produced in the governed areas and in Liaojiqing of Annan were better. Based on *Chorography of Guizhou Province*, paper produced in Shiqian County is thick and smooth, good enough for calligraphy. While paper made in Zhenfeng was even better in whiteness and durability. *Local Chronicles of Guanling County During the Republican Era of China* recorded that white paper produced in Shaying was in good quality and sold well. Straw paper was produced in Shaying, Yalong and Banjiujing. In the book, under the commercial entry, it is recorded that white paper produced in Shaying was sold outside the county, especially to the Yunnan Province. All these words lead to a conclusion that white paper produced in Shaying was popular then. Guanling County was the strategic pass from Guizhou Province to Yunnan Province, and this advantage in transportation made the white bast paper produced locally conveniently sold to Yunnan Province. Moreover, bast paper produced in the areas of Duyun, Changshun and Liuzhi even served as tribute to the imperial court, or was used in imperial examinations.

The development of handmade papermaking practice in Guizhou Province is marked by another feature: innovation. In 1950s, Duyun Paper Factory, in response to the market demands, produced Duyun Wax Paper, which obtained countrywide attention. The Shiqiao Village of Danzhai County visited Jingxian County of Anhui Province for Xuan Papermaking techniques and started their trial production of the traditional Chinese painting paper in the year of 1978.

With development of handmade paper techniques and papermaking industry, papermaking sheds increasing influence on the local customs and daily life of the area. Besides the traditional funeral customs or folk customs, it is connected also to some very special cultural events or innovative uses.

For example, Wang Yangming, grand master of "Yangming's Theory of Mind", composed a poem, *Young Servant Making Paper Lantern*, when he was in Longchang in exile, which vividly depicted the rich celebrating holidays with lanterns in Guizhou Province, with the cost of one lantern amounting to ten mid-income families' annual income.

From the title of the poem we can tell that handmade paper was used and was a luxury that not accessible to the common families in the Ming Dynasty.

In many areas of Guizhou Province, handmade paper was recorded as a part of local culture. Based on our field investigation in Changshun County, local papermakers enjoyed this papermaking song in Wenggui:

Paper mulberry trees bloom in Longjingkan, papermaking troughs were built near Longtan.

We count on Cai Lun for blessings, we make paper enjoying praises.

Paper mulberry trees grow in Longjingkan, papermaking mills were built in Wenggui Village.

The Book of *Heavenly Creations* inspired our practice, white paper we made was tribute to the emperor.

Papermaking techniques, tools and raw materials could be traced in local cultural heritage like the papermaking song, passed down in the southeast areas of Guizhou Province, e.g., Zhouxi Village and Kaitang Village of Kaili City, Shiqiao Village of Nangao Town in Danzhai County, Shidongkou Village of Jianhe County.

Handmade paper was important in local sacrificial ceremonies. Some interesting changes have occurred recently (e.g., discrimination between "real money" and "fake money").

Nowadays, machine-made paper is sold for sacrificial ceremonies in many areas in Guizhou Province. But in areas of Yinjiang, Jinsha, Cengong, only "real money" (handmade paper) could be used, which when burned, would be accepted by ancestors and used in heaven as money. While "fake money" (machine-made paper) would be useless high in heaven. Thus, one's ancestors would reject machine-made joss paper.

Traditional handmade papermaking is time-consuming and labor-intensive. Therefore, many papermakers in Guizhou Province worshiped and valued the paper they made. Since the 1980s, industrialization and urbanization caused many handmade papermakers to cease production. However, many retired papermakers still keep the paper they made and some papermaking tools and equipment, including papermakers from Tangkou Village and Liudong Village of Shazipo Town in Yinjiang County, from Jianglong Town of Zhenning County, from Bandang Town of Ziyun County, Langdai Town of Liuzhi Special Area, and Longchang Town of Zhenfeng County. Based on our investigation, these papermakers kept some of the handmade paper and tools they previously used because they hoped that someday in the future handmade paper might flourish again. Some kept self-made paper as a memorial and other elder natives kept their self-made paper for their own future funerals. There is a local belief that only handmade paper can be used as "real money" when in heaven. From a cultural perspective, this sort of love and attachment towards the handmade paper among the local papermakers in the early 21st century that lead to the preservation and development of the handmade papermaking techniques and culture in Guizhou Province.

2 Current Production Status of Handmade Paper in Guizhou Province

2.1 Impact of Geographic Environment on Handmade Paper in Guizhou Province

Guizhou is located in Yunnan-Guizhou plateau. The topography is characterized by its isolated location and is difficult to access. A local ballad vividly depicts the geographic and economic status of Guizhou: "You can never see sun continuously for more than three days; we have rough roads only; both you and me can barely make both ends meet." Geographic features also shed influences on the population, distribution of minority groups, communication pattern, ways of living and manufacturing, religious beliefs and local customs.

Guizhou and neighboring provinces as well as some small areas inside Guizhou, are restricted by tall mountains and deep valleys, and are also isolated from the outside world. As a result, the exchange and communication is rare. As for external forces during Hongwu Reign and Yongle Reign of the Ming Dynasty, large-scaled migration caused a surge of immigrants in a short time and brought in outsiders and new culture. However, as time went on, the outsiders gradually formed their own relatively closed regional groups, communities and culture, such as the famous phenomenon of "the Tunbao people" (distinctive Han people who lived in enclosed groups and kept traditional habits) . Ultimately a very obvious characteristic of Guizhou residents and ethnic distribution "living compactly in a small community but dispersed in a large area" took shape.

The isolated environment took its toll to the economic and cultural interactions with the outside world. Yet, these "disadvantages" contributed to the preservation of the diverse cultural customs and cultural heritage, and these cultural features are fully presented in the papermaking techniques of the area.

The research group began their field investigation of Guizhou Province in December 2008, and finished in March 2016. Based on the investigation, in the early 21st century, there are about 40 handmade papermaking mills in the province still involved in the papermaking practice. Despite the modern social background featured by open communication and convenient transportation, the handmade paper practice in Guizhou Province still sticks to its original style, namely, independent and enclosed. This is typically demonstrated by our own experiences: when we first contacted the administration bureau of the government in charge of protecting intangible cultural heritage, they could only provide less than enough papermaking sites. City governments also failed us and even internet, which is quite informative nowadays, could only provide limited sources of useful messages.

Papermaking is only practiced during the agricultural slack season, meeting the neighboring rural market demands in Guizhou Province. Papermakers use the money to improve their living, and are not highly motivated to expand their production. Usually the papermakers are family-based, but a few such as papermakers in the areas of Shiqiao Village of Nangao Town, Danzhai County in Qiandongnan Miao and Dong Autonomous Prefecture, which are usually areas featured by tourism. In these areas, papermaking factories have been large-scaled, so as to meet the tourists' needs of handmade paper products. In these villages which are strongly influenced by marketization, papermaking factories own employers more than just family members and they produce paper regularly instead of in slack season. They even have a division of groups involving in development, procurement, production and wholesale and retail marketing. Take Danzhai County for example, the local government set up Shiqiao Handmade Papermaking Centre, which after Pacific Asia Travel Association (PATA) paid a visit in January 2006, successfully raised a fund of 200 thousand yuan.

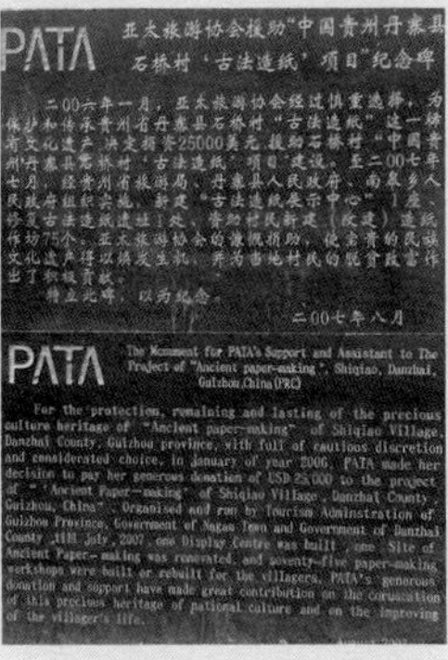

Monument in memory of Pacific Asia Travel Association's funding for the traditional papermaking techniques in Shiqiao Village of Danzhai County

From December 2008 to March 2016, over 30 researchers exhausted their visit of existing papermaking sites in Guizhou Province, including the remote villages, with systematic investigations, interviews, and literature review. Papermaking practices, including techniques, marketing data, heritages and beliefs, of early 21st century, in about 48 villages and towns belonging to 31 counties (districts, cities) were recorded through recordings, videos, photographs and words. Our work will contribute to the complete and detailed description of handmade paper in Guizhou Province, which will be presented to the world through our efforts.

2.2 Diversity: A Typical Feature of Guizhou Handmade Paper

Over the course of a-thousand-years, the various minority groups lived in small-scaled gatherings and swiftly adapted to the mountainous and isolated geographical conditions. These minority groups' handmade paper practices illustrate the following features: taking advantage of local natural resources and making paper with typical ethnic characteristics with diverse papermaking techniques.

2.2.1 Diverse Distribution of Papermaking Sites

Based on our field investigation and literature review, the handmade papermaking industry is challenged nowadays and withering. However, in Guizhou Province, due to its enclosed natural environment and small-scaled gathering, handmade paper is thriving. The practice can be traced almost everywhere in every county and every city of the province. Some of the papermaking sites are: Sanyuan Village of Wuma Town and Sangshuwan Village of Luban Town in Renhuai City; Xiangshuyuan Village of Tangshan Town, Xialinba Village of Pingdichang Town and Guandaotu Village of Dashaba Town in Shiqian County; Xingwang Village of Heshui Town in Liudong Village and Tangkou Village of Shazipo Town in Yinjiang County; Gaohua Village of Cuili Town, Dage Village of Xiutang Town, Xiaohuang Village and Zhanli Village of Gaozeng Town in Congjiang County. Papermaking sites can also be found in

Duyun City, Jinsha, Cengong, Zhenfeng, Guanling and Ziyun Counties. This is quite different from the withering state as a whole in China.

2.2.2 Papermakers of Different Ethnic Groups

Based on our field investigation and literature review, various ethnic groups, including the Han, Miao, Yao, Dong, Gelo, Tujia, Bouyei, Yi, Li etc. were involved in papermaking.

Handmade papermaking techniques were passed from the central areas of China to southwest areas by the Han people, and they consistently practiced papermaking. These techniques were passed on to other ethnic groups in Guizhou Province, and they contributed more ethnic features and culture into this practice, so as to preserve and well inherit the heritage. For instance, bamboo paper made by the Bouyei and the Tujia Ethnic Groups of Cengong County is still popular with people living in Qiandongnan Miao and Dong Autonomous Prefecture and west area of Hunan Province. Bast paper made by the Dong Ethnic Group in Rongjiang County, and by the Yao Ethnic Group in Congjiang County both vividly reflect local customs and cultural traditions.

It is worth to mentioning that according to the genealogy, many papermakers were descendants of Han soldiers and immigrants from Jiangxi Province and Hunan Province. Nowadays, they were classified as minorities. For instance, Yang Zaixiang, a papermaker in Guidong Village of Bagong Town in Sansui County, is now a Dong people, but according to genealogy, his ancestors came from Jiangxi Province. Another example, white bast papermakers in Xingwang Village of Heshui Town in Yinjiang County are now Tujia Ethnic Group, yet based on their family records, especially of the Cais and the Lus, they immigrated to the place from Jiangxi Province in Hongwu Reign of the Ming Dynasty and the late years of Ming Dynasty respectively, and their ancestors were supposed to be Han people. Of course, we can explain the lack of connection: one possibility is that they were influenced by Han people's ancestor worship, so their genealogy or family memories might have been biased in this direction. Another possibility is that they were immigrants from Jiangxi Province and Hunan Province, and their ancestors were Han people. Their status as a minority group was a result of mis-registration. One more possibility is that these papermakers were actually descendants of Han people from the northern China marry local people of minority groups, i.e., descendants of multi-groups. This example also illustrates the intermixing state of various groups in the area.

2.2.3 Diverse Papermaking Raw Materials

The raw materials used in the area, based on our field investigation and literature review, are mostly paper mulberry bark and bamboo. Moreover, in some villages, cotton and sabaigrass are also employed. Wax paper in Duyun City is a unique local product, using a kind of hemp called "Ye Meng Hua Ma".

Paper mulberry tree, *Broussonetia L' Herit.*

Wild paper mulberry trees

ex Vent., serves as a qualified raw material for papermaking. Early in the Wei, Jin and Southern & Northern Dynasties, it had been used for papermaking and became quite popular in the Sui Dynasty. Jia Sixie of the Northern Wei Dynasty, depicted planting of paper mulberry tree and papermaking in his work *Qimin Yaoshu*, volume 5: "It's better for paper mulberry trees to be planted in valleys. In fall, ripe fruits were harvested, washed clean, and dried in sunshine""It should be planted with hemp, so when winter arrived, the paper mulberry tree wouldn't succumb to the frost." Yet, it was quite different for Guizhou Province, where paper mulberry bark used for papermaking was mostly wild, due to its warm and moist climate. This natural advantage saves a

Phyllostachys bambussoides cv.Tanakae

lot of money and energy. Moreover, the prosperity of the paper mulberry trees also depended on using burned ashes as fertilizer in each January (lunar calendar). When were the trees got chopped down should also be considered. December (lunar calendar) would be the best choice, then April (lunar calendar). In other months of the year, the tree might die if lopped down.

Although the local papermakers have never heard of Jia Sixie and his instructions on paper mulberry tree planting in his book, they were basically practicing following similar techniques and tips. For instance, Wang Piquan, papermaker in Yuxiazhai Natural Village of Guosong Administrative Village in Guihua Town of Jinsha County in Bijie City of Guizhou Province, told the researchers in March 2014 in an interview, that in the lunar calendar, March and April are best for lopping paper mulberry trees for papermaking. A forest fire of the previous year usually leads to the prosperity of paper mulberry trees in the following year.

Nowadays, paper made from paper mulberry bark is widely used for ceremonial and elegant purposes. For instance, it is primarily used for painting and calligraphy, or for copying religious scriptures or genealogy. Moreover, it is also used for sacrificial offerings, packaging, folk arts

Neosinocalamus affinis

and crafts, and household items like paper fans, lanterns or umbrellas, etc.

Bamboo paper made from bamboo is another important type of paper produced in Guizhou Province. Guizhou Province enjoys diverse types of bamboo, e.g., *Neosinocalamus affinis, Fargesia semicoriacea* Yi, *PhyllostachysacutaChuetChao*, *Phyllostachys heteroclada* Oliver, *Phyllostachys bambussoides* cv.Tanakae, *Phyllostachys parvifolia, B.blumeana* Schult.f, *Phyllostachys pubescens* Mazel ex H. de Leh., *Bambusa intermedia* Hsueh et Yi, Man bamboo, *Pleioblastus amarus* (Keng) keng f., *Dendrocalamus tsiangii* (McClure) Chia et H. L. Fung, etc. Among them, *Neosinocalamus affinis* is most frequently used in papermaking. *Dendrocalamus tsiangii* (McClure) Chia et H. L. Fung was originally found in Guizhou Province, while now, it can also be found in Guangdong Province and Guangxi Zhuang Autonomous Region. It is roughly 5 to 6 meters high, and 2 to 4 centimeters in diameter. It is also refered to as fishing bamboo, because it is perfect for fishing, as it is thin and hard, pretty and practical.

In comparison to bast paper, bamboo paper made in Guizhou Province is limited in its usage. It is usually used for sacrificial offerings, e.g., joss paper or firecrackers. Bamboo paper is rough and of yellowish grey color. When the researchers visited the area, machine-made paper boomed and greatly challenged the market demands for handmade paper. Only in rural areas, handmade paper as sacrificial offering is still used while in cities and towns, machine-made paper has become trendy.

It is unavoidable that with urbanization, more and more people will join the trend and use machine-made paper. Moreover, burning handmade paper in large amount is believed to be harmful in terms of environment protection and low-carbon living. These facts make it urgent for handmade bamboo paper to enhance its quality and outlook, in order to expand its usage.

In terms of raw materials in Guizhou Province, wax paper in Duyun City is the peculiar one. It uses a certain "Ye Meng Hua Ma" (a kind of hemp) as its raw material.

Wikstroemia delavayi

Based on our research,"Ye Meng Hua Ma" is a type of *Wikstroemia*. Shen Jihui in his paper "*White Bast Paper in Duyun City: Paper with A Long History*", *Wikstroemia* in Zhejiang and Jiangxi Provinces are called "Shan Mian Pi"; in the minority groups, it is called "Ban Lan", or "Di Mian Gen", or "Shan Dou Liao", while in Japan, it is called "Yan Pi". *Wikstroemia* is classified as *Thymelaeaceae*, erect shrubs, with a height of 30 to 90 centimeters, grows in moist areas, such as beside a stream, a rock seam or valleys. The branches are thin and filamentous, and the bark is smooth, with symmetric leaves and egg-shaped fruits. At least 3~7 centimeters of stump should be kept while chopping down the tree for raw materials. *Wikstroemia* for wax paper in Duyun City usually comes from Zunyi City of Guizhou Province. It is similar to other raw papermaking materials, such as *Wikstroemia delavayi Lecomte* for Dongba paper by the Naxi Ethnic Group of Yunnan Province, *Gelsemium elegans* for Tibetan paper in Dege County of Sichuan Province and Nimu County of Tibet Autonomous Region.

After 1990s, some paper factories in Guizhou Province added in machine-made paper pulp, to offset costs and increase profits. Paper made in this way is called Huilong paper (circulated paper). It is true that paper made in this way could be more competitive in terms of price, yet this is a deviation of the handmade paper tradition. It is hard to say that it will do any good for the preservation and development of handmade paper.

Most handmade paper made in Guizhou Province is of a singular raw material, while some used mixed materials. For example, paper made in Shiqiao Village of Danzhai County used paper mulberry bark and *Juncus effuses*, with a fixed proportion. Only in this way, paper can be qualified for calligraphy and painting. This is similar to Xuan paper, which is demanding for the proportion of raw materials of *Wingceltis* bark and straw grown in the sands. In Zhenfeng County, papermakers use bark and bamboo to make paper for painting and calligraphy, while papermakers of Renhuai City use bark and cotton as raw materials.

Alkaline agents and mucilage are usually used as catalyst in papermaking. Alkaline agents used in Guizhou Province usually include wood ash, lime and caustic soda. While mucilage includes cactus, Chinese gooseberry, cedar pine root, *Abelmoschus manihot (L.) Medicus* leaves and polyacrylamide. Use of chemical materials like caustic soda and polyacrylamide shows that traditional handmade papermaking techniques have been challenged by the development of modern chemistry.

Papermaking places, papermaker's ethnic groups and raw materials in Guizhou Province are listed in Table 1.3.

Table 1.3 Papermaking places, papermaker's ethnic groups and raw materials in Guizhou Province

Paper type	Raw material	Papermaker's ethnic group	Place of production
Bast paper in Nayong County	Paper mulberry bark	Han people	Shabao Town of Nayong County in Bijie City, northwest area of Guizhou Province
Bast paper in Jinsha County	Paper mulberry bark	Han people	Guihua Town, Chengguan Town and Shatu Town of Jinsha County in Bijie City, northwest area of Guizhou Province
Bamboo paper in Jinsha County	*Neosinocalamus affinis*	Han people	Chayuan Town of Jinsha County in Bijie City, northwest area of Guizhou Province
Bast paper in Renhuai City	Paper mulberry bark	Han people	Wuma Town of Renhuai City in Zunyi City, north area of Guizhou Province

(Continued)

Paper type	Raw material	Papermaker's ethnic group	Place of production
Bamboo paper in Renhuai City	*Neosinocalamus affinis*	Han people	Luban Town of Renhuai City in Zunyi City, north area of Guizhou Province
Bast paper by the Gelo Ethnic Group in Wuchuan County	Paper mulberry bark	Gelo Ethnic Group	Fengle Town of Wuchuan Gelo and Miao Autonomous County in Zunyi City, north area of Guizhou Province
Bamboo paper by the Gelo Ethnic Group in Wuchuan County	*Neosinocalamus affinis*	Gelo Ethnic Group	Huangdu Town of Wuchuan Gelo and Miao Autonomous County in Zunyi City, north area of Guizhou Province
Bamboo paper in Zheng'an County	*Neosinocalamus affinis*, *B.blumeana* Schult.f	Han people	Fengyi Town and Hexi Town of Zheng'an County in Zunyi City, north area of Guizhou Province
Bast paper in Zheng'an County	*Salix guebriantiana* bark	Han people	Fengyi Town of Zheng'an County in Zunyi City, north area of Guizhou Province
Bamboo paper in Yuqing County	*Neosinocalamus affinis*	Han people	Dawujiang Town of Yuqing County in Zunyi City, north area of Guizhou Province
Bamboo paper in Wudang District	*Pleioblastus amarus* (Keng) Keng f., *Neosinocalamus affinis*	Han people, Bouyei Ethnic Group	Xinbao Bouyei Town of Wudang District in Guiyang City, central area of Guizhou Province
Bast paper by the Gelo Ethnic Group in Shiqian County	Paper mulberry bark	Gelo Ethnic Group	Tangshan Town of Shiqian County in Tongren City, northeast area of Guizhou Province
Bamboo paper by the Gelo Ethnic Group in Shiqian County	*Bambusa intermedia* Hsueh et Yi, *Phyllostachys bambussoides* cv. Tanakae, *Phyllostachys parvifolia*, *Pleioblastus amarus* (Keng) Keng f.	Gelo, Dong Ethnic Groups	Pingdichang Gelo and Dong Town and Dashaba Town of Shiqian County in Tongren City, northeast area of Guizhou Province
Bast paper in Heshui Town of Yinjiang County	Paper mulberry bark	Han people, Tujia Ethnic Group	Heshui Town of Yinjiang Tujia and Miao Autonomous County in Tongren City, northeast area of Guizhou Province
Bast paper by the Tujia Ethnic Group in Shazipo Town of Yinjiang County	Paper mulberry bark	Tujia Ethnic Group	Shazipo Town of Yinjiang Tujia and Miao Autonomous County in Tongren City, northeast area of Guizhou Province
Bamboo paper by the Tujia Ethnic Group in Shazipo Town of Yinjiang County	*Neosinocalamus affinis*	Tujia Ethnic Group	Shazipo Town of Yinjiang Tujia and Miao Autonomous County in Tongren City, northeast area of Guizhou Province
Bamboo paper by the Tujia Ethnic Group in Jiangkou County	*Neosinocalamus affinis, Shan bamboo*	Tujia Ethnic Group	Nuxi Tujia and Miao Town and Taiping Tujia and Miao Town of Jiangkou County in Tongren City, northeast area of Guizhou Province
Bast paper by the Miao Ethnic Group in Danzhai County	Paper mulberry bark	Miao Ethnic Group	Nangao Town of Danzhai County in Qiandongnan Miao and Dong Autonomous Prefecture, southeast area of Guizhou Province

(Continued)

Paper type	Raw material	Papermaker's ethnic group	Place of production
Bamboo paper by the Dong Ethnic Group in Sansui County	*Fargesia semicoriacea* Yi	Dong Ethnic Group	Bagong Town of Sansui County in Qiandongnan Miao and Dong Autonomous Prefecture, southeast area of Guizhou Province
Bamboo paper by the Miao Ethnic Group in Kaili City	*Neosinocalamus affinis*	Miao Ethnic Group	Wanshui Town of Kaili City in Qiandongnan Miao and Dong Autonomous Prefecture, southeast area of Guizhou Province
Bast paper by the Dong Ethnic Group in Rongjiang County	Paper mulberry bark	Dong, Miao Ethnic Groups	Leli Town and Jihua Town of Rongjiang County in Qiandongnan Miao and Dong Autonomous Prefecture, southeast area of Guizhou Province
Bast paper by the Yao Ethnic Group in Gaohua Village of Cuili Town in Congjiang County	Paper mulberry bark	Yao, Dong Ethnic Groups	Cuili Town of Congjiang County in Qiandongnan Miao and Dong Autonomous Prefecture, southeast area of Guizhou Province
Bast paper by the Dong Ethnic Group in Xiaohuang Village and Zhanli Village of Gaozeng Town in Congjiang County	Paper mulberry bark	Yao, Dong Ethnic Groups	Gaozeng Town of Congjiang County in Qiandongnan Miao and Dong Autonomous Prefecture, southeast area of Guizhou Province
Straw paper by the Yao Ethnic Group in Gaohua Village of Cuili Town in Congjiang County	Glutinous rice straw	Yao Ethnic Group	Cuili Town of Congjiang County in Qiandongnan Miao and Dong Autonomous Prefecture, southeast area of Guizhou Province
Bamboo paper by the Yao Ethnic Group in Dage Village of Xiutang Town in Congjiang County	*Phyllostachys pubescens* Mazel ex H. de Leh.	Yao Ethnic Group	Xiutang Town of Congjiang County in Qiandongnan Miao and Dong Autonomous Prefecture, southeast area of Guizhou Province
Bast paper by the Dong Ethnic Group in Liping County	Paper mulberry bark	Dong Ethnic Group	Maogong Town of Liping County in Qiandongnan Miao and Dong Autonomous Prefecture, southeast area of Guizhou Province
Bamboo paper by the Miao Ethnic Group in Huangping County	Doujiu bamboo (local dialect)	Miao Ethnic Group	Wengping Town of Huangping County in Qiandongnan Miao and Dong Autonomous Prefecture, southeast area of Guizhou Province
Bamboo paper by the Dong Ethnic Group in Cengong County	*Bambusa intermedia* Hsueh et Yi	Dong, Tujia Ethnic Groups	Shuiwei Town and Yangqiao Tujia Town of Cengong County in Qiandongnan Miao and Dong Autonomous Prefecture, southeast area of Guizhou Province
Bamboo paper in Longli County	*Dendrocalamus tsiangii* (McClure) Chia et H. L. Fung	Han people	Longshan Town of Longli County in Qiannan Bouyei and Miao Autonomous Prefecture, south area of Guizhou Province
Bast paper in Duyun City	Paper mulberry bark	Han people	Chengguan Town of Duyun City in Qiannan Bouyei and Miao Autonomous Prefecture, south area of Guizhou Province

(Continued)

Paper type	Raw material	Papermaker's ethnic group	Place of production
Wax paper in Duyun City	"Ye Meng Hua Ma"	Han people	Chengguan Town of Duyun City in Qiannan Bouyei and Miao Autonomous Prefecture, south area of Guizhou Province
Bast paper in Changshun County	Paper mulberry bark	Han people	Baiyunshan Town of Changshun County in Qiannan Bouyei and Miao Autonomous Prefecture, south area of Guizhou Province
Bamboo paper in Huishui County	*Phyllostachys heteroclada* Oliver, *Cephalostachyum pergracile* Munro, *Pleioblastus amarus* (Keng) Keng f.	Han people	Lushan Town of Huishui County in Qiannan Bouyei and Miao Autonomous Prefecture, south area of Guizhou Province
Bamboo paper by the Bouyei Ethnic Group in Libo County	*Dendrocalamus tsiangii* (McClure) Chia et H. L. Fung	Bouyei Ethnic Group	Yongkang Shui Town of Libo County in Qiannan Bouyei and Miao Autonomous Prefecture, south area of Guizhou Province
Bast paper in Pu'an County	Paper mulberry bark	Han people	Baisha Town of Pu'an County in Qianxinan Bouyei and Miao Autonomous Prefecture, southwest area of Guizhou Province
Bamboo paper by the Bouyei Ethnic Group in Anlong County	Man bamboo, *B.blumeana* Schult.f	Bouyei Ethnic Group	Wanfenghu Town of Anlong County in Qianxinan Bouyei and Miao Autonomous Prefecture, southwest area of Guizhou Province
Bast paper in Zhenfeng County	Paper mulberry bark	Han people	Xiaotun Town of Zhenfeng County in Qianxinan Bouyei and Miao Autonomous Prefecture, southwest area of Guizhou Province
Bamboo paper by the Bouyei Ethnic Group in Zhenfeng County	*Bambusa intermedia* Hsueh et Yi, *Neosinocalamus affinis, Phyllostachys parvifolia*	Bouyei Ethnic Group	Longchang Town of Zhenfeng County in Qianxinan Bouyei and Miao Autonomous Prefecture, southwest area of Guizhou Province
Bast paper in Guanling County	Paper mulberry bark	Han people	Shaying Town of Guanling Bouyei and Miao Autonomous County in Anshun City, west area of Guizhou Province
Bamboo paper by the Bouyei and Miao Ethnic Groups in Guanling County	*Bumbusa intermedia* Hsueh et Yi, *Pleioblastus amarus* (Keng) Keng f., *Phyllostachys bambussoides* cv. Tanakae	Bouyei, Miao Ethnic Groups, Han people	Yongning Town of Guanling Bouyei and Miao Autonomous County in Anshun City, west area of Guizhou Province
Bast paper in Zhenning County	Paper mulberry bark	Han people	Jianglong Town of Zhenning Bouyei and Miao Autonomous County in Anshun City, west area of Guizhou Province
Bast paper in Ziyun County	Paper mulberry bark	Han people, Miao, Bouyei Ethnic Groups, etc.	Bandang Town of Ziyun Bouyei and Miao Autonomous County in Anshun City, west area of Guizhou Province

(Continued)

Paper type	Raw material	Papermaker's ethnic group	Place of production
Bamboo paper in Ziyun County	*Phyllostachys bambussoides* cv.Tanakae, *Cephalostachyum pergracile* Munro, *Dendrocalamus tsiangii* (McClure) Chia et H. L. Fung	Han people, Miao, Bouyei Ethnic Groups, etc.	Bandang Town of Ziyun Bouyei and Miao Autonomous County in Anshun City, west area of Guizhou Province
Bast paper in Panxian County	Paper mulberry bark	Han people	Yangchang Bouyei, Bai and Miao Town of Panxian County in Liupanshui City, west area of Guizhou Province
Bamboo paper in Panxian County	*Phyllostachys nuda*, *Phyllostachys parvifolia*	Han people	Laochang Town of Panxian County in Liupanshui City, west area of Guizhou Province
Bast paper by the Yi, Miao and Gelo Ethnic Groups in Liuzhi Special Area	Paper mulberry bark	Yi, Miao and Gelo Ethnic Groups	Langdai Town and Zhongzhai Town of Liuzhi Special Area in Liupanshui City, west area of Guizhou Province

2.2.4 Diverse Papermaking Techniques

Diverse papermaking techniques are involved in Guizhou Province, e.g., papermaking with raw or boiled materials, different ways of stirring paper pulp, different ways of getting paper out of water, different ways of drying paper. Bast paper produced in Guizhou Province usually use boiled materials, while bamboo paper uses boiled or raw materials. In terms of beating, local papermakers use bulls, or their bare foot, a foot pestle, waterpower or machine. They get paper out of the papermaking trough either with a fixed screen or movable one. They make paper alone or with help. But generally, they use movable screen to make paper, while a fixed screen is used in Congjiang, Rongjiang and Liping of Qiandongnan Miao and Dong Autonomous Prefecture, and papermakers from Zhenfeng County and Danzhai County also use fixed screen when making flower paper. In terms of drying, they dry paper by fire, in the sun or shade.

All the previously mentioned procedures and papermaking techniques are still in use and inherited in the peculiar Guizhou style: independent, special, yet inherently connected, as a component of Chinese intangible cultural heritage.

One man making paper in Longjiao Village of Xinbao Bouyei Town in Wudang District

Trimming papermaking materials: one procedure in making white bast paper by the Gelo Ethnic Group in Xiangshuyuan Village of Tangshan Town in Shiqian County

2.2.5 Different Paper Counting Ways

Papermakers of Guizhou Province use various ways to count the paper they made. Most frequently used methods are in units of 10, 20 or 50. While in Shaying Town of Guanling County, papermakers counted in unit of 11, which is quite inconvenient yet for the reason unknown to the researchers. In Ziyun County, the papermakers use the straw core to count the paper made, and the straw core they use is processed through special procedures.

2.2.6 Diverse Types and Usages

Paper made in Guizhou Province can be used for task such as production and daily life, calligraphy, painting, sacrificial ceremonies, and religious celebration. They use the paper to copy family genealogy, and use it as offerings to be burnt and therefore it will become "real money" for their ancestors in heaven. They make paper for sacrificial purposes, and use paper to copy scriptures, to paste window, and to make kites, fans and umbrellas. Handmade paper has actually penetrated every aspect of human life and production in Guizhou Province.

2.3 Diverse Customs, Taboos and Cultural Phenomena of Handmade Paper in Guizhou Province

Here are some examples to show the diverse cultural customs regarding handmade paper in Guizhou Province.

2.3.1 Worshipping the Originator of Papermaking, Cai Lun

In many papermaking sites in Guizhou Province, legends about the originator of papermaking Cai Lun. Specifically, there are stories about how he dreamed of using papermaking mucilage to split paper layers. One papermaker in Zhenning County vividly described the story to a researcher: when paper was made, Cai Lun was unable to split the paper layers no matter how hard he tried. Out of anger, he threw the bundle of paper away. Then a pig came and pushed the bundle with the nose and paper was able to be separated. When Cai Lun saw this, he exclaimed that the pig was even smarter than he was. Therefore, pig was entitled "Mr. Pig". Though in Zhenning County papermaking was declining after 1960s, the local papermakers still worshipped Cai Lun and Mr. Pig. This worshipping tradition has become proof of the papermaking history of the area, according to the local papermakers.

As the originator of papermaking recognized by all the papermakers in Guizhou Province, Cai Lun has been worshipped in many places in Guizhou Province. Usually, the papermakers will use a separate room for his memorial tablet or to house his portrait. In some areas, such as Duyun City and Longli County, they even built up Cai Lun Temple. While in Liuzhi Special Area, local people attached his portrait in Xuanyuan (Yellow Emperor) Palace. Of all the similarities in worshipping Cai Lun, there are also quite a lot of differences in terms of time to worship and rituals, which are worth further exploration and research.

2.3.2 Papermaking Songs

Han people in Shiqiao Village of Nangao Town in Danzhai County practiced

Cai Lun and Mr. Pig are worshipped together with gods and ancestors

papermaking in the first place, yet when the white bast papermaking techniques spread to the place, Miao people learned and mastered the techniques, and the techniques gradually became a part of their culture. Among the Miao Ethnic Group in Danzhai County, Jianhe County and Kaili City, Papermaking Song, Song of Looking for Paper were still popular, which depicted how the ancestors made bamboo paper, Mian paper (white bast paper), and the papermaking tools and techniques they used (which has already been depicted in the previous part and will just be omitted here). From these facts, we can see that handmade papermaking has become a unique part of local ethnic culture, especially in their dancing and singing culture.

2.3.3 Paper Worshipping

Papermaking is time-consuming and laborious, and this along generates a special respect for civilization, people respect and value paper. This tradition can also be traced to Guizhou Province. For instance, in Zheng'an County, there are at least two Paper-Burning Towers. One is located in Fanrong Villagers' Group in Hezuo Village of Jianping Town, and another one is located nearby the Danping Elementary School of Banzhu Town. These towers are used to burn paper, so that the used paper wouldn't get abandoned or decayed. This practice shows the respect for paper, which in turn represents civilization and education, and it is still valued and respected today which is quite significant for the preservation of ancient cultural classics.

2.3.4 Taboos Concerning Papermaking

It is obvious that in some areas of Guizhou Province, only men are allowed to implement the major procedures of papermaking, while women are only allowed to help with unimportant procedures. In other areas, women are allowed to learn the papermaking techniques and bring the techniques to their new home after marriage.

However, in other areas, things are quite different. For example, among the Dong and Miao Ethnic Groups in Benli Village and Jiuqiu Village of Rongjiang County, Yao Ethnic Group in Gaohua Village of Cuili Town in Congjiang County, Dong Ethnic

Paper-Burning Tower in Zheng'an County

Group in Xiaohuang Village and Zhanli Village of Gaozeng Town, and Dong Ethnic Group in Dimen Village of Maogong Town in Liping County, only female members of a family can be allowed to learn bast papermaking techniques, while men only help with simple, unimported steps. Moreover, their techniques are not limited to only family members, neighbors in the same village, even relatives from other villages are allowed to learn papermaking techniques.

Why in these areas, women are dominant in papermaking practice? We proposed the assumption: it is found that the above mentioned areas are also the places where fixed papermaking screens are employed. Therefore, it is quite possible that papermakers believe women are more capable completing such demanding tasks, which need scrupulosity and calmness just as in weaving or embroidery. Of course, there may be other causes, or many factors may be playing together.

A comprehensive investigation and survey leads to the following conclusion concerning the status of papermaking in Guizhou Province: with the development of society, economy, technology and culture in a world featured by industrialization and urbanization, handmade paper in Guizhou Province has been challenged by the prosperity of machine-made paper and has endured a gradual decline in production. Dramatic changes from agriculture-oriented society to a newly industrialized society result in differences of the traditions through different generations. Through out history, Guizhou Province once boasted of diverse handmade paper in abundance, while nowadays, it has become an unavoidable trend for some papermakers to cease production. Currently, even remote papermaking places with inconvenient transportation are facing extinction of their papermaking techniques and are in urgent need of preservation as an intangible cultural heritage.

3 Current Preservation and Researches of Handmade Paper in Guizhou Province

3.1 Characteristics of Existing Handmade Paper Heritage Resources in Guizhou Province

Based on literature review and multiple field investigations, researchers of *Library of Chinese Handmade Paper: Guizhou* conclude that the inheritance and distribution of handmade paper practice in Guizhou Province is in a "bittersweet" state. The specific characteristics to be observed are as follows.

3.1.1 Diverse Paper Types

Needless to say that industrialization, urbanization and development of information technology (mobile communications), have challenged the traditions of handmade papermaking all over the country. In less than half a century, more than half of papermaking mills that used to be widely distributed throughout China just disappeared. Most Chinese papermaking villages are now faced with a difficult situation, and some places used to be famous have ceased papermaking practice, except two places: Jingxian County of Anhui Province and Jiajiang County of Sichuan Province, the two capitals of Chinese painting and calligraphy paper, are still quite prosperous. However, Guizhou Province is not completely immune to this trend and its papermaking practice is also shrinking. Famous local paper such as white bast paper in Duyun City, iron pen wax paper in Duyun City, bast paper in Zhenning County have disappeared, and other papermakers just ceased the production of bamboo paper in Jinsha County and bamboo paper in Zhenfeng County. These have all spurred great concerns.

Meanwhile, the investigation team was impressed by the diverse variety of handmade paper resources in Guizhou Province.

(1) High-quality and the number of paper types recognized as the intangible cultural heritage. Among the famous papermaking places such as Jingxian, Jiajiang, Fuyang, Lijiang, Qian'an, Moyu, Dege, none locates in Guizhou Province. But unexpectedly, in 2006 the Ministry of Culture announced the national intangible cultural heritage protection list, 3 papermaking counties of Guizhou Province were included, namely bast paper by the Miao Ethnic Group in Shiqiao Village of Danzhai County, bamboo paper by the Bouyei Ethnic Group in Xiangzhigou Village of Wudang District in Guiyang City, and bast paper in Longjing Village of Zhenfeng County. This level of government protection towards intangible cultural heritage is unprecedented in China (see Table 1.4).

A female Dong papermaker pouring paper pulp on the papermaking screen in Benli Village of Rongjiang County

Table 1.4 Distribution of paper preserved by the national intangible cultural heritage program in China

Grant No.	Protected papermaking techniques	Location
V111-65	Xuan papermaking techniques	Jingxian County of Anhui Province
V111-66	Liansi papermaking techniques in Yanshan County	Yanshan County of Jiangxi Province
V111-67	Bast papermaking techniques	Zhenfeng County, Danzhai County and Guiyang City of Guizhou Province
V111-68	Handmade papermaking techniques of the Dai Ethnic Group and Naxi Ethnic Group	Shangri-la County and Lincang City of Yunnan Province
V111-69	Tibetan papermaking techniques	Tibet Autonomous Region
V111-70	Paper mulberry bark papermaking techniques of the Uygur Ethnic Group	Turpan District of Xinjiang Uygur Autonomous Region
V111-71	Bamboo papermaking techniques	Jiajiang County of Sichuan Province and Fuyang City of Zhejiang Province

(2) Diverse ethnic cultures are illustrated. Various ethnic groups living in Guizhou Province, e.g., Bouyei, Yao, Dong, Tujia, Gelo, Miao, Yi, Li, stick to their own cultural features in the papermaking practices. Even within one ethnic group, due to different locations, distinct features are visible which is in accordance with Guizhou's typical feature of small communities and enjoying unique local culture and papermaking techniques. For instance, bamboo paper by the Bouyei Ethnic Group in different places such as Bapan Village of Anlong County in southwest area of Guizhou Province, Xiangzhigou Village of Wudang District in Guiyang City of central Guizhou Province and Yaogu Village of Libo County in south area of Guizhou Province, all showed specific features in techiques and culture.

3.1.2 Wide Distribution of Pristine Papermaking Techniques

Starting from the late 20th century, machine-made and half-machine-made paper became quite popular and challenged the market of handmade paper by replacing it through out China, especially for the paper in the mid-to-low end market lacking specific or exclusive usage.
For instance, a considerable number of papermakers gathered together in a trend of mechanization, such as the ones that enjoy great reputation: the bast papermaking gathering area in Qian'an City, Hebei Province, and bamboo papermaking gathering area in Huili County of Sichuan Province and Fuyang City in Zhejiang Province. In these areas, handmade paper has been completely replaced by machine-made and half-machine-made paper.

Abandoned hydraulic grinder in Chayuan Town of Jinsha County

However, the fine pristine handmade papermaking techniques in Guizhou Province is still impressive.

(1) When we first started our field investigation of Guizhou Province, we were impressed by the wide distribution, popularity and large quantity of papermaking sites there. While on our seven and a half years field investigation was indeed thorough, there is still much more work to be done. There are 44 villages involved in papermaking practice that are listed in *Library of Chinese Handmade Paper: Guizhou* , distributed in 31 counties (districts, cities) and cover 9 prefecture-level cities (prefectures). This status and distribution of the existing resources is quite unique in China.

(2) It was found that in the 44 villages under investigation most papermaking villages still maintain the traditional practice, namely, papermaking is family-based, but some that have ceased production. Even in areas where there are many rural papermaking households such as Cengong County, Danzhai County of Qiandongnan Miao and Dong Autonomous Prefecture, and Nayong County of Bijie City, papermaking practices still remain in the traditional self-sufficient way, and few of them turn to the machines or employ helpers. This is quite different from any other papermaking sites in China. Why this traditional and pristine way has been preserved widely is worthy of serious and deep consideration.

A row of handmade papermaking mills in Cengong County

3.1.3 Protection of the Surviving Papermaking Techniques Greatly Challenged

From the interviews and local literature obtained in our field investigation, it has been revealed that the pressure of inheriting handmade papermaking techniques in Guizhou Province comes from the following three aspects:

(1) There are mainly two kinds of paper in Guizhou Province: one is bast paper using paper mulberry bark as a raw material, and the other is bamboo paper made from bamboo, and the latter dominates. Judging from the statistical data, there are 47 kinds of paper listed in the *Library of Chinese Handmade Paper: Guizhou*, among which there are 21 kinds of bamboo paper, accounting for 44.7%. However, bamboo paper in Guizhou Province is mostly used as joss paper which is quite low in economic value. Almost 16 kinds of bamboo paper out of 21 are mainly used as joss paper, including: bamboo paper in Panxian County, bamboo paper by the Bouyei Ethnic Group in Anlong County, bamboo paper in Zhenfeng County, bamboo paper by the Bouyei and Miao Ethnic Groups in Guanling County, bamboo paper in Ziyun County, bamboo paper in Huishui County, bamboo paper in Longli County, bamboo paper by the Bouyei Ethnic Group in Libo County, bamboo paper in Jinsha County, bamboo paper by the Dong Ethnic Group in Sansui County, bamboo paper by the Tujia Ethnic Group in Yinjiang County, bamboo paper by the Miao Ethnic Group in Kaili City, bamboo paper by the Yao Ethnic Group in Xiutang Town of Congjiang County, bamboo paper by the Gelo Ethnic Group in Wuchuan County, bamboo paper in Yuqing County, bamboo paper by the Miao Ethnic Group in Huangping County, accounting for 76.2% of all. Though it is documented that bamboo paper in Guizhou Province once enjoyed medium or even significant usage, Guizhou handmade paper never developed significant usage for Chinese calligraphy and painting in the

Gaohua Village surrounded by mountains

same way bamboo paper in Jiajiang County or bamboo paper in Fuyang County did.

The direct pressure the researchers found in the investigation was that low-price joss paper was overwhelmed by machine-made paper in market, which put the cost-performance ratio of bamboo paper at a disadvantage in the contemporary market.

(2) Another pressure is that the development of modern modes is not easy. Due to the natural geography and ethnic distribution, many handmade paper villages are distributed in some remote areas with poor living conditions, inconvenient traffic, and limited communication with the outside world. Therefore, it is difficult to strengthen exchange and circulation, bring in modern advanced knowledge or develop tourism in the area, following other intangible cultural heritage preservation and development modes.

For instance, during many field investigations,

the members of survey team couldn't reach the final destinations because of the severe road conditions. One example was our trip to Dage Village of Xiutang Zhuang Town, Congjiang County for example, the research group's first attempt to visit failed due to the heavy rain. In the second try (one year later), in continuous sunny days, it took about 3-hour motorcycle ride to get to the destination from the town government. While for Baishui Village of Cengong County and Gaohua Village of Congjiang County, the researchers had to take SUV or even farm vehicles to get there.

(3) The above two pressures lead to a third pressure: few or no one wants to carry on the torch of handmade paper, while inheritance is critical for protecting handmade paper. Since the reforming and opening in 1978, industrialization and urbanization encourage many younger and middle-aged people to leave their rural work for the city. Luxurious way of life, money, and a brighter future in the prosperous cities attract them in an irresistible way.

When the investigation team first visited Guizhou Province, it had been a common phenomenon for many countrymen to leave their hometown for cities. Papermaking is very tedious work, from morning to night, day after day. Moreover, most papermaking villages locate in remote mountains, do not have interaction and communication with outside world; thus, they are limited to small communities (since they devote themselves to the arduous papermaking), while the economic disadvantage served as the last straw for the young who lose the motivation to learn and pass down the craftsmanship. In the survey of 44 villages, it is rare to find young papermakers, but middle-aged or older ones. For instance, Chen Changcai, the inheritor of bamboo paper in Jinsha County, Wang Piquan, the inheritor of bast paper, He Lianqing and Dong Huaxiang, the inheritors of bamboo paper in Panxian County, their offspring all go out to work in cities. Therefore, the current state of papermaking is that either the middle-aged and old people struggled to sustain the practice all by themselves, or the elderly are too old to continue the job and ceased the practice. The young generation are no longer able to make paper, even for those who had learned the techniques eventually went out and were unwilling to take up the hard work again. The chance of passing down the techniques is very slim.

It is true that some places in Guizhou Province developed a positive tradition of papermaking, yet this imbalanced status is quite disturbing.

3.2 Efforts to Protect Handmade Paper in Guizhou Province

3.2.1 Achievements in Policies and Regulations

Efforts of the Guizhou government in protecting intangible cultural heritage are quite eminent, and have achieved remarkable results. Guizhou Province has developed well in the multi-ethnic cultural heritage protection themed as "Colorful Guizhou".

The representative progress can be summarized as:

1. Laws and Regulations

(1) On March 30, 2012, the NPC Standing Committee of Guizhou Province formally approved the *Regulations on the Protection of Intangible Cultural Heritage of Guizhou Province*. On May 1, in the same year, the Regulations in the form of local law was officially issued and implemented in the whole province. Guizhou is the first province that promotes the protection of Intangible Cultural Heritage in the form of regional laws. Before that, the draft had been revised for many times.

(2) On June13, 2014, the Guizhou Provincial Office, General Office of Provincial Government officially issued and implemented *Plan for Developing and Protecting Intangible Cultural Heritage in Guizhou Province* (*2014~2020*), which is the first comprehensive plan released by provincial government in China. The Plan clearly and specifically maps out eight key objectives:

Conducting a thorough survey on Intangible Cultural Heritage and building up a database;

Protecting the Intangible Cultural Heritage resources, specifically the production;

Training talents in protecting Intangible Cultural Heritage;

Do more theoretical research on Intangible Cultural Heritage;

Cultivating Intangible Cultural Heritage properly;

Promoting the creation of works on Intangible Cultural Heritage;

Organizing more activities and building brands on Intangible Cultural Heritage;

Constructing demonstration platforms for Intangible Cultural Heritage.

It is the first time that the Chinese cultural authorities of the government treated the protection of intangible cultural heritage as an official and public activity. Even the heads of the Provincial Party Committee and Guizhou government were involved and led the activities. This is an important milestone for the protection of intangible cultural heritage in Guizhou Province and in China. There is no doubt that the multi-ethnic handmade papermaking techniques will benefit more from these activities aiming to protect intangible cultural heritage.

2. Construction of Protecting Mechanism

(1) The four-level protection system of Intangible Cultural Heritage has been established in Guizhou Province which takes a leading role in China. National-level and province-level intangible cultural heritage protection list has been released successively four times, and so for the recommendation and awarding of intangible cultural heritage inheritors. Altogether, there are 74 projects with 125 items with 57 inheritors are listed on the national-level intangible cultural heritage list; 440 projects and 568 items with 301 inheritors are listed into the province-level intangible cultural heritage. One national ecological protection experimental zone of intangible cultural heritage has been established in southeast area of Guizhou. Three national demonstrative bases of ethnic cultures are built up, including handmade paper of Shiqiao Village, Danzhai County (The Shiqiao Qianshan traditional cooperative papermaking technique was listed in 2011). The Dong Ethnic Song is enlisted into "Masterpieces of the Intangible Cultural Heritage".

At the same time, the county or city-level application for intangible cultural heritage is proceeding in an orderly way. The city (prefecture) and county-level programs and inheritors database are being enriched and improved with the help from relevant protective institutions and groups.

(2) Achievements are made in handmade paper and papermaking inheritor included in Intangible Cultural Heritage Protection List. At the province-level, by December 2014, 9 counties' papermaking techniques were listed as the intangible cultural heritage, including Danzhai County (Shiqiao traditional papermaking techniques), Zhenfeng County (Xiaotun Baimian papermaking techniques), Wudang District of Guiyang City (Wudang handmade papermaking techniques), Sansui County (Traditional papermaking techniques), Panxian County (Traditional papermaking techniques), Huishui County (Traditional papermaking techniques), Changshun County (Traditional papermaking techniques), Cengong County (Local Huo papermaking techniques) and Zheng'an County (Local paper folding techniques). There are 7 handmade papermakers who were acknowledged as experts in the field, namely Wang Xingwu (Danzhai County), You Xinglun (Zhenfeng County), Luo Shouquan (Wudang District), Yang Zaixiang (Sansui County), He Lianqing (Panxian County), Li Fatian (Changshun County) and Pan Yuhua (Danzhai County).

In terms of handmade paper, the effect to protect handmade paper through laws and regulations has proven to be significant.

3.2.2 Exploitation of Resources and Implementation of Policies

1. Financial Support

(1) Guizhou government plays a dominant role in funding the protection of intangible cultural heritage. For instance, direct funding from the provincial government, in 2005 was 1 million yuan per year. In 2009, the fund increased to 11.5 million yuan per year and in 2012 to 16.3 million yuan per year. Since 2014, 10 million more yuan was funded per year. This input on intangible cultural heritage protection put Guizhou Province on the leading role in China, based on an interview in September 2009 of head of the Department of Culture in Guizhou Province.

Since 2009, Guizhou government consistently awards the national-level, province-level and city (prefecture)-level inheritor of intangible cultural heritage 8 000, 5 000 and 3 000 yuan prize money respectively. This is strong evidence of support from the government considering the financial state of Guizhou Province.

(2) Danzhai County serves as a typical example in handmade paper field. Handmade paper in Shiqiao Village of Danzhai County is a national Intangible Cultural Heritage, which also boasts a national productive protection brand of handmade paper. The high-quality resources in the county are abundant, e.g., the sacrificial dancing and singing using local musical instruments Mangtong and Lusheng. Every year, Danzhai government allocates 1% of the real estate development fund to the preservation and development of intangible cultural heritage, 5 million yuan on the Intangible Cultural Heritage Development Fund, 2 million yuan from local fiscal funds on the Cultural Heritage Protection, and registered 6 culture-related trademarks, e.g., "Shiqiao traditional papermaking techniques".

Besides Danzhai County, Yinjiang County is also remarkable in handmade paper protection. Though it is not listed the national or province-level intangible cultural heritage protection list, the handmade paper practice there is actually enjoying a long history and wide coverage. In November 2010, Yinjiang County started the "Heshui traditional handmade papermaking mills" maintenance project, and 250 thousand yuan was invested, in order to repair or rebuild 76 handmade papermaking mills located in Xiazhai and Caijiawan Natural Village of Xingwang Administrative and Caijiawan Natural Village, and Qiaotou Natural Village of Mula Administrative Village in Heshui Town. Altogether, 45 mills were repaired and 31 were rebuilt, and 231 family-based papermakers were involved, amounting to 2 950.74 square meters in coverage. This is a significant supporting effort to the basic papermaking platform.

2. Promoting the Public Understanding of the Techniques

(1) Establishment and modification of the promotion platform is a significant sign of these efforts. In 2003, *Guizhou Folk Culture Protection Regulations* was issued. In July 2004, Folk Culture Protection Committee of Guizhou Province was founded and in September 2006, the committee was renamed the Guizhou Intangible Cultural Heritage Protection Committee, and subsidiaries were built up in cities and prefectures. In 2009, Expert Committee of Guizhou Intangible Cultural Heritage was founded, and over 50 experts in the field of cultural heritage were appointed then.

The first Ethnic Cultural Eco-Museum in Asia, located in Suoga Village of Liuzhi Special Area

Since 1995, co-sponsored by the National Administration of Cultural Heritage and the Norwegian Government, Guizhou Province has built quite a few cultural eco-museums, such as Suoga Cultural Eco-Museum in Liuzhi Special Area, Zhenshan Bouyei Cultural Eco-Museum in Huaxi Town, Tang'an Dong Cultural Eco-Museum in Liping County, and Longli Cultural Eco-Museum in Jinping County. In December 2011, 14 provincial intangible cultural heritage protection bases were established.

In 2007, the project "Guizhou Ethnic Folk Cultural Resources Information Network Construction" succeeded in competing for the small grants from the World Bank, which (website: http://www.gzfefax.com) was soon activated and invited volunteers to contribute relevant information, which became an important public platform for the protection of the intangible cultural heritage.

(2) The local publicity work is full of vigor. As early as September 1982, when Chinese Traditional Techniques Troupe visited Canada, a papermaker from Shiqiao of Danzhai County, Yang Dawen, displayed Chinese handmade papermaking techniques in the world famous Toronto Ontario Science Centre. In the year of 2010, a series of books, *Guizhou Intangible Cultural Heritage*: *Collection of Local Civilizations* were started to promote the publicity of the Intangible Cultural Heritage. By now, book on *Mawei Embroidery of Shui Ethnic Group* was released, and books on Guizhou handmade papermaking techniques were being worked on.

In August 2009, Kaili College of Qiandongnan Miao and Dong Autonomous Prefecture issued the inaugural *Journal of Pristine Ethnic Culture* and textbook series on the same subject. They also officially started the preparation of an undergraduate

Papermaking scenes in Shiqiao Village captured in a movie

program on ethnic cultures.

In the towns and counties of Guizhou Province, various means have been taken to promote the publicity. For instance, the Radio, Film and Television Administration of Zhenfeng County released book series of *Intangible Cultural Heritages in Zhenfeng County* including 4 nation-level, 12 province-level and over 50 city- or county-level intangible cultural heritages, among them Zhenfeng bast papermaking techniques and culture are highlighted. A movie funded by Danzhai County named "Sun Above the Cloud" vividly presented Shiqiao traditional papermaking techniques to the world and the movie was awarded Best Foreign Language Film, Best Cinematography, Best Film (nomination) by the 17th Sedona International Film Festival. These awards greatly promoted the world publicity of this unique papermaking technique.

(3) Promoting the public understanding of Guizhou cultural heritages to students in elementary schools and middle schools is an important task. In October 2002, the Guizhou Education Department and Ethnic Affair Commission co-issued a regulation named *Promoting Ethnic Folk Culture Education in Schools of the Province*, and made it clear that the government should take the lead in promoting the education of cultural heritage throughout various places and schools in the province, and by now it has been quite effective.

Take Liping County of Qiandongnan Dong and Miao Autonomous Prefecture for example, Liping embraces the intangible cultural heritages of the Dong Ethnic Group, including Dong ethnic songs and handmade papermaking techniques of Dimen Village. In 2005, the county government issued a brief on famous scenic spots in Liping, and in 2007, a series of textbooks on local ethnic culture (4 volumes) were published, and handed out to all the students in the county for free. Meanwhile, the county office in charge of education even released the guidelines on the details of promoting education of ethnic culture in Liping County, such as each week each school must organize extracurricular activities on ethnic culture. No.1 Middle School of Liping County, Dimen Elementary School of Maogong Town and Yandong Middle School were set as research and teaching bases of ethic culture, and every two years a symposium should be held.

3.2.3 Concerns on Preservation and Development of Intangible Cultural Heritage

(1) Measures taken by Guizhou government on intangible cultural heritage protection are quite effective. However, there is a problem on how to balance the development of pristine culture and newly emerging culture. This is a complicated issue in protecting intangible cultural heritage.

Local scholars of Guizhou Province have proposed solutions. For example, Cai Qun in his paper argued that intangibility is a typical feature with intangible cultural heritage. So its protection should consider the following factors: firstly, help the current inheritors and train the more skilled ones; secondly, develop the endangered techniques in new areas through recording, describing and transferring; thirdly, remodel or rebuild the present existing environment to make it friendly to the techniques.

Based on the international practice, the modern ways of protection include: being family-based, resorting to school education, developing as a tourist spot, protecting the whole ecosystem, converting to digital forms or making relevant videos and other means.

(2) Typical problems concerning the perservation of intangible cultural heritage, considering the current status of handmade paper development in Guizhou Province, are: how to motivate the youth to inherit the techniques, developing tourism may harm the local pristine environment, remote papermaking mills are hard to reach, promoting environment of protection and controlling pollution in gathered production, low-end paper is disappearing fast because most rescuing efforts (such as videoing) are focusing on high-end paper, homogenization of low-end or medium paper fails to cater to the various needs of education, enforced measures by the government fail to consider the varieties, inheritors of the techniques are not educated enough nor do they know how to take advantage of new marketing means and techniques.

3.3 Current Overview of Handmade Papermaking Researches in Guizhou Province

Guizhou Province has accumulated a solid cultural foundation on handmade paper. However, research on the subject has been delayed and are not systematic. Untill now, not even one monograph has been published that systematically studies the handmade paper in Guizhou Province. The other typical feature with researches on handmade paper in Guizhou Province is that, they only focus on a select few papermaking sites, such as Danzhai, Zhenfeng, Wudang, Changshun or Liping, and newly-developed sites like Pu'an, Huishui, Shiqian and Liuzhi are only studied by our researchers based on our field investigation and literature review of *Library of Chinese Handmade Paper*. In reality, most papermaking sites are not studied enough.

Here comes a brief introduction to the researches on handmade paper in Guizhou Province:

(1) In 1984, Zhu Dazhen, from the History Museum of China, published a paper, *White Bast Papermaking Techniques in Danzhai County of Guizhou Province*, based on field investigations, in which the raw materials, tools and procedures were specifically depicted concerning bast paper made by the Miao Ethnic Group in Shiqiao Village of Danzhai County. Cleaning bag and bamboo drying methods exclusively used in the area were also introduced. In 1996, another paper written by Zhu Dazhen, *Traditional Papermaking Techniques and Their Influence on Guizhou Ethnic Groups*, introduced Shiqiao paper of Danzhai County again, as well as the techniques used in bamboo papermaking in Laochang Town of Panxian County. His works initiated the modern research on handmade paper in Guizhou Province.

(2) In 1991, Liang Taihe et al. of the

Guizhou Meseum finished their field investigation report of *Research on Guizhou Traditional Techniques*, based on their investigation over the years. In the report, various papermaking sites in Guizhou Province and relevant papermaking techniques during 1980s were recorded. Though this is only a small portion of the report, it failed to give the details or cover all the papermaking areas. However, it still serves as the first systematic field investigation on handmade paper in Guizhou Province which makes it a meaningful resource.

(3) Wang Shiwen, veteran handmade paper history researcher, published *Dictionary of Chinese Handmade Paper*, which keeps an account of handmade paper from the origin to the raw materials, place of production, techniques, usage and current status. It aims to cover China as a whole, and Guizhou Province is just briefly depicted and far from being enough.

(4) Wu Zhengguang in his paper *Papermaking Culture in Guizhou Provinc*e gives a brief introduction to handmade papermaking in Shiqiao Village of Danzhai County and papermaking practice in other places including Longjiao Village of Wudang District, Laochang Village of Panxian County, Longtan Village of Liuzhi Special Area, Heshui Village of Yinjiang County, Sandaogou of Xingren County, Shuijing Village of Pu'an County, and Wenggui Village of Guangshun County. Among the 8 papermaking sites, some of them have now ceased production.

(5) In *Disappearing Techniques*: *Notes on Handmade Papermaking Mills in Wenggui Village of Changshun County*, Zu Ming reviews the history of handmade papermaking and usage in Guizhou Province, and illustrates field investigation results of handmade paper made in Wenggui Village of Changshun County in terms of history, inheritance, raw materials, tools and techniques. This was the first systematic and thorough research paper on Wenggui paper.

(6) In *Handmade Paper in Ethnic Areas in China*, Liu Renqing introduced four traditional handmade paper types made by the ethnic groups, including bast paper made by the Miao Ethnic Group in Danzhai County.

(7) Wei Lai et al. explore family-based papermaking practices on white bast paper in Shiqiao Village, in terms of its current status, features, obstacles, development mode and strategies in his paper *Development of Family-based Handicraft in Southeast Guizhou Province*: *An Investigation on Two Villages in Danzhai County*.

(8) Wei Dengliang in his paper *Introduction to Bast Papermaking Inheritor*: *Luo Shouquan* presents a descriptive introduction to Luo Shouquan, inheritor of national intangible cultural heritage, and bast papermaking techniques of the Bouyei Ethnic Group of Longjiao Village, Wudang District of Guiyang City. However, there is a discrepancy between the title which attempts to introduce the bast paper included in the national intangible cultural heritage list, he actually introduced bamboo paper made in Longjiao Village. This may be due to the fact that this collection is supposed to focus on the biographies of inheritors of national intangible culture heritage in Guizhou Province.

(9) Deng Guocheng in his paper *Inheritor of Baimian Papermaking Techniques in Xiaotun Town of Zhenfeng County*: *Liu Shiyang* introduced Baimian paper produced in Xiaotun Town of Zhenfeng County, in terms of its papermaking techniques and tools, and discussed the inheritor of the techniques with descriptions and pictures.

(10) In Dai Cong's paper *Preservation and Development of Intangible Cultural Heritage in Ethnic Areas*: *A Case Study of Danzhai County in Qiandongnan Miao and Dong Autonomous Prefecture*, he explores into bast papermaking techniques and its preservation in Shiqiao Village of Danzhai County. In 2007, Dai Cong was a volunteer college student teaching there, and accumulated field investigation data on the intangible cultural heritage. Later, he collaborated with Yu Shiming and Zhang Weina, and published a paper *Status Survey of Intangible Cultural Heritage in Ethnic Areas*: *A Case Study of Danzhai County in Qiandongnan Miao and Dong Autonomous Prefecture* in *Journal of Guizhou University* (*Social Science*) , No. 3 of 2010, in which bast paper made in Shiqiao Village was depicted but many parts in this paper just copied the previous one.

(11) Zeng Yun wrote a paper named *Development and Preservation of Intangible Cultural Heritage in Ethnic Areas*: *A Case Study of Traditional Papermaking Techniques of Shiqiao Village in Guizhou Province*, and in it the quality, market, utility, developing conditions and paths, and protection modes of bast paper made by the Miao Ethnic Group in Shiqiao Village were discussed and suggestions proposed.

(12) Chen Hongli in his *Review on Handmade Paper Research in Southwest Ethnic Areas* elaborates on the researches on the handmade papermaking techniques and current status in ethnic areas of southwestern China. However, Guizhou handmade paper is not included, specifically related and some important research was not included, potentially due to the fact that Guizhou Province failed to concentrate on the research in this field while other areas are more focused.

(13) Chen Biao et al. in *Investigation on Bast Paper Made in Katang Village of Pu'an County in Guizhou Province* probes into Pu'an bast papermaking techniques, raw materials, tools and inheritance based on the field investigation analysis.

(14) Chen biao et al. in their paper *White Bast Paper Made by the Gelo Ethnic Group in Xiangshuyuan Village of Shiqian County*, depicted in details the making techniques, techniques, sales, profits and inheritance of white bast paper made by the Gelo Ethnic Group based on the field investigation.

(15) Tian Maowang in his papers *Survey on Traditional Handmade Papermaking Techniques in Baishuihe Village of Guizhou Province* and *Protection of Traditional Handmade Papermaking in Baishuihe Village of Guizhou Province* explore into the raw materials, procedures, tools of the papermaking, and analyzes the status and causes of the shrinking paper practice in Baishuihe Village of Xinbao Bouyei Town in Wudang District of Guiyang City, and proposes various suggestion on methods of protection.

(16) Ma Yingna in her paper *Handmade Paper Made by the Dong Ethnic Group of Dimen Village* reports the field investigation results of handmade papermaking status of the Dong Ethnic Group in Dimen Dong Town of Liping County. She further elaborates on the techniques and features, usage development and causes, social network of papermakers, etc., using sociology and anthropology methods.

(17) Thesis of Li Jinhai for master's degree, *Developments of Manual Papermaking Technology and Its Influences on Paper Properties in Zhongzhai Town of Liuzhi Special Area in Guizhou Province*, focuses on the papermaking techniques,

performance testing and technique development of bast paper in Liuzhi Special Area. Technical measures were taken to test the features of the paper which makes it a systematic research.

(18) *Sources of Handmade Papermaking Techniques of Ethnic Groups in China: A Case Study of Southwest Ethnic Areas* written by Qin Ying works on the distribution, techniques and application of handmade paper in the ethnic areas in Yunnan Province, Guangxi Zhuang Autonomous Region and Guizhou Province. Using geographic information system analysis tools, systematic regional distribution of technique sources and the technique-transmitting route are studied. This report focuses on Yunnan Province, while handmade paper in Guizhou Province has seized less attention and failed to present the panorama of handmade paper in Guizhou Province in terms of its current status and technique development.

(19) Fan Shengjiao took the economic perspective in the paper *The Pathway from Intangible Cultural Heritage to Intangible Economic Industry: A Case Study of Shiqiao Traditional Papermaking in Guizhou Province*, which studies on the bast paper in Shiqiao Village features and its industrialization.

(20) Zhang Jianshi wrote a book named *Protection of Handmade Paper Traditions of Ethnic Groups in Southwest China*, in which representative papermaking sites and technique development of various ethnic groups, e.g., Tibetan, Dai, Naxi, Bai, Yi, Yao, Dong, Miao, Bouyei and Tujia Ethnic Groups are reviewed. Paper made by the Miao Ethnic Group of Shiqiao Village is specifically introduced, and paper made in Zhenfeng and Liping is also briefly introduced.

(21) Yang Zhengwen et al. in their paper *Handmade Papermaking Technique Development of Ethnic Groups in Guizhou Province of China* depicts the raw materials, papermaking mucilage, procedures and techniques of paper made in Shiqiao Village of Danzhai County by the Miao Ethnic Group, Dimen Village of Liping County by the Dong Ethnic Group, and Xiaotun Village of Zhenfeng County by the Bouyei Ethnic Group. The research is standard; yet only 3 papermaking sites in Guizhou Province are introduced, which is far from being enough.

In the above part, we mainly reviewed the research published in journals, while three other sources of research on handmade paper in Guizhou Province deserve attention:

① Local chronicles, which not only include the relevant information taken from ancient chronicles, but also new progress in papermaking practices. Comparatively, they are more like a panorama, covering papermaking details of various cities (prefectures), counties and towns. Such chronicles include *Chronicle of Zunyi City, Chronicle of Guanling County, Chronicle of Laochang Town*, etc.

② Literature and History Committee of the Political Consultative Conference of various levels works on the local cultural and historical resources, and among them papermaking techniques, folklores and history were recorded. These articles are usually based on personal experiences or field investigations, which is direct and very valuable. For instance, *Cultural and Historical Resources of Qiannan Bouyei and Miao Autonomous Prefecture*, *Cultural and Historical Resources of Zhenfeng County*.

③ Local newspaper or website also publish articles on handmade paper culture for the purpose of attracting tourists. These articles are usually written by journalists, and lack the research value.

An overview of researches on handmade paper reveals the weakness of Guizhou Province in this field compared to other southwest areas of China, e.g., Yunnan Province, Sichuan Province and Guangxi Zhuang Autonomous Region. For the existing researches, the focus is usually towards bast paper by the Miao Ethnic Group in Shiqiao Village of Danzhai County, while other paper types such as bamboo paper by the Bouyei Ethnic Group in Xinbao Town of Wudang District, bast paper by the Dong Ethnic Group in Dimen Town of Liping County are barely touched upon. These examples are far from being enough in terms of research coverage or balance.

It took us about seven and a half years field investigation to present a panorama of handmade paper practice in Guizhou Province in the early 21st century in *Library of Chinese Handmade Paper: Guizhou*. We have also followed a standard research paradigm, in order to present detailed information on all the accessible papermaking villages and practices in Guizhou Province.

A researcher interviewing Liang Taihe (left) on the distribution of handmade paper in Guizhou Province

图目
Figures

章节	图中文名称	图英文名称
第一章	贵州省手工造纸概述	Chapter I Introduction to Handmade Paper in Guizhou Province
第一节	贵州省手工造纸业的历史沿革	Section 1 History of Handmade Paper in Guizhou Province
	王阳明像	Statue of Wang Yangming
	修文县龙场镇"阳明玩易窝"	Yangming Wanyiwo stone cave monument in Longchang Town of Xiuwen County
	长顺县翁贵村光绪年间的"永垂定例"碑	Monument of "Permanent Stipulation" erected during Guangxu Reign of the Qing Dynasty, in Wenggui Village of Changshun County
	贵州省1950~1988年手工纸产量变化示意图	Variation trend of handmade paper output in Guizhou Province from1950 to 1988
	务川县造纸塘村存手抄本《卢氏经单薄》	*History of the Lus*, handwritten version kept in Zaozhitang Village of Wuchuan County

章节	图中文名称	图英文名称
第二节	贵州省手工造纸的当代生产现状	Section 2 Current Production Status of Handmade Paper in Guizhou Province
	亚太旅游协会援助丹寨石桥古法造纸项目纪念碑	Monument in memory of Pacific Asia Travel Association's funding for the traditional papermaking techniques in Shiqiao Village of Danzhai County
	野生构树林	Wild paper mulberry trees
	慈竹	*Neosinocalamus affinis*
	斑竹	*Phyllostachys bambusiodes* cv. Tanakae (spotted leopard bamboo)
	荛花	*Wikstroemia delavayi*
	乌当区新堡布依族乡陇脚村单人抄纸	One man making paper in Longjiao Village of Xinbao Bouyei Town of Wudang District
	石阡汤山镇香树园村仡佬族白皮纸切料工序	Trimming the papermaking materials: one procedure in making white bast paper by the Gelo Ethnic Group in Xiangshuyuan Village of Tangshan Town in Shiqian County
	与祭祀天地君亲师一起祭祀蔡伦和猪拱先师	Cai Lun and Mr. Pig are worshipped together with gods and ancestors
	正安县的字库塔	Paper-Burning Tower in Zheng' an County
	榕江县本里村侗族女性在浇纸	A female Dong papermaker pouring paper pulp on the papermaking screen in Benli Village of Rongjiang County

章节	图中文名称	图英文名称
第三节	贵州省手工造纸的保护与研究现状	Section 3 Current Preservation and Researches of Handmade Paper in Guizhou Province
	金沙县茶园乡废弃的水碾房	Abandoned hydraulic grinder in Chayuan Town of Jinsha County
	岑巩乡间连排的手工纸槽坊	A row of handmade papermaking mills in Cengong County
	深山处的高华村	Gaohua Village surrounded by mountains
	六枝梭戛村亚洲第一座民族文化生态博物馆	The first Ethnic Cultural Eco-Museum in Asia, located in Suoga Village of Liuzhi Special Area
	电影中的石桥造纸场景	Papermaking scenes in Shiqiao Village captured in a movie
	调查组成员与梁太鹤(左)交流贵州手工纸分布	A researcher interviewing Liang Taihe (left) on the distribution of handmade paper in Guizhou Province

章节	图中文名称	图英文名称
第二章	六盘水市	Chapter II Liupanshui City
第一节	盘县皮纸	Section 1 Bast Paper in Panxian County
	村口背料小景	A villager carrying a bundle of paper mulberry bark
	盘县山地风光	Landscape of Panxian County
	普安州文庙	Local Confucian Temple in Pu'an Prefecture
	进入羊场乡的乡间公路	Country road leading to Yangchang Town
	调查组成员采访郑福玉等人	Researchers interviewing Zhen Fuyu and other papermakers
	调查组成员与造纸人交流	A researcher communicating with a papermaker
	泡楮皮	Soaking paper mulberry bark
	造纸人背料回村	A papermaker carrying the bark back to the village
	捆好的干楮皮	Bundles of dried paper mulberry bark
	泡楮皮	Soaking paper mulberry bark
	浆楮皮	Fermenting paper mulberry bark in limewater
	上甑	Putting the papermaking materials in the kiln
	甑锅	Wok for steaming the papermaking materials
	蒸料	Steaming the papermaking materials
	摆石灰皮	Cleaning the fermented bark
	清皮	Cleaning the bark
	榨皮子	Pressing wet papermaking materials to squeeze water out
	蒸地灰皮	Steaming the papermaking materials with alkali
	洗料	Cleaning the papermaking materials
	踩干	Squeezing water out by stamping the papermaking materials
	抄纸	Scooping and lifting the papermaking screen out of water and turning it upside down on the board
	榨纸	Pressing the paper
	晒纸	Drying the paper
	拆纸	Peeling the paper down
	理纸	Sorting the paper
	捆纸	Binding the paper
	甑子	Kiln for steaming the papermaking materials
	木榨	Wooden presser
	纸帘	Papermaking screen
	下午村皮纸纤维形态图(10×)	Fibers of bast paper in Xiawu Village (10× objective)
	下午村皮纸纤维形态图(20×)	Fibers of bast paper in Xiawu Village (20× objective)
	旧日的文书	Old documents written on bast paper
	纸甑旁的祭祀小龛	Niche for worship beside the papermaking kiln
	村中正在造纸的老年人	Old ladies making paper
	废弃烘纸房里的造纸老人	An old papermaker in an abandoned drying room
	村里随处可见的造纸设施	Papermaking tools scattered everywhere in the village
	下午村皮纸透光摄影图	A photo of bast paper in Xiawu Village seen through the light

章节	图中文名称	图英文名称
第二节	盘县竹纸	Section 2 Bamboo Paper in Panxian County
	老厂村小景	View of Laochang Village
	老厂民居	Residences in Laochang Town
	老厂镇人民政府	Local government of Laochang Town
	老厂村造纸作坊群街区旧址	Former papermaking mills in Laochang Village
	老厂优质煤	High quality coal produced in Laochang Town
	《董氏族谱》封面	The cover of *Genealogy of the Dongs*
	非物质文化遗产代表性传承人董华祥	Dong Huaxiang, a representative inheritor of intangible cultural heritage
	贵州省非物质文化遗产代表性传承人何联庆与其妻子赵本美	He Lianqing, a representative inheritor of intangible cultural heritage in Guizhou Province, and his wife Zhao Benmei
	记述老厂造纸历史的乡土文献	Local literature on papermaking history in Laochang Village
	何联庆忆造纸旧事	He Lianqing recalling the old days of papermaking
	何联庆抄纸的小纸坊	He Lianqing's papermaking mill
	国营盘县造纸厂的旧车间外景	Former workshop of state-run Panxian Papermaking Factory
	老厂村旁的新竹	Bamboo springing up by Laochang Village
	煮料用的造纸窑	Papermaking kiln for steaming the papermaking materials
	何联庆家踩竹麻用的布质钉鞋	Spiked cloth shoes for stamping the papermaking materials (owned by He Lianqing)
	老厂村旧窑边的洗料池	Pool for cleaning the papermaking materials near an old kiln in Laochang Village
	废弃的老厂煮料窑	Abandoned boiling kiln in Laochang Village
	调查组成员试踩何联庆家的脚碓	A researcher trying the foot pestle in He Lianqing's house
	老厂镇黑土坡村的纸槽	Papermaking trough belonging to Heitupo Village of Laochang Town
	何联庆正在抄纸	He Lianqing making paper
	何联庆家待榨的湿纸垛	Wet paper pile to be pressed and dried in He Lianqing's house
	赵本美正在揭纸上墙	Zhao Benmei peeling the paper down and pasting it on the wall for drying
	烘纸的火墙	Wall for drying the paper
	打浆的脚碓	Foot pestle for beating the papermaking materials
	老厂镇何联庆家的木榨	Wooden presser in He Lianqing's house in Laochang Town
	老厂村的废弃窑孔	Abandoned papermaking kiln in Laochang Village
	废弃的窑孔内部	Inner view of an abandoned papermaking kiln
	钉鞋	Spiked shoes for stamping the papermaking materials
	背纸板	Board for carrying the wet paper
	老厂镇竹纸纤维形态图 (10×)	Fibers of bamboo paper in Laochang Town (10× objective)
	老厂镇竹纸纤维形态图 (20×)	Fibers of bamboo paper in Laochang Town (20× objective)
	何联庆家中待售的成捆竹纸	Bundles of bamboo paper for sale in He Lianqing's house
	正忆旧日辉煌的董华祥	Dong Huaxiang recalling the past glorious days of papermaking
	荒废倒塌的老厂房	Abandoned papermaking factory
	调查组成员与何联庆交流销售情况	Researchers talking with He Lianqing about the sales of handmade paper
	调查组成员练习揭纸	A researcher practicing the procedure of peeling the paper down
	调查组成员与董华祥考察老纸厂车间	A researcher and Dong Huaxiang visiting a former papermaking workshop
	在废弃不久的旧车间门口的董华祥老人	Dong Huaxiang standing by a door of newly abandoned papermaking workshop
	随处可见的废弃造纸旧迹	Abandoned papermaking sites can be seen everywhere
	老厂镇编印的宣传材料	Propaganda brochure of Laochang Town
	废弃的造纸车间	Abandoned papermaking workshop
	国营盘县造纸厂旧大门	Gate of the former state-run Panxian Papermaking Factory
	老厂镇竹纸透光摄影图	A photo of bamboo paper in Laochang Town seen through the light

章节	图中文名称	图英文名称
第三节	六枝彝族苗族仡佬族皮纸	Section 3 Bast Paper by the Yi, Miao and Gelo Ethnic Groups in Liuzhi Special Area
	郎岱镇旧日门上的标语	Slogans on a gate in Langdai Town
	老王山风光	Landscape of Laowang Mountain
	牂牁江风光	Scenery of Zangke River
	郎岱古镇一角	A section of Langdai Town
	长角苗人居住的村落	A village where Longhorn Miao people live
	民国年间的皮纸印本书籍	A copy of book printed on bast paper during the Republican Era of China
	20世纪60年代使用手工纸抄写的文契	Contract written on handmade paper in 1960s
	火碱	Caustic soda (sodium hydroxide)
	泡构皮	Soaking paper mulberry bark
	滤水	Withdrawing and airing soaked paper mulberry bark
	打皮	Hammering the bark with a foot pestle
	洗料	Cleaning the papermaking materials
	抄纸	Scooping and lifting the papermaking screen out of water and turning it upside down on the board
	拆纸	Peeling the paper down
	理纸	Sorting the paper
	皮甑	Kiln for steaming the papermaking materials
	纸帘	Papermaking screen
	计数器	Paper counting apparatus
	脚碓座	Stone board under the foot pestle
	脚碓	Foot pestle
	纸槽和纸榨	Papermaking trough and pressing device
	火坑村皮纸纤维形态图 (10×)	Fibers of bast paper in Huokeng Village (10× objective)
	火坑村皮纸纤维形态图 (20×)	Fibers of bast paper in Huokeng Village (20× objective)
	郎岱镇皮纸纤维形态图 (10×)	Fibers of bast paper in Langdai Town (10× objective)
	郎岱镇皮纸纤维形态图 (20×)	Fibers of bast paper in Langdai Town (20× objective)
	皮纸抄本	Transcript on bast paper
	壁纸	Paintings on bast paper
	造纸户堂屋供奉的牌位	Memorial tablets in a papermaker's house
	观音阁	Guanyin (Goddess of Mercy) Pavilion
	艰难生存的小造纸槽坊	A small papermaking mill that barely survived
	调查组成员与中寨造纸村民在纸甑旁	Researchers and papermakers standing by a papermaking kiln in Zhongzhai Town
	郎岱老街	An old street in Langdai Town
	郎岱镇皮纸透光摄影图	A photo of bast paper in Langdai Town seen through the light
	火坑村黄金纸透光摄影图	A photo of golden paper in Huokeng Village seen through the light
	火坑村白绵纸透光摄影图	A photo of Baimian paper in Huokeng Village seen through the light

章节	图中文名称	图英文名称
第三章	黔西南布依族苗族自治州	Chapter III Qianxinan Bouyei and Miao Autonomous Prefecture
第一节	普安皮纸	Section 1 Bast Paper in Pu'an County
	乌蒙山风光	Landscape of Wumeng Mountain
	野生古茶树王	Ancient wild tea tree
	出土的古茶籽化石	Unearthed ancient tea seed fossil
	进入白沙乡的乡间公路	Country road leading to Baisha Town
	龙溪石砚	Longxi inkstone
	通往白沙乡的山道	Mountain road leading to Baisha Town
	普安县白沙乡自然风光	Landscape of Baisha Town in Pu'an County
	砍楮枝示意	Showing how to lop paper mulberry branches

章节	图中文名称	图英文名称
	剥楮皮	Stripping paper mulberry bark
	干楮皮	Dried paper mulberry bark
	摆二道皮示意	Showing how to double clean the bark
	打　皮	Hammering the bark with a foot pestle
	洗　料	Cleaning the papermaking materials
	挤　水	Squeezing water out of papermaking materials
	堵住出水口	Using cloth to block the water outlet
	抄　纸	Scooping and lifting the papermaking screen out of water and turning it upside down on the board
	揭　纸	Peeling the paper down
	敲打纸	Flattening the paper with a wooden ruler
	纸帘及纸帘架	Papermaking screen and its supporting frame
	料槽及滑缸	Papermaking trough and the vat for holding the papermaking mucilage
	皮　甑	Kiln for steaming the papermaking materials
	河沟头村皮纸纤维形态图（10×）	Fibers of bast paper in Hegoutou Village (10× objective)
	河沟头村皮纸纤维形态图（20×）	Fibers of bast paper in Hegoutou Village (20× objective)
	保存在竹篓里的卡塘白绵纸	Baimian paper in Katang Village stored in a bamboo basket
	用卡塘白绵纸抄写的《崔氏记录》	*Records of the Cuis* written on Baimian paper in Katang Village
	用卡塘白绵纸创作的书法作品	Calligraphy written on Baimian paper in Katang Village
	蔡伦先师牌位	Memorial tablets in memory of Cai Lun, the originator of papermaking
	经书印版	Scripture printing plates
	装木榨	Fixing the wooden presser
	河沟头村白绵纸透光摄影图	A photo of Baimian paper in Hegoutou Village seen through the light
	河沟头村皮纸透光摄影图	A photo of bast paper in Hegoutou Village seen through the light

章节	图中文名称	图英文名称
第　二　节	安龙布依族竹纸	Section 2 Bamboo Paper by the Bouyei Ethnic Group in Anlong County
	坝盘村边的南盘江风光	Scenery of Nanpan River alongside the Bapan Village
	笃山溶洞	Dushan Karst Cave
	天生桥水电站坝区	Dam area of Tianshengqiao Hydropower Station
	十八先生墓	Tomb of the Eighteen Loyalists (Ming Dynasty)
	兴义府试院	Imperial Examination Academy of Xingyi Fu
	招堤荷花	Lotus pond alongside Zhaodi Causeway
	穿民族服饰的布依农妇	Bouyei women wearing ethnic clothing
	香车河制香场景	Producing incense in Xiangchehe Village
	“八音坐唱”表演	Eight Musical Instruments performance
	万峰湖风光	Scenery of Wanfeng Lake
	坝盘村的民居	Local residences in Bapan Village
	在纸坊内小憩的王济周	Papermaker Wang Jizhou taking a rest in the papermaking mill
	纸药槽中浸泡的仙人掌	Soaking cactus in a papermaking mucilage trough
	坝盘村边的小竹林	Small bamboo forest alongside Bapan Village
	竹料浸泡池	Pool for soaking the bamboo materials
	浸泡发酵完竹料的浸泡池	Pool for soaking the fermented bamboo materials
	王济周家打浆用的脚碓	Wang Jizhou's foot pestle for hammering the papermaking materials
	正在打槽的王济周	Papermaker Wang Jizhou stirring the papermaking materials
	添加纸药仙人掌汁水	Adding in cactus extract as papermaking mucilage
	舀纸的四个环节	Four procedures of papermaking using a movable papermaking screen
	王济周在榨纸	Wang Jizhou pressing the paper
	屋内晾纸	Drying the paper inside a house
	捆纸成挑	Binding the paper
	王济周使用的纸帘	Papermaking screen used by Wang Jizhou
	王济周家的纸槽	Wang Jizhou's papermaking trough
	纸　榨	Presing device
	坝盘村竹纸纤维形态图（10×）	Fibers of bamboo paper in Bapan Village (10× objective)
	坝盘村竹纸纤维形态图（20×）	Fibers of bamboo paper in Bapan Village (20× objective)
	调查组成员在王济周家	Researchers in Wang Jizhou's house
	一挑竹纸	A bundle of bamboo paper
	坝盘村中的神树	Divine tree in Bapan Village
	坝盘村布依族木屋	Bouyei ethnic cabin in Bapan Village
	调查组成员在纸坊调研	Researchers visiting a papermaking mill
	坝盘村竹纸透光摄影图	A photo of bamboo paper in Bapan Village seen through the light

章节	图中文名称	图英文名称
第　三　节	贞丰皮纸	Section 3 Bast Paper in Zhenfeng County
	洛帆河风光	Scenery of Luofan River
	双乳奇峰	Breasts-shaped mountains
	布依浪哨	Bouyei ethnic courtship ritual
	龙井村寨风景图	View of Longjing Village
	造纸作坊群	Papermaking mills
	龙井村的《龙氏家谱》	*Genealogy of the Longs* in Longjing Village
	龙氏来黔始祖石碑	Stone monument in memory of the ancestors of the Longs immigrants to Guizhou Province
	泡仙人掌	Soaking the cactus
	剥　皮	Stripping the bark
	构　皮	Paper mulberry bark
	泡　料	Soaking the papermaking materials
	捆　料	Binding the papermaking materials
	浆　料	Fermenting the papermaking materials
	蒸　料	Steaming the papermaking materials
	蒸汽锅炉	Steaming boiler
	出　甑	Picking the materials out of papermaking kiln
	揉　料	Rubbing the papermaking materials
	用脚踩料	Stamping the papermaking materials
	洗　料	Cleaning the papermaking materials
	加漂精	Adding in bleach
	漂　料	Soaking the bark for bleaching
	拣　料	Picking out the impurities
	打　浆	Beating the papermaking materials with a beating machine
	淘　料	Sieving the papermaking materials
	打　槽	Stirring the papermaking materials
	添　滑	Adding in cactus extract as papermaking mucilage
	抄　纸	Scooping and lifting the papermaking screen out of water and turning it upside down on the board
	小算盘	Paper counting apparatus
	榨　纸	Pressing the paper
	刮　纸	Trimming the paper
	揭　纸	Peeling the paper down
	理　纸	Sorting the paper
	包　装	Packing the paper
	传统木甑、石甑群	Traditional wooden and stone kilns
	木　甑	Wooden kiln
	打碓示意	Showing how to beat the papermaking materials
	纸槽及一帘一纸的纸帘	Papermaking trough and papermaking screen that can make one piece of paper
	一帘二纸的纸帘	Papermaking screen that can make two pieces of paper simultaneously

章节	图中文名称	图英文名称
	纸焙	Drying wall
	龙井村白绵纸纤维形态图 (10×)	Fibers of Baimian paper in Longjing Village (10× objective)
	龙井村白绵纸纤维形态图 (20×)	Fibers of Baimian paper in Longjing Village (20× objective)
	龙井村灯笼纸纤维形态图 (10×)	Fibers of Denglong (lantern) paper in Longjing Village (10× objective)
	龙井村灯笼纸纤维形态图 (20×)	Fibers of Denglong (lantern) paper in Longjing Village (20× objective)
	书法创作现场	Calligraphy performance
	用龙井白绵纸画的古画	Ancient painting on Baimian paper in Longjing Village
	用龙井白绵纸书写的民国契约	Contract during the Republican Era of China written on Baimian paper in Longjing Village
	经书	Taoist classics
	1945年的手抄经书	Taoist classics transcript written in 1945
	印经文	Printing Taoist classics
	印神马	Printing divine horse
	神马	Divine horse
	20世纪50年代的小学生作业本	Pupil's exercise book used in the 1950s
	纸绳	Paper robe
	打青	Making grave marker
	青	Grave marker
	花纸	Flower paper
	祭蔡伦	Worshipping Cai Lun, the originator of papermaking
	国家级"非遗"传承人刘仕阳（证书上写成"刘世阳"）	Liu Shiyang, an inheritor of National Intangible Cultural Heritage, and his Certificate of Honor
	龙井村的"非遗"标志	Logo of Intangible Cultural Heritage in Longjing Village
	清代用龙井白绵纸抄写的经书	Taoist classics written on Baimian paper in Longjing Village (Qing Dynasty)
	龙井村灯笼纸透光摄影图	A photo of lantern paper in Longjing Village seen through the light
	龙井村白绵纸透光摄影图	A photo of Baimian paper in Longjing Village seen through the light
	龙井村黄筋纸透光摄影图	A photo of Huangjin paper in Longjing Village seen through the light
	龙井村黑夹纸透光摄影图	A photo of Heijia paper in Longjing Village seen through the light
	龙井村皮纸透光摄影图	A photo of bast paper in Longjing Village seen through the light

章节	图中文名称	图英文名称
第四节	贞丰布依族竹纸	Section 4 Bamboo Paper by the Bouyei Ethnic Group in Zhenfeng County
	坡柳村口的土窑	An adobe kiln at the entrance of Poliu Village
	龙场镇小景	Landscape of Longchang Town
	龙场古镇夜景	Night scene of Longchang Town
	造纸人黄敬德一家	Papermaker Huang Jingde's family
	《黄氏家谱》	*Genealogy of the Huangs*
	牌位	Memorial tablets
	纸帘及纸帘架	Papermaking screen and its supporting frame
	牛皮榨索	Cowhide rope
	棕刷	Coir brush
	刷纸	Pasting the paper on the wall
	捆纸	Binding the paper
	料塘	Pond for storing the papermaking materials
	槽缸及滑缸	Papermaking trough and the vat for holding the papermaking mucilage
	坡柳村竹纸纤维形态图 (10×)	Fibers of bamboo paper in Poliu Village (10× objective)
	坡柳村竹纸纤维形态图 (20×)	Fibers of bamboo paper in Poliu Village (20× objective)
	龙场镇的民居	Residence in Longchang Town
	坡柳村竹纸透光摄影图	A photo of bamboo paper in Poliu Village seen through the light

章节	图中文名称	图英文名称
第四章	安顺市	Chapter IV Anshun City
第一节	关岭布依族苗族竹纸	Section 1 Bamboo Paper by the Bouyei and Miao Ethnic Groups in Guanling County
	关岭花江大峡谷	Huajiang Canyon in Guanling County
	关索古驿道	Guansuo Ancient Courier Route
	神秘的红岩天书	Mysterious holy book inscribed on red rock
	花江铁索桥	Huajiang Chain Bridge
	紫山村自然环境	Natural environment of Zishan Village
	紫山村竹林中的竹子	Local bamboo in Zishan Village
	竹麻	Bamboo materials
	打浆把	Binding the dried bamboo
	煮料的窑	Kiln for boiling the papermaking materials
	抄纸帘子及帘上纸膜	Papermaking screen with a piece of paper newly made on it
	榨纸工具	Device for pressing the paper
	晒纸	Drying the paper
	揭纸	Peeling the paper down
	捆纸	Binding the paper
	存放的成品竹纸	Final product of bamboo paper
	槽棍	Stick for stirring paper pulp
	帘架	Frame for supporting the papermaking screen
	盖板	Wooden boards for pressing the paper
	窑口	Entrance of the papermaking kiln
	槽子	Papermaking trough
	槽坎	Wooden board by the papermaking trough
	纸帘	Papermaking screen
	紫山村竹纸纤维形态图 (10×)	Fibers of bamboo paper in Zishan Village (10× objective)
	紫山村竹纸纤维形态图 (20×)	Fibers of bamboo paper in Zishan Village (20× objective)
	现场制作纸钱过程	Procedures of making joss paper
	网上销售的关岭布依族苗族竹纸	Bamboo paper by the Bouyei and Miao Ethnic Groups in Guanling County sold online
	紫山村竹纸透光摄影图	A photo of bamboo paper in Zishan Village seen through the light
	紫山村竹纸透光摄影图	A photo of bamboo paper in Zishan Village seen through the light

章节	图中文名称	图英文名称
第二节	关岭皮纸	Section 2 Bast Paper in Guanling County
	沙营乡的一处造纸作坊群	Papermaking mills in Shaying Town
	美丽的蕹叶芸香	Beautiful herbal medicine named *C. Nitidissima*
	抽旱烟的布依族老人	An old Bouyei man smoking
	纸厂村边的造纸皮甑	Papermaking kiln by Zhichang Village
	《易氏家谱》(手抄本)	*Genealogy of the Yis* (handwritten version)
	砍树	Lopping paper mulberry trees
	剥皮	Stripping the bark
	装运构皮的竹篓	Bamboo baskets for carrying the bark
	驮运构皮的马	A horse carrying the bark
	晒皮	Drying the bark
	绑皮	Binding the bark
	皮甑	Kiln for steaming the papermaking materials
	木榨	Wooden presser
	扳干后的构皮	Pressed paper mulberry bark
	包袱 (塑料网)	Baofu (plastic net for sieving and cleaning the bark)
	滑水	Papermaking mucilage
	搅槽	Stirring the papermaking materials
	安帘子	Fixing the papermaking screen
	摇水	Stirring paper pulp
	翻扣上板	Turning the papermaking screen upside down on the board

章节	图中文名称	图英文名称
	垛　子	Wooden presser
	晒　纸	Drying the paper
	揭　纸	Peeling the paper down
	折　纸	Folding the paper
	理纸和修纸边	Sorting the paper and trimming the deckle edges
	包　装	Packing the paper
	纸厂村皮纸纤维形态图（10×）	Fibers of bast paper in Zhichang Village (10× objective)
	纸厂村皮纸纤维形态图（20×）	Fibers of bast paper in Zhichang Village (20× objective)
	纸　钱	Joss paper
	皮　纸	Bast paper
	关岭皮纸手抄本	Manuscript on bast paper in Guanling County
	计数器	Paper counting apparatus
	坟头挂青的钱串	Grave marker
	纸厂村皮纸透光摄影图	A photo of bast paper in Zhichang Village seen through the light

章节	图中文名称	图英文名称
第　三　节	镇宁皮纸	Section 3 Bast Paper in Zhenning County
	江龙镇俯瞰图	Aerial view of Jianglong Town
	茅草坡山	Maocaopo (thatch) Mountain
	黄果树瀑布	Huangguoshu Waterfall
	竹新村风光	Landscape of Zhuxin Village
	毛氏家谱	Genealogy of the Maos
	蔡应芳（左）与毛万忠（右）于旧永宁州和安顺府交界处	Cai Yingfang (left) and Mao Wanzhong (right) at the border of Yongning Prefecture and Anshun Fu
	舀纸示意	Showing how to scoop and lift the papermaking screen out of water
	压榨示意	Showing how to press the paper
	药　槽	Trough for holding the papermaking mucilage
	碓	Stone pestle
	碓　杈	V-shaped stone pestle
	纸　帘	Papermaking screen
	焙　架	Supporting frame for drying the paper
	帘　架	Frame for supporting the papermaking screen
	窑上村制作的油纸伞	Oil-paper umbrella in Yaoshang Village
	同治年间的地契	Land certificate during Tongzhi Reign of the Qing Dynasty
	民国时期的地契	Land certificate during the Republican Era of China
	1950年的地契	Land certificate in 1950
	经　文	Scripture of Taoist classics
	对　联	Chinese couplets
	祭蔡伦与猪拱先师	Worshipping Cai Lun, the originator of papermaking, and Mr. Pig, who inspired Cai Lun in the papermaking procedures
	镇宁波波糖	Bobo Candy in Zhenning County
	镇宁县内的“竹王后裔”	Descendants of “Bamboo King” in Zhenning County
	徐霞客黄果树探险纪念碑	Monument in memory of Xu Xiake at Huangguoshu Waterfall

章节	图中文名称	图英文名称
第　四　节	紫云手工纸	Section 4 Handmade Paper in Ziyun County
	板当翠河风光	Scenery of Cuihe River in Bandang Town
	板当夹山风光	Scenery of Jiashan Mountain in Bandang Town
	板当纸厂废弃厂房	Abandoned Bandang Papermaking Factory
	紫云县板当纸厂公章	Official seal of Bandang Papermaking Factory in Ziyun County
	废弃的旧纸槽	Abandoned papermaking trough
	废弃的窑子	Abandoned papermaking kiln
	废弃的水碾	Abandoned hydraulic grinder
	纸　帘	Papermaking screen
	用板当白纸抄的经书	Scripture written on white paper in Bandang Town
	做法事时烧纸	Burning handmade paper on religious ceremonies

章节	图中文名称	图英文名称
第　五　章	黔南布依族苗族自治州	Chapter V Qiannan Bouyei and Miao Autonomous Prefecture
第　一　节	长顺皮纸	Section 1 Bast Paper in Changshun County
	白云山	Baiyun Mountain
	白云山寺院大殿	Great hall of Baiyun Mountain Temple
	翁贵古亭	Ancient pavilion in Wenggui Village
	泡料池	Soaking pool
	甑锅灶口	Entrance of the steaming wok
	打　槽	Stirring the papermaking materials
	抄　纸	Scooping and lifting the papermaking screen out of water and turning it upside down on the board
	压　榨	Pressing the paper
	外墙晒纸	Pasting the paper on the wall
	晒　纸	Drying the paper
	揭　纸	Peeling the paper down
	成品纸	Final product of paper
	翁贵村双层皮纸纤维形态图（10×）	Fibers of double-layered bast paper in Wenggui Village (10× objective)
	翁贵村双层皮纸纤维形态图（20×）	Fibers of double-layered bast paper in Wenggui Village (20× objective)
	翁贵纸厂旧址	Former site of Wenggui Papermaking Factory
	“永垂定例”碑	Monument of “Permanent Stipulation”
	现存生产厂房	Papermaking factory still in use
	广顺古镇	Guangshun Ancient Town
	翁贵村皮纸透光摄影图	A photo of bast paper in Wenggui Village seen through the light

章节	图中文名称	图英文名称
第　二　节	惠水竹纸	Section 2 Bamboo Paper in Huishui County
	芦山镇	Lushan Town
	竹　料	Bamboo materials
	舂竹麻	Beating the bamboo materials
	窑　口	Entrance of the papermaking kiln
	窑上部	Top of the papermaking kiln
	碾竹麻	Grinding the bamboo materials
	水碾的外部	Outside of the hydraulic grinder
	打　槽	Stirring the papermaking materials
	抄　纸	Scooping and lifting the papermaking screen out of water and turning it upside down on the board
	晾　干	Drying the paper
	切　纸	Trimming the paper
	打　孔	Punching holes in the paper
	芦山镇竹纸纤维形态图（10×）	Fibers of bamboo paper in Lushan Town (10× objective)
	芦山镇竹纸纤维形态图（20×）	Fibers of bamboo paper in Lushan Town (20× objective)
	芦山镇上出售惠水竹纸的小商店	Bamboo paper in Huishui County sold in a small store in Lushan Town
	芦山镇竹纸透光摄影图	A photo of bamboo paper in Lushan Town seen through the light

章节	图中文名称	图英文名称
第　三　节	龙里竹纸	Section 3 Bamboo Paper in Longli County
	龙里山地风光	Landscape of mountains in Longli County
	钓鱼竹原料	*Dendrocalamus tsiangii* (McClure) Chia et H. L. Fung, as the raw material
	煮好的竹麻	Boiled bamboo materials
	碾竹麻	Grinding the bamboo materials
	抄　纸	Scooping and lifting the papermaking screen out of water and turning it upside down on the board
	压纸垛	Pressing a pile of paper
	压　榨	Pressing the paper

章节	图中文名称	图英文名称
	晾 纸	Drying the paper
	成品纸	Final product of paper
	纸厂村竹纸纤维形态图 (10×)	Fibers of bamboo paper in Zhichang Village (10× objective)
	纸厂村竹纸纤维形态图 (20×)	Fibers of bamboo paper in Zhichang Village (20× objective)
	蔡伦庙遗址	Cai Lun Temple relics
	调查组成员和造纸户合影	Researchers and papermakers
	纸厂村竹纸透光摄影图	A photo of bamboo paper in Zhichang Village seen through the light

章节	图中文名称	图英文名称
第 四 节	荔波布依族竹纸	Section 4 Bamboo Paper by the Bouyei Ethnic Group in Libo County
	尧古村中立着的茂兰景区简介牌	Introduction to Maolan Scenic Spot in Yaogu Village
	尧古村远眺图	Overlook of Yaogu Village
	村里的造纸作坊	Papermaking mill in the village
	身着布依族服饰的少女	A maid in Bouyei ethnic clothing
	清代尧古纸遗品	Yaogu paper of the Qing Dynasty
	"古纸坊"宣传牌	Billboard of Ancient Papermaking Mill
	荔波尧古村手工纸成品	Final product of handmade paper in Yaogu Village of Libo County
	覃自凡造纸老人	Qin Zifan, an old papermaker
	晒干的糯叶	Dried leaf, a local raw material of papermaking mucilage
	泡 竹	Soaking the bamboo in limewater
	取 料	Picking out the papermaking materials with a wooden rake
	砍 料	Cutting the materials into fixed length
	碾 料	Grinding the materials
	放 料	Transferring the materials to the papermaking trough
	煮糯叶粉	Boiling the papermaking mucilage
	放 药	Adding in papermaking mucilage
	打 槽	Stirring the papermaking materials
	抄 纸	Scooping and lifting the papermaking screen out of water and turning it upside down on the board
	压 纸	Pressing the paper
	揭 纸	Peeling the paper down
	晒 纸	Drying the paper
	打纸钱	Making joss paper
	纸 槽	Papermaking trough
	滑 槽	Trough for holding the papermaking mucilage
	纸 帘	Papermaking screen
	帘 架	Frame for supporting the papermaking screen
	钉纸架	Shelf for making joss paper
	打纸刀	Nail for making joss paper
	打纸锤	Hammer for making joss paper
	尧古村竹纸纤维形态图 (10×)	Fibers of bamboo paper in Yaogu Village (10× objective)
	尧古村竹纸纤维形态图 (20×)	Fibers of bamboo paper in Yaogu Village (20× objective)
	纸 钱	Joss paper
	供神灵、祖先	Sacrificial offerings to the gods and ancestors
	尧古傩戏	Nuo Opera in Yaogu Village
	神龛台	Niche table
	阁楼上的覃自凡、覃万恒父子	Qin Zifan and his son, Qin Wanheng sitting in the attic
	茂兰景区	Maolan Scenic Spot
	尧古村竹纸透光摄影图	A photo of bamboo paper in Yaogu Village seen through the light

章节	图中文名称	图英文名称
第 五 节	都匀皮纸	Section 5 Bast Paper in Duyun City
	用都匀皮纸创作的绘画作品	Painting on bast paper in Duyun City
	正在寻找旧厂址的调查组成员	Researchers looking for the former papermaking factory
	斗篷山风景区小景	Scenery of Doupeng Mountain Scenic Spot
	石板街入口	Entrance to Shiban Street
	水家布制作的服饰	Shui ethnic clothing
	调查组成员在调查现场进行调查	Researchers investigating the field
	造纸老人们：左、右分别为章氏、简氏后人，中为肖明远	Old papermakers, Mr. Zhang (left), Mr. Jian (right) and Xiao Mingyuan
	肖明远向调查组成员演示工艺	Xiao Mingyuan demonstrating the papermaking procedures to a researcher
	都匀地方志上记录的都匀皮纸洗料工艺	Procedures of cleaning the papermaking materials as recorded in the local annals
	拣 料	Picking out the impurities
	打 碓	Beating the papermaking materials
	都匀国画纸纤维形态图 (10×)	Fibers of Chinese painting paper in Duyun City (10× objective)
	都匀国画纸纤维形态图 (20×)	Fibers of Chinese painting paper in Duyun City (20× objective)
	皮纸上的"抱石皴"	Painting by Fu Baoshi on local bast paper
	原匀阳造纸厂厂房	Former Yunyang Papermaking Factory
	现遗弃在路边的皮纸生产器具	Abandoned apparatus used to make bast paper

章节	图中文名称	图英文名称
第 六 节	都匀蜡纸	Section 6 Wax Paper in Duyun City
	"文革"时期的蜡纸油印本	Book mimeographed on wax paper during the Chinese Cultural Revolution
	野梦花	Ye Meng Hua, *Daphne tangutica* Maxim var. *wilsonii*
	打字蜡纸	Wax paper for mimeograph
	都匀蜡纸厂留守处的工人与调查组成员	Workers and researchers standing in front of the Duyun Wax Papermaking Liaison Office
	都匀蜡纸厂蜡纸透光摄影图	A photo of wax paper in Duyun Wax Papermaking Factory seen through the light

章节	图中文名称	图英文名称
第 六 章	毕节市	Chapter VI Bijie City
第 一 节	纳雍皮纸	Section 1 Bast Paper in Nayong County
	流经沙包乡的启河	Qihe River flowing through Shabao Town
	纳雍"穿青人"	Chuanqing people (people who wear cyan clothes) in Nayong County
	纳雍苗族"滚山珠"演出照	Local show performed by the Miao Ethnic Group in Nayong County
	晒构皮	Drying paper mulberry bark
	蒸构皮	Steaming paper mulberry bark
	洗构皮	Cleaning paper mulberry bark
	打 碓	Beating the papermaking materials
	洗 料	Cleaning the papermaking materials
	拥 水	Squeezing water out of the papermaking materials
	滑 缸	Vat for holding the papermaking mucilage
	铲 槽	Stirring paper pulp
	舀 纸	Scooping and lifting the papermaking screen out of water and turning it upside down on the board
	榨 纸	Pressing the paper
	用棕刷将纸刷到墙上	Pasting the paper on the wall with a coir brush
	扯 纸	Peeling the paper down
	晒纸——墙晾	Drying the paper on a wall
	晒纸——挂晾	Drying the paper on a string
	两人合作撕纸	Two people working together to peel the paper down
	用海绵蘸水浸湿纸的边缘	Moisturizing the paper edge with a sponge (to avoid damage to the paper when peeling it down)
	理 纸	Sorting the paper
	槽 子	Papermaking trough
	帘 子	Papermaking screen

	窑 子	Kiln for steaming the papermaking materials
	石墩打	Beating the papermaking materials with a stone mallet
	打浆机	Beating machine
	大寨村皮纸纤维形态图（10×）	Fibers of bast paper in Dazhai Village (10× objective)
	大寨村皮纸纤维形态图（20×）	Fibers of bast paper in Dazhai Village (20× objective)
	旧皮纸家谱	Old genealogy written on bast paper
	祭蔡伦	Worshipping Cai Lun, the originator of papermaking
	大寨村皮纸透光摄影图	A photo of bast paper in Dazhai Village seen through the light

章节	图中文名称	图英文名称
第 二 节	金沙竹纸	Section 2 Bamboo Paper in Jinsha County
	红土坡村风光	Landscape of Hongtupo Village
	金沙县城新修建的黄河大道	Yellow River Road newly built in Jinsha County
	冷水河风景区	Lengshui (cold water) River Scenic Spot
	敖家古墓石刻	Tombstones of the Ao's Family Graveyard, now protected as cultural relics
	茶园万寿宫戏楼	Theatre Building named Longevity Palace in Chayuan Town
	造纸人陈昌财	Chen Changcai, a local papermaker
	红土坡村外景环境	Landscape of Hongtupo Village
	半荒废的水碾现场照片	Nearly abandoned hydraulic grinder
	陈昌财的老母亲	Chen Changcai's mother
	已泡上竹料的泡料池	Soaking pool with bamboo materials in it
	从山上引流下来的山泉水pH测试结果	pH testing result of the local spring
	已砍的造纸原料慈竹段	Chopped bamboo (*Neosinocalamus affinis*) as the raw material of papermaking
	泡满竹料的泡料池	Soaking pool with bamboo materials
	再泡料	Soaking the materials for the second time
	水 碾	Hydraulic grinder
	用槽棍搅拌纸药液	Using a stirring stick to mix paper pulp and the papermaking mucilage
	抄 纸	Scooping and lifting the papermaking screen out of water and turning it upside down on the board
	榨 水	Procedures of pressing the paper
	屋内晾晒的竹纸	Drying the bamboo paper in the room
	展开的长纸纸样	An unfolded sheet of paper sample
	纸帘及纸帘架	Papermaking screen and its supporting frame
	竹纸帘	Bamboo papermaking screen
	纸帘架	Frame for supporting the papermaking screen
	纸药槽	Trough for holding the papermaking mucilage
	抄纸槽	Papermaking trough
	纸 榨	Pressing device
	红土坡村竹纸纤维形态图（10×）	Fibers of bamboo paper in Hongtupo Village (10× objective)
	红土坡村竹纸纤维形态图（20×）	Fibers of bamboo paper in Hongtupo Village (20× objective)
	已裁剪好并打完钱孔的竹纸	Trimmed bamboo paper with holes on it
	调查组成员与陈昌财等造纸人交流	Researchers interviewing Chen Changcai and other papermakers
	陈昌财家已浸泡数年的竹料	Bamboo materials soaked for years in Chen Changcai's house
	陈昌财家在小坡上的纸坊	Chen Changcai's papermaking mill on a hill
	红土坡村河边的水碾房	Hydraulic grinding mill by a river in Hongtupo Village
	红土坡村竹纸透光摄影图	A photo of bamboo paper in Hongtupo Village seen through the light
	红土坡村竹纸透光摄影图	A photo of bamboo paper in Hongtupo Village seen through the light

章节	图中文名称	图英文名称
第 三 节	金沙皮纸	Section 3 Bast Paper in Jinsha County
	果松村山中丰富的构树资源	Abundant paper mulberry trees in Guosong Village
	果松村的山溪水	Mountain stream flowing through Guosong Village
	2014年春天调查组成员在王丕全家做访谈	Researchers interviewing papermaker Wang Piquan in the spring of 2014
	兑好的纸药液（左槽）	Processed papermaking mucilage (left trough)
	晒干并捆扎好备用的构树皮	Dried and bounded paper mulberry bark
	王丕全家的打浆机	Beating machine in Wang Piquan's house
	舀 纸	Scooping and lifting the papermaking screen out of water and turning it upside down on the board
	晒 纸	Drying the paper
	撕 纸	Peeling the paper down
	蒸料窑锅	Kiln for steaming the papermaking materials
	泡皮槽	Trough for soaking the bark
	抄池、帘架和抄台	Papermaking trough, frame for supporting the papermaking screen, and the board for piling wet paper
	雨下寨村皮纸纤维形态图（10×）	Fibers of bast paper in Yuxiazhai Village (10× objective)
	雨下寨村皮纸纤维形态图（20×）	Fibers of bast paper in Yuxiazhai Village (20× objective)
	用手工皮纸所抄家谱（封面）	Family genealogy transcribed on handmade bast paper (cover)
	用手工皮纸所抄家谱（内页）	Family genealogy transcribed on handmade bast paper (inside page)
	雨下寨村皮纸透光摄影图	A photo of bast paper in Yuxiazhai Village seen through the light
	雨下寨村皮纸透光摄影图	A photo of bast paper in Yuxiazhai Village seen through the light
	联盟村皮纸透光摄影图	A photo of bast paper in Lianmeng Village seen through the light
	河透底村皮纸透光摄影图	A photo of bast paper in Hetoudi Village seen through the light

章节	图中文名称	图英文名称
第 七 章	贵阳市	Chapter VII Guiyang City
第 一 节	乌当竹纸	Section 1 Bamboo Paper in Wudang District
	香纸沟古法造纸博物馆	Traditional Papermaking Museum in Xiangzhigou Scenic Spot
	洪水后废弃的造纸作坊	Abandoned papermaking mill after flood
	白水河村造纸现场	Papermaking site in Baishuihe Village
	白水河村小景	View of Baishuihe Village
	香纸沟小景	View of Xiangzhigou Scenic Spot
	调查组成员采访汪长伦	A researcher interviewing the local papermaker Wang Changlun
	"越国汪公彭氏宗祖之位"牌位	Memorial tablets of Wang Hua, lord of Yue State, and the ancestor of the Pengs in ancient China
	香纸沟的抄纸展演	Papermaking show in Xiangzhigou Scenic Spot
	香纸沟土法造纸工艺流程展板	Flowchart of traditional papermaking techniques in Xiangzhigou Scenic Spot
	砍 竹	Lopping the bamboo
	修竹叶	Cutting off the bamboo leaves
	破 竹	Beating the bamboo with a wooden mallet
	晾 竹	Drying the bamboo
	打石灰水	Soaking the bamboo in limewater
	碾竹麻	Grinding the bamboo materials with a grinder
	拉开水板	Removing the wooden water barrier
	搅 料	Stirring the papermaking materials
	加清水	Adding in water
	滑 根	Papermaking mucilage as adhesive
	搅纸浆	Stirring the paper pulp
	抄 纸	Scooping and lifting the papermaking screen out of water and turning it upside down on the board
	石头压榨	Pressing the paper with stones
	石头上的红记	Red mark on stone
	打红用的竹笔	Bamboo stick for marking the paper
	晾 纸	Drying the paper
	理 纸	Sorting the paper

章节	图中文名称	图英文名称
	裁　纸	Trimming the paper edge
	造纸人用比子测量香纸的大小	A papermaker using a ruler to measure the size of Xiang paper
	晾纸耙	Rake for drying the paper
	纸　帘	Papermaking screen
	陇脚村竹纸纤维形态图（10×）	Fibers of bamboo paper in Longjiao Village (10× objective)
	陇脚村竹纸纤维形态图（20×）	Fibers of bamboo paper in Longjiao Village (20× objective)
	打钱眼	Perforating the paper
	“越国汪公”及蔡伦的牌位	Memorial tablets of Wang Hua, lord of Yue State, and Cai Lun the originator of papermaking
	香纸沟的纸坊	Papermaking mill in Xiangzhigou Scenic Spot
	陇脚村竹纸透光摄影图	A photo of bamboo paper in Longjiao Village seen through the light

章节	图中文名称	图英文名称
第八章	遵义市	Chapter VIII Zunyi City
第一节	仁怀手工纸	Section 1 Handmade Paper in Renhuai City
	五马河边的造纸村落——三元村	Papermaking sites in Sanyuan Village alongside Wuma River
	仁怀市云仙洞遗址出土的陶缸、陶圆底壶、陶杯、陶杯盖及石器	Earthenwares and stonewares unearthed from Yunxian Cave in Renhuai City
	五马古街	Ancient Street in Wuma Town
	五马乡村文化节	Rural Cultural Festival in Wuma Town
	堂屋正面陈设	Furnishings of the living room
	双魁田四合院内院落	Shuangkuitian Courtyard
	五马河	Wuma River
	仁怀市五马镇手工造纸协会公章	Official seal of Handmade Papermaking Association in Wuma Town of Renhuai City
	手工造纸协会会员证	Membership card of Handmade Papermaking Association
	三元村支书葛光远指认朱怀顺墓地遗址	Tomb site of Zhu Huaishun, a brilliant local papermaker during the Qing Dynasty (pointed out by the local secretary Ge Guangyuan of Sanyuan Village)
	当地野生构树	Wild paper mulberry tree
	已砍下的构树枝	Lopped paper mulberry branches
	待蒸煮的皮料	Papermaking materials ready for steaming
	桑　皮	Paper mulberry bark
	打浆机加工的皮纸料	Papermaking materials processed by electronic beater
	搅料子	Stirring the paper pulp
	舀　纸	Scooping and lifting the papermaking screen out of water and turning it upside down on the board
	吊垛子	Pressing the paper with stones
	退垛子	Removing the stones
	理垛子	Sorting the paper
	刷米浆	Pasting rice pulp on the paper as adhesive
	揭纸角	Peeling the paper down from the upper corner
	揭　纸	Peeling the paper down
	刷纸上墙	Pasting the paper on a wall for drying
	撕　纸	Peeling the paper down
	抖　纸	Shaking the paper
	理　平	Flattening the paper
	对　折	Folding the paper
	捆　纸	Binding the paper
	运　输	Transporting the paper
	纸帘及纸帘架	Papermaking screen and its supporting frame
	泡料池	Pool for soaking the papermaking materials
	废弃的蒸灶	Abandoned kiln for steaming the papermaking materials
	火　焙	Drying wall
	三元村皮纸纤维形态图（10×）	Fibers of bast paper in Sanyuan Village (10× objective)
	三元村皮纸纤维形态图（20×）	Fibers of bast paper in Sanyuan Village (20× objective)
	荨　麻	Nettle as adhesive
	造纸工杨存志老人讲述竹纸制作工艺	Yang Cunzhi, an old papermaker introducing the papermaking procedures
	废弃的泡料塘	Abandoned soaking pool
	掩映在草丛中的平面碾	Grinder covered by grasses
	废弃的纸槽	Abandoned papermaking trough
	废弃的捞纸棚	Abandoned papermaking shed
	纸　榨	Pressing device
	裁　刀	Sickle for trimming the paper
	调查组成员与杨存志合影	A researcher and Yang Cunzhi
	鲁班镇竹纸纤维形态图（10×）	Fibers of bamboo paper in Luban Town (10× objective)
	鲁班镇竹纸纤维形态图（20×）	Fibers of bamboo paper in Luban Town (20× objective)
	书写在仁怀皮纸上的《见病知方汤头歌》	Medical book written on bast paper in Renhuai City
	在仁怀皮纸上进行绘画创作	Painting on bast paper in Renhuai City
	用仁怀皮纸抄写的账簿	Account book on bast paper in Renhuai City
	用仁怀皮纸包装的地方白酒	Local liquor bottle wrapped by bast paper in Renhuai City
	1972年用仁怀皮纸包装的茅台酒	Maotai bottle wrapped by bast paper in Renhuai City in 1972
	仁怀皮纸检验报告	Test report of bast paper in Renhuai City
	第2届中国国际茶叶及茶文化（深圳）博览会参展申请表	Application form of the Second International Tea Culture Exposition (Shenzhen)
	蔡伦先师牌位	Memorial tablet in memory of Cai Lun, the originator of papermaking
	感谢支持传统造纸工艺标语	Slogan thanking the government's support for traditional papermaking technique
	呼吁自动拆除造纸作坊的标语	Slogan advocating removal of papermaking mills
	五马镇取缔岩头小造纸指挥部	Head office for abolishing papermaking mills in Wuma Town
	拆除后的造纸作坊	A demolished papermaking site
	三元村皮纸（加厚）透光摄影图	A photo of bast paper (extra thick) in Sanyuan Village seen through the light
	三元村皮纸（带丝）透光摄影图	A photo of bast paper (with threads) in Sanyuan Village seen through the light
	三元村皮纸透光摄影图	A photo of bast paper in Sanyuan Village seen through the light
	鲁班镇竹纸透光摄影图	A photo of bamboo paper in Luban Town seen through the light

章节	图中文名称	图英文名称
第二节	正安手工纸	Section 2 Handmade Paper in Zheng'an County
	清代正安竹纸制成的土坪镇郑氏家谱	Genealogy of the Zhengs in Tuping Town written on bamboo paper in Zheng'an County during the Qing Dynasty
	尹珍像	Statue of Yin Zhen
	瑞溪镇三把车村李红全老人指认竹原料	Choosing the papermaking materials by the local papermaker Li Hongquan in Sanbache Village of Ruixi Town
	造纸作坊的水源	Water source for papermaking
	和溪镇杉木坪村蒋礼昌老人在舀纸	Jiang Lichang, the local papermaker in Shanmuping Village of Hexi Town, lifting the papermaking screen out of water
	目前使用的纸帘示意图	Schematic diagram of the papermaking screen currently in use
	凤仪镇梨坝行政村大田坝村民组屋内晾纸	Drying the paper in a room in Datianba Villagers' Group of Liba Administrative Village in Fengyi Town
	和溪镇杉木坪村的舀纸棚	Papermaking sheds in Shanmuping Village of Hexi Town
	三把车行政村堡上村民组的舀纸池	Papermaking trough in Baoshang Villagers' Group of Sanbache Administrative Village
	梨坝村现在使用的纸帘及纸帘架	Papermaking screen and its supporting frame currently in use in Liba Village
	梨坝村过去使用的纸帘及其示意图	Papermaking screen and its schematic diagram formerly used in Liba Village

章节	图中文名称	图英文名称
	梨坝村使用的捅耙	Rake for stirring paper pulp used in Liba Village
	梨坝村的纸筋刷	Sticks for picking out the impurities in Liba Village
	三把车行政村堡上村民组李红全老人介绍池子的使用方法	Li Hongquan, the local papermaker in Baoshang Villagers' Group of Sanbache Administrative Village, introducing the use of soaking pool
	碾竹材用的碾盘	Grinder for grinding bamboo materials
	梨坝村竹纸纤维形态图（10×）	Fibers of bamboo paper in Liba Village (10× objective)
	梨坝村竹纸纤维形态图（20×）	Fibers of bamboo paper in Liba Village (20× objective)
	纸焙示意图	Schematic diagram of the drying wall
	梨坝村道士用皮纸书写的《永言孝思》	*Book in Memory of the Deceased* written by Taoist priest on bast paper in Liba Village
	用正安皮纸绘制的神像	Buddha paintings on bast paper in Zheng'an County
	用正安手工纸印制的道家经书《金光神咒》	Taoist classics named *Golden Light Mantra* on handmade paper in Zheng'an County
	祭祀牌位前的火纸	Huo paper in front of memorial tablets
	灵屋和花圈	Paper house and wreath used for sacrificial ritual
	糍粑灯笼（周信绘画，1999年版《正安县志》）	Lantern made of bast paper in Zheng'an County (drawn by Zhou Xin, included in *The Annals of Zheng'an County*, 1999)
	正安班竹乡的旧字库塔	Ancient Paper-Burning Tower in Banzhu Town of Zheng'an County
	创修字库碑	Memorial monument of the ancient Paper-Burning Tower
	和溪镇杉木坪村的泡料池群	Soaking pools in Shanmuping Village of Hexi Town
	梨坝村竹纸透光摄影图	A photo of bamboo paper in Liba Village seen through the light

章节	图中文名称	图英文名称
第三节	务川仡佬族皮纸	Section 3 Bast Paper by the Gelo Ethnic Group in Wuchuan County
	村里的造纸作坊	Local papermaking mills
	造纸水碾	Hydraulic grinder for papermaking
	造纸塘村旧藏《卢氏经单薄》	*Book of the Lus* collected in Zaozhitang Village
	砍构树	Lopping paper mulberry tree
	剥构皮	Stripping paper mulberry bark
	阴干构皮	Drying paper mulberry bark
	泡构皮	Soaking paper mulberry bark
	泡好的构皮	Soaked paper mulberry bark
	蒸毛料	Steaming the papermaking materials
	打皮板	Beating the papermaking materials
	皮板	Folded papermaking materials
	切皮板	Trimming the papermaking materials
	淘料	Cleaning the papermaking materials
	打槽	Stirring the papermaking materials
	掺滑水	Adding in papermaking mucilage
	滑石粉	Talcum powder
	胚纸	Preliminary paper
	榨纸	Pressing the paper
	撕纸	Peeling the paper down
	刷纸	Pasting the paper on the wall for drying
	揭角角	Peeling the paper down from the upper corner
	揭纸	Peeling the paper down
	理纸和捆纸	Sorting and binding the paper
	甑子	Utensil for steaming the papermaking materials
	锅炉	Wok for steaming the papermaking materials
	打碓示意	Showing how to beat the papermaking materials with a pestle
	一帘三纸纸帘	Papermaking screen that can make three pieces of paper simultaneously
	纸焙	Drying wall
	造纸塘村皮纸纤维形态图（10×）	Fibers of bast paper in Zaozhitang Village (10× objective)
	造纸塘村皮纸纤维形态图（20×）	Fibers of bast paper in Zaozhitang Village (20× objective)
	门神案图	Painting of Door Gods
	过桥图	Local traditional painting
	清代邹氏族谱	Genealogy of the Zous in the Qing Dynasty
	民国契约	Contract during the Republican Era of China
	清光绪三十二年（1906年）抄经	Scriptures during Guangxu Reign of the Qing Dynasty (1906)
	清咸丰十年（1861年）抄经	Scriptures during Xianfeng Reign of the Qing Dynasty (1861)
	周相清老人抄写的花灯书	Book of Lantern Fesitival transcribed by Zhou Xiangqing
	民国时期的派款通知	Levy during the Republican Era of China
	"青"纸	Grave marker
	民国三十五年（1946年）申氏经单簿	Genealogy of the Shens during the Republican Era of China (1946)
	笑和尚、孙猴子面具	Masks of smiling monk and Sun Wukong
	花轿	Palanquin
	祭"将军柱"示意	Offering sacrifices to the General Pillars
	清末民国时期的旧栈房	Ancient house during the late Qing Dynasty and the Republican Era of China
	"非遗"传承人卢朝辉（左四）、卢朝松（右三）全家合影	A family photo of Lu Zhaohui (fourth from the left) and Lu Zhaosong (third from the right), inheritors of intangible cultural heritage
	新建的造纸厂房	Newly built papermaking factory
	新建的水碾	Newly built hydraulic grinder
	访谈造纸传人	Researchers interviewing papermakers
	洪渡河	Hongdu River
	造纸塘村皮纸透光摄影图	A photo of bast paper in Zaozhitang Village seen through the light
	造纸塘村皮纸透光摄影图	A photo of bast paper in Zaozhitang Village seen through the light

章节	图中文名称	图英文名称
第四节	务川仡佬族竹纸	Section 4 Bamboo Paper by the Gelo Ethnic Group in Wuchuan County
	学堂坡村远景	View of Xuetangpo Village
	学堂坡村民居小景	View of residences in Xuetangpo Village
	四川引种的滑药植株黄蜀葵	Papermaking mucilage (*Abelmoschus manihot*) introduced from Sichuan Province
	已用过的滑药植物残草	Used plant for making papermaking mucilage
	村寨边的慈竹丛	*Neosinocalamus affinis* forest alongside the village
	泡料	Soaking the papermaking materials
	熟料	Fermented papermaking materials
	碾料	Grinding the papermaking materials
	碾好的料	Processed papermaking materials
	槽棍	Sticks for stirring the papermaking materials
	挤滑	Squeezing the papermaking mucilage
	舀"头帘水"	Scooping the papermaking screen for the first time
	舀"二帘水"	Scooping the papermaking screen for the second time
	待榨的纸	Paper to be pressed
	榨纸过程	Procedures of pressing the paper
	榨好的纸	Pressed paper
	晒纸	Drying the paper
	理纸	Sorting the paper
	捆好的纸	Bundles of paper
	纸槽	Papermaking trough
	一帘二纸纸帘	Papermaking screen that can make two pieces of paper simultaneously
	木榨	Wooden presser
	石碾	Stone grinder for grinding the papermaking materials
	学堂坡村竹纸纤维形态图（10×）	Fibers of bamboo paper in Xuetangpo Village (10× objective)

章节	图中文名称	图英文名称
	学堂坡村竹纸纤维形态图（20×）	Fibers of bamboo paper in Xuetangpo Village (20× objective)
	山坡上的造纸作坊	Papermaking mills on a hill
	学堂坡村竹纸透光摄影图	A photo of bamboo paper in Xuetangpo Village seen through the light

章节	图中文名称	图英文名称
第五节	余庆竹纸	Section 5 Bamboo Paper in Yuqing County
	构皮滩水电站	Goupitan Hydropower Station
	乌江村边的泡料塘	Soaking pool alongside Wujiang Village
	调查组成员在料塘边访谈	Researchers interviewing papermakers by a soaking pool
	泡竹麻	Soaking bamboo materials
	打浆机碾料	Grinding the papermaking materials with a machine
	打 槽	Stirring the papermaking materials
	沙根滑	Shagenhua, papermaking mucilage as adhesive
	抄 纸	Scooping and lifting the papermaking screen out of water
	纸帘示意图	Schematic diagram of papermaking screen
	压榨示意	Showing how to press the paper
	刮 纸	Scraping the paper
	拆 纸	Spliting the paper layers
	打纸钱	Making joss paper
	纸钱架	Frame for making joss paper
	乌江村竹纸纤维形态图（10×）	Fibers of bamboo paper in Wujiang Village (10× objective)
	乌江村竹纸纤维形态图（20×）	Fibers of bamboo paper in Wujiang Village (20× objective)
	乌江村景	View of Wujiang Village
	乌江村竹纸透光摄影图	A photo of bamboo paper in Wujiang Village seen through the light

章节	图中文名称	图英文名称
第九章	铜仁市	Chapter IX Tongren City
第一节	石阡仡佬族皮纸	Section 1 Bast Paper by the Gelo Ethnic Group in Shiqian County
	尧上文化村祭祀活动	Sacrificial ceremony in Yaoshang Cultural Village
	石阡茶灯表演	Shiqian Tea Show during the Lantern Festival
	搭在梁架上的皮料	Papermaking materials hanging on the beams
	选 料	Picking out the impurities
	舂 料	Beating the papermaking materials
	切 料	Trimming the papermaking materials
	舂 融	Beating the papermaking materials until they're melted
	洗 料	Cleaning the papermaking materials
	纸 筋	Processed papermaking material ball
	打 槽	Stirring the papermaking materials
	加滑根水	Adding in papermaking mucilage
	抄 纸	Scooping and lifting the papermaking screen out of water and turning it upside down on the board
	压 榨	Pressing the paper
	揭 纸	Peeling the paper down
	撕 纸	Peeling the paper down
	理 纸	Sorting the paper
	帘架与头子	Frame and sticks for supporting the papermaking screen
	纸 帘	Papermaking screen
	香树园村皮纸纤维形态图（10×）	Fibers of bast paper in Xiangshuyuan Village (10× objective)
	香树园村皮纸纤维形态图（20×）	Fibers of bast paper in Xiangshuyuan Village (20× objective)
	用石阡皮纸制成的旧日"礼尚往来"簿	Family gift money book with "Courtesy Demands Reciprocity" written on the cover, made of bast paper in Shiqian County
	访谈结束后的告别	Saying goodbye to local papermakers
	造纸户的院墙外	Outside of a papermaker's courtyard
	香树园村皮纸透光摄影图	A photo of bast paper in Xiangshuyuan Village seen through the light

章节	图中文名称	图英文名称
第二节	石阡仡佬族竹纸	Section 2 Bamboo Paper by the Gelo Ethnic Group in Shiqian County
	关刀土村的纸坊在榨纸	Pressing the paper in a papermaking mill in Guandaotu Village
	石阡山区冬景	Winter view of the mountain area in Shiqian County
	下林坝村的侗族民居	Dong ethnic residence in Xialinba Village
	坪地场山中的小溪	Stream in the mountain of Pingdichang Town
	龙塘坑造纸作坊的山路上十几年前曾是繁华市场	Country road to the Longtangkeng Papermaking Mill (used to be a prosperous market)
	下林坝村造纸作坊的水源	Water source of the local papermaking mills in Xialinba Village
	盛石灰水的石槽	Stone trough holding limewater
	踩料用的槽	Trough for stamping the papermaking materials
	洗料的河水	River for cleaning the papermaking materials
	依山而建的窑子	Papermaking kiln built in the mountain area
	取出料子	Lifting the papermaking materials
	在踩槽中踩料	Stamping the papermaking materials in the trough
	用拱耙打散料子	Stirring the paper pulp with a rake
	用捞筋棍搅匀	Stirring the paper pulp with a lifting stick
	捞 纸	Scooping and lifting the papermaking screen out of water
	取下捞好湿纸的纸帘	Removing the papermaking screen from its supporting frame
	把纸帘上的湿纸翻扣在竹席上	Turning the papermaking screen upside down on the bamboo matt
	揭 帘	Removing the papermaking screen
	压榨用的工具	Pressing device
	窑 子	Papermaking kiln
	窑子入口	Entrance to the papermaking kiln
	窑子顶部支起煮料大锅的地方	Place for holding the boiling wok on the top of the papermaking kiln
	踩 槽	Trough for stamping the papermaking materials
	料 池	Pool for holding the papermaking materials
	木 榨	Wooden presser
	纸 帘	Papermaking screen
	帘 架	Frame for supporting the papermaking screen
	拱 耙	Stirring rake
	捞筋棍（放在槽子沿上）	Lifting stick (put on the edge of the trough)
	关刀土村竹纸纤维形态图（10×）	Fibers of bamboo paper in Guandaotu Village (10× objective)
	关刀土村竹纸纤维形态图（20×）	Fibers of bamboo paper in Guandaotu Village (20× objective)
	石阡山里人走山路	Local residents walking on a country road
	访谈中热情的造纸老人	Hospitable old papermakers
	下林坝村的造纸农家宅居	Papermakers' residence in Xialinba Village
	关刀土村竹纸透光摄影图	A photo of bamboo paper in Guandaotu Village seen through the light

章节	图中文名称	图英文名称
第三节	江口土家族竹纸	Section 3 Bamboo Paper by the Tujia Ethnic Group in Jiangkou County
	云舍村口农家生活小景	Rural view of Yunshe Village
	怒溪乡河口村的民宅	Local residence in Hekou Village of Nuxi Town
	怒溪乡的山景	Landscape of Nuxi Town
	云舍村边的小河	River alongside Yunshe Village
	云舍村的造纸棚	Papermaking shed in Yunshe Village
	调查组成员访谈云舍村造纸人	A researcher interviewing a papermaker in Yunshe Village
	调查组成员访谈河口村造纸人	A researcher interviewing a papermaker in Hekou Village

章节	图中文名称	图英文名称
	云舍村的连片纸坊	Papermaking mills in Yunshe Village
	进入云舍村的桥	Bridge leading to Yunshe Village
	拌 料	Stamping the papermaking materials
	用撸筋棍搅拌	Stirring the papermaking materials with a lifting stick
	加 滑	Adding in papermaking mucilage
	由外往里舀水	Scooping the papermaking screen from far to near
	由右往左挖水	Scooping the papermaking screen from right to left
	由左往右送水	Scooping the papermaking screen from left to right
	盖 纸	Turning the papermaking screen upside down on the wet paper
	去纸边	Trimming the deckle edges
	盖榨板	Putting on the pressing board
	手榨压榨	Pressing the paper with hand presser
	大榨压榨	Pressing the paper with large presser
	二次压榨	Pressing the paper for the second time
	纸 垛	A pile of paper
	纸帘及纸帘架	Papermaking screen and its supporting frame
	纸帘架	Frame for supporting the papermaking screen
	云舍村竹纸纤维形态图（10×）	Fibers of bamboo paper in Yunshe Village (10× objective)
	云舍村竹纸纤维形态图（20×）	Fibers of bamboo paper in Yunshe Village (20× objective)
	待泡的竹料	Bamboo materials to be fermented
	浸泡池	Soaking pool
	搅 料	Stirring the papermaking materials
	加 滑	Adding in papermaking mucilage
	抄纸和扣纸	Scooping and lifting the papermaking screen out of water and turning it upside down on the board
	榨 纸	Pressing the paper
	揭 纸	Peeling the paper down
	在家门外的地上晒纸	Drying the paper on the ground
	捆 纸	Binding the paper
	可销售的成品竹纸	Bamboo paper ready for sale
	河口村打纸钱的老人	An old papermaker making joss paper in Hekou Village
	河口村的山道	Country road in Hekou Village
	云舍村口的纸坊	Papermaking mill in Yunshe Village
	云舍村的旅游宣传牌	Tourism billboard in Yunshe Village
	调查组成员在通往云舍村的山道上	A researcher walking on the country road to Yunshe Village
	云舍村竹纸透光摄影图	A photo of bamboo paper in Yunshe Village seen through the light
第 四 节	印江合水镇皮纸	Section 4 Bast Paper in Heshui Town of Yinjiang County
	兴旺村以草棚为主的造纸作坊群	Thatched papermaking mills in Xingwang Village
	成排的蒸料窑	Row of steaming kilns
	兴旺村以瓦房为主的造纸作坊群	Tiled papermaking mills in Xingwang Village
	印江的山水	Mountains and rivers in Yinjiang County
	路边的大型印江文化宣传牌	Cultural billboard of Yinjiang County by the road
	调查组成员进入兴旺村	Researchers arriving at Xingwang Village
	印江河畔的造纸现场	Papermaking site alongside Yinjiang River
	采下来的松蒿	*Phtheirospermum japonicum* used as papermaking mucilage
	调查组成员向造纸人认真核实工艺	A researcher confirming the papermaking technique with the local papermakers
	泡构皮	Soaking paper mulberry bark
	浆构皮	Fermenting paper mulberry bark
	堆构皮	Piling paper mulberry bark
	盖稻草	Covering the papermaking materials with dried straw
	结草绳	Twisting straw robe
	钩料上窑	Lifting the papermaking materials onto the kiln
	捶 窑	Hammering the papermaking materials
	盖 窑	Covering the kiln
	窑中蒸料	Steaming the papermaking materials in the kiln
	造纸人正在观察蒸锅里的水量	A papermaker checking water volume in the steaming wok
	流水冲洗皮料	Cleaning the papermaking materials in the running river
	漂花皮	Cleaning the spotted papermaking materials
	踩花皮	Stamping the spotted papermaking materials
	晾料子	Drying the papermaking materials
	翻晒料子	Turning the papermaking materials for drying
	捆料子	Binding the papermaking materials
	挑料子	Carrying the papermaking materials
	踩料子	Stamping the papermaking materials
	揭窑扔料	Lifting the papermaking materials from the kiln to the ground
	挑料至河中	Carrying the papermaking materials to the river
	漂料子	Cleaning the papermaking materials
	拆料子	Picking out the impurities
	脚 碓	Foot pestle
	水 碓	Hydraulic pestle
	切皮板	Trimming the papermaking materials
	踩踏洗料	Stamping the papermaking materials for cleaning
	舂料粑子	Beating the papermaking materials
	搅 槽	Stirring the papermaking materials
	抽松蒿	Adding in *Phtheirospermum japonicum* as papermaking mucilage
	搅纸浆	Stirring the paper pulp
	坎 垛	First ten pieces of paper
	洗纸帘及帘架	Cleaning the papermaking screen and its supporting frame
	舀 纸	Scooping and lifting the papermaking screen out of water and turning it upside down on the board
	再次搅槽	Stirring the papermaking materials for the second time
	加纸浆	Adding in paper pulp
	压 垛	Pressing the paper
	千斤顶压垛	Pressing the paper with a lifting jack
	撕 垛	Peeling the paper down
	刷 贴	Pasting the paper on a wall for drying
	捆 纸	Binding the paper
	甑 子	Papermaking kiln for steaming
	灰 池	Fermenting pool
	料 槌	Mallet for beating the papermaking materials
	捞 钩	Lifting hook
	石碓与舂料棒	Stone pestle and pestle stick
	纸 帘	Papermaking screen
	纸帘架	Frame for supporting the papermaking screen
	拱 耙	Papermaking rake
	捞纸槽	Papermaking trough
	蒿 槽	Trough for holding *Phtheirospermum japonicum*
	计数器	Counting apparatus
	水动力石碾	Hydraulic grinder
	兴旺村皮纸纤维形态图（10×）	Fibers of bast paper in Xingwang Village (10× objective)
	兴旺村皮纸纤维形态图（20×）	Fibers of bast paper in Xingwang Village (20× objective)
	光绪年间用白皮纸书写的卖纸槽地基文契	Contract for selling papermaking field written on white bast paper during Guangxu Reign of the Qing Dynasty
	道光年间用白皮纸书写的契约	Contract written on white bast paper during Daoguang Reign of the Qing Dynasty
	咸丰年间用白皮纸书写的契约	Contract written on white bast paper during Xianfeng Reign of the Qing Dynasty
	同治年间用白皮纸书写的卖山土文契	Contract for selling mountain field written on white bast paper during Tongzhi Reign of the Qing Dynasty

章节	图中文名称	图英文名称
	民国时期用白皮纸书写的卖水田文契	Contract for selling paddy field written on white bast paper during the Republican Era of China
	薄如蝉翼的白皮纸	Thin and transparent white bast paper
	老太太向调查组成员赠送鞋屉	An old papermaker giving her shoemaking kit paper bag to a researcher
	鞋屉外观	Outside of the shoemaking kit paper bag
	鞋屉部分内部空间	Inside of the shoemaking kit paper bag
	用白皮纸制作的鞋样	Shoe sample made of white bast paper
	供奉蔡伦造纸先师	Worshipping Cai Lun, the originator of papermaking
	用兴旺村白皮纸写的书法条幅	Calligraphy written on white bast paper in Xingwang Village
	造纸村民展示皮纸	Local papermakers showing bast paper
	兴旺村连片的纸坊	Papermaking mills in Xingwang Village
	河边的老水车	Waterwheel by the river
	兴旺村皮纸透光摄影图	A photo of bast paper in Xingwang Village seen through the light

章节	图中文名称	图英文名称
第 五 节	印江土家族手工纸	Section 5 Handmade Paper by the Tujia Ethnic Group in Yinjiang County
	六洞村湮没在荒草中的捞纸槽	Abandoned papermaking trough in Liudong Village
	高山上的六洞村	Liudong Village on a high mountain
	六洞村里的传承访谈	Interviewing the papermaking inheritors in Liudong Village
	调查组成员与昔日造纸人的合影	Researchers and former papermakers
	残留的纸帘与帘架	Broken papermaking screen and its supporting frame
	荒弃的捞纸槽	Abandoned papermaking trough
	皮　刀	Iron knife for trimming the paper
	钩竹麻示意	Showing how to pick out the bamboo materials with a hook
	踩竹麻示意	Showing how to stamp the bamboo materials
	抄纸示意	Showing how to scoop and lift the papermaking screen out of water and turn it upside down on the board
	纸　帘	Papermaking screen
	纸帘架	Frame for supporting the papermaking screen
	踩竹料用的钉鞋	Spiked shoes for stamping the papermaking materials
	双捞钩	Stick with double lifting hooks
	六洞村土家族皮纸制作、书写的斋事礼簿	Praying signature book made of bast paper by the Tujia Ethnic Group in Liudong Village
	六洞村土家族皮纸制作、书写的新婚礼簿	Wedding signature book made of bast paper by the Tujia Ethnic Group in Liudong Village
	仅存的竹纸	Last pieces of bamboo paper kept by an old papermaker
	六洞村外小河边的废纸槽	Abandoned papermaking trough by a river in Liudong Village
	在造纸老人家中夜谈	Interviewing an old papermaker at night
	六洞村皮纸透光摄影图	A photo of bast paper in Liudong Village seen through the light

章节	图中文名称	图英文名称
第 十 章	黔东南苗族侗族自治州	Chapter X Qiandongnan Miao and Dong Autonomous Prefecture
第 一 节	岑巩侗族竹纸	Section 1 Bamboo Paper by the Dong Ethnic Group in Cengong County
	进入白水村的乡村公路	Country road leading to Baishui Village
	调查组成员与村民在白水村纸坊中交流	A researcher communicating with villagers in a papermaking mill in Baishui Village
	流经水尾镇的龙鳌河	Long'ao River flowing through Shuiwei Town
	传说中的陈圆圆墓	Tomb of Chen Yuanyuan according to the local legend
	思州砚作坊	Ink slab mill in Sizhou Prefecture (ancient name of Cengong County)
	神奇的思州傩特技	Magical Nuo Show in Sizhou Prefecture (ancient name of Cengong County)
	龙鳌河沿岸竹纸作坊	Bamboo papermaking mill along the Long'ao Riverside
	水稻田中的造纸作坊	Papermaking mill in a paddy field
	纸坊中的黄姓纸工	Local papermaker Mr. Huang, in a papermaking mill
	白水村小山坡上的煮料窑	Boiling kiln on a hill in Baishui Village
	岩山脚村的一处造纸作坊群	Papermaking mills in Yanshanjiao Village
	阳山竹	Local Yangshan bamboo (*Bambusa intermedia* Hsueh et Yi)
	纸坊中的姚菊莲	Local papermaker Yao Julian in the papermaking mill
	堆放的竹麻原料	Piled raw material of bamboo materials
	料塘中沤竹	Fermenting the bamboo in a pool
	煮竹麻	Boiling the bamboo materials
	打　碓	Beating the papermaking materials
	舀　纸	Scooping and lifting the papermaking screen out of water and turning it upside down on the board
	榨　纸	Pressing the paper
	分　纸	Splitting the paper layers
	晒　纸	Drying the paper
	捆　纸	Binding the paper
	于河村的纸窑	Papermaking kiln in Yuhe Village
	中寨村的纸碓	Beating pestle in Zhongzhai Village
	岩山脚村的纸槽	Papermaking trough in Yanshanjiao Village
	纸　帘	Papermaking screen
	岩山脚村的纸榨	Pressing device in Yanshanjiao Village
	白水村竹纸纤维形态图（10×）	Fibers of bamboo paper in Baishui Village（10× objective）
	白水村竹纸纤维形态图（20×）	Fibers of bamboo paper in Baishui Village（20× objective）
	优质的白水村竹纸	Quality bamboo paper in Baishui Village
	纸　钱	Joss paper
	重点访谈对象黄俊有	Local papermaker Huang Junyou, an important interviewee of the research
	中寨村打纸钱的老人	An old man making joss paper in Zhongzhai Village
	打纸钱	Making joss paper
	包纸钱寄给祖先示意	Showing how to wrap joss paper as sacrificial offerings to ancestors
	送调查组成员去于河村的小船	The boat taking the researchers to Yuhe Village
	岩山脚村的连排纸坊	A row of papermaking mills in Yanshanjiao Village
	于河村正在烧窑的造纸人	A papermaker boiling the papermaking materials
	满池待洗的竹麻料	Bamboo materials to be cleaned
	山脚下的料塘与竹料加工现场	Soaking pool and the locale for processing the bamboo materials at the foot of a mountain
	水尾镇陪同调研的年轻女孩正在学习揭纸	A young girl from Shuiwei Town learning how to peel the paper down
	白水村竹纸透光摄影图	A photo of bamboo paper in Baishui Village seen through the light

章节	图中文名称	图英文名称
第 二 节	三穗侗族竹纸	Section 2 Bamboo Paper by the Dong Ethnic Group in Sansui County
	贵洞村的乡土建筑	Local residences in Guidong Village
	贵洞村寨风景图	View of Guidong Village
	贵洞村造纸作坊群	Papermaking mills in Guidong Village
	《贵州省三穗县地名志》书影	*The Annals of Place Names in Sansui County of Guizhou Province*
	调查组成员采访"非遗"传承人杨再祥（中）	Researchers interviewing local intangible cultural heritage inheritor, Yang Zaixiang (middle)
	敲猕猴桃根	Beating Chinese gooseberry root to make mucilage
	泡　麻	Soaking the papermaking materials
	沤　麻	Fermenting the papermaking materials
	碾麻示意	Showing how to grind the papermaking materials
	碾好的竹料	Processed bamboo materials
	捞　筋	Picking out the impurities
	拱　麻	Lifting the papermaking materials
	加　玄	Adding in papermaking mucilage
	抄　纸	Scooping and lifting the papermaking screen out of water and turning it upside down on the board
	刮　纸	Scraping the paper

章节	图中文名称	图英文名称
	压盖板	Pressing the paper with a cover board
	榨　纸	Pressing the paper
	石　碾	Stone grinder
	纸　槽	Papermaking trough
	踩　槽	Trough for stamping the papermaking materials
	纸　帘	Papermaking screen
	帘　架	Frame for supporting the papermaking screen
	贵洞村竹纸纤维形态图（10×）	Fibers of bamboo paper in Guidong Village（10× objective）
	贵洞村竹纸纤维形态图（20×）	Fibers of bamboo paper in Guidong Village（20× objective）
	亲戚们帮忙将嫁妆送至新郎家	Relatives carrying the dowery to bridegroom's house
	烧纸钱	Burning the joss paper
	告诉祖先小辈喜结良缘	Informing the ancestors of the offspring's wedding
	贵洞村边的造纸作坊群	Papermaking mills alongside the Guidong Village
	废弃的纸槽房	Abandoned papermaking mill
	还在使用的碾坊	Grinding workshop still in use
	贵洞村竹纸透光摄影图	A photo of bamboo paper in Guidong Village seen through the light

章节	图中文名称	图英文名称
第　三　节	黄平苗族竹纸	Section 3 Bamboo Paper by the Miao Ethnic Group in Huangping County
	黄平境内的山地河流	Mountains and rivers in Huangping County
	黄平县的苗族大歌表演	Miao ethnic polyphonic music show in Huangping County
	热闹非凡的革家哈戎节	Harong Festival celebrated by the Gejia people
	别具一格的革家蜡染服饰	Gejia ethnic batik clothing
	革家妇女与孩童	Gejia women and kid
	王启光夫妇	Wang Qiguang and his wife
	纸药树原料	Raw material of papermaking mucilage
	晒　竹	Drying the bamboo
	泡竹塘	Pool for soaking the bamboo
	正在碾竹料的碾子	Grinder for grinding the bamboo materials
	初步碾碎的竹料	Processed bamboo materials
	放到簸箕里的纸药原料	Raw material of the papermaking mucilage in a bamboo basket
	捞　纸	Scooping and lifting the papermaking screen out of water and turning it upside down on the board
	石头压纸	Pressing the paper with a stone
	放　水	Opening sluice to let water out
	纸浆沉淀	Precipitate of the paper pulp
	晾　纸	Drying the paper
	收　纸	Collecting the paper
	晒干的一提纸	A pile of dried paper
	叠　纸	Folding the paper
	用凿纸担压纸	Pressing the paper with a carrying pole
	凿　纸	Perforating the paper to make joss paper
	砍　纸	Trimming the paper
	打纸棒	Beating stick
	纸　帘	Papermaking screen
	凿纸桩	Trimming stake
	砍纸刀、砍纸卡	Trimming knife and trimming ruler
	凿纸槌	Mallet for perforating the paper
	凿纸担	Carrying pole for fixing the paper
	满溪村竹纸纤维形态图（10×）	Fibers of bamboo paper in Manxi Village (10× objective)
	满溪村竹纸纤维形态图（20×）	Fibers of bamboo paper in Manxi Village (20× objective)
	打纸钱	Making joss paper
	成品纸钱	Final product of joss paper
	满溪村外景	Landscape of Manxi Village
	翁坪乡的革家寨	Gejia ethnic residence in Wengping Town
	翁坪乡的犀牛河	Xiniu River in Wengping Town
	满溪村竹纸透光摄影图	A photo of bamboo paper in Manxi Village seen through the light

章节	图中文名称	图英文名称
第　四　节	凯里苗族竹纸	Section 4 Bamboo Paper by the Miao Ethnic Group in Kaili City
	流经凯里的重安江	Chong'an River flowing through Kaili City
	进入湾水造纸村落的吊桥	Suspension bridge leading to Wanshui Papermaking Village
	造纸人姜文安夫妇和儿子姜明春	Papermakers Jiang Wen'an with his wife and son Jiang Mingchun
	调查组成员与姜明亮合影	Researchers and the local papermaker Jiang Mingliang
	药　树	Raw material of papermaking mucilage
	采访造纸村民	Interviewing papermakers
	砍　竹	Lopping the bamboo
	捶　竹	Hammering the bamboo
	头道挖纸示意	Showing how to scoop the papermaking screen for the first time
	二道挖纸示意	Showing how to scoop the papermaking screen for the second time
	装　架	Fixing the shelf
	打钱眼	Perforating the paper
	打好钱眼的纸	Paper with holes on it
	切　纸	Trimming the paper
	姜明亮及其家人	Jiang Mingliang and his family members
	碾　子	Stone grinder
	水　车	Waterwheel
	纸　塘	Papermaking trough
	一帘三纸的帘子	Papermaking screen that can make three pieces of paper simultaneously
	有制帘人和造纸人姓名的帘子	Papermaking screen with the names of screen-maker and papermaker on it
	帘子架	Frame for supporting the papermaking screen
	带提手的帘子架	Papermaking frame with handles
	木　榨	Wooden presser
	砍纸刀和钱鋻	Trimming knife and chisel
	切纸凳	Trimming table
	纸钱架	Frame for perforating the paper
	切板与切刀	Trimming board and trimming knife
	湾水镇竹纸纤维形态图（10×）	Fibers of bamboo paper in Wanshui Town (10× objective)
	湾水镇竹纸纤维形态图（20×）	Fibers of bamboo paper in Wanshui Town (20× objective)
	湾水镇集市上的纸钱交易	Joss paper trade in a market in Wanshui Town
	湾水镇集市上销售纸钱场景	Scene of joss paper trade in a market in Wanshui Town
	烧纸钱	Burning the joss paper
	三行钱与四行钱	Joss paper with different numbers of holes（3×7, 4×9）
	江口村的旧纸坊	Old papermaking mill in Jiangkou Village
	湾水镇苗族村民的酒席	Miao people's feast in Wanshui Town
	湾水镇竹纸透光摄影图	A photo of bamboo paper in Wanshui Town seen through the light

章节	图中文名称	图英文名称
第　五　节	丹寨苗族皮纸	Section 5 Bast Paper by the Miao Ethnic Group in Danzhai County
	石桥山崖下的造纸作坊	Papermaking mill at the foot of mountain in Shiqiao Village
	杨大文于加拿大抄纸展演旧照	Photo of papermaking show performed by Yang Dawen in Canada
	丹寨县的苗族斗牛盛会	Miao Ethnic Bullfighting Pageant in Danzhai County
	天然石桥	Natural stone bridge
	石桥村入口介绍牌	Introduction to Shiqiao Village
	造纸胜地——大岩脚石壁	Papermaking scenic spot named Dayanjiao Cliff
	亚太旅游协会援助项目碑	Monument in memory of Pacific Asia Travel Association's assistance to local papermaking program

章节	图中文名称	图英文名称
	黔山古法造纸合作社	Papermaking cooperative employing Qianshan traditional papermaking techniques
	易兴古法造纸合作社	Papermaking cooperative employing Yixing traditional papermaking techniques
	河水中泡料	Soaking the papermaking materials in a river
	浆 料	Fermenting the papermaking materials
	挑 煤	Carrying the coal to heat papermaking materials
	煮 料	Boiling the papermaking materials
	选 料	Picking out the impurities
	泡 料	Soaking the papermaking materials
	打 浆	Beating the papermaking materials with a machine
	抄 纸	Scooping and lifting the papermaking screen out of water and turning it upside down on the board
	压 榨	Pressing the paper
	揭纸角	Peeling the paper down from the upper corner
	撕 纸	Peeling the paper down
	晒 纸	Drying the paper
	揭 纸	Peeling the paper down
	捆 纸	Binding the paper
	村民自豪加自负的造纸标语	Couplets showing the local papermakers' pride and arrogance of their papermaking techniques
	浇纸法新品种	New paper type employing the fixed screen
	石桥村白皮纸纤维形态图（10×）	Fibers of white bast paper in Shiqiao Village (10× objective)
	石桥村白皮纸纤维形态图（20×）	Fibers of white bast paper in Shiqiao Village (20× objective)
	石桥村花纸纤维形态图（10×）	Fibers of flower paper in Shiqiao Village (10× objective)
	石桥村花纸纤维形态图（20×）	Fibers of flower paper in Shiqiao Village (20× objective)
	白皮纸糊窗	Pasting a sheet of white bast paper over a lattice window
	制作花草纸	Making flower-grass paper
	花草纸工艺品	Crafts of flower-grass paper
	杨大文于加拿大展演旧照	Photo of Yang Dawen's papermaking show in Canada
	祭蔡伦的手写牌位	Handwritten tablet in memory of Cai Lun
	各种新的纸制品	Various new paper products
	国外游客参观抄纸过程	Foreign tourists watching papermaking show
	国内外游客观纸留言	Notes left by tourists from China and abroad
	中国石桥古纸上海书画名家笔会	Exhibition of paintings and calligraphy on bast paper in Shiqiao Village was held in Shanghai City
	石桥村纸毯透光摄影图	A photo of paper blanket in Shiqiao Village seen through the light
	石桥村花纸透光摄影图	A photo of flower paper in Shiqiao Village seen through the light
	石桥村染色纸透光摄影图	A photo of dyed paper in Shiqiao Village seen through the light
	石桥村白皮纸透光摄影图	A photo of white bast paper in Shiqiao Village seen through the light
	石桥村书画纸透光摄影图	A photo of calligraphy and painting paper in Shiqiao Village seen through the light
章节	图中文名称	图英文名称
第 六 节	榕江侗族皮纸	Section 6 Bast Paper by the Dong Ethnic Group in Rongjiang County
	皮纸制作的"时髦"针线包	"Modish" sewing kit wrappers made of bast paper
	洁白漂亮的本里皮纸	White and smooth bast paper in Benli Village
	九秋村侗寨的民宅	Local Dong ethnic residences in Jiuqiu Village
	本里村的侗家村寨吊脚楼	Dong ethnic Diaojiao (suspended) buildings in Benli Village
	旧日用皮纸抄写的风水书	Old Fengshui book transcribed on bast paper
	本里村造纸的侗族女村民	Female Dong papermakers in Benli Village
	本里村中"话构麻"	Enquiring papermaking techniques in Benli Village
	构树采样	Sample of paper mulberry tree
	"藤本构树"采样	Sample of *Broussonetia kaempfori Sieb.* var. *australis* Suzuki
	野生猕猴桃树	Wild Chinese gooseberry tree
	剥 皮	Stripping the bark
	晒干的构树皮	Dried paper mulberry bark
	煮 料	Boiling the papermaking materials
	捞 料	Picking out the papermaking materials
	洗 料	Cleaning the papermaking materials
	捶猕猴桃树藤	Hammering Chinese gooseberry vine
	过滤猕猴桃树藤汁	Filtering Chinese gooseberry sap
	搅 料	Stirring the papermaking materials
	浇 纸	Pouring paper pulp onto the papermaking screen
	晾 纸	Drying the paper
	撕 纸	Peeling the paper down
	理 纸	Sorting the paper
	垫石和棒槌	Stone beating pad and wooden mallet
	立 框	Frame for supporting the papermaking screen
	草木灰盆、火钳和煮料锅	Plant ash container, tongs and boiling wok
	本里村皮纸纤维形态图（10×）	Fibers of bast paper in Benli Village (10× objective)
	本里村皮纸纤维形态图（20×）	Fibers of bast paper in Benli Village (20× objective)
	九秋村皮纸纤维形态图（10×）	Fibers of bast paper in Jiuqiu Village (10× objective)
	九秋村皮纸纤维形态图（20×）	Fibers of bast paper in Jiuqiu Village (20× objective)
	油纸斗笠	Hat made of oil paper
	剪 纸	Paper cutting
	七十二寨侗族古典芦笙曲谱	Music score of traditional Dong ethnic musical instrument, Lusheng
	九秋村河边洗料的造纸人	A papermaker cleaning the papermaking materials in a river in Jiuqiu Village
	进入九秋村的山道	Country road leading to Jiuqiu Village
	山环水绕的榕江	Rongjiang County surrounded by mountains and rivers
	正在浇纸的侗族女人	A Dong woman making paper
	珍贵的国家一级保护树种	Trees under First-Grade State Protection
	辛苦的侗族造纸老人	An old Dong papermaker
	本里村皮纸透光摄影图	A photo of bast paper in Benli Village seen through the light
	九秋村皮纸透光摄影图	A photo of bast paper in Jiuqiu Village seen through the light
章节	图中文名称	图英文名称
第 七 节	从江秀塘瑶族竹纸	Section 7 Bamboo Paper by the Yao Ethnic Group in Xiutang Town of Congjiang County
	前往打格村的山路	Country road leading to Dage Village
	进入秀塘乡的壮族风格门楼	Unique Zhuang ethnic gate entrance to Xiutang Town
	南竹林环抱的打格村民居	Local residences in Dage Village; surrounded by *Phyllostachys pubescens* Mazel ex H. de Leh.
	正在上课的打格小学学生	Students of Dage Elementary School having classes
	高山上的瑶族村寨	Yao ethnic residences on high mountains
	打格村瑶民烹饪食物	A Yao people cooking food in Dage Village
	落满灰土的20年前的纸帘	Papermaking screen covered by dirt which hadn't been used for 20 years
	穿瑶族盛装服饰少女旧照	Old photo of maids wearing Yao ethnic clothing
	访谈赵金学夫妇	Interviewing local papermakers Zhao Jinxue and his wife
	打格瑶族竹纸的原料——南竹	Local papermaking raw material, *Phyllostachys Pubescens* Mazel ex H. de Leh.
	打格村旧日沿河造纸的山溪	Stream formerly used for cleaning the papermaking materials in Dage Village
	溪水的pH约为6.2	pH value of the stream was about 6.2
	赵金成讲述当年砍竹	Zhao Jincheng relating the bamboo-lopping experiences
	赵金学示意抄纸动作	Zhao Jinxue showing how to scoop and lift the papermaking screen out of water
	揭纸动作示意	Showing how to peel the paper down
	晾纸示意	Showing how to dry the paper

章节	图中文名称	图英文名称
	旧抄纸帘与抄纸架	Old papermaking screen and its supporting frame
	溪边遗存的石臼窝	Stone mortar alongside the stream
	打格村竹纸纤维形态图 (10×)	Fibers of bamboo paper in Dage Village (10× objective)
	打格村竹纸纤维形态图 (20×)	Fibers of bamboo paper in Dage Village (20× objective)
	打纸钱	Making joss paper
	打格村的纸钱	Joss paper in Dage Village
	神龛旁的祭物	Sacrificial offerings (bamboo basket and joss paper) by the niche
	打格度戒仪式旧照	Old photo of the Coming-of-Age Ceremony in Dage Village
	破旧的"盘王节"手抄祭本	Transcribed sacrificial book used for King Pan Festival
	打格村长讲述"盘王节"盛况	Head of Dage Village relating the grand occasion of King Pan Festival
	告别造纸村民时的合影	Photo of researchers and local papermakers before the departure time
	讲述造纸往事的赵金学	Zhao Jinxue telling the papermaking stories
	祭　台	Worshipping niche
	打格村竹纸透光摄影图	A photo of bamboo paper in Dage Village seen through the light

章节	图中文名称	图英文名称
第　八　节	从江翠里瑶族手工纸	Section 8 Handmade Paper by the Yao Ethnic Group in Cuili Town of Congjiang County
	高华村田里收割完的糯稻秆	Glutinous rice stalk in Gaohua Village
	流经从江县城的都柳江	Duliu River flowing through Congjiang County
	侗塔与古树	Dong ethnic tower and ancient trees
	茂林环抱的高华村	Gaohua Village surrounded by dense forest
	高华村的吊脚楼	Diaojiao (suspended) buildings in Gaohua Village
	柴灶铁锅熬煮的药浴药水	Boiling the herbs in an iron wok for bathing
	盛满药水的洗浴木桶	Bath tub for holding liquid herbs
	访谈正在吃饭的赵成富	Interviewing papermaker, Zhao Chengfu, who was having dinner
	访谈赵金妹等造纸人	Interviewing Zhao Jinmei and other papermakers
	已剥好的构树皮	Processed paper mulberry bark
	芙蓉树根	Root of *Actinidia chinensis* (Planch.)
	吊脚楼旁的构树	Paper mulberry tree by the Diaojiao (suspended) building
	晒好的干构树皮和皮纸	Dried paper mulberry bark and bast paper
	煮构皮	Boiling paper mulberry bark
	捶　纸	Hammering the paper
	摇纸料	Stirring the papermaking mucilage
	浇　纸	Pouring paper pulp onto the papermaking screen
	烤　纸	Drying the paper by fire
	晒　纸	Drying the paper
	扯　纸	Peeling the paper down with an ox bone
	叠　纸	Folding the paper
	木　槌	Wooden mallet
	牛　骨	Ox bone for peeling the paper down
	里　白	White cloth as a sieve on the papermaking screen
	纸帘架	Frame for supporting the papermaking screen
	高华村皮纸纤维形态图 (10×)	Fibers of bast paper in Gaohua Village (10× objective)
	高华村皮纸纤维形态图 (20×)	Fibers of bast paper in Gaohua Village (20× objective)
	5把干糯稻秆	Five bundles of dried glutinous rice stalk
	纸药芙蓉树叶	Papermaking mucilage, leaves of *Actinidia chinensis* (Planch.)
	未收割的糯稻田与收割后的糯稻田	Glutinous rice field to be reaped and after reaping
	赵掌珠为调查组成员示范如何搅拌	Zhao Zhangzhu showing how to stir the papermaking materials
	赵掌珠正在示范浇纸	Zhao Zhangzhu showing how to pour paper pulp onto the papermaking screen
	屋外晾纸	Drying the paper outside the room
	赵掌珠正在示范剥纸	Zhao Zhangzhu showing how to peel the paper down
	煮料所用的大锅	Iron wok for boiling the papermaking materials
	用于挑料的"粪挑"	Bamboo basket and shoulder pole for carrying the papermaking materials
	木　舂	Wooden pestle
	用于盛放纸浆的桶	Vat for holding paper pulp
	葫芦瓢	Gourd ladle
	用于筛选火灰的簸箕	Sieve for winnowing ashes
	纸帘架	Frame for supporting the papermaking screen
	牛　骨	Ox bone for peeling the paper down
	里　白	White cloth as a sieve on the papermaking screen
	高华村草纸纤维形态图 (10×)	Fibers of straw paper in Gaohua Village (10× objective)
	高华村草纸纤维形态图 (20×)	Fibers of straw paper in Gaohua Village (20× objective)
	赵成富演示打纸钱	Zhao Chengfu showing how to make joss paper
	赵成富在家中的神龛前祭祖	Zhao Chengfu performing sacrificial ritual in front of the family niche
	展开的成品皮纸	Unfolded bast paper
	展开的成品草纸	Unfolded straw paper
	赵成富家中用皮纸抄写的书	Books transcribed on bast paper in Zhao Chengfu's house
	调查组成员正在向赵金妹买测试纸样	Researchers buying paper samples from papermaker Zhao Jinmei
	瑶族女人正在浇纸	A Yao woman pouring paper pulp onto the papermaking screen
	高华村村民家中的神龛	Niche in a papermaker's house in Gaohua Village
	用草纸做的篮子	Basket made of straw paper
	赵成富正在用草纸和竹子做幡	Zhao Chengfu making sacrificial flag by straw paper and bamboo
	高华村的一处药浴传习基地	Herbal Bath Training Base in Gaohua Village
	调查组成员与赵掌珠老人	A researcher and local papermaker Zhao Zhangzhu
	高华村皮纸透光摄影图	A photo of bast paper in Gaohua Village seen through the light
	高华村草纸透光摄影图	A photo of straw paper in Gaohua Village seen through the light

章节	图中文名称	图英文名称
第　九　节	从江小黄侗族皮纸	Section 9 Bast Paper by the Dong Ethnic Group in Xiaohuang Village of Congjiang County
	类似"藤本构树"的构树种类	Local paper mulberry tree, similar to *Broussonetia kaempferi Sieb.* var. *australis* Suzuki
	高增乡的侗寨	Dong ethnic residence in Gaozeng Town
	小黄鼓楼前的侗族大歌表演	Dong ethnic polyphonic music show persented in front of Xiaohuang drum tower
	男女老少齐合唱	Chorus of local residents including all ages and both genders
	小黄村民歌手汇演旧照	A photo of local singers in a show (Xiaohuang Village)
	侗歌表演中的诙谐小品	Funny show performed in Dong ethnic polyphonic music show
	小黄晒谷风俗	Unique custom of drying millets in Xiaohuang Village
	小黄村头迎客的风雨桥	Wind-Rain Bridge in Xiaohuang Village
	鼓楼底层玩石子游戏的男人们	Men playing stone games on the ground floor of the drum tower
	吴奶保卫与小孙女	Wu Baowei's mother (mothers are named after their son, based on local tradition), an old papermaker, and her granddaughter
	气宇轩昂的85岁老爷子	Wu Baowei's 85-year-old father, still in good spirits
	访谈正在采构皮的吴奶保卫	Interviewing Wu Baowei's mother, who was collecting paper mulberry bark
	剥　皮	Stripping the bark
	干构皮	Dried paper mulberry bark
	洗料水池	Pool for cleaning the papermaking materials
	捏料团	Balling up the papermaking materials
	捶料团	Hammering the papermaking material balls
	浸泡好的纸药液	Soaked papermaking mucilage
	浇　纸	Pouring paper pulp onto the papermaking screen
	折　纸	Folding the paper

章节	图中文名称	图英文名称
	帘 子	Papermaking screen
	木 槌	Wooden mallet
	煮料锅	Wok for boiling the papermaking materials
	小黄村皮纸纤维形态图（10×）	Fibers of bast paper in Xiaohuang Village (10× objective)
	小黄村皮纸纤维形态图（20×）	Fibers of bast paper in Xiaohuang Village (20× objective)
	村里旧日传下的皮纸手抄风水书	Old Fengshui book transcribed on local bast paper
	有皮纸夹层的绣花鞋	Embroidered shoes with bast paper inside
	针线盒外包装	Sewing kit wrapped by bast paper
	展开的精美针线盒	Exquisite unfolded sewing kit wrapper
	“女人的活计”	Papermaking practice limited to female papermakers only
	祭祀焚烧的纸钱	Joss paper used in sacrificial ritual
	调查组成员在买皮纸	A researcher buying bast paper from the papermakers
	在吴奶保卫家聚会的造纸老人们	Old papermakers gathering in Wu Baowei's mother's house
	从江黄金周旅游海报	Tourism poster for the Golden Week in Congjiang County
	小黄小学的“非遗”传习基地	Intangible cultural heritage training base set in Xiaohuang Elementary School
	调查组成员向造纸老人敬酒	A researcher making a toast to the old papermakers
	小黄村皮纸透光摄影图	A photo of bast paper in Xiaohuang Village seen through the light

章节	图中文名称	图英文名称
第十节	从江占里侗族皮纸	Section 10 Bast Paper by the Dong Ethnic Group in Zhanli Village of Congjiang County
	占里村每户门前贴的人口档案卡	Household population file card attached on each house gate in Zhanli Village
	从江深山里的小村寨	A small village in mountain area of Congjiang County
	村中心的文化广场——侗家鼓楼	Dong ethnic drum tower at the cultural square in the village centre
	安宁的占里侗族小街	Peaceful Dong ethnic alley in Zhanli Village
	占里村雨中连排的晒谷架	Rows of drying rack in Zhanli Village in the rain
	秀美富饶的占里侗寨	Rich and beautiful Dong ethnic residence in Zhanli Village
	占里村的夜色	Night scene of Zhanli Village
	调查组成员与吴奶卫和吴奶妹合影	Researchers with Wu Naiwei and Wu Naimei
	造纸使用的干构皮	Dry paper mulberry bark for papermaking
	造纸使用的灶灰	Stove ash for papermaking
	实测山泉水pH	pH value of the local spring water
	吴奶妹在剥黑皮	Wu Naimei stripping the bark
	吴奶妹拿出存放的干构皮	Dry bark for papermaking stored by Wu Naimei
	吴奶妹示范煮皮	Wu Naimei showing how to boil the bark
	洗料的竹编箩筐	Bamboo basket for cleaning the papermaking materials
	吴奶妹在示范舂料	Wu Naimei showing how to beat the papermaking materials
	吴奶妹在示范捶料	Wu Naimei showing how to hammer the papermaking materials
	浇纸装帘	Fixing the papermaking screen
	折 纸	Folding the paper
	装纸帘示意	Showing how to fix the papermaking screen
	吴奶妹家的石板和槌	Wu Naimei's stone board and mallet
	石 舂	Stone pestle
	占里村皮纸纤维形态图（10×）	Fibers of bast paper in Zhanli Village (10× objective)
	占里村皮纸纤维形态图（20×）	Fibers of bast paper in Zhanli Village (20× objective)
	占里村流行的住家门头驱鬼用纸习俗	Paper hanging on door for expelling ghosts, a local custom in Zhanli Village
	吴奶妹卖纸前数纸	Wu Naimei counting the paper before it was sold
	“生意”达成	A deal was made
	吴奶卫展示浇纸帘	Wu Naiwei showing the papermaking screen
	提倡控制生育的占里古歌	Zhanli ballad on birth control
	占里鼓楼底层戏耍的男人	Men enjoying their time on the ground floor of Zhanli drum tower
	占里村风雨桥头	One end of the Wind-Rain Bridge in Zhanli Village
	造纸老人吴奶卫	Old papermaker Wu Naiwei
	占里村皮纸透光摄影图	A photo of bast paper in Zhanli Village seen through the light

章节	图中文名称	图英文名称
第十一节	黎平侗族皮纸	Section 11 Bast Paper by the Dong Ethnic Group in Liping County
	浇纸的老人	An old woman pouring paper pulp onto the papermaking screen
	黎平侗乡小景	Landscape of Dong ethnic residence in Liping County
	肇兴侗寨	Dong ethnic residence in Zhaoxing Town
	鼓楼内部仰望	Inside view of the drum tower
	茅贡乡间的桥	Bridge in Maogong Town
	侗寨花桥	Flower Bridge in Dong ethnic residence
	调查组成员与吴永峰	A researcher and local papermaker Wu Yongfeng
	地扪村水塘边的民居	Local residences by a pond in Dimen Village
	杨桃藤原料	*Averrhoa carambola* vine, raw material of papermaking mucilage
	捶纸药料	Hammering the papermaking mucilage
	揉出纸药液	Squeezing out the papermaking mucilage sap
	剥构树外皮	Stripping paper mulberry bark
	干 皮	Dried bark
	浸泡好的纸药液	Soaked papermaking mucilage
	煮 皮	Boiling the bark
	洗 皮	Cleaning the bark
	拣 皮	Picking out the impurities
	捏 团	Balling up the papermaking materials
	捶 皮	Hammering the bark
	搅纸浆	Stirring paper pulp
	加杨桃藤汁	Adding in the sap of *Averrhoa carambola* vine
	摇 纸	Shaking the papermaking screen to even paper pulp
	晒 纸	Drying the paper
	剥纸和折纸	Peeling the paper down and folding it
	木 槌	Wooden mallet
	葫芦瓢	Gourd ladle
	撮 箕	Winnowing basket
	地扪村皮纸纤维形态图（10×）	Fibers of bast paper in Dimen Village (10× objective)
	地扪村皮纸纤维形态图（20×）	Fibers of bast paper in Dimen Village (20× objective)
	地扪手抄皮纸书	Book transcribed on bast paper in Dimen Village
	粉红色绣领下的刺绣纸	Embroidery paper covered under the pink neckline
	皮纸针线盒	Sewing kit wrapper made of bast paper
	包装用纸	Packing paper
	地扪吸引了大批艺术家	Dimen Village appeals to the artists
	鬼娃娃剪纸	Ghost kid paper cutting for protecting kids in the family
	皮纸长串串	A string of bast paper
	竹纸钱	Bamboo joss paper
	穿侗装的造纸老太与调查组成员	A researcher and an old Dong papermaker wearing ethnic clothing
	河边挂钩上摇纸	Shaking the papermaking screen with a pothook by the river
	黏稠的纸药液	Sticky papermaking mucilage
	地扪村的风雨桥文化	Wind-Rain Bridge culture in Dimen Village
	地扪村的旅游宣传牌	Tourism billboard in Dimen Village
	地扪村皮纸透光摄影图	A photo of bast paper in Dimen Village seen through the light

表目
Tables

表中英文名称

表1.1 贵州省当代手工造纸点历史溯源表

Table1.1 History of current handmade papermaking sites in Guizhou Province

表1.2 民国中后期贵州各县产纸概况表

Table1.2 Paper output of various counties in Guizhou Province in the mid-late period of the Republican Era of China

表1.3 贵州省手工造纸文化的地域、民族和原料多样性

Table1.3 Papermaking places, papermaker's ethnic groups and raw materials in Guizhou Province

表1.4 第一批"非遗"全国纸项目分布表

Table1.4 Distribution of paper preserved by the National Intangible Cultural Heritage program in China

表2.1 下午村皮纸的相关性能参数

Table2.1 Performance parameters of bast paper in Xiawu Village

表2.2 老厂镇竹纸的相关性能参数

Table2.2 Performance parameters of bamboo paper in Laochang Town

表2.3 六枝特区历史上手工纸生产的地点及现状

Table2.3 Handmade papermaking sites in history and their current status in Liuzhi Special Area

表2.4 1950~1964年六枝手工纸生产的基本情况

Table2.4 Handmade papermaking data in Liuzhi Special Area between 1950 and 1964

表2.5 传统和现代六枝皮纸生产工艺对照表

Table2.5 Contrast of traditional and modern papermaking techniques of bast paper in Liuzhi Special Area

表2.6 火坑村皮纸的相关性能参数

Table2.6 Performance parameters of bast paper in Huokeng Village

表2.7 郎岱镇皮纸的相关性能参数

Table2.7 Performance parameters of bast paper in Langdai Town

表3.1 河沟头村皮纸的相关性能参数

Table3.1 Performance parameters of bast paper in Hegoutou Village

表3.2 王家良家造纸一年原材料及各种辅料费用表

Table3.2 Wang Jialiang's annual papermaking cost of raw materials and other auxiliary materials

表3.3 坝盘村竹纸的相关性能参数

Table3.3 Performance parameters of bamboo paper in Bapan Village

表3.4 传统和现代龙井白绵纸生产工艺对照表

Table3.4 Contrast of traditional and modern papermaking techniques of Baimian paper in Longjing Village

表3.5 龙井村白绵纸的相关性能参数

Table3.5 Performance parameters of Baimian paper in Longjing Village

表3.6 龙井村灯笼纸的相关性能参数

Table3.6 Performance parameters of Denglong (lantern) paper in Longjing Village

表3.7 龙井村普通造纸户一年造纸平均开支表

Table3.7 Papermakers' annual papermaking cost in Longjing Village

表3.8 坡柳村竹纸的相关性能参数

Table3.8 Performance parameters of bamboo paper in Poliu Village

表4.1 紫山村竹纸的相关性能参数

Table4.1 Performance parameters of bamboo paper in Zishan Village

表4.2 纸厂村皮纸的相关性能参数

Table4.2 Performance parameters of bast paper in Zhichang Village

表5.1 翁贵村皮纸的相关性能参数

Table5.1 Performance parameters of bast paper in Wenggui Village

表5.2 芦山镇竹纸的相关性能参数

Table5.2 Performance parameters of bamboo paper in Lushan Town

表5.3 晚清、民国年间龙里竹纸和草纸历史产量统计

Table5.3 Output of bamboo paper and straw paper in Longli County during the late Qing Dynasty and the Republican Era of China

表5.4 纸厂村竹纸的相关性能参数

Table5.4 Performance parameters of bamboo paper in Zhichang Village

表5.5 龙里竹纸销售情况表

Table5.5 Sales status of bamboo paper in Longli County

表5.6 尧古村竹纸的相关性能参数

Table5.6 Performance parameters of bamboo paper in Yaogu Village

表5.7 都匀造纸厂皮纸的相关性能参数

Table5.7 Performance parameters of bast paper in Duyun Papermaking Factory

表6.1 大寨村皮纸的相关性能参数

Table6.1 Performance parameters of bast paper in Dazhai Village

表6.2 红土坡村竹纸的相关性能参数

Table6.2 Performance parameters of bamboo paper in Hongtupo Village

表6.3 生产7 000刀成品竹纸所需的成本

Table6.3 Costs of papermaking raw materials to produce 7 000 dao bamboo paper

表6.4 雨下寨村皮纸的相关性能参数

Table6.4 Performance parameters of bast paper in Yuxiazhai Village

表7.1 陇脚村竹纸的相关性能参数

Table7.1 Performance parameters of bamboo paper in Longjiao Village

表8.1 1949~1988年仁怀市构皮收购量与手工纸产量的变化

Table8.1 Paper mulberry bark purchase and handmade paper output from 1949 to 1988 in Renhuai City

表8.2 仁怀皮纸制作的传统工艺和20世纪80年代以后逐渐采用的新工艺的对比

Table8.2 Contrast of traditional and modern papermaking techniques after 1980s of bast paper in Renhuai City

表8.3 三元村皮纸的相关性能参数

Table8.3 Performance parameters of bast paper in Sanyuan Village

表8.4 鲁班镇竹纸的相关性能参数

Table8.4 Performance parameters of bamboo paper in Luban Town

表8.5 三元村普通造纸户一年造纸开支表

Table8.5 Papermakers' annual papermaking cost in Sanyuan Village

表8.6 2009年仁怀市取缔小造纸作坊前五马河沿岸手工造纸户基本情况

Table8.6 Basic information of papermakers alongside the Wuma River before papermaking mills are banned in 2009

表8.7 正安历史上手工纸生产的地点及现状

Table8.7 Handmade papermaking sites in history and their current status in Zheng'an County

表8.8 1982~1990年正安县乡镇企业土纸及石灰产量变化情况

Table8.8 Local handmade paper and lime output in Zheng'an County from 1982 to 1990

表8.9 梨坝村竹纸的相关性能参数

Table8.9 Performance parameters of bamboo paper in Liba Village

表8.10 1952~1990年正安县部分年度全县手工纸销售数量

Table8.10 Annual sales of handmade paper in Zheng'an County between 1952 and 1990 (parts of the data)

表8.11 传统和现代造纸塘皮纸生产工艺对照表

Table8.11 Contrast of traditional and modern papermaking techniques of bast paper in Zaozhitang Village

表8.12 造纸塘村皮纸的相关性能参数

Table8.12 Performance parameters of bast paper in Zaozhitang Village

表8.13 传统和现代学堂坡村竹纸生产工艺对照表

Table8.13 Contrast of traditional and modern papermaking techniques of bamboo paper in Xuetangpo Village

表8.14 学堂坡村竹纸的相关性能参数

Table8.14 Performance parameters of bamboo paper in Xuetangpo Village

表8.15 张明芳家一年造纸开支表

Table8.15 Zhang Mingfang's annual papermaking cost

表8.16 乌江村竹纸的相关性能参数

Table8.16 Performance parameters of bamboo paper in Wujiang Village

表9.1 香树园村皮纸的相关性能参数

Table9.1 Performance parameters of bast paper in Xiangshuyuan Village

表9.2 石阡县部分竹株的基本特征

Table9.2 Basic information of some varieties of bamboo in Shiqian County

表9.3 石阡县部分竹种造纸性能指标

Table9.3 Papermaking performance parameters of some varieties of bamboo in Shiqian County

表9.4 关刀土村竹纸的相关性能参数

Table9.4 Performance parameters of bamboo paper in Guandaotu Village

表9.5 云舍村竹纸的相关性能参数

Table9.5 Performance parameters of bamboo paper in Yunshe Village

表9.6 兴旺村皮纸的相关性能参数

Table9.6 Performance parameters of bast paper in Xingwang Village

表9.7 杨边畅户五代传承图

Table9.7 Papermaker Yang Bianchang's genealogy of papermaking inheritors

表10.1 白水村竹纸的相关性能参数

Table10.1 Performance parameters of bamboo paper in Baishui Village

表10.2 贵洞村竹纸的相关性能参数

Table10.2 Performance parameters of bamboo paper in Guidong Village

表10.3 满溪村竹纸的相关性能参数

Table10.3 Performance parameters of bamboo paper in Manxi Village

表10.4 湾水镇竹纸的相关性能参数

Table10.4 Performance parameters of bamboo paper in Wanshui Town

表10.5 石桥村白皮纸的相关性能参数

Table10.5 Performance parameters of white bast paper in Shiqiao Village

表10.6 石桥村花纸的相关性能参数

Table10.6 Performance parameters of flower paper in Shiqiao Village

表10.7 本里村皮纸的相关性能参数

Table10.7 Performance parameters of bast paper in Benli Village

表10.8 九秋村皮纸的相关性能参数

Table10.8 Performance parameters of bast paper in Jiuqiu Village

表10.9 打格村竹纸的相关性能参数

Table10.9 Performance parameters of bamboo paper in Dage Village

表10.10 赵掌珠家五代造纸人传承表

Table10.10 Papermaker Zhao Zhangzhu's genealogy of papermaking inheritors

表10.11 高华村皮纸的相关性能参数

Table10.11 Performance parameters of bast paper in Gaohua Village

表10.12 高华村草纸的相关性能参数

Table10.12 Performance parameters of straw paper in Gaohua Village

表10.13 高华村皮纸、草纸销售价格表

Table10.13 Price list of bast paper and straw paper in Gaohua Village

表10.14 吴奶保卫家四代造纸人传承表

Table10.14 Papermaker Wu Baowei's mother genealogy of papermaking inheritors

表10.15 小黄村皮纸的相关性能参数

Table10.15 Performance parameters of bast paper in Xiaohuang Village

表10.16 吴奶卫和吴奶妹的家谱

Table10.16 Genealogy of Wu Naiwei and Wu Naimei

表10.17 占里村皮纸的相关性能参数

Table10.17 Performance parameters of bast paper in Zhanli Village

表10.18 地扪村皮纸的相关性能参数

Table10.18 Performance parameters of bast paper in Dimen Village

表10.19 不同年份地扪侗族构皮纸价格

Table10.19 Price list of bast paper by the Dong Ethnic Group in Dimen Village in different years

术语
Terminology

地理名 Places

汉语术语 Term in Chinese	英语术语 Term in English
安龙县	Anlong County
安顺市	Anshun City
八弓镇	Bagong Town
坝盘行政村	Bapan Administrative Village
白沙乡	Baisha Town
白水行政村	Baishui Administrative Village
白云山镇	Baiyunshan Town
百花村	Baihua Village
班竹乡	Banzhu Village
板当镇	Bandang Town
本里村	Benli Village
毕节市	Bijie City
蔡家湾	Caijiawan
岑巩县	Cengong County
茶园乡	Chayuan Town
长顺县	Changshun County
城关镇	Chengguan Town
从江县	Congjiang County
翠里瑶族壮族乡	Cuili Yao and Zhuang Town
打格行政村	Dage Administrative Village
大沙坝仡佬族侗族乡	Dashaba Gelo and Dong Town
大乌江镇	Dawujiang Town
丹寨县	Danzhai County
旦坪村	Danping Village
地扪行政村	Dimen Administrative Village
都匀市	Duyun City
堕却乡	Duoque Town
丰乐镇	Fengle Town
凤仪镇	Fengyi Town
桴焉坪村	Fuyanping Village
桴焉乡	Fuyan Town
高华村	Gaohua Village
高增乡	Gaozeng Town
关刀土村	Guandaotu Village
关岭布依族苗族自治县	Guanling Bouyei and Miao Autonomous County
广顺州	Guangshun Prefecture

贵洞行政村	Guidong Administrative Village
贵阳市	Guiyang City
贵州省	Guizhou Province
桂花乡	Guihua Town
合水镇	Heshui Town
和溪镇	Hexi Town
河沟头村民组	Hegoutou Villagers' Group
红岩村	Hongyan Village
黄都镇	Huangdu Town
黄平县	Huangping County
惠水县	Huishui County
计划乡	Jihua Town
俭坪乡	Jianping Town
江口村	Jiangkou Village
江口县	Jiangkou County
江龙镇	Jianglong Town
金沙县	Jinsha County
九秋村	Jiuqiu Village
卡塘行政村	Katang Administrative Village
凯里市	Kaili City
腊岩行政村	Layan Administrative Village
烂溽河村	Lanruhe Village
郎岱造纸厂	Langdai Papermaking Factory
郎岱镇	Langdai Town
老厂镇	Laochang Town
乐里镇	Leli Town
梨坝村	Liba Village
黎平县	Liping County
荔波县	Libo County
流渡镇	Liudu Town
六洞村	Liudong Village
六盘水市	Liupanshui City
六枝特区	Liuzhi Special Area
龙井村	Longjing Village
龙里县	Longli County
龙山镇	Longshan Town
龙潭村	Longtan Village
陇脚行政村	Longjiao Administrative Village
芦山镇	Lushan Town
鲁班镇	Luban Town
满溪行政村	Manxi Administrative Village
茅贡乡	Maogong Town
纳雍县	Nayong County
南皋乡	Nangao Town
怒溪土家族苗族乡	Nuxi Tujia and Miao Town
盘县	Panxian County
坪地场仡佬族侗族乡	Pingdichang Gelo and Dong Town
坡柳村	Poliu Village
普安县	Pu'an County
黔北区域	North area of Guizhou Province
黔东南苗族侗族自治州	Qiandongnan Miao and Dong Autonomous Prefecture
黔南布依族苗族自治州	Qiannan Bouyei and Miao Autonomous Prefecture
黔南区域	South area of Guizhou Province
黔西北区域	Northwest area of Guizhou Province
黔西南布依族苗族自治州	Qianxinan Bouyei and Miao Autonomous Prefecture
黔西区域	West area of Guizhou Province
黔中区域	Middle area of Guizhou Province
仁怀市	Renhuai City
瑞溪镇	Ruixi Town
三把车村	Sanbache Village
三合行政村	Sanhe Administrative Village
三穗县	Sansui County
三元村	Sanyuan Village
桑树湾	Sangshuwan
沙包乡	Shabao Town
沙营乡	Shaying Town
沙子坡镇	Shazipo Town
杉木坪村	Shanmuping Village
石阡县	Shiqian County
石桥行政村	Shiqiao Administrative Village
市坪村	Shiping Village
水尾镇	Shuiwei Town
太平土家族苗族乡	Taiping Tujia and Miao Town
汤山镇	Tangshan Town
塘口村	Tangkou Village
桐梓坪村	Tongziping Village
铜仁市	Tongren City
湾水镇	Wanshui Town
万峰湖镇	Wanfenghu Town
翁贵村	Wenggui Village
翁贵镇	Wenggui Town
翁坪乡	Wengping Town
乌当区	Wudang District
乌江行政村	Wujiang Administrative Village
五马镇	Wuma Town
务川仡佬族苗族自治县	Wuchuan Gelo and Miao Autonomous County
下林坝村	Xialinba Village
下午行政村	Xiawu Administrative Village
香树园村	Xiangshuyuan Village
香纸沟风景区	Xiangzhigou Scenic Spot
小黄行政村	Xiaohuang Administrative Village
小屯乡	Xiaotun Town
新堡布依族乡	Xinbao Bouyei Town
新场村	Xinchang Village
兴旺村	Xingwang Village
秀塘壮族乡	Xiutang Zhuang Town
岩脚镇	Yanjiao Town
羊场布依族白族苗族乡	Yangchang Bouyei, Bai and Miao Town
尧古村	Yaogu Village
印江土家族苗族自治县	Yinjiang Tujia and Miao Autonomous County
营盘乡	Yingpan Town
永康水族乡	Yongkang Shui Town
永宁镇	Yongning Town
余庆县	Yuqing County
云舍村	Yunshe Village
占里村	Zhanli Village
贞丰县	Zhenfeng County
镇宁布依族苗族自治县	Zhenning Bouyei and Miao Autonomous County
正安县	Zheng'an County
郑家河自然村	Zhengjiahe Natural Village

纸厂村	Zhichang Village
中寨乡	Zhongzhai Town
中寨乡火坑造纸厂	Huokeng Papermaking Factory in Zhongzhai Town
竹新村	Zhuxin Village
紫山村	Zishan Village
紫云苗族布依族自治县	Ziyun Miao and Bouyei Autonomous County
遵义市	Zunyi City

纸品名 Paper names

汉语术语	英语术语
Term in Chinese	Term in English
安龙布依族竹纸	Bamboo paper by the Bouyei Ethnic Group in Anlong County
岑巩侗族竹纸	Bamboo paper by the Dong Ethnic Group in Cengong County
长顺皮纸	Bast paper in Changshun County
从江翠里瑶族草纸	Straw paper by the Yao Ethnic Group in Cuili Town of Congjiang County
从江翠里瑶族皮纸	Bast paper by the Yao Ethnic Group in Cuili Town of Congjiang County
从江小黄侗族皮纸	Bast paper by the Dong Ethnic Group in Xiaohuang Village of Congjiang County
从江秀塘瑶族竹纸	Bamboo paper by the Yao Ethnic Group in Xiutang Town of Congjiang County
从江占里侗族皮纸	Bast paper by the Dong Ethnic Group in Zhanli Village of Congjiang County
丹寨苗族皮纸	Bast paper by the Miao Ethnic Group in Danzhai County
都匀蜡纸	Wax paper in Duyun City
都匀皮纸	Bast paper in Duyun City
关岭布依族苗族竹纸	Bamboo paper by the Bouyei and Miao Ethnic Groups in Guanling County
关岭皮纸	Bast paper in Guanling County
黄平苗族竹纸	Bamboo paper by the Miao Ethnic Group in Huangping County
惠水竹纸	Bamboo paper in Huishui County
江口土家族竹纸	Bamboo paper by the Tujia Ethnic Group in Jiangkou County
金沙皮纸	Bast paper in Jinsha County
金沙竹纸	Bamboo paper in Jinsha County
凯里苗族竹纸	Bamboo paper by the Miao Ethnic Group in Kaili City
黎平侗族皮纸	Bast paper by the Dong Ethnic Group in Liping County
荔波布依族竹纸	Bamboo paper by the Bouyei Ethnic Group in Libo City
六枝彝族苗族仡佬族皮纸	Bast paper by the Yi, Miao and Gelo Ethnic Groups in Liuzhi Special Area
龙里竹纸	Bamboo paper in Longli County
纳雍皮纸	Bast paper in Nayong County
盘县皮纸	Bast paper in Panxian County
盘县竹纸	Bamboo paper in Panxian County
普安皮纸	Bast paper in Pu'an County
仁怀手工纸	Handmade paper in Renhuai City
榕江侗族皮纸	Bast paper by the Dong Ethnic Group in Rongjiang County
三穗侗族竹纸	Bamboo paper by the Dong Ethnic Group in Sansui County
石阡仡佬族皮纸	Bast paper by the Gelo Ethnic Group in Shiqian County
石阡仡佬族竹纸	Bamboo paper by the Gelo Ethnic Group in Shiqian County
乌当竹纸	Bamboo paper in Wudang District
务川仡佬族皮纸	Bast paper by the Gelo Ethnic Group in Wuchuan County
务川仡佬族竹纸	Bamboo paper by the Gelo Ethnic Group in Wuchuan County
印江合水镇皮纸	Bast paper in Heshui Town of Yinjiang County
印江土家族手工纸	Handmade paper by the Tujia Ethnic Group in Yinjiang County
余庆竹纸	Bamboo paper in Yuqing County
贞丰布依族竹纸	Bamboo paper by the Bouyei Ethnic Group in Zhenfeng County
贞丰皮纸	Bast paper in Zhenfeng County
镇宁皮纸	Bast paper in Zhenning County
正安手工纸	Handmade paper in Zheng'an County
紫云手工纸	Handmade paper in Ziyun County

原料与相关植物名 Raw materials and plants

汉语术语	英语术语
Term in Chinese	Term in English
白竹	*Fargesia semicoriacea* Yi
斑竹	*Phyllostachys bambusoides* cv. Tanakae
草木灰	Wood ash
楮树	Paper mulberry tree
楮树皮	Paper mulberry bark
慈竹	*Neosinocalamus affinis*
刺竹	*B.blumeana* Schult.f
钓鱼竹（黔竹）	*Dendrocalamus tsiangii* (McClure) Chia et H.L Fung
构树皮	Paper mulberry bark
滑液	Papermaking mucilage
滑（纸药）	Hua (papermaking mucilage)
金竹	*Phyllostachys parvifolia*
聚丙烯酰胺	Polyacrylamide
苦竹	*Pleioblastus amarus* (Keng) Keng f.
蛮竹	Man bamboo
煤	Coal
绵竹	*Bambusa intermedia* Hsueh et Yi
糯稻杆	Glutinous rice stalk
青竹	*PhyllostachysacutaChuetChao*
杉松树根（沙根子）	Cedar pine root
山棉皮	Shan Mian Pi
山泉水	Spring water
烧碱	Caustic soda
石灰	Lime
水	Water
水竹	*Phyllostachys heteroclada* Oliver
洗衣粉	Washing powder
仙人掌	Cactus
野梦花麻	Ye Meng Hua Ma
野生构树	Wild paper mulberry tree
灶灰	Stove ash
竹子	Bamboo

工艺技术和工艺设备 Techniques and tools

汉语术语	英语术语
Term in Chinese	Term in English
摆二道皮	Double cleaning the bark
摆石灰皮	Cleaning the fermented bark
摆头道皮	Cleaning the bark
拌料	Stirring the papermaking materials
绑皮	Binding the bark
棒槌	Wooden mallet
包装	packing
焙架	Frame for drying the paper
焙纸	Drying the paper
比子	Bizi, a ruler for measuring the paper

剥构皮	Stripping paper mulberry bark
剥皮	Stripping the bark
簸箕	Bamboo basket
裁刀	Sickle for trimming the paper
裁纸	Cutting paper
踩槽	Trough for stamping the papermaking materials
踩料	Stamping the papermaking materials
踩、洗竹麻	Stamping and cleaning the bamboo materials
踩竹	Stamping the bamboo
槽缸	Papermaking trough
槽棍	Stirring stick
拆构皮	Splitting the papermaking mulberry bark
拆料子	Splitting the papermaking materials
拆纸（扯纸）	Peeling the paper down
抄纸槽	Papermaking trough
抄纸帘	Papermaking screen
舂料	Beating the papermaking materials with a foot pestle
舂竹麻	Beating the bamboo materials
出甑	Picking the materials out of the kiln
捶纸	Hammering the paper
槌	Mallet
搓料子	Rubbing the materials
打碓	Beating the papermaking materials
打浆耙	Stirring rake
打料	Beating the materials
打皮	Beating the bark
打纸棒	Beating stick
打纸锤	Beating hammer
打纸钱	Making joss paper
大盖板	Cover board for pressing the paper
叠纸	Folding the paper
钉鞋	Spiked shoes
钉纸架	Papermaking frame with nail
抖皮	Shaking the bark
抖纸	Shaking the paper
堆构皮	Piling paper mulberry bark
碓	Pestle
碓杈	V-shaped stone pestle
二次搅槽	Stirring the papermaking materials for the second time
二次压榨	Pressing the paper for the second time
二次煮竹麻	Boiling the bamboo materials for the second time
发塘	Fermenting pool
放滑	Adding in papermaking mucilage
分纸	Spliting the paper layers
盖方	Cover board
拱耙	Rake
构皮	Paper mulberry bark
刮构皮	Stripping paper mulberry bark
刮料	Stripping the materials
葫芦瓢	Gourd ladle
划竹	Cleaving the bamboo
滑缸	Vat for holding the papermaking mucilage
灰池（石灰池）	Lime pool
计数器	Paper counting appratus
加滑	Adding in papermaking mucilage
加清水	Adding in water
加纸药	Adding in papermaking mucilage
拣料	Picking out the impurities
浆皮	Soaking and fermenting the bark
浆竹麻	Soaking and fermenting the bamboo materials in the lime water
浇纸	Pouring the paper pulp onto the papermaking screen
搅料	Stirring the materials
搅纸浆	Stirring the paper pulp
揭纸	Peeling the paper down
浸泡稻草	Soaking the straw
砍	Lopping
砍构树	Lopping paper mulberry bark
砍料	Lopping the materials
砍树	Lopping the tree
砍纸	Trimming the paper
砍纸刀	Trimming knife
砍纸卡	Trimming ruler
砍竹麻	Lopping the bamboo materials
砍竹子	Lopping the bamboo
炕纸	Drying the paper
烤纸架	Drying frame
捆稻草	Binding the straw
捆构皮	Binding paper mulberry bark
捆料	Binding the materials
捆皮	Binding the bark
捆纸	Binding the paper
捆竹	Binding the bamboo
捞钩	Hook for picking up the papermaking materials
捞筋	Picking out the impurities
捞料	Picking out the materials
捞纸槽	Papermaking trough
捞纸棚和纸榨	Papermaking shed and presser
捞竹	Picking the bamboo out of water
捞竹麻	Picking the bamboo materials out of water
理构皮	Sorting paper mulberry bark
理皮	Sorting the bark
理纸	Sorting the paper
帘架	Frame for supporting the papermaking screen
帘子	Papermaking screen
晾干	Drying
晾构皮	Drying paper mulberry bark
晾料	Drying the materials
晾纸	Drying the paper
晾竹	Drying the bamboo
晾竹麻	Drying the bamboo materials
料锤	Hammer for beating materials
料兜	Bag for holding the papermaking materials
料坑	Pool for soaking the bark
料塘	Soaking pool
滤料	Filtering the materials
买构皮	Buying paper mulberry bark
民间火纸制作技艺	Folk papermaking techniques of Huo paper
木槌	Wooden mallet

木 碓	Wooden pestle
木 甑	Wooden utensil
木 榨	Wooden presser
碾稻草	Grinding the straw
碾 料	Grinding the materials
碾 竹	Grinding the bamboo
碾竹麻	Grinding the bamboo materials
碾 子	Grinder
牛 骨	Ox bone
沤构皮	Soaking paper mulberry bark
沤 麻	Soaking the materials
沤 竹	Soaking the bamboo
泡稻草	Soaking the straw
泡构皮	Soaking paper mulberry bark
泡 料	Soaking the materials
泡料塘	Soaking pool
泡 皮	Soaking the bark
泡 窑	Soaking kiln
泡 竹	Soaking the bamboo
泡竹麻	Soaking bamboo materials
皮 甑	Kiln for steaming the materials
皮纸制作技艺	Papermaking techniques of bast paper
瓢	Gourd ladle
漂构皮	Bleaching paper mulberry bark
漂 皮	Bleaching the bark
漂 洗	Bleaching
平面碾（石碾）	Stone grinder
破 竹	Beating the bamboo with a wooden mallet
齐 纸	Sorting the paper
钱 錾	Trimming chisel
切 板	Cutting board
切皮板	Trimming the bark
清 皮	Cleaning the bark
揉二道皮	Rubbing the bark for the second time
揉 料	Rubbing the materials
揉 皮	Rubbing the bark
晒	Drying
晒楮皮	Drying paper mulberry bark
晒料子	Drying the materials
晒 皮	Drying the bark
晒 竹	Drying the bamboo
晒竹麻	Drying bamboo materials
上 窑	Putting the papermaking materials into the kiln
上 甑	Putting the papermaking materials into the utensil
绳 索	Rope
石碓与舂料棒	Stone pestle and beating stick
石桥古法造纸技艺	Traditional papermaking techniques in Shiqiao Village
石 甑	Stone utensil
双捞钩	Double hooks for picking up the materials
水 碓	Hydraulic pestle
撕 纸	Peeling the paper down
淘 料	Cleaning the materials
挑 料	Carrying the materials
挑竹麻	Carrying the bamboo materials

土法造纸工艺	Traditional papermaking techniques
乌当手工土纸制作工艺	Papermaking techniques of handmade paper in Wudang District
捂料子	Fermenting the materials
捂 皮	Fermenting the bark
洗	Cleaning
洗 池	Cleaning pool
洗稻草	Cleaning the straw
洗构皮	Cleaning paper mulberry bark
洗 料	Cleaning the materials
洗 竹	Cleaning the bamboo
洗竹麻	Cleaning the bamboo materials
小屯白绵纸造纸工艺	Papermaking techniques of Baimian Paper in XiaotunVillage
修 竹	Trimming the bamboo
选 料	Choosing the materials
压 干	Pressing and squeezing the paper
压 纸	Pressing the paper
摇	Shaking
摇 纸	Shaking the paper
舀 纸	Lifting the papermaking screen out of water and turning it upside down on the board
舀纸棚	Papermaking shed
药 槽	Trough for holding the papermarking mucilage
印 刷	Printing
凿 纸	Perforating the paper to make joss paper
凿纸锤	Perforating hammer
凿纸担	Carrying pole for pressing the paper
凿纸桩	Trimming stack
甑 子	Utensil for steaming the papermaking materials
榨 料	Pressing the materials
榨 纸	Pressing the paper
折	Folding
折 纸	Folding the paper
蒸构皮	Steaming paper mulberry bark
蒸 料	Steaming the materials
蒸 皮	Steaming the bark
蒸 灶	Steaming stove
蒸 煮	Steaming
纸 焙	Drying wall
纸 槽	Papermaking trough
纸 架	Papermaking frame
纸筋刷	Sticks for picking out the impurities
纸 帘	Papermaking screen
纸帘架	Frame for supporting the papermaking screen
纸 塘	Papermaking pool
竹 棍	Bamboo stick
煮	Boiling
煮构皮	Boiling paper mulberry bark
煮 料	Boiling the materials
煮料锅	Pot for boiling the materials
煮 皮	Boiling the bark
煮 竹	Boiling the bamboo
煮竹麻	Boiling the bamboo materials
装 池	Putting the materials in the pool

历　史　文　化 History and culture

汉语术语 Term in Chinese	英语术语 Term in English
八音坐唱	"Eight Musical Instruments" performance
不吃竹笋为造纸	" Do not eat bamboo shoots and save those for papermaking"
草纸祭俗	Worshipping with straw paper
岑巩火纸	Huo paper in Cengong County
打经簿	Dajingbu, a book in memory of the dead
调北填南	North-to-South Migration for the purpose of occupying
调北征南	North-to-South Migration for the purpose of conquering
侗族大歌	Dong ethnic polyphonic music show
都匀爱国皮纸厂	Aiguo Bast Paper Factory in Duyun City
都匀蜡纸厂	Wax Paper Factory in Duyun City
都匀皮纸联营社	Bast Paper Joint Agency in Duyun City
都匀市匀阳造纸厂	Yunyang Papermaking Factory in Duyun City
度　戒	Coming-of-Age Ceremony
盖公纸(皮纸)	Gaigong paper (bast paper)
工分制	Distribution system in the period of People's Commune
供奉蔡伦造纸先师	Worshipping Cai Lun, the originator of papermaking
谷冰造纸厂	Gubing Papermaking Factory
国翁造纸社	Guoweng Papermaking Agency
裹脚纸	Paper used to wrap women's feet
过世送纸	Burning joss paper
哈戎节	Harong Festival
糊墙壁	Pasting the paper on a wall
花溪镇山布依文化生态馆	The Bouyei Ethnic Cultural Eco-Museum in Zhenshan Village of Huaxi District
惠水利民造纸厂	Limin Papermaking Factory in Huishui County
火纸、火枚纸	Huo paper, Huomei paper
祭蔡伦	Worshipping Cai Lun
祭"将军柱"	Offering sacrifices to the General Pillars
祭祀"盘王"	Worshipping King Pan
锦屏隆里文化生态馆	Cultural Eco-Museum in Longli Ancient City of Jingping County
开天门、传弟子	A ceremony to pass on the papermaking tools from the dead papermaker to the inheritor
开窑祭	The sacrificial ceremonies before starting papermaking
黎平堂安侗文化生态馆	The Dong Ethnic Cultural Eco-Museum in Tang'an Village of Liping County
留下买路钱	Paying the road toll
六枝梭戛民族文化生态博物馆	Ethnic Cultural Eco-Museum in Suoga Village of Liuzhi Special Area
《龙氏家谱》	*Genealogy of the Longs*
龙溪石砚	Longxi inkstone
梅子沟纸厂	Meizigou Paper Factory
冥镪 (纸钱)	Joss paper
藕金谷公私纸厂	Oujingu Public-Private Paper Factory
盘王节	King Pan Festival
青　纸	Joss paper
烧钱化纸，以缅阳人	To burn joss paper to the dead, so as to bless the living
烧　纸	Burning joss paper
神奇纸袋	Magical paper bag
思州火纸	Huo paper in Sizhou Prefecture
心　学	Theory of Mind
夜挑灯笼去造纸	Making paper even at night
衣服夹层	Interlayer of clothes
"永垂定例"碑	Monument of "Permanent Stipucation"
越国公汪华	Wang Hua, lord of Yue State
"真钱"与"假钱"	"Real money" and "fake money"
纸　钱	Joss paper
猪拱先师	Mr. Pig
竹廉纸 (竹纸)	Zhulian paper (bamboo paper)

后 记

《中国手工纸文库·贵州卷》作为整个文库起步较早的工作，不包括团队成员更早的田野工作积累，从2008年12月至2016年3月，田野研究历经七年半时间。其间，项目组深入贵州省各手工造纸点的调查采样以及一次又一次的补充调查和求证贯穿始终，直到2016年3月，这种意在求准确、求完善的田野采集与实验分析仍在进行中。

由于田野调查和文献梳理基本上是多位成员协同完成的，且前后多次的补充修订也并非始终出自一人之手，因而即便有田野调查工作规范、撰稿的标准及示范的样稿，全卷的信息采集方式和表述风格依然存在多样性。针对这一状况，初稿合成后，统稿工作小组从2014年6月到2016年8月一共进行了7轮统稿，最终定稿。虽然仍感觉全卷有诸多可进一步完善之处，但《中国手工纸文库·贵州卷》的调查研究及书稿撰写已历时太久，行业同仁和团队成员均有出版传播的期待，因此若干的未尽之义只能暂时心怀遗憾，以待来日有修订缘分时再尽心完善了。

本卷书稿的完成有赖团队所有成员全心全意的投入与持续不懈的努力，除了对所有参与成员表示衷心感谢外，特在后记中对各位同仁的工作做如实的记述。

Epilogue

Library of Chinese Handmade Paper: *Guizhou* is among the earliest of our serial handmade paper studies. From December 2008 to March 2016, seven and a half years have passed since our earliest fieldwork efforts. The research members explored into papermaking sites in Guizhou Province over these years. Up to March 2016, our papermaking odyssey was still on.

Modification of the book was cooperated by many research members, therefore, the writing style and information collection method may vary due to the fact that the fieldwork and literature surveys were undertaken by different groups of researchers. Although the investigation rules, writing norms and format were enacted, we may make amends for the possible deviation. From June 2014 to August 2016, seven rounds of modification efforts have contributed to this version for publication, though the version can never claim perfection. Research and writing of *Library of Chinese Handmade Paper*: *Guizhou* last several years and finally go to the verge of publication as heartily expected by our team members and friends in the field. We still harbor expectation for further and deeper exploration and modification.

This volume acknowledges the consistent efforts and great contributions of the following researchers:

一、田野调查与文稿撰写部分

市/州	县/区	乡村/厂	纸名(种类)	撰写人	调查人
六盘水市	盘县	羊场布依族白族苗族乡下午行政村	皮纸	陈　彪、张义忠	陈　彪、张义忠
		老厂镇	竹纸	汤书昆	汤书昆、蓝　强、吴明卫
	六枝特区	中寨苗族彝族布依族乡火坑行政村	皮纸	陈　彪、张义忠、张美丽	陈　彪、张义忠、张美丽
黔西南布依族苗族自治州	普安县	白沙乡卡塘行政村河沟头村民组	皮纸	陈　彪、张义忠	陈　彪、张义忠
	安龙县	万峰湖镇坝盘行政村	竹纸	汤书昆	汤书昆、蓝　强、陈　彪
	贞丰县	小屯乡龙井行政村	皮纸	陈　彪	陈　彪、王　祥
		龙场镇坡柳行政村	竹纸	陈　彪	陈　彪、王　祥
安顺市	关岭县	永宁镇紫山行政村	竹纸	陈　彪	陈　彪、王　祥
		沙营乡纸厂行政村	皮纸	王　祥	王　祥、陈　彪
	镇宁县	江龙镇竹新行政村	皮纸	陈　彪、祝秀丽	陈　彪、王　祥
	紫云县	板当镇	皮纸、竹纸	陈　彪、陈　佳	陈　彪、王　祥
黔南布依族苗族自治州	长顺县	白云山镇翁贵行政村	皮纸	黄飞松、陈敬宇	陈　彪、黄飞松、张义忠
	惠水县	芦山镇	竹纸	黄飞松	黄飞松、陈　彪、张义忠
	龙里县	龙山镇纸厂行政村	竹纸	黄飞松	陈　彪、张义忠、黄飞松
	荔波县	永康水族乡尧古行政村	竹纸	汤书昆、万　丽	蓝　强、陈　彪、杨洋、张　娟、万　丽、王红莲
	都匀市	斗篷山景区、石板街	皮纸	黄飞松、陈　彪、汤书昆	陈　彪、张义忠、黄飞松、祖明
		斗篷山景区、石板街	蜡纸	黄飞松、陈　彪、汤书昆	陈　彪、张义忠、黄飞松、祖明
毕节市	纳雍县	沙包乡大寨行政村	皮纸	汤书昆、谢起慧、陈　彪	陈　彪、蓝　强
	金沙县	茶园乡新桥行政村红土坡自然村	竹纸	汤书昆	汤书昆、蓝　强、王　霄、杜加文、吕　华
		沙土镇天星行政村河透底自然村、桂花乡果松行政村雨下寨自然村、城关镇联盟行政村	皮纸	汤书昆、陈　彪、王　研、朱安达、朱　赟、孙　舰	汤书昆、陈　彪、蓝　强、王　霄、杜加文、吕　华
贵阳市	乌当区	新堡布依族乡陇脚行政村	竹纸	王　祥	王　祥、陈　彪、汪常明
遵义市	仁怀市	五马镇三元行政村	皮纸	李宪奇、陈　彪、祝秀丽	陈　彪、李宪奇
		鲁班镇	竹纸	李宪奇、陈　彪	李宪奇、陈　彪
	正安县	凤仪镇黎坝行政村、河溪镇杉木坪行政村	皮纸、竹纸	李宪奇	汤书昆、陈　彪、李宪奇、蓝　强
	务川县	丰乐镇新场行政村造纸塘村民组	皮纸	邹进扬、陈　彪、汤书昆	汤书昆、陈　彪、李宪奇、邹进扬、蓝　强
		黄都镇三河行政村学堂坡村民组	竹纸	邹进扬、陈　彪、汤书昆	汤书昆、陈　彪、李宪奇、邹进扬、蓝　强
	余庆县	大乌江镇乌江行政村	竹纸	陈　彪、陈敬宇	陈　彪

铜仁市	石阡县	汤山镇香树园行政村	皮纸	陈　彪、刘　靖、祝秀丽	陈　彪、刘　靖
		大沙坝仡佬族侗族乡关刀土行政村、坪地场仡佬族侗族乡下林坝行政村	竹纸	陈　彪、祝秀丽	陈　彪、刘　靖
	江口县	怒溪土家族苗族乡河口行政村、太平土家族苗族乡云舍行政村	竹纸	陈　彪、汪常明	陈　彪、刘　靖
	印江县	合水镇兴旺行政村	皮纸	刘　靖	陈　彪、刘　靖
		沙子坡镇六洞行政村与塘口行政村	皮纸、竹纸	刘　靖	陈　彪、刘　靖
黔东南苗族侗族自治州	岑巩县	水尾镇白水行政村与腊岩行政村、羊桥土家族乡龙统行政村	竹纸	汤书昆	汤书昆、蓝　强
	三穗县	八弓镇贵洞行政村	竹纸	陈　彪、孙　舰	陈　彪
	黄平县	翁坪乡满溪行政村	竹纸	陈　彪	陈　彪、吴寿仁
	凯里市	湾水镇	竹纸	陈　彪	陈　彪、吴寿仁
	丹寨县	南皋乡石桥行政村	皮纸	陈　彪	陈　彪、汤书昆、刘　靖、祖　明
	榕江县	乐里镇本里行政村、计划乡九秋行政村	皮纸	杨　洋、张　娟、李　昂、汤书昆	蓝　强、陈　彪、杨　洋、张　娟、吴寿仁、黄飞松、翟剑锋
	从江县	秀塘壮族乡打格行政村	竹纸	汤书昆、高洁恒	汤书昆、蓝　强、高洁恒
		翠里瑶族壮族乡高华行政村	皮纸、草纸	汤书昆、陈　彪、李淑丹、朱安达、李　昂、宋彩龙、胡文英	汤书昆、蓝　强、陈　彪、李淑丹、宋彩龙、王　妍、吴寿仁、漆小芳、宜苗苗、王　霄、郭帅辉
		高增乡小黄行政村	皮纸	汤书昆、陈　彪	汤书昆、陈　彪、蓝　强、漆小芳、王　霄、宜苗苗、郭帅辉
		高增乡占里行政村	皮纸	汤书昆、高洁恒	汤书昆、蓝　强、高洁恒、方建霞
	黎平县	茅贡乡地扪行政村	皮纸	陈　彪	陈　彪、吴寿仁

二、纸样测试分析与拍摄部分

主测试：朱赟、郑久良、刘伟　数据分析：朱赟、郑久良、刘伟　分析统筹：汤书昆、朱赟　实物纸样拍摄：黄晓飞

此外，本卷的测试得到了安徽省泾县中国宣纸集团公司罗鸣、黄立新、赵梦君、王钟玲、宋福星的支持和帮助，其中赵梦君在具体测试中给予了较多的协助。

三、英文翻译部分

领衔主译：合肥工业大学外国语学院　方媛媛

感谢美国罗格斯大学亚洲语言与文化系教授Richard V. Simmons，以及罗格斯大学Lucas Richards同学对书稿英文版提出的修改建议。合肥工业大学翻译硕士刘婉君和朱丽君参与了本书部分章节的翻译。

四、总序、编撰说明、概述与附录部分

总序撰写	汤书昆
编撰说明撰写	汤书昆、朱赟、陈　彪
概述撰写	第一节：陈敬宇、陈　彪 第二节：孙　舰、陈　彪、汤书昆 第三节：汤书昆、陈　彪 概述由陈　彪起草初稿，由汤书昆、陈敬宇、陈　彪统稿审定
附录	表目整理编制：郑久良、刘伟 图目整理编制：刘伟 术语整理编制：李昂、朱安达 后记撰写：汤书昆

1. Field Investigation and Writing of the Book

City/Prefecture	County/District	Village/Factory	Paper type	Author	Investigator
Liupanshui City	Panxian County	Xiawu Administrative Village of Yangchang Bouyei, Bai and Miao Town	Bast paper	Chen Biao, Zhang Yizhong	Chen Biao, Zhang Yizhong
		Laochang Town	Bamboo paper	Tang Shukun	Tang Shukun, Lan Qiang, Wu Mingwei
	Liuzhi Special Area	Huokeng Administrative Village of Zhongzhai Miao, Yi and Bouyei Town	Bast paper	Chen Biao, Zhang Yizhong, Zhang Meili	Chen Biao, Zhang Yizhong, Zhang Meili
Qianxinan Bouyei and Miao Autonomous Prefecture	Pu'an County	Hegoutou Villagers' Group of Katang Administrative Village in Baisha Town	Bast paper	Chen Biao, Zhang Yizhong	Chen Biao, Zhang Yizhong
	Anlong County	Bapan Administrative Village of Wanfenghu Town	Bamboo paper	Tang Shukun	Tang Shukun, Lan Qiang, Chen Biao
	Zhenfeng County	Longjing Administrative Village of Xiaotun Town	Bast paper	Chen Biao	Chen Biao, Wang Xiang
		Poliu Administrative Village of Longchang Town	Bamboo paper	Chen Biao	Chen Biao, Wang Xiang
Anshun City	Guanling County	Zishan Administrative Village of Yongning Town	Bamboo paper	Chen Biao	Chen Biao, Wang Xiang
		Zhichang Administrative Village of Shaying Town	Bast paper	Wang Xiang	Wang Xiang, Chen Biao
	Zhenning County	Zhuxin Administrative Village of Jianglong Town	Bast paper	Chen Biao, Zhu Xiuli	Chen Biao, Wang Xiang
	Ziyun County	Bandang Town	Bast paper, Bamboo paper	Chen Biao, Chen Jia	Chen Biao, Wang Xiang
Qiannan Bouyei and Miao Autonomous Prefecture	Changshun County	Wenggui Administrative Village of Baiyunshan Town	Bast paper	Huang Feisong, Chen Jingyu	Chen Biao, Huang Feisong, Zhang Yizhong
	Huishui County	Lushan Town	Bamboo paper	Huang Feisong	Huang Feisong, Chen Biao, Zhang Yizhong
	Longli County	Zhichang Administrative Village of Longshan Town	Bamboo paper	Huang Feisong	Chen Biao, Zhang Yizhong, Huang Feisong
	Libo County	Yaogu Administrative Village of Yongkang Shui Town	Bamboo paper	Tang Shukun, Wan Li	Lan Qiang, Chen Biao, Yang Yang, Zhang Juan, Wan Li, Wang Honglian
	Duyun City	Doupeng Mountain Scenic Spot, Shiban Street	Bast paper	Huang Feisong, Chen Biao, Tang Shukun	Chen Biao, Zhang Yizhong Huang Feisong, Zu Ming
		Doupeng Mountain Scenic Spot, Shiban Street	Wax paper	Huang Feisong, Chen Biao, Tang Shukun	Chen Biao, Zhang Yizhong, Huang Feisong, Zu Ming
Bijie City	Nayong County	Dazhai Administrative Village of Shabao Town	Bast paper	Tang Shukun, Xie Qihui, Chen Biao	Chen Biao, Lan Qiang
		Hongtupo Natural Village of Xinqiao Administrative Village in Chayuan Town	Bamboo paper	Tang Shukun	Tang Shukun, Lan Qiang, Wang Xiao, Du Jiawen, Lv Hua
	Jinsha County	Hetoudi Natural Village of Tianxing Administrative Village in Shatu Town, Yuxiazhai Natural Village of Guosong Administrative Village in Guihua Town, Lianmeng Administrative Village of Chengguan Town	Bast paper	Tang Shukun, Chen Biao, Wang Yan, Zhu Anda, Zhu Yun, Sun Jian	Tang Shukun, Chen Biao, Lan Qiang, Wang Xiao, Du Jiawen, Lv Hua
Guiyang City	Wudang District	Longjiao Administrative Village of Xinbao Bouyei Town	Bamboo paper	Wang Xiang	Wang Xiang, Chen Biao, Wang Changming

Zunyi City	Renhuai City	Sanyuan Administrative Village of Wuma Town	Bast paper	Li Xianqi, Chen Biao, Zhu Xiuli	Chen Biao, Li Xianqi
		Luban Town	Bamboo paper	Li Xianqi, Chen Biao	Li Xianqi, Chen Biao
	Zheng'an County	Liba Administrative Village of Fengyi Town, Shanmuping Administrative Village of Hexi Town	Bast paper, Bamboo paper	Li Xianqi	Tang Shukun, Chen Biao, Li Xianqi, Lan Qiang
	Wuchuan County	Zaozhitang Villagers' Group of Xinchang Administrative Village in Fengle Town	Bast paper	Zou Jinyang, Chen Biao, Tang Shukun	Tang Shukun, Chen Biao, Li Xianqi, Zou Jinyang, Lan Qiang
		Xuetangpo Villagers' Group of Sanhe Administrative Village in Huangdu Town	Bamboo paper	Zou Jinyang, Chen Biao, Tang Shukun	Tang Shukun, Chen Biao, Li Xianqi, Zou Jinyang, Lan Qiang
	Yuqing County	Wujiang Administrative Village of Dawujiang Town	Bamboo paper	Chen Biao, Chen Jingyu	Chen Biao
Tongren City	Shiqian County	Xiangshuyuan Administrative Village of Tangshan Town	Bast paper	Chen Biao, Liu Jing, Zhu Xiuli	Chen Biao, Liu Jing
		Guandaotu Administrative Village of Dashaba Gelo and Dong Town, Xialinba Administrative Village of Pingdichang Gelo and Dong Town	Bamboo paper	Chen Biao, Zhu Xiuli	Chen Biao, Liu Jing
	Jiangkou County	Hekou Administrative Village of Nuxi Tujia and Miao Town, Yunshe Administrative Village of Taiping Tujia and Miao Town	Bamboo paper	Chen Biao, Wang Changming	Chen Biao, Liu Jing
	Yinjiang County	Xingwang Administrative Village of Heshui Town	Bast paper	Liu Jing	Chen Biao, Liu Jing
		Liudong Administrative Village and Tangkou Administrative Village of Shazipo Town	Bast paper, Bamboo paper	Liu Jing	Chen Biao, Liu Jing
Qiandongnan Miao and Dong Autonomous Prefecture	Cengong County	Baishui Administrative Village and Layan Administrative Village of Shuiwei Town, Longtong Administrative Village of Yangqiao Tujia Town	Bamboo paper	Tang Shukun	Tang Shukun, Lan Qiang
	Sansui County	Guidong Administrative Village of Bagong Town	Bamboo paper	Chen Biao, Sun Jian	Chen Biao
	Huangping County	Manxi Administrative Village of Wengping Town	Bamboo paper	Chen Biao	Chen Biao, Wu Shouren
	Kaili City	Wanshui Town	Bamboo paper	Chen Biao	Chen Biao, Wu Shouren
	Danzhai County	Shiqiao Administrative Village of Nangao Town	Bast paper	Chen Biao	Chen Biao, Tang Shukun, Liu Jing, Zu Ming
	Rongjiang County	Benli Administrative Village of Leli Town, Jiuqiu Administrative Village of Jihua Town	Bast paper	Yang Yang, Zhang Juan, Li Ang, Tang Shukun	Lan Qiang, Chen Biao, Yang Yang, Zhang Juan, Wu Shouren, Huang Feisong, Zhai Jianfeng
	Congjiang County	Dage Administrative Village of Xiutang Zhuang Town	Bamboo paper	Tang Shukun, Gao Jieheng	Tang Shukun, Lan Qiang, Gao Jieheng
		Gaohua Administrative Village of Cuili Yao and Zhuang Town	Bast paper, Straw paper	Tang Shukun, Chen Biao, Li Shudan, Zhu Anda, Li Ang, Song Cailong, Hu Wenying	Tang Shukun, Lan Qiang, Chen Biao, Li Shudan, Song Cailong, Wang Yan, Wu Shouren, Qi Xiaofang, Yi Miaomiao, Wang Xiao, Guo Shuaihui
		Xiaohuang Administrative Village of Gaozeng Town	Bast paper	Tang Shukun, Chen Biao	Tang Shukun, Chen Biao, Lan Qiang, Qi Xiaofang, Wang Xiao, Yi Miaomiao, Guo Shuaihui
		Zhanli Administrative Village of Gaozeng Town	Bast paper	Tang Shukun, Gao Jieheng	Tang Shukun, Lan Qiang, Gao Jieheng, Fang Jianxia
	Liping County	Dimen Administrative Village of Maogong Town	Bast paper	Chen Biao	Chen Biao, Wu Shouren

2. Technical Analysis and Photographer of the Paper Samples

The test was mainly undertaken by Zhu Yun, Zheng Jiuliang and Liu Wei. Data analysis was undertaken by Zhu Yun, Zheng Jiuliang and Liu Wei. Tang Shukun and Zhu Yun took charge of analyzing and co-ordinating. Sample images were photographed by Huang Xiaofei.

Moreover, Luo Ming, Huang Lixin, Zhao Mengjun, Wang Zhongling and Song Fuxing from China Xuan Paper Co., Ltd. (located in Jingxian County of Anhui Province) helped with the tests, among whom, Zhao Mengjun provided most assistance.

3. Translation

Translation was mainly undertaken by Fang Yuanyuan from the School of Foreign Studies, HFUT, together with her MTI students Liu Wanjun and Zhu Lijun. We should also thank Richard V. Simmons, professor of Asian Language & Culture at Rutgers, the State University of New Jersey, and his student, Lucas Richards, for their valuable suggestions in modifying the English parts of the book.

4. Preface, Introduction to the Writing Norms, Introduction and Appendices

Preface	Tang Shukun
Introduction to the Writing Norms	Tang Shukun, Zhu Yun, Chen Biao
Introduction	Section one: Chen Jingyu, Chen Biao Section two: Sun Jian, Chen Biao, Tang Shukun Section three: Tang Shukun, Chen Biao First version of the introduction was written by Chen Biao, and then modified by Tang Shukun, Chen Jingyu and Chen Biao.
Appendices	Tables: Zheng Jiuliang, Liu Wei Figures: Liu Wei Terminology: Li Ang, Zhu Anda Epilogue: Tang Shukun

在历时两年多的多轮修订增补与统稿工作中，汤书昆、陈彪、方媛媛、朱赟、刘靖、李宪奇、陈敬宇、祝秀丽、黄飞松、郭延龙等作为过程主持人或重要内容模块修订的负责人，对文稿内容、图片与示意图的修订增补、技术分析数据、英文翻译、文献注释考订、表述格式的规范性与准确性核实做了大量基础性的工作，这是全卷书稿能够以今天的面貌和质量展现不容忽视的工作。

在贵州卷的田野调查过程中，先后得到原贵州省博物馆馆长梁太鹤先生、贵州省文化厅非遗处处长张诗莲女士等多位贵州非物质文化遗产研究与保护专家的帮助与指导，在《中国手工纸文库·贵州卷》出版之际，谨对各位支持者真诚致谢！

汤书昆

2017年12月于中国科学技术大学

Tang Shukun, Chen Biao, Fang Yuanyuan, Zhu Yun, Liu Jing, Li Xianqi, Chen Jingyu, Zhu Xiuli, Huang Feisong and Guo Yanlong, et al., who were in charge of the writing and modification for more than two years, all contributed their efforts to the completion of this book. Their meticulous efforts in writing, drawing or photographing, mapping, technical analysis, translating, modifying format, noting and proofreading should be recognized and eulogized in the achievement of the high-quality work.

We owe thanks to Liang Taihe, former curator of Guizhou Provincial Museum and Zhang Shilian who is in charge of Intangible Cultural Heritage Office of Department of Culture in Guizhou government, and all other experts in the field of intangible cultural heritage research and protection who helped in our field investigations. At the time of publication of *Library of Chinese Handmade Paper*: *Guizhou*, sincere gratitude should go to all those who supported and recognized our efforts!

Tang Shukun
University of Science and
Technology of China
December 2017